# GENETICS, GENOMICS AND BREEDING OF POPLAR

# Genetics, Genomics and Breeding of Crop Plants

*Series Editor*
Chittaranjan Kole
Department of Genetics and Biochemistry
Clemson University
Clemson, SC
USA

## Books in this Series:

# GENETICS, GENOMICS AND BREEDING OF POPLAR

*Editors*

## Chandrashekhar P. Joshi

School of Forest Resources and Environmental Science
Michigan Technological University
Houghton, MI
USA

## Stephen P. DiFazio

Department of Biology
West Virginia University
Morgantown, West Virginia
USA

## Chittaranjan Kole

Department of Genetics and Biochemistry
Clemson University
Clemson, SC
USA

**CRC Press**
Taylor & Francis Group
an **informa** business
www.crcpress.com

6000 Broken Sound Parkway, NW
Suite 300, Boca Raton, FL 33487
270 Madison Avenue
New York, NY 10016
2 Park Square, Milton Park
Abingdon, Oxon OX 14 4RN, UK

**Science Publishers**
Enfield, New Hampshire

Published by Science Publishers, P.O. Box 699, Enfield, NH 03748, USA
An imprint of Edenbridge Ltd., British Channel Islands

E-mail: _info@scipub.net_                                   Website: _www.scipub.net_

*Marketed and distributed by:*

CRC Press
Taylor & Francis Group
an **informa** business
www.crcpress.com

6000 Broken Sound Parkway, NW
Suite 300, Boca Raton, FL 33487
270 Madison Avenue
New York, NY 10016
2 Park Square, Milton Park
Abingdon, Oxon OX 14 4RN, UK

Copyright reserved © 2011

ISBN 978-1-57808-714-3

**Library of Congress Cataloging-in-Publication Data**

Genetics, genomics and breeding of poplar / editors,
          Chandrashekhar P.
Joshi, Stephen P. DiFazio, Chittaranjan Kole. -- 1st ed.
      p. cm. -- (Genetics, genomics and breeding of crop
          plants)
  Includes bibliographical references and index.
  ISBN 978-1-57808-714-3 (hardcover : alk. paper)
1.  Poplar--Genetics. 2.  Poplar--Genome mapping. 3.
Poplar--Breeding.  I. Joshi, Chandrashekhar P. II. DiFazio,
          Stephen P.
III. Kole, Chittaranjan. IV. Series: Genetics, genomics and
          breeding
of crop plants.
  QK495.S16G46 2011
  634.9'723--dc22
                                                    2010051279

Printed in the United States of America

# Dedication

**Dr. Gopi Krishna Podila**

September 14, 1957–February 12, 2010

This book is dedicated to the loving memory of Dr. Gopi Krishna Podila who was a stellar scientist, excellent teacher, kind-hearted mentor, enthusiastic colleague, and dear friend to so many of us in the poplar community.

# Preface to the Series

Genetics, genomics and breeding has emerged as three overlapping and complimentary disciplines for comprehensive and fine-scale analysis of plant genomes and their precise and rapid improvement. While genetics and plant breeding have contributed enormously towards several new concepts and strategies for elucidation of plant genes and genomes as well as development of a huge number of crop varieties with desirable traits, genomics has depicted the chemical nature of genes, gene products and genomes and also provided additional resources for crop improvement.

In today's world, teaching, research, funding, regulation and utilization of plant genetics, genomics and breeding essentially require thorough understanding of their components including classical, biochemical, cytological and molecular genetics; and traditional, molecular, transgenic and genomics-assisted breeding. There are several book volumes and reviews available that cover individually or in combination of a few of these components for the major plants or plant groups; and also on the concepts and strategies for these individual components with examples drawn mainly from the major plants. Therefore, we planned to fill an existing gap with individual book volumes dedicated to the leading crop and model plants with comprehensive deliberations on all the classical, advanced and modern concepts of depiction and improvement of genomes. The success stories and limitations in the different plant species, crop or model, must vary; however, we have tried to include a more or less general outline of the contents of the chapters of the volumes to maintain uniformity as far as possible.

Often genetics, genomics and plant breeding and particularly their complimentary and supplementary disciplines are studied and practiced by people who do not have, and reasonably so, the basic understanding of biology of the plants for which they are contributing. A general description of the plants and their botany would surely instill more interest among them on the plant species they are working for and therefore we presented lucid details on the economic and/or academic importance of the plant(s); historical information on geographical origin and distribution; botanical origin and evolution; available germplasms and gene pools, and genetic and cytogenetic stocks as genetic, genomic and breeding resources; and basic information on taxonomy, habit, habitat, morphology, karyotype, ploidy level and genome size, etc.

Classical genetics and traditional breeding have contributed enormously even by employing the phenotype-to-genotype approach. We included detailed descriptions on these classical efforts such as genetic mapping using morphological, cytological and isozyme markers; and achievements of conventional breeding for desirable and against undesirable traits. Employment of the in vitro culture techniques such as micro- and megaspore culture, and somatic mutation and hybridization, has also been enumerated. In addition, an assessment of the achievements and limitations of the basic genetics and conventional breeding efforts has been presented.

It is a hard truth that in many instances we depend too much on a few advanced technologies, we are trained in, for creating and using novel or alien genes but forget the infinite wealth of desirable genes in the indigenous cultivars and wild allied species besides the available germplasms in national and international institutes or centers. Exploring as broad as possible natural genetic diversity not only provides information on availability of target donor genes but also on genetically divergent genotypes, botanical varieties, subspecies, species and even genera to be used as potential parents in crosses to realize optimum genetic polymorphism required for mapping and breeding. Genetic divergence has been evaluated using the available tools at a particular point of time. We included discussions on phenotype-based strategies employing morphological markers, genotype-based strategies employing molecular markers; the statistical procedures utilized; their utilities for evaluation of genetic divergence among genotypes, local landraces, species and genera; and also on the effects of breeding pedigrees and geographical locations on the degree of genetic diversity.

Association mapping using molecular markers is a recent strategy to utilize the natural genetic variability to detect marker-trait association and to validate the genomic locations of genes, particularly those controlling the quantitative traits. Association mapping has been employed effectively in genetic studies in human and other animal models and those have inspired the plant scientists to take advantage of this tool. We included examples of its use and implication in some of the volumes that devote to the plants for which this technique has been successfully employed for assessment of the degree of linkage disequilibrium related to a particular gene or genome, and for germplasm enhancement.

Genetic linkage mapping using molecular markers have been discussed in many books, reviews and book series. However, in this series, genetic mapping has been discussed at length with more elaborations and examples on diverse markers including the anonymous type 2 markers such as RFLPs, RAPDs, AFLPs, etc. and the gene-specific type 1 markers such as EST-SSRs, SNPs, etc.; various mapping populations including $F_2$, backcross, recombinant inbred, doubled haploid, near-isogenic and pseudotestcross; computer software including MapMaker, JoinMap, etc. used; and different types of genetic maps including preliminary, high-resolution, high-density, saturated, reference, consensus and integrated developed so far.

Mapping of simply inherited traits and quantitative traits controlled by oligogenes and polygenes, respectively has been deliberated in the earlier literature crop-wise or crop group-wise. However, more detailed information on mapping or tagging oligogenes by linkage mapping or bulked segregant analysis, mapping polygenes by QTL analysis, and different computer software employed such as MapMaker, JoinMap, QTL Cartographer, Map Manager, etc. for these purposes have been discussed at more depth in the present volumes.

The strategies and achievements of marker-assisted or molecular breeding have been discussed in a few books and reviews earlier. However, those mostly deliberated on the general aspects with examples drawn mainly from major plants. In this series, we included comprehensive descriptions on the use of molecular markers for germplasm characterization, detection and maintenance of distinctiveness, uniformity and stability of genotypes, introgression and pyramiding of genes. We have also included elucidations on the strategies and achievements of transgenic breeding for developing genotypes particularly with resistance to herbicide, biotic and abiotic stresses; for biofuel production, biopharming, phytoremediation; and also for producing resources for functional genomics.

A number of desirable genes and QTLs have been cloned in plants since 1992 and 2000, respectively using different strategies, mainly positional cloning and transposon tagging. We included enumeration of these and other strategies for isolation of genes and QTLs, testing of their expression and their effective utilization in the relevant volumes.

Physical maps and integrated physical-genetic maps are now available in most of the leading crop and model plants owing mainly to the BAC, YAC, EST and cDNA libraries. Similar libraries and other required genomic resources have also been developed for the remaining crops. We have devoted a section on the library development and sequencing of these resources; detection, validation and utilization of gene-based molecular markers; and impact of new generation sequencing technologies on structural genomics.

As mentioned earlier, whole genome sequencing has been completed in one model plant (Arabidopsis) and seven economic plants (rice, poplar, peach, papaya, grapes, soybean and sorghum) and is progressing in an array of model and economic plants. Advent of massively parallel DNA sequencing using 454-pyrosequencing, Solexa Genome Analyzer, SOLiD system, Heliscope and SMRT have facilitated whole genome sequencing in many other plants more rapidly, cheaply and precisely. We have included extensive coverage on the level (national or international) of collaboration and the strategies and status of whole genome sequencing in plants for which sequencing efforts have been completed or are progressing currently. We have also included critical assessment of the impact of these genome initiatives in the respective volumes.

Comparative genome mapping based on molecular markers and map positions of genes and QTLs practiced during the last two decades of the last century provided answers to many basic questions related to evolution, origin and phylogenetic relationship of close plant taxa. Enrichment of genomic resources has reinforced the study of genome homology and synteny of genes among plants not only in the same family but also of taxonomically distant families. Comparative genomics is not only delivering answers to the questions of academic interest but also providing many candidate genes for plant genetic improvement.

The 'central dogma' enunciated in 1958 provided a simple picture of gene function—gene to mRNA to transcripts to proteins (enzymes) to metabolites. The enormous amount of information generated on characterization of transcripts, proteins and metabolites now have led to the emergence of individual disciplines including functional genomics, transcriptomics, proteomics and metabolomics. Although all of them ultimately strengthen the analysis and improvement of a genome, they deserve individual deliberations for each plant species. For example, microarrays, SAGE, MPSS for transcriptome analysis; and 2D gel electrophoresis, MALDI, NMR, MS for proteomics and metabolomics studies require elaboration. Besides transcriptome, proteome or metabolome QTL mapping and application of transcriptomics, proteomics and metabolomics in genomics-assisted breeding are frontier fields now. We included discussions on them in the relevant volumes.

The databases for storage, search and utilization on the genomes, genes, gene products and their sequences are growing enormously in each second and they require robust bioinformatics tools plant-wise and purpose-wise. We included a section on databases on the gene and genomes, gene expression, comparative genomes, molecular marker and genetic maps, protein and metabolomes, and their integration.

Notwithstanding the progress made so far, each crop or model plant species requires more pragmatic retrospect. For the model plants we need to answer how much they have been utilized to answer the basic questions of genetics and genomics as compared to other wild and domesticated species. For the economic plants we need to answer as to whether they have been genetically tailored perfectly for expanded geographical regions and current requirements for green fuel, plant-based bioproducts and for improvements of ecology and environment. These futuristic explanations have been addressed finally in the volumes.

We are aware of exclusions of some plants for which we have comprehensive compilations on genetics, genomics and breeding in hard copy or digital format and also some other plants which will have enough achievements to claim for individual book volume only in distant future. However, we feel satisfied that we could present comprehensive deliberations on genetics, genomics and breeding of 30 model and economic plants, and their groups in a few cases, in this series. I personally feel also happy that I could work with many internationally celebrated scientists who edited the book volumes on the leading plants and plant groups and included chapters authored by many scientists reputed globally for their contributions on the concerned plant or plant group.

We paid serious attention to reviewing, revising and updating of the manuscripts of all the chapters of this book series, but some technical and formatting mistakes will remain for sure. As the series editor, I take complete responsibility for all these mistakes and will look forward to the readers for corrections of these mistakes and also for their suggestions for further improvement of the volumes and the series so that future editions can serve better the purposes of the students, scientists, industries, and the society of this and future generations.

Science publishers, Inc. has been serving the requirements of science and society for a long time with publications of books devoted to advanced concepts, strategies, tools, methodologies and achievements of various science disciplines. Myself as the editor and also on behalf of the volume editors, chapter authors and the ultimate beneficiaries of the volumes take this opportunity to acknowledge the publisher for presenting these books that could be useful for teaching, research and extension of genetics, genomics and breeding.

Chittaranjan Kole

# Preface to the Volume

Members of the genus *Populus* have long fascinated researchers, land managers, funding agencies, and the general public. Dr. Ernst J. Schreiner's early hybridization studies, sponsored by the Oxford Paper Company in the early 1920s, represent one of the first large-scale efforts to genetically improve trees through breeding. Spontaneous aspen polyploids captured the attention of one of the fathers of modern population genetics, Dr. Henrik Nilsson-Ehle in the early 20th century in Sweden, and this fascination with *Populus* polyploids, both ancient and recent, continue to intrigue researchers to this day. *Populus* has consistently garnered substantial attention from researchers, even as its commercial importance in the forest industry has waxed and waned. This interest has spiked in recent years, as genomics tools have rapidly developed, driven in large part by the sequencing of the genome and the rapid development of genomics technologies. The time is therefore ripe for a compendium of reviews that provides an overview of *Populus* biology, with particular focus on the genetic and genomic breakthroughs that have occurred in recent years.

The purpose of this book is therefore to highlight the development of *Populus* as a model organism for plant genomics. The first several chapters set the stage by enumerating the tools and resources that have been developed for *Populus* in recent years, including an introduction to poplar (Chapter 1), an overview of the genetic and genomic resources that have been created (Chapters 2), discussion on various methods for haplotyping complex traits (Chapter 3), a description of the whole genome sequence in poplar (Chapter 4), and reviews of the state-of-the art of genomics (Chapter 5), proteomics (Chapter 6) and metabolomics (Chapter 7) in *Populus*.

The second part of the book describes applications of genomics resources to the understanding of key aspects of the biological features of *Populus*, with particular focus on those features that set this organism apart from other plant model species. Chapter 8 provides an extensive overview of transcription factors that are involved in growth and development in *Populus*, with a focus on aspects of growth that are unique to trees, including crown architecture and secondary growth resulting in the development of wood. The role of the phytohormone auxin in regulating these aspects of growth and development is reviewed in detail in Chapter 9, which covers major aspects of auxin biosynthesis and signal transduction, as experimentally determined in *Populus*. Chapter 10 further explores another aspect that is specific to perennial plants, the annual dormancy cycle. Chapter 11 focuses on yet another key feature of the dormancy and developmental process in perennial plants: the timing of flowering, both on an annual basis and as a maturation process over the lifespan of an organism. Another economically and scientifically important aspect of *Populus* biology is the production of secondary cell walls. Chapter 12 characterizes pathways leading to the production of phenylpropanoids and their derivatives, which are essential components of the secondary cell wall matrix, and which play an important role in determining the commercial and ecological characteristics of *Populus* genotypes. The final contributed chapter (13) looks beyond the confines of poplar trees to examine interactions with other organisms. This chapter examines the intricate association between *Populus* and the arbuscular mycorrhizal fungus *Glomus*, and how this association changes at the molecular level upon exposure to elevated carbon dioxide and ozone.

This volume brings together contributions from 51 authors representing 24 institutions and six countries. We believe that their thorough reviews and analyses of *Populus* genomics and genetics research will provide insights into future discoveries about the basic biology of this fascinating genus, and pave the way for advances in applied breeding and genetic improvement. These are indeed exciting times to be working in this field, and we feel privileged to be part of such a collaborative community of talented scientists. We thank the authors for their hard work and patience in producing this volume. Expert editorial assistance provided by Ms. Sandra Hubscher, and editorial direction by Dr. Chittaranjan Kole is also gratefully acknowledged.

Chandrashekhar P. Joshi
Stephen P. DiFazio
Chittaranjan Kole

# Contents

# List of Contributors

**Victoria Allison**
Ministry of Agriculture and Forestry, Auckland, New Zealand.
Tel: +64 09-909-3510
Fax: +64 09-909-3505
Email: *victoria.allison@maf.govt.nz*

**Manojit M. Basu**
Environmental Sciences Division, PO BOX 2008, Oak Ridge National Laboratory, TN 37831, USA.
Tel: 865-576-3918
Fax: 865-574-0133
Email: *basumm@ornl.gov*

**Frank Bedon**
INRA, UMR1202 BIOGECO, 69 Route d'Arcachon, Cestas F-33612, France.
Tel: +33 5-57-12-28-89
Fax: +33 5-57-12-28-81
Email: *bedon@pierroton.inra.fr*

**Eric P. Beers**
Department of Horticulture, Virginia Polytechnic Institute and State University, Blacksburg, VA 24061-0327, USA.
Tel: 540-231-3210
Fax: 540-231-3083
Email: *ebeers@vt.edu*

**Catherine I. Benedict**
School of Forest Resources and Conservation, Genetics Institute, University of Florida, PO Box 110410, Gainesville, FL 32611, USA.
Tel: 352-273-8193
Fax: 352-846-1277
Email: *cibenedict@ufl.edu*

**Ludovic Bonhomme**
UMR de GENETIQUE   VEGETALE, INRA/Univ Paris-Sud/CNRS/
AgroParisTech, Ferme du Moulon, Gif-sur-Yvette, F-91190, France.
Tel: +33 1-69-33-23-65
Fax: +33 1-69-33-23-40
Email: *bonhomme@moulon.inra.fr*

**Amy M. Brunner**
Department of Forest Resources and Environmental Conservation, Virginia
Polytechnic Institute and State University, Blacksburg, VA 24061-0324,
USA.
Tel: 540-231-3165
Fax: 540-231 3698
Email: *abrunner@vt.edu*

**Leland J. Cseke Ph.D.**
Dept. of Biological Sciences, University of Alabama in Huntsville,
Huntsville, AL 35899, USA.
Tel: 256-824-6774
Fax : 256-824-6305
Email: *csekel@uah.edu*

**Yuehua Cui**
Department of Statistics and Probability, Michigan State University, East
Lansing, MI 48824, USA.
Tel: 517-432-7098
Fax: 517-432-1405
Email: *cui@stt.msu.edu*

**Stephen P. DiFazio**
Department of Biology, West Virginia University, Morgantown, West
Virginia 26506-6057, USA.
Tel: 304-293-5201 ext 31512
Fax: 304- 293-6363
Email: *spdifazio@mail.wvu.edu*

**Carl J. Douglas**
Department of Botany, University of British Columbia, Vancouver BC V6T
1Z4, Canada.
Tel: 604-822-2618
Fax : 604-822-6089
Email: *cdouglas@interchange.ubc.ca*

**Derek R. Drost**
Graduate Program in Plant Molecular and Cellular Biology, School of Forest Resources and Conservation Genetics Institute, University of Florida, PO Box 110410, Gainesville, FL 32611, USA.
Tel: 352-273-8193
Fax: 352-846-1277
Email: *ddrost@ufl.edu*

**Sébastien Duplessis**
Interactions Arbres/Micro-organismes, Centre INRA de Nancy, UMR1136 INRA/Université Nancy, Champenoux 54280, France.
Tel: +33 3-83-39-40-13
Fax: +33 3-83-39-40-69
Email: *duplessi@nancy.inra.fr*

**Jürgen Ehlting**
Centre for Forest Biology and Department of Biology, University of Victoria, Victoria BC V8W 3N5, Canada.
Tel: 250-472-5091
Fax: 250-721-6611
Email: *je@uvic.ca*

**Kyung-Hwan Han**
Department of Forestry, Michigan State University, East Lansing, MI 48824-1222, USA
and
Department of Bioenergy Science and Technology, Chonnam National University, 300 Yongbong-Dong, Bukgu, Gwangju 500-757, Korea.
Tel: 517-353-4751
Fax: 517-432-1143
Email: *hanky@msu.edu*

**Scott A. Harding**
Warnell School of Forestry, University of Georgia, Athens, GA 30602-2152, USA.
Tel: 706-542-1239
Fax : 706-542-3910
Email: *sharding@uga.edu*

**Chuan-Yu Hsu**
Department of Forestry, Mississippi State University, PO Box 9681, Mississippi State, MS 39762, USA.
Tel: 662-325-8726
Fax: 662-325 8726
Email: *ch11@msstate.edu*

**Sara S. Jawdy**
Environmental Sciences Division, PO BOX 2008, Oak Ridge National Laboratory, TN 37831, USA.
Tel: 865-574-7833
Fax: 865-574-0133
Email: *jawdys@ornl.gov*

**Johann Joets**
UMR de GENETIQUE VEGETALE, INRA/Univ. Paris XI/CNRS/INA PG, Ferme du Moulon, Gif-sur-Yvette F-91190, France.
Tel: +33 1-69-33-23-78
Fax: +33 1-69-33-23-40
Email: *joets@moulon.inra.fr*

**Virgil E. Johnson**
School of Forestry and Natural Resources, University of Georgia, Athens, Georgia 30602, USA.
Tel: 706-542-1271
Fax : 706-542-3910
Email : *vedjohns@uga.edu*

**Yves Jolivet**
UMR1137 Ecologie et Ecophysiologie Forestières, INRA/UHP, Nancy-Université BP239, Vandoeuvre-lès-Nancy Cedex F-54506, France.
Tel: +33 3-83-68-42-43
Fax: +33 3-83-68-42-40
Email: *jolivet@scbiol.uhp-nancy.fr*

**Chandrashekhar P. Joshi**
School of Forest Resources and Environmental Science, Michigan Technological University, 1400 Townsend Drive, Houghton, MI 49931, USA.
Tel: 906-487-3480
Fax: 906-487-2915
Email: *cpjoshi@mtu.edu*

**Udaya C. Kalluri**
Biosciences Division, PO BOX 2008, Oak Ridge National Laboratory, TN 37831, USA.
Tel: 865-576-9495
Fax: 865-576-9939
Email: *kalluriudayc@ornl.gov*

**Daniel E. Keathley**
Department of Forestry, Michigan State University, East Lansing, MI 48824-1222, USA.
Tel: 517-355-5191 x 1374
Fax : 517-353-0890
Email : *keathley@msu.edu*

**Matias Kirst**
School of Forest Resources and Conservation and Genetics Institute, Genetics Institute, University of Florida, PO Box 110410, Gainesville, FL 32611, USA.
Tel: 352-846-0900
Fax: 352-846-1277
Email: *mkirst@ufl.edu*

**Jae-Heung Ko**
Department of Plant & Environmental New Resources, Kyung Hee University, Yongin, 446-701, Republic of Korea.
Tel: 82-31-201-3863
Fax: 82-10-2687-5365
Email: *ko@msu.edu*

**Christer Larsson**
Department of Biochemistry, Lund University, Box 124, Lund S-22100, Sweden.
Tel: +46 46 222 8885
Fax: +46 46 222 4116
Email: *Christer.Larsson@plantbio.lu.se*

**Yao Li**
Dept. of Statistics, P.O. Box 6330, West Virginia University, Morgantown, WV 26506, USA.
Tel: 304-293-3607 ext. 1063
Fax: 304-293-2272
Email: *yli@stat.wvu.edu*

**Tian Liu**
Human Genetics Group, Genome Institute of Singapore, 60 Biopolis Street, Singapore 138672.
Tel: +65 6808-8000
Fax: +65 6808-8292
Email: *liut2@gis.a-star.edu.sg*

**Shawn D. Mansfield**
Department of Wood Science, Faculty of Forestry, University of British Columbia, 4030-2424 Main Mall, Vancouver, BC, V6T 1Z4, Canada.
Tel: 1-604-822-0196
Fax: 1-604-822-9104
Email: *shawn.mansfield@ubc.ca*

**R. Michael Miller**
Biosciences Division, 9700 South Cass Avenue, Argonne National Laboratory, Argonne, IL 60439, USA.
Tel:  630-252-3395
Fax: 630-252-8895
Email: *rmmiller@anl.gov*

**Domenico Morabito**
UFR-Faculté des Sciences, Laboratoire de Biologie des Ligneux et des Grandes Cultures, Université d'Orléans, UPRES EA 1207, rue de Chartres, BP 6759, Orléans Cedex 45067 02, France.
Tel: +33 2-38-41-72-35
Fax: +33 2-38-49-40-89
Email: *domenico.morabito@univ-orleans.fr*

**Robert Nilsson**
Department of Forest Genetics and Plant Physiology, Swedish University of Agricultural Sciences, Umeå Plant Science Centre, Umeå, 90183, Sweden.
Tel: +46 90-786-8326
Fax: +46 90-786-5901
Email: *Robert.Nilsson@genfys.slu.se*

**Evandro Novaes**
School of Forest Resources and Conservation, Genetics Institute, University of Florida, PO Box 110410, Gainesville, FL 32611, USA.
Tel: 352-273-8193
Fax: 352-846-1277
Email: *evandro@ufl.edu*

**Sunchung Park**
Department of Forestry, Michigan State University, East Lansing, MI 48824-1222, USA.
Tel: 517-353-3205
Fax: 517-432-1143
Email: *parksu16@msu.edu*

**Olga Pechanova**
Department of Biochemistry and Molecular Biology, Mississippi State University, Starkville, MS 39762, USA.
Tel: 662-325-2640
Fax: 662-325-8664
Email: *op2@msstate.edu*

**Christophe Plomion**
INRA, UMR1202 BIOGECO, 69 Route d'Arcachon, Cestas F-33612, France.
Tel: +33 5-57-12-28-38
Fax: +33 5-57-12-28-81
Email: *plomion@pierroton.inra.fr*

**G.K. Podila**
Dept. of Biological Sciences, University of Alabama, Huntsville, AL 35899, USA.
Tel: 256-824-6263
Fax: 256-824-6305
Email: *podilag@uah.edu*
Note: Dr. Podila passed away on February 12, 2010.

**Priya Ranjan**
Biosciences Division, Oak Ridge National Laboratory, Oak Ridge, Tennessee 37831, USA.
Tel: 865-574-5870
Fax: 865-576-9939
Email: *ranjanp@ornl.gov*

**Jenny Renaut**
Centre de Recherche Public - Gabriel Lippmann, Department of Environment and Agrobiotechnologies (EVA), Proteomics Platform, 41 rue du Brill, Belvaux L-4422, Luxembourg.
Tel: +3 52 470261 860
Fax: +3 52 470264
Email: *renaut@lippmann.lu*

**Andrew R. Robinson**
Department of Wood Science, Faculty of Forestry, University of British Columbia, 4030-2424 Main Mall, Vancouver, BC, V6T 1Z4, Canada.
Tel: 1-604-375-4389
Fax: 1-604-822-9104
Email: *andrewrobinsonnz@gmail.com*

**Gancho T. Slavov**
Department of Biology, Morgantown, West Virginia 26506-6057, USA.
Tel: 304-293-5201 ext 31480
Fax: 304-293-6363
Email: *gancho.slavov@mail.wvu.edu*

**Steven H. Strauss**
Department of Forests and Society, Oregon State University, Corvallis, OR 97331-5752, USA.
Tel: 541-760-7357
Fax: 541-737-1393
Email: *steve.strauss@oregonstate.edu*

**Chunfa Tong**
Center for Statistical Genetics, Pennsylvania State University, Hershey, PA 17033, USA.
Tel: 717-531-0008 ext 289580
Fax: 717-531-0480
Email: *ctong@hmc.psu.edu*

**Chung-Jui Tsai**
School Forestry and Natural Resources, and Department of Genetics, University of Georgia, 170 Green Street, Athens, Georgia 30602, USA.
Tel: 706-542-1271
Fax: 706-542-3910
Email: *cjtsai@warnell.uga.edu*

**Gerald A. Tuskan**
Biosciences Division, Oak Ridge National Laboratory, Oak Ridge, Tennessee 37831, USA.
and
Laboratory Science Program, Joint Genome Institute, Walnut Creek, California 94598, USA.
Tel: 865-576-8141
Fax: 865-576-9939
Email: *tuskanga@ornl.gov*

**Delphine Vincent**
INRA, UMR1202 BIOGECO, 69 Route d'Arcachon, Cestas F-33612, France.
Tel: +33 5 57 12 28 19
Fax: +33 5 57 12 28 81
Email: *vincent@pierroton.inra.fr*

**Holly L. White**
DIATHERIX Laboratories, Inc., Huntsville, AL 35806.
Tel: 256-327-9442
Fax : 256-327-5296
Email: *holly.white@diatherix.com*


**Gunnar Wingsle**
Department of Forest Genetics and Plant Physiology, Swedish University of
Agricultural Sciences, Umeå Plant Science Centre, Umeå 90183, Sweden.
Tel: +46 90 786 8326
Fax: +46 90 786 5901
Email: *Gunnar.Wingsle@genfys.slu.se*

**Jiasheng Wu**
School of Forestry and Biotechnology, Zhejiang Forestry University, Lin'an,
Zhejiang 311300 People's Republic of China.
Tel: 86-571-63743858
Fax: 86-571-63732738
Email: *jswu2005@hotmail.com*

**Rongling Wu**
Center for Statistical Genetics, Pennsylvania State University, Hershey, PA
17033, USA.
Tel: 717-531-2037
Fax: 717-531-0480
Email: *rwu@hes.hmc.psu.edu*

**Xiaohan Yang**
Biosciences Division, Oak Ridge National Laboratory, Oak Ridge, TN
37831, USA.
Tel: 865-241-6895
Fax: 865-576-9939
Email: *yangx@ornl.gov*

**Cetin Yuceer**
Department of Forestry, PO Box 9681, Mississippi State University,
Starkville, MS 39762, USA.
Tel: 662-325-2795
Fax: 662-325-8726
Email: *mcy1@msstate.edu*

**Yanru Zeng**
School of Forestry and Biotechnology, Zhejiang Forestry University, Lin'an,
Zhejiang 311300 People's Republic of China.
Tel: 86-571-63743858
Fax: 86-571-63732738
E-mail: *zengyr@hotmail.com* and *yrzeng@zjfc.edu.cn*

**Bo Zhang**
The Poplar Research Institute, Key Laboratory of Forest Genetics &
Biotechnology, Nanjing Forestry University, Nanjing, Jiangsu 210037
People's Republic of China.
Tel: 86-25-85427412
Fax: 86-25-85427412
Email: *bo.zhang@plantphys.umu.se*

# Abbreviations

| | |
|---|---|
| 1-DE | one-dimensional gel electrophoresis |
| 2-DE | Two-dimensional gel electrophoresis |
| 2-ME | 2-Mercaptoethanol |
| 4-CL | 4-Coumarate-CoA ligase |
| ABP | Auxin binding protein |
| ADT | Arogenate dehydratase |
| AFLP | Amplified fragment length polymorphism |
| AGF1 | AT-HOOK PROTEIN OF GA FEEDBACK |
| *AIL5* | *ANT-LIKE5* |
| AM | Arbuscular mycorrhizal fungi |
| *ANT* | *AINTEGUMENTA* |
| *AP1* | *APETALA1* |
| *AP2* | *APETALA2* |
| *APL* | *ALTERED PHLOEM* |
| ARF | Auxin response factor |
| ATP | adenosine tri-phosphate |
| AUX | Auxin/Indole Acetic Acid transcriptional regulator |
| BAC | Bacterial artificial chromosome |
| *BAN* | *BANYULS* |
| bHLH | Basic Helix-Loop-Helix transcription factor |
| BHT | Butylated hydroxytoluene |
| *Bl* | *BLIND* |
| *BP/KNAT1* | *BREVIPEDICELLUS* |
| *BRC1* | *BRANCHED1* |
| bZIP | Basic Leucine Zipper transcription factor |
| C3H | 4-Coumaroyl-shikimate/quinate-3-hydroxlase |
| C4H | Cinnamate-4-hydroxylase |
| CA | carrier-ampholyte |
| CAD | Cinnamyl alcohol dehydrogenase |
| CADL | Cinnamyl alcohol dehydrogenase-like |
| CaMV | Cauliflower mosaic virus |
| CBB | coomassie brilliant blue |
| CCA | Canonical correlation analysis |
| CCoAOMT | Caffeoyl-CoA O-methyltransferase |

| | |
|---|---|
| CCR | Cinnamyl CoA reductase |
| CCRL | Cinnamyl CoA reductase-like |
| CHAPS | 3-[(3-Cholamidopropyl)dimethylammonio]-1-propanesulfonate |
| CHI | Chalcone isomerase |
| CHS | Chalcone synthase |
| CM | Chorismate mutase |
| *CNA* | *CORONA* |
| *CO* | *CONSTANS* |
| COMT | Caffeic acid O-methyltransferase |
| COW | Correlation optimized warping |
| *CPC* | *CAPRICE* |
| *CRY1* | *CRYPTOCHROME 1* |
| *CRY2* | *CRYPTOCHROME 2* |
| CS | Chorismate synthase |
| CT | Condensed tannins |
| CYP | Cytochrome P450 monooxygenase |
| DB | database |
| DELLA | A motif characterizing a subgroup of the GRAS family |
| *DFR* | Dihydroflavonol 4-reductase |
| DHQD | 3-Dehydroquinate dehydratase |
| DHQD | 3-Dehydro-quinate dehydratase/shikimate dehydrogenase |
| DHQS | 3-Dehydroquinate synthase |
| DHS | 3-Deoxy-arabino-heptulosonate-7-phosphate synthase |
| DIGE | Difference gel electrophoresis |
| DOC | Sodium deoxycholate |
| DOF | DNA-binding with One Finger domain transcription factor |
| DTT | Dithiothreitol |
| E4P | Erythrose-4-phosphate |
| ECM | Ectomycorrhizal fungi |
| EDTA | ethylenediaminetetraacetic acid |
| $eCO_2$ | Elevated atmospheric concentrations of carbon dioxide |
| EI | Electron ionization |
| $eO_3$ | Ozone |
| EPSPS | 5-Enolpyruvylshikimate-3-phosphate synthase |
| *ERF* | *ETHYLENE RESPONSE FACTOR* |
| ESI | Electrospray ionization |
| ESI-QTOF | electrospray ionization-quadrupole time of flight |
| EST | Expressed sequence tag |
| F5H | Ferulate-5-hydroxylase |

| | |
|---|---|
| FACE | Free-air carbon dioxide enrichment |
| | *FACTOR1* |
| FDA | Food and Drug Administration |
| *FLC* | *FLOWERING LOCUS C* |
| FLD | FLOWERING LOCUS D |
| FLK | *FLOWERING LATE KH MOTIF* |
| FRI | *FRIGIDA* |
| *FT* | *FLOWERING LOCUS T* |
| FT | Fourier transform |
| FT1 | *FLOWERING LOCUS T1* |
| FT2 | *FLOWERING LOCUS T2* |
| FUL | FRUITFULL |
| G | Guaiacyl (G) subunits |
| G × E | Genotype × environment interaction |
| GA | Gibberellins |
| *GA20ox* | *GA20-oxidase* |
| GAI | GA INSENSITIVE |
| GARP | GARP group of Myb-related transcription factors |
| GC | Gas chromatography |
| GEO | Gene Expression Omnibus |
| *GI* | *GIGANTEA* |
| GMD | Gölm Metabolite Database |
| GRAS | Plant-specific gene family named after the first three members: *GIBBERELLIC-ACID INSENSITIVE*, *REPRESSOR of GAI* and *SCARECROW* |
| GSH | glutation |
| GxE | genotype by environment interaction |
| HCA | Hierarchical cluster analysis |
| HCQ | Hydroxycinnamyl-quinate transferase, |
| *HFR1* | *LONG HYPOCOTYL IN FAR-RED1* |
| HILC | Hydrophilic interaction chromatography |
| HPLC | High pressure liquid chromatography |
| HSP | heat shock protein |
| IAA | Indole-3-acetic acid |
| IAM | Indole-3-acetamide |
| IAOx | Indole-3-acetaldoxime |
| IEF | Isoelectric focusing |
| *IFL1* | *INTERFASCICULAR FIBER1* |
| InDel | Insertion/deletion |
| INRA | institute national de la recherche agronomique |
| IPA | Indole-3-pyruvic acid |
| IPG | Immobilized pH gradient |
| IPGC | International *Populus* Genome Consortium |

| | |
|---|---|
| IPCC | intergovernmental panel on climate change |
| iTRAQ | Isobaric multiplex tagging |
| JGI | Joint Genome Institute |
| *KAN1* | *KANADI1* |
| LAX | Like-AUX |
| LC/MS | Liquid-chromatograph Mass Spectrometry |
| LCM | Laser capture microdissection |
| *LD* | *LUMINIDEPENDENS* |
| LD | Linkage disequilibrium |
| LDC | Long-day and cold conditions |
| LDD | Long-day drought |
| LDs | Long daylengths |
| LDW | Long-day warm grown plants |
| *LFY* | *LEAFY* |
| *LMI2* | *LATE MERISTEM IDENTITY2* |
| LRR | Leucine-rich repeat |
| *Ls* | *LATERAL SUPPRESSOR* |
| MADS | MADS box transcription factor |
| MALDI | Matrix assisted laser desorption ionization- |
| MALDI-TOF-MS | matrix-assisted laser desorption ionization-time of flight-mass spectrometry |
| MAQC | Microarray quality control |
| MDA | Multiple discriminant analysis |
| MDR | Multi-drug resistance |
| miRNA | MicroRNA |
| mRNA | messenger RNA |
| MLE | Maximum likelihood estimate |
| MS | Mass spectrometry |
| MS | Murashige and Skoog |
| MS/MS | tandem MS |
| *MS35* | *MYB26/MALESTERILE35* |
| MSTFA | N-Methyl-N-trimethylsilyltrifluoroacetamide |
| MW | Molecular weight |
| MYB | Myb transcription factor |
| NAC | NAC domain transcription factor |
| NADP | nicotinamide adenine dinucleotide phosphate |
| nat-siRNA | Natural antisense RNA |
| NBS-LRR | nucleotide-binding site leucine-rich repeat |
| NCBI | National Center for Biotechnology Information |
| nLC | Nanoscale liquid chromatography |
| nLC-ESI-MS/MS | nanoscale liquid chromatography-electrospray ionization-mass spectrometry |
| NLFA | Neutral lipid fatty acid |

| | |
|---|---|
| NMR | Nuclear Magnetic Resonance |
| NP | not provided |
| NPA | 1-Naphthylphthalamic acid |
| NPGI | National Plant Genome Initiative |
| NPP | Net primary production |
| *NPR1* | *NONEXPRESSOR OF PR GENES1* |
| NRC | National Research Council (Canada) |
| Nr NCBI | non redundant database of the NCBI |
| *NST1* | *NAC SECONDARY WALL THICKENING PROMOTING* |
| O2PLS | Orthogonal 2 partial least squares |
| *OBP2* | *OBF BINDING PROTEIN2* |
| ORNL | Oak Ridge National Laboratory (TN, USA) |
| PAGE | Polyacrylamide gel electrophoresis |
| PAL | Phenylalanine ammonia-lyase |
| PAT | Polar auxin transport |
| PCA | Principal components analysis |
| PDA | Photodiode array |
| PEP | Phosphoenolpyruvate |
| PG | Phenolic glycosides |
| PGP | p-glycoproteins |
| *PHB* | *PHABULOSA* |
| *PHV* | *PHAVOLUTA* |
| | $p$-hydroxyphenyl (H) subunits |
| PHYA | *PHYTOCHROME A* |
| PHYB | *PHYTOCHROME A* |
| pI | isoelectric point |
| PIF3 | PHYTOCHROME INTERACTING FACTOR3 |
| PIN | PINFORMED mutant, also the Auxin efflux carrier |
| PLFA | Phospholipid fatty acid |
| PLSR | Partial least squares regression |
| PM | Plasma membrane |
| PMF | Peptide mass fingerprinting |
| PMFQ | peptide mass fingerprint data quality |
| PMSF | phenylmethanesulphonylfluoride |
| PNNL | Pacific Northwest National Laboratory (Richland, WA, USA) |
| PNT | Prephenate aminotransferase |
| PR | Pathogenesis-related |
| PSII | photosystem II |
| PTM | Post-translational modification |
| PVP | polyvinyl pyrrolidone |
| QTL | Quantitative trait loci |
| QTOF | Quadrupole time of flight |

| | |
|---|---|
| RAPD | Random amplified polymorphic DNA |
| *RAX* | *REGULATOR OF AXILLARY MERISTEMS* |
| *REV* | *REVOLUTA* |
| RFLP | Restriction fragment length polymorphisms |
| *RGA* | *REPRESSOR OF GA1-3* |
| ROS | reactive oxygen species |
| RSG | REPRESSION OF SHOOT GROWTH |
| SAM | Shoot apical meristem |
| *SAUR* | *Small auxin-induced RNA* |
| SD | Short-day |
| SDC | Short-day and cold conditions |
| SDH | Shikimate dehydrogenase |
| SDS | Sodium dodecyl sulphate |
| SDW | Short-day and warm |
| siRNA | Small interfering RNA |
| SK | Shikimate kinase |
| SNP | Single nucleotide polymorphism |
| *SOC1* | *SUPPRESSOR OF OVEREXPRESSION OF CONSTANS* |
| SOD | super oxyde dismutase |
| SP | SwissProt |
| SPY | *SPINDLY* |
| SSR | Simple sequence repeats |
| S | Syringyl (S) subunits |
| TAM | Tryptamine |
| tasiRNA | Trans-acting RNA |
| *TB1* | *TEOSINTE BRANCHED1* |
| TCA | Trichloroacetic acid |
| *TCP* | TCP domain transcription factor |
| TF | Transcription factors |
| *TFL1* | *TERMINAL FLOWER1* |
| TM | trans membrane |
| TOF | Time of flight |
| TrEMBL | translated European Molecular Biology Laboratory |
| *TSF* | *TWIN SISTER of FT* |
| *TT8* | *TRANSPARENT TESTA8* |
| U-HPLC | Ultra-high pressure liquid chromatography |
| UV | Ultraviolet |
| VIN3 | *VERNALIZATION INSENSITIVE 3* |
| *VND1* | *VASCULAR NAC DOMAIN1* |
| WRKY | WRKY transcription factor |
| *XND1* | *XYLEM NAC DOMAIN1* |

# 1

# *Populus*: A Premier Pioneer System for Plant Genomics

*Stephen P. DiFazio,*[1,a,]* *Gancho T. Slavov*[1,b] and
*Chandrashekhar P. Joshi*[2]

## ABSTRACT

The genus *Populus* has emerged as one of the premier systems for studying multiple aspects of tree biology, combining diverse ecological characteristics, a suite of hybridization complexes in natural systems, an extensive toolbox of genetic and genomic tools, and biological characteristics that facilitate experimental manipulation. Here we review some of the salient biological characteristics that have made this genus such a popular object of study. We begin with the taxonomic status of *Populus*, which is now a subject of ongoing debate, though it is becoming increasingly clear that molecular phylogenies are accumulating. We also cover some of the life history traits that characterize the genus, including the pioneer habit, long-distance pollen and seed dispersal, and extensive vegetative propagation. In keeping with the focus of this book, we highlight the genetic diversity of the genus, including patterns of differentiation among populations, inbreeding, nucleotide diversity, and linkage disequilibrium for species from the major commercially-important sections of the genus. We conclude with an overview of the extent and rapid spread of global *Populus* culture, which is a testimony to the growing economic importance of this fascinating genus.

**Keywords:** *Populus*, SNP, population structure, linkage disequilibrium, taxonomy, hybridization

[1]Department of Biology, West Virginia University, Morgantown, West Virginia 26506-6057, USA;
[a]e-mail: *spdifazio@mail.wvu.edu*
[b]e-mail: *gancho.slavov@mail.wvu.edu*
[2]School of Forest Resources and Environmental Science, Michigan Technological University, 1400 Townsend Drive, Houghton, MI 49931, USA; e-mail: *cpjoshi@mtu.edu*
*Corresponding author

## 1.1 Introduction

The genus *Populus* is full of contrasts and surprises, which combine to make it one of the most interesting and widely-studied model organisms. *Populus* seeds are among the smallest produced by North American trees (Hewitt 1998), yet these tiny propagules ultimately yield some of the fastest-growing and largest angiosperm trees in the temperate regions (Dickmann 2001), and some genotypes reach astounding sizes by spreading vegetatively across the landscape (Mock et al. 2008). These tiny seeds carry very little endosperm and therefore require nearly optimal conditions for establishment, yet this tree occurs in some of the harshest environments in the world, including nutrient-poor sand bars that are subject to frequent flooding and scouring, boreal landscapes with severe winters and short growing seasons, and even harsh desert climates with xeric, saline, and alkaline soils.

*Populus* is also renowned (and reviled) for being almost completely intolerant of competition from other species, and for being highly susceptible to numerous pathogens and herbivores, thereby acquiring a reputation among land managers as a *"prima donna "* (Dickmann et al. 1987). Consequently, *Populus* is often confined to early successional stages, and is therefore an ephemeral presence in some systems. However, *Populus* is also eminently adapted for a pioneer life history, and it thrives on the cycles of cataclysmic disturbance that are typical of riparian ecosystems, continuously forming nearly monotypic stands on freshly deposited substrates. Furthermore, some species can attain dominance on a landscape scale, and individual clones can persist for hundreds or even thousands of years in the same locales (Kemperman and Barnes 1976; Mock et al. 2008).

The genus appears to have relatively low species diversity compared to some other tree genera, with only 30 to 40 species recognized by most taxonomists (Eckenwalder 1996). However, this relative simplicity is deceptive, because hybridization is rampant across the genus, and many species form intergrading hybrid swarms, often involving multiple species mixing in all possible combinations (Eckenwalder 1984a; Floate 2004). Closely-related sympatric species, therefore, have variable levels of introgression, which creates challenges and opportunities for systematists and evolutionary biologists interested in mechanisms of species isolation and persistence. These hybrid swarms are also zones of tremendous genetic diversity layered upon substantial environmental heterogeneity. This has created a setting in which remarkably strong genetic effects are exerted by the host tree onto the complex communities of associated organisms, which has helped to spur the development of the new field of community genetics (Whitham et al. 2006). All of this and more has grown from the humble beginnings of a tiny, wind-blown seed. In this chapter we will provide a broad overview of the biological characteristics of this genus, with a particular focus on those features that make *Populus* a pioneer model organism in comparative, functional, and ecological genomics research.

## 1.2 Systematics and Evolution

### 1.2.1 *Taxonomic Position of the Genus* Populus

*Populus* and *Salix* (willows) have traditionally been considered the only two genera in the Salicaceae family (Eckenwalder 1996). However, a number of genera formerly included in the Flacourtiaceae are now assigned to Salicaceae *sensu lato*, within the Malpighiales order of the "Eurosid I" clade (Chase et al. 2002; Angiosperm Phylogeny Group 2003). The closest genera to *Populus* and *Salix* within the Salicaceae *sensu lato* are mostly woody tropical species, including the genera *Chosenia, Idesia, Itoa, Carrierea, Bennettiodendrion*, and *Poliothyrsis*, all native to the Asian subcontinent (Cronk 2005), which is apparently the center of diversity for the Salicaceae.

Recent molecular phylogenetic studies in the Salicaceae (Leskinen and Alstrm-Rapaport 1999; Cervera et al. 2005; Hamzeh et al. 2006) have shown that *Populus* and *Salix* clearly form two separate groups. Interestingly, in one of these studies the presumably most ancient species of *Populus* (*P. mexicana*; Eckenwalder 1996) showed higher similarity to *Salix* than to any other species of *Populus* (Cervera et al. 2005). It still remains an open question whether *Populus* and *Salix* are truly monophyletic.

### 1.2.2 *Classification at the Species Level*

There is broad disagreement about the number of species in the genus *Populus*, with some taxonomists recognizing as few as 22 species and others enumerating as many as 85 species, including 53 in China alone (Eckenwalder 1996; Dickmann 2001). The difficulties in taxonomy arise because of the extensive phenotypic variation observed within broadly-distributed *Populus* species, as well as the existence of many hybrids, which blur the lines between some species, and which themselves are sometimes misclassified as separate species (Eckenwalder 1996). The classification scheme most commonly used today is that of Eckenwalder (1996), who recognized 29 species subdivided into six sections based on relative morphological similarity and crossability (Table 1-1). A consensus cladogram from the 840 most parsimonious trees built based on 76 morphological characters (Fig. 6 in Eckenwalder 1996) provided evidence that all sections except for *Tacamahaca* are monophyletic. Section *Tacamahaca* was split into two monophyletic groups, one comprised of "typical balsam poplars" (e.g., *P. balsamifera* and *P. trichocarpa*) and the other comprised of "narrow-leaved, thin-twigged" species (e.g., *P. angustifolia* and *P. simonii*).

**Table 1-1** Eckenwalder's (1996) classification of the genus *Populus*.

| Section [synonym] | Species | Distribution |
|---|---|---|
| *Abaso* Eckenwalder | *Populus mexicana* Wesmael | Mexico |
| *Turanga* Bunge | *P. euphratica* Olivier | Africa, Asia |
| | *P. ilicifolia* (Engler) Rouleau | Africa |
| | *P. pruinosa* Schrenk | Asia |
| *Leucoides* Spach | *P. glauca* Haines *sl*[b] | China |
| | *P. heterophylla* L. | N America |
| | *P. lasiocarpa* Olivier | China |
| *Aigeiros* Duby | *P. deltoides* Marshall *sl*[c] | N America |
| | *P. fremontii* S. Watson | N America |
| | *P. nigra* L. | Eurasia, N Africa |
| *Tacamahaca* Spach | *P. angustifolia* James | N America |
| | *P. balsamifera* L. | N America |
| | *P. ciliata* Royle | Himalayas |
| | *P. laurifolia* Ledebour | Eurasia |
| | *P. simonii* Carrière | E Asia |
| | *P. suaveolens* Fischer *sl*[d] | China, Japan |
| | *P. szechuanica* Schneider | Eurasia |
| | *P. trichocarpa* Torrey & A. Gray | N America |
| | *P. yunnanensis* Dode | Eurasia |
| *Populus*[a] | *P. adenopoda* Maximowicz | China |
| | *P. alba* L. | Eurasia, Africa |
| | *P. gamblei* Haines | Eurasia |
| | *P. grandidentata* Michaux | N America |
| | *P. guzmanantlensis* Vazques & Cuevas | Mexico |
| | *P. monticola* Brandegee | Mexico |
| | *P. sieboldii* Miquel | Japan |
| | *P. simaroa* Rzedowski | Mexico |
| | *P. tremula* L. | Eurasia, Africa |
| | *P. tremuloides* Michaux | N America |

[a]Synonymous with *Leuce* Duby.
[b]*Sensu lato.* Synonymous with *P. wilsonii* Schneider.
[c]Synonymous with *P. sargentii* Dode and *P. wislizenii* Sargent.
[d]Synonymous with *P. cathayana* Rehder, *P. koreana* Rehder, and *P. maximowiczii* A. Henry.

## *1.2.3 Origin and Evolution of* Populus

*Populus* appears in the fossil record after the Eocene (40 million years ago (MYA)), and probably as early as the late Paleocene (58 MYA; Eckenwalder 1996; Dickmann and Kuzovkina 2008). A representative of the *Abaso* section related to modern *P. mexicana* is the oldest clearly identifiable *Populus* fossil specimen. The swamp poplars of the section *Leucoides* appeared in the late Eocene period (40 MYA) in temperate regions of North America and Asia. Fossils from the Oligocene period (30 MYA) representing ancestors of the sections *Tacamahaca* and *Aigeiros* have been found in both the new and old worlds. Representatives of section *Populus*, however, did not appear in fossil records until the early Miocene (20 MYA; Eckenwalder 1996; Cronk 2005).

Combining fossil records with information from phylogenetic analyses, Eckenwalder (1996) speculated that the genus originated in North America or tropical Asia during the Paleocene, and that the three "advanced" sections of the genus, *Populus*, *Aigeiros*, and *Tacamahaca*, evolved rapidly in their distinctive habitats during the Miocene. Recent studies based on molecular markers mostly support the delineations of sections and their evolutionary relationships (Cervera et al. 2005; Hamzeh et al. 2006). However, numerous questions remain to be resolved. For example, analyses based on chloroplast DNA clearly group *P. nigra* with section *Populus* (Smith and Sytsma 1990; Hamzeh and Dayanandan 2004), yet analyses based on nuclear DNA and morphology clearly place *P. nigra* in section *Aigeiros* (Eckenwalder 1996; Hamzeh and Dayanandan 2004; Cervera et al. 2005). There is also some ambiguity about the position of section *Populus* (also called *Leuce*), and the taxonomic status of *P. mexicana*, which may be divergent enough to be removed from the genus (Cervera et al. 2005). Also, *P. angustifolia* groups more closely with Asian species from section *Tacamahaca* rather than with the sympatric *P. trichocarpa* and *P. balsamifera* in some molecular phylogenies (Cervera et al. 2005), raising the possibility that the current sympatry with those two species is due to secondary contact following intercontinental migration. Many of these questions are likely to be resolved as molecular data accumulate throughout this genus.

### 1.2.4 Natural Hybridization

Many problems in the taxonomy of *Populus* may prove intractable due to the extensive occurrence of hybridization and the possibility that reticulate evolution has played a role in the evolution of the modern taxa (Eckenwalder 1996; Hamzeh and Dayanandan 2004; Cervera et al. 2005; Hamzeh et al. 2006). For example, based on variation in leaf morphology across the extensive range of *P. tremuloides* and similarity to fossil aspen taxa, Barnes (1967) and Eckenwalder (1996) hypothesized that the modern species is actually a "compilospecies" derived from hybridization among multiple ancestral species. Hybridization occurs freely within sections and also between closely-related sections of the genus (Fig. 1-1), and natural hybrids have been observed in most locations where compatible species co-occur (Eckenwalder 1984a, b, c; Rood et al. 1986; Campbell et al. 1993; Martinsen et al. 2001; Floate 2004; Lexer et al. 2005; Hamzeh et al. 2007). For example, most of the five North American species from sections *Aigeiros* and *Tacamahaca* overlap in their ranges (Fig. 1-2), and most hybrid combinations are found in the wild (Eckenwalder 1984a), including some complex hybrids involving three species (Floate 2004). Interestingly, many *Populus* hybrid zones show nonrandom patterns of introgression, with crossing often occurring preferentially in the direction of one of the parental species (Keim et al. 1989; Floate 2004; Lexer et al. 2005). Furthermore, not all portions of the genome introgress equally: some genome segments introgress more than expected based on overall rates of hybridization, and some portions seem to be inhibited in their introgression (Martinsen et al. 2001; Lexer and van Loo 2006).

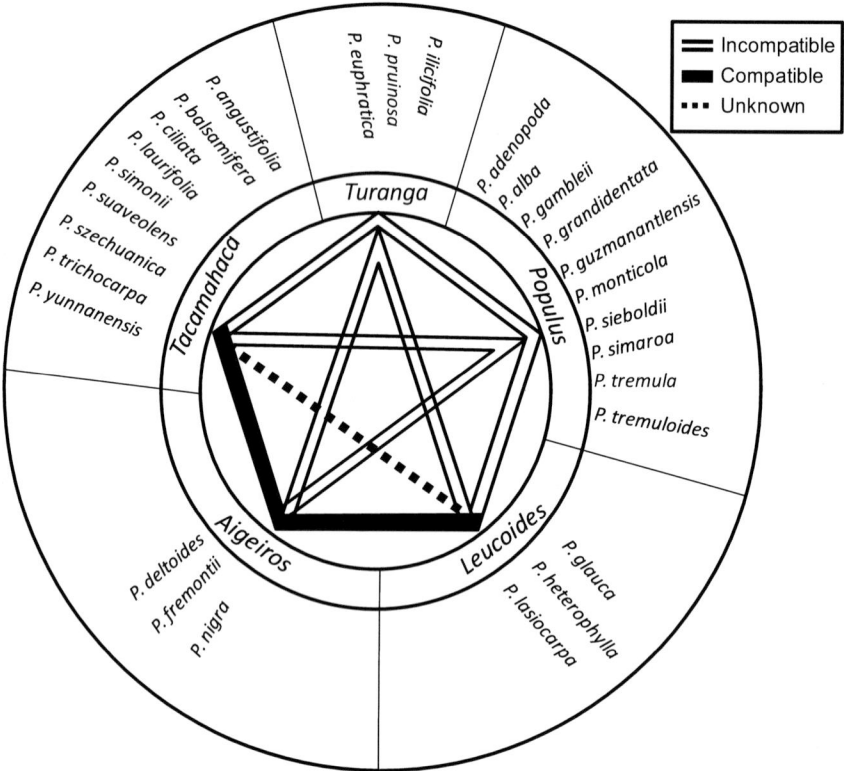

**Figure 1-1** Crossing relationships for *Populus* species. Solid lines connecting sections indicate that species from these sections intercross freely. Open lines indicate that crosses do not normally occur spontaneously, and can only be accomplished using specialized procedures. The dashed line indicates that insufficient data exist to definitively characterize compatibility of species from sections *Leucoides* and *Tacamahaca*. After Willing and Pryor (1976).

*Populus* hybrids zones have proven to be fertile areas for ecological research because of their high levels of genetic variation and a diverse and sometimes unique assemblage of dependent organisms, including arthropods (Wimp et al. 2005), microbes (Schweitzer et al. 2005), and fungi (Bailey et al. 2005). Furthermore, the biotic community is apparently determined in part by the genotypic composition of the *Populus* hybrids, which has led to the development of concepts like community heritability and even community evolution (Whitham et al. 2006). *Populus* hybrids

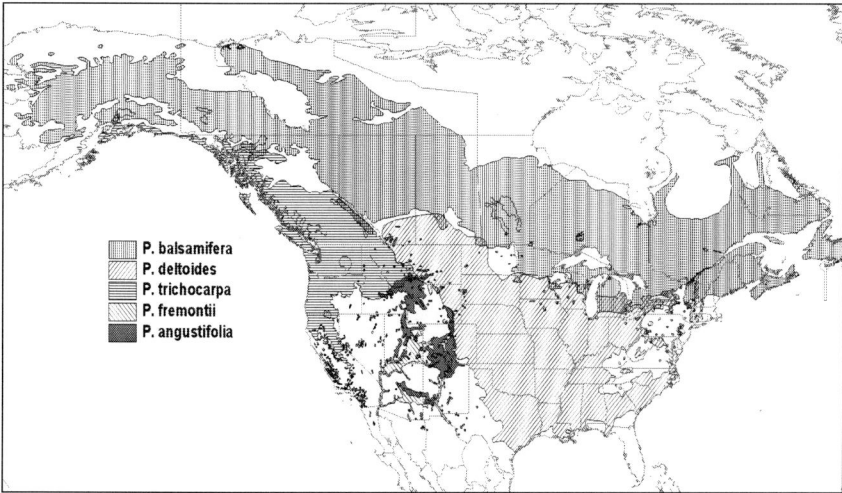

**Figure 1-2** Ranges of the North American species from sections *Aigeiros* and *Tacamahaca*. Overlap occurs among most pairs of species in some portion of their range, and hybridization zones exist in many of these zones of overlap. Range maps were created from GIS layers obtained from the USGS Earth Surface Processes web site (*http://esp.cr.usgs.gov/data/atlas/ little/*).

seem to have unique characteristics that set them apart from either of the parental species. For example, relict stands of hybrids often occur far away from the current distribution of one or both of the presumed species, suggesting that hybrids are adapted in locations where the parental species are not (Woolbright et al. submitted). Hybrid zones have therefore yielded numerous insights about the nature of species boundaries and the structure and functioning of communities and ecosystems, and the application of genomic tools in these natural laboratories is likely to accelerate the pace of discovery in many fields (Whitham et al. 2008).

## 1.3 Life History

### 1.3.1 Habitat

The supreme adaptability of *Populus* is revealed by its extensive and varied range. Although *Populus* species are mostly confined to the Northern Hemisphere, individual species span entire continents across a startling range of environments. For example, *P. tremuloides* is the most widely distributed tree in North America, spanning 111° of longitude and 48° of latitude, stretching from the west to east coasts and from northern Alaska to central Mexico (Perala 1990). *Populus* species also span a remarkable range of habitats. The cottonwoods, species from sections *Aigeiros* and *Tacamahaca*, are typically riparian (Fig. 1-3), thriving on the fresh sediments deposited by flooding (Fig. 1-4) and capitalizing on establishment opportunities created by ice scouring and even fire (Braatne et al. 1996; Rood et al. 2007). However, even within these sections there is substantial variation in habitats, with cottonwoods occurring on upland sites in northern regions

**Figure 1-3** Photos of *Populus* habitats and reproductive structures. (A) A typical *P. trichocarpa* stand on the Willamette River near Salem, Oregon. (B) Stands of *P. tremuloides* in Grand Teton National Park, Wyoming. Different genets can be distinguished based on variation in fall leaf colors. All photos by S. DiFazio.

*Color image of this figure appears in the color plate section at the end of the book.*

where desiccation is not excessive (Dickmann 2001), and even as isolated patches of trees associated with warm springs in vast stretches of arctic tundra (Zasada and Phipps 1990).

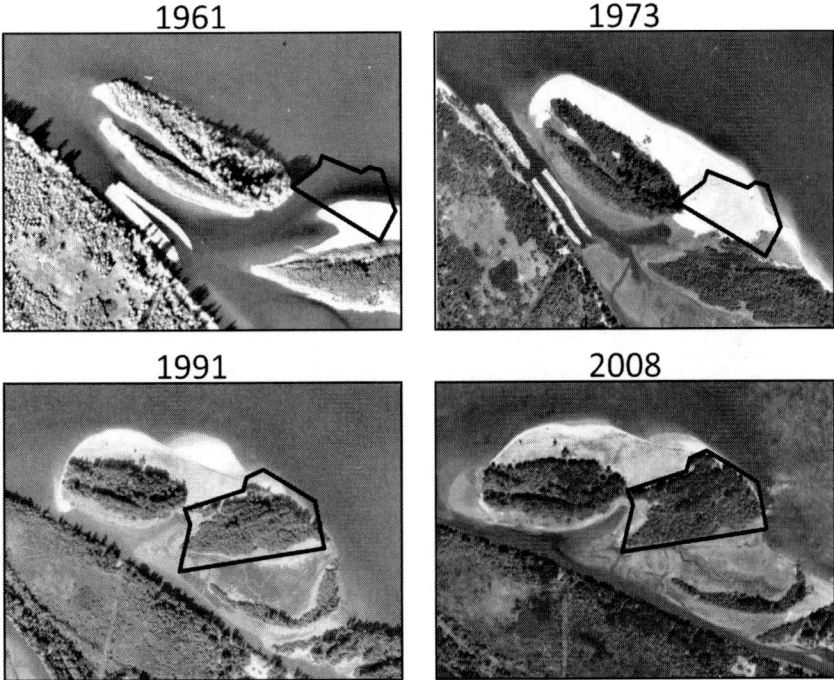

**Figure 1-4** Chronosequence of air photos depicting establishment of *P. trichocarpa* populations on Gull Island on the Columbia River near Clatskanie, Oregon. A sand bar was deposited between 1961 and 1973, and a *P. trichocarpa* stand became established between 1973 and 1983 (indicated by a black outline in each photograph). Photos courtesy of U.S. Army Corps of Engineers, except 2008 photo, which is from Google Earth.

Poplars from section *Populus* show much broader ranges of habitats. For example, *P. tremuloides* and the closely-related *P. tremula* occur in mountainous or upland habitats and mixed conifer forests as well as in nearly pure stands in riparian areas (Perala 1990; Lexer et al. 2005). Other species of this section have more restricted habitats. For example, *P. alba* is almost exclusively riparian in some parts of its range in Europe (Lexer et al. 2005), whereas it is found on a much broader range of sites in other parts of its extensive range (Dickmann 2001). *P. grandidentata*, the North American big-toothed aspen, has a more restricted range than *P. tremuloides*, and tends to occur on sites with better drainage than *P. tremuloides* in their zones of overlap (Laidly 1990; Dickmann 2001). This ecological differentiation is perhaps one of the factors that maintains the various members of section *Populus* as separate species in Europe and North America despite extensive potential for hybridization (Lexer et al. 2005).

The genus also includes several species with distinctive habitats and distributions. Section *Leucoides* presents an interesting case because this was the first section to appear in the North American fossil record after *Abaso* (Eckenwalder 1996; Cronk 2005), perhaps indicating that it is ancestral to sections *Populus, Aigeiros*, and *Tacamahaca*. However, this section is now represented by only a single species in North America, *P. heterophylla*, which has a very limited distribution, restricted to permanent swamps of the eastern US (Johnson 1990). Perhaps the most ecologically idiosyncratic section of the genus is *Turanga*, which includes *P. euphratica*, a species that grows in extremely hot and dry environments with water tables as deep as 10–13 m (Ma et al. 1997; Hukin et al. 2005). This species shows high tolerance of salinity, surviving doses approaching sea water (300 mM; Chen et al. 2002).

### 1.3.2 Sexual Reproduction

*Populus* species are mostly dioecious, with separate male and female sexes, although the occurrence of cosexual trees has been reported in multiple species (Stettler 1971; Rottenberg et al. 2000; Rowland et al. 2002; Cronk 2005; Slavov et al. 2008). *Populus* flowers occur in pendent catkins with strongly reduced perianths (Boes and Strauss 1994; Eckenwalder 1996). *Populus* trees reach reproductive maturity within 4–8 years in intensively managed plantations and within 10–15 years under favorable conditions in natural populations (Stanton and Villar 1996). Flowering usually occurs before leaf emergence in early spring (Braatne et al. 1996; Eckenwalder 1996). Individual trees flower for 1–2 weeks (Stanton and Villar 1996), but the pollination period in a population can exceed one or even two months (Braatne et al. 1996). The timing of flowering depends in part on temperature, with populations at higher elevations, more northern latitudes, and more continental climates flowering later (DeBell 1990; Perala 1990; Zasada and Phipps 1990; Braatne et al. 1996). Pollen is dispersed by wind, and effective pollination distances can be extensive (Tabbener and Cottrell 2003; Lexer et al. 2005; Pospíšková and Šálková 2006; Vanden Broeck et al. 2006; Slavov et al. 2009). Fertilization occurs within 24 hours after a viable pollen grain has landed on a receptive stigma (Braatne et al. 1996). Capsules typically dehisce 4–6 weeks after fertilization, but seed development can occur in as little as 2–3 weeks or as long as 3–5 months in some species and locations.

The timing of seed dispersal tends to coincide with annual flooding, which creates favorable sites for seed establishment for many species (Braatne et al. 1996; Stella et al. 2006). Seeds are produced in great numbers (> 25 million per tree per year; Braatne et al. 1996), and their small size and cotton-like appendages facilitate dispersal over large distances by wind and water (Johnson 1994; Braatne et al. 1996; Karrenberg et al. 2002).

Seeds normally retain viability for only 1–2 weeks in natural systems, and germination occurs within 24 hours under warm, moist conditions (Braatne et al. 1996; Karrenberg et al. 2002). Seedlings can establish in great numbers on bare, moist mineral substrates (e.g., up to 4,000 m$^{-2}$), but mortality in the first year can be quite high due to the severe conditions found in most establishment sites (Braatne et al. 1996; Karrenberg et al. 2002; Dixon 2003; Dixon and Turner 2006).

## 1.3.3 Vegetative Propagation

The extensive occurrence of vegetative propagation (also called asexual reproduction) sets *Populus* apart from most other dominant temperate trees (Dickmann 2001). Vegetative propagation is one of the traits that enable *Populus* to occupy tumultuous habitats along river banks and to persist long-term in landscapes that are frequently reset by large-scale fires. Vegetative propagation is also one of the traits that make *Populus* such an attractive model organism and one of the features that allows rapid deployment of genetically improved materials due to an almost unlimited capacity for replicating high-performing genotypes, thus capturing both additive and nonadditive genetic variation.

The extent and modes of reproduction vary tremendously among the different sections of the genus. Some species from section *Populus* have extensive vegetative propagation, as typified by the North American quaking aspen (*P. tremuloides*; Fig. 1-3). This species spreads vegetatively through the production of vegetative sprouts from adventitious buds on shallow lateral roots, a process known as "suckering" (Perala 1990). Suckering is believed to have been the predominant means of reproduction of *P. tremuloides* in the western part of its range in the recent past, with sexual reproduction from seeds occurring only rarely (Romme et al. 1997, 2005). Some aspen clones have reached a remarkable size, and have even been described as the largest organisms known (Kemperman and Barnes 1976; DeWoody et al. 2008; Mock et al. 2008). These large clone sizes and putative rarity of sexual reproduction have led to speculation that aspen clones could be thousands of years old (Barnes 1975; Mitton and Grant 1996). However, relatively high genotypic diversities have been observed in most genetic studies of aspen clones (Hyun et al. 1987; Jelinski and Cheliak 1992; Lund et al. 1992; Liu and Furnier 1993; Yeh et al. 1995; Namroud et al. 2005), suggesting that sexual reproduction may be more frequent and/or its impact on the genetic structure of aspen populations may be more persistent than previously assumed. Observations of large numbers of somatic mutations in vegetative tissues of extensive aspen clones suggests that advanced ages are possible for individual genets, but accurate estimation of clone age remains a substantial technical challenge (Ally et al. 2008; Mock et al. 2008).

Trees from section *Tacamahaca*, including the North American cottonwoods *P. balsamifera*, *P trichocarpa*, and *P. angustifolia*, commonly spread vegetatively through rooting of shoots from broken branches or entire tree trunks that have been toppled during storms and floods and then buried in sediment (Braatne et al. 1996; Rood et al. 2003; Barsoum et al. 2004; Rood et al. 2007; Smulders et al. 2008). One unusual mode of vegetative propagation in this section is cladoptosis, the continuous self-pruning of live short-shoots, which then act as propagules that can be dispersed by wind or water and become established (Galloway and Worrall 1979; Dewit and Reid 1992). However, in some settings, the breakage and subsequent transport of large branches by flood waters is the most important mode of vegetative propagation in this section, and cladoptosis contributes little to vegetative spread (Rood et al. 2003). Stem coppicing and root sprouting also occur in *Tacamahaca* cottonwoods, but these are usually not as important as branch and stem sprouting for vegetative spread (Rood et al. 1994; Rood et al. 2003) except when shallow roots are disturbed or the main stem is destroyed by fire or breakage (Rood et al. 1994; Gom and Rood 1999a, b).

Trees from section *Aigeiros* typically have less extensive vegetative propagation than species from *Populus* or *Tacamahaca* (Braatne et al. 1996; Bradshaw and Strauss 2001; Rood et al. 2007). For example, the North-American plains cottonwood (*P. deltoides*) and Fremont cottonwood (*P. fremontii*) have relatively rare vegetative propagation, which occurs primarily via coppicing of broken or killed shoots (Braatne et al. 1996; Gom and Rood 1999b; Schweitzer et al. 2002; Rood et al. 2003), although there are reports of extensive root sprouting in *P. fremontii* in some settings (Howe and Knopf 1991). *P. nigra* shows higher levels of vegetative propagation than North American cottonwoods from this section, sprouting extensively from branches, roots, and broken stems (Legionnet et al. 1997; Arens et al. 1998; Barsoum et al. 2004).

### 1.3.4 Sex Ratios

Sex ratios in natural *Populus* populations are highly variable, with many studies revealing male-biased ratios, some studies showing no deviation from the 1:1 expectation, and a few studies showing a female bias (reviewed by Braatne et al. 1996; Farmer 1996; Stanton and Villar 1996; see also Gom and Rood 1999b; Rottenberg et al. 2000; Rowland et al. 2001; Hultine et al. 2007). Although no consistent pattern has emerged, several of these studies suggest that biases in sex ratio may be driven by differential responses of the sexes to environmental conditions. For example, several studies have shown that males are more common and have superior growth at high elevations, as well as in warmer, drier, and more extreme environments, while females predominate where moisture is higher and resources are more abundant (Grant and Mitton 1979; Sakai and Burris 1985; Rottenberg et al. 2000). Controlled physiological studies have recently supported the contention that male and female *P. cathayana* (*P. suaveolens sensu lato*) have differential drought responses, with males showing smaller growth reductions than females (Xu et al. 2008b), and greater photosynthetic adjustment under drought treatments (Xu et al. 2008a). Similarly, male clones of *P. tremuloides* showed higher basal photosynthetic rates and greater responsiveness to elevated $CO_2$ compared to male clones (Wang and Curtis 2001). There are therefore intriguing possibilities for sexual selection in dioecious species like *Populus*, especially given the possible existence of a sex chromosome with reduced recombination (Yin et al. 2008). However, future studies on natural populations will need to take into account developmental differences between the two genders (e.g., male trees may reach reproductive maturity before female trees, thus possibly skewing sex ratios; Stanton and Villar 1996) and the effects of clonality, which can greatly reduce genotypic sampling and therefore yield apparently skewed sex ratios due to sampling error.

## 1.4 Genetic Variation

### 1.4.1 Neutral Diversity and Differentiation

Numerous studies have characterized patterns of variation in *Populus* using putatively neutral molecular markers like allozymes, Restriction Fragment Length Polymorphisms (RFLP), and microsatellites (Simple Sequence Repeat, SSR). Although there is substantial variation among species and marker types, the overall conclusion is that *Populus* species show high levels of variation within populations and low levels of differentiation among populations (Table 1-2), as is expected for obligately outcrossing organisms with large effective population sizes and long-distance dispersal of pollen and seeds (Hamrick et al. 1992). Levels of polymorphism and expected heterozygosity are higher for *Populus* than for plants in general ($A = 1.5$, $H_e = 0.11$), but close to the mean values for long-lived woody species ($A = 1.8$, $H_e = 0.15$; Hamrick et al. 1992), with SSR markers showing substantially higher allelic diversity and heterozygosity than allozymes and RFLP, as expected. Multiple studies show a deficiency of heterozygotes compared to Hardy-Weinberg expectations, which is probably due in part to undetected substructure in population samples (Hedrick 2005), and/ or null alleles (Ewen et al. 2000). Population differentiation as measured by $F_{ST}$ (Wright 1965) is quite low in most studies, with differences among populations accounting for only 1 to 12% of the total genetic variation (Table 1-2). In comparison, the mean $F_{ST}$ for long-lived woody species is 0.084, and that for plants in general is 0.228 (Hamrick et al. 1992). The relatively

**Table 1-2** Molecular diversity and differentiation in *Populus* based on allozymes (Allo.), RFLP, and SSR markers.

| Section | Species | Marker | $N_{loci}$ | $N_{pop}$ | $N$ | $A$ | $H_o$ | $H_e$ | $F_{IS}$ | $F_{ST}$ | Reference |
|---|---|---|---|---|---|---|---|---|---|---|---|
| *Aigeiros* | *P. deltoides* | Allo. | 33 | 9 | 84 | 1.7 | 0.06 | - | - | - | Rajora et al. 1991 |
| | | Allo. | 22 | 21 | - | 1.5 | - | 0.08 | - | 0.064 | Marty 1984 |
| | *P. fremontii* | RFLP | 36 | 4 | 47 | 1.5 | 0.18 | 0.15 | -0.175 | 0.074 | Martinsen et al. 2001 |
| | *P. nigra* | Allo. | 8 | 3 | 146 | - | - | 0.16 | 0.113 | 0.063 | Legionnet and Lefvre 1996 |
| | | SSR | 6 | 22 | 574 | - | 0.78 | 0.73 | -0.077 | 0.047 | Imbert and Lefevre 2003 |
| | | SSR | 7 | 17 | 921 | - | 0.74 | 0.76 | 0.027 | 0.081 | Smulders et al. 2008 |
| *Tacamahaca* | *P. angustifolia* | RFLP | 36 | 10 | 281 | 1.4 | 0.10 | 0.08 | -0.236 | 0.022 | Martinsen et al. 2001 |
| | *P. balsamifera* | Allo. | 17 | 5 | 248 | - | - | 0.04 | 0.061 | 0.014 | Farmer et al. 1988 |
| | *P. trichocarpa* | Allo. | 18 | 10 | 456 | 1.2 | - | 0.09 | - | 0.063 | Weber and Stettler 1981 |
| | | SSR | 10 | 2 | 282 | 17.5 | 0.71 | 0.77 | 0.058 | - | Slavov et al. 2008 |
| | | SSR | 9 | 47 | 372 | 6.1 | 0.60 | 0.80 | 0.293 | 0.078 | M. Ismail, unpublished data |
| *Populus* | *P. alba* | SSR | 19 | 1 | 169 | 6.4 | 0.37 | 0.38 | 0.027 | - | van Loo et al. 2008 |
| | *P. grandidentata* | Allo. | 14 | - | 96 | 1.4 | 0.07 | 0.08 | 0.125 | - | Liu and Furnier 1993 |
| | | RFLP | 37 | - | 75 | 1.8 | 0.12 | 0.13 | 0.077 | - | Liu and Furnier 1993 |
| | *P. tremula* | SSR | 9 | 3 | 113 | - | 0.35 | 0.41 | 0.120 | 0.117 | Suvanto and Latva-Karjanmaa 2005 |
| | | SSR | 25 | 12 | 116 | - | 0.50 | 0.62 | 0.197 | 0.015 | Hall et al. 2007 |
| | *P. tremuloides* | Allo. | 13 | - | 118 | 2.8 | 0.19 | 0.25 | 0.240 | - | Liu and Furnier 1993 |
| | | Allo. | 17 | 6 | 156 | 2.4 | 0.32 | 0.29 | -0.102 | 0.030 | Jelinski and Cheliak 1992 |
| | | Allo. | 15 | 8 | 200 | 2.7 | 0.13 | 0.24 | 0.462 | 0.068 | Hyun et al. 1987 |
| | | Allo. | 26 | 7 | 222 | 2.3 | 0.52 | 0.42 | -0.238 | - | Cheliak and Dancik 1982 |
| | | Allo. | 10 | 9 | 347 | 2.6 | 0.22 | 0.22 | 0.017 | 0.003 | Lund et al. 1992 |
| | | RFLP | 41 | - | 91 | 2.7 | 0.21 | 0.25 | 0.160 | - | Liu and Furnier 1993 |
| | | SSR | 16 | 11 | 189 | 4.9 | 0.41 | 0.45 | 0.093 | 0.045 | Cole 2005 |
| | | SSR | 4 | - | 266 | 8.8 | 0.47 | 0.67 | 0.300 | - | Namroud et al. 2005 |
| *Turanga* | *P. euphratica* | Allo. | 20 | 3 | 85 | 1.8 | 0.10 | 0.24 | 0.592 | - | Rottenberg et al. 2000 |
| | Median | Allo. | 17 | 7.5 | 156 | 2.1 | 0.16 | 0.22 | 0.113 | 0.063 | |
| | | RFLP | 36.5 | 7.0 | 83 | 1.7 | 0.15 | 0.14 | -0.049 | 0.048 | |
| | | SSR | 9.0 | 11.5 | 266 | 6.4 | 0.50 | 0.67 | 0.093 | 0.063 | |

$N_{loci}$ is the number of loci used; $N_{pop}$ is the number of populations sampled; $N$ is the number of genets (or trees) analyzed; $A$ is the average number of alleles per locus detected in each population; $H_e$ is the expected heterozygosity (Nei 1973); $F_{IS}$ is the fixation index as reported in the study or calculated as $F_{IS} = (H_e - H_o)/H_e$; $F_{ST}$ is the among-population differentiation (Wright 1965).

weak differentiation among *Populus* populations is probably a reflection of extensive gene flow by pollen, which is corroborated by numerous direct studies based on paternity analysis (Tabbener and Cottrell 2003; Pospíšková and Šálková 2006; Vanden Broeck et al. 2006; Slavov et al. 2009).

It is intriguing that trembling aspen species (*P. tremula* and *P. tremuloides*) have consistently higher numbers of alleles per locus and expected heterozygosity for allozyme and RFLP markers compared to other *Populus* species (Table 1-2). The median gene diversity from six studies of *P. tremuloides* ($H_e$ = 0.25) is comparable to that from two studies of its "sister" species *P. tremula* in Europe ($H_e$ = 0.20), and is more than two times higher than that for other species of *Populus* ($H_e$ = 0.09). This is in agreement with the general trend in woody plants (mean $H_e$ = 0.25 for species with both asexual and sexual reproduction versus $H_e$ = 0.14 for species that only reproduce sexually; Hamrick et al. 1992). One possible explanation lies in the extensive occurrence of vegetative propagation in aspens, as described above. Highly clonal organisms typically show higher allelic diversity and heterozygosity compared to organisms with similar life histories due to the accumulation of mutations (Balloux et al. 2003; Halkett et al. 2005; de Meeûs et al.2007). Another possible explanation is that some of the sampled trees were polyploids, an occurrence that seems to be relatively common in section *Populus* (Mock et al. 2008; Zhu et al. 1998; DiFazio et al., unpublished). Polyploidy could cause an apparent excess in allelic diversity and heterozygosity due to sampling of extra chromosomes (Krieger and Keller 1998; Ridout 2000). A final possibility is that this increased allelic diversity is a result of the hypothesized hybrid origin of these species, as described above (Barnes 1967). It will be interesting to see if this phenomenon is upheld by rangewide studies of genetic variation in microsatellite and SNP markers, and if further insights are gained into the mechanisms of elevated genetic diversity in aspen.

### 1.4.2 Nucleotide Diversity

The whole-genome sequence of *P. trichocarpa* provides an excellent resource for enhancing our understanding of the population genetics and genomics of the genus. Levels of nucleotide diversity in *Populus* are highly variable (Table 1-3), but generally comparable to those in other tree species (González-Martínez et al. 2006b; Savolainen and Pyhäjärvi 2007). Interestingly, nucleotide diversity in trees does not seem to be substantially higher than in other plants, including the nearly completely selfing annual *Arabidopsis thaliana* (Savolainen and Pyhäjärvi 2007). Presumably, this is due to the long generation times of trees, and the possible existence of genetic bottlenecks in the evolutionary history of many trees (Savolainen and Pyhäjärvi 2007; Ingvarsson 2008). Although data on among-population differentiation for

**Table 1-3** Nucleotide diversity, linkage disequilibrium, and differentiation in *Populus*.

| Species | $N_{genes}$ | $N_{pop}$ | $N_{hap}$ | $S_{kb}$ | $\pi$ | $\pi_S$ | $\pi_N$ | $LD_{0.2}$ | $F_{ST}$ | Reference |
|---|---|---|---|---|---|---|---|---|---|---|
| *P. nigra* | 9 | - | 48 | 38.7 | 0.0070 | 0.0107 | 0.0046 | 300 | - | Chu et al. 2009 |
| *P. trichocarpa* | 9 | - | 78 | 7.7 | 0.0018 | 0.0029 | - | - | - | Gilchrist et al. 2006 |
| *P. balsamifera* | 11 | 7 | 185 | 18.7 | 0.0020 | 0.0030 | 0.0008 | - | 0.053 | Keller et al. 2010 |
| *P. balsamifera* | 460 | 15 | 30 | 10.5 | 0.0026 | 0.0045 | 0.0012 | - | - | Olson et al. 2010 |
| *P. angustifolia* | 3 | 8 | 64 | 17.5 | 0.0024 | - | - | 120 | 0.086 | Slavov et al., unpublished data |
| *P. tremula* | 5 | 4 | 48 | - | 0.0111 | 0.0220 | 0.0059 | <100 | 0.117 | Ingvarsson 2005b |
| | 6 | 4 | 44 | 68.0 | 0.0144 | 0.0207 | 0.0117 | - | 0.107 | Ingvarsson 2005a |
| | 1 | 4 | 48 | 39.3 | 0.0061 | 0.0085 | 0.0030 | - | 0.045 | Ingvarsson et al. 2006 |
| | 76 | - | 24-38 | 19 | 0.0042 | 0.0120 | 0.0017 | <100 | - | Ingvarsson 2008 |

$N_{genes}$ is the number of genes sequenced; $N_{pop}$ is the number of populations analyzed; $N_{hap}$ is the number of haplotypes sequenced; $S_{kb}$ is the number of Single Nucleotide Polymorphisms (SNP) per 1 kb of sequence; $\pi$ is the overall nucleotide diversity; $\pi_S$ is the nucleotide diversity for synonymous sites; $\pi_N$ is the nucleotide diversity for nonsynonymous sites; $LD_{0.2}$ is the distance (bp) at which mean linkage disequilibrium between pairs of loci ($r^2$) decays to 0.2; $F_{ST}$ is the average among-population differentiation based on SNP.

SNP markers are still extremely limited in *Populus*, the few available values of $F_{ST}$ appear slightly higher but generally comparable with those based on other types of codominant markers (Tables 1-2 and 1-3).

In contrast to nucleotide diversity, the rate of decay of linkage disequilibrium (LD) with distance among sites appears to be extraordinarily high in some *Populus* populations. For example, average LD (as measured by the $r^2$ statistic) has been observed to decay below 0.2 within several hundred bp in *P. nigra*, *P. tremula*, and *P. angustifolia*, but other studies in *P. trichocarpa* and *P. balsamifera* have found that decay occurs over much greater distances (Table 1-3). In contrast, LD declines to background levels within 0.5 to 1.5 kb in conifers (Neale and Savolainen 2004; Krutovsky and Neale 2005; Heuertz et al. 2006; González-Martinez et al. 2006a), and within 10 kb in Arabidopsis thaliana. The reasons for this discrepancy among Populus species is far from clear, but it appears to be driven in part by population structure and population history (Invarsson 2008; Olsen et al. 2010) as well as substantial variation in recombination rates across the genome (Slavov et al. in preparation).

The levels of LD have major implications for gene discovery and characterization using association genetics in natural populations of *Populus*. On the one hand, low LD makes candidate gene association studies very accurate because polymorphisms underlying an association are expected to be within <1 kb of the SNP used to detect it. On the other hand, however, low LD renders whole-genome approaches unfeasible in the near future because millions of markers would be needed to scan a tree genome. The creative combination of plant materials characterized by high LD (e.g., sib families or hybrids) with natural populations characterized by very low LD, as well as the development of appropriate study designs and analytical approaches is a major short-term priority for population and quantitative geneticists working with *Populus* and other forest trees (Lexer and van Loo 2006; Lexer et al. 2007).

### 1.4.3 Adaptive Traits

*Populus* trees typically have high levels of adaptive genetic variation both within and among populations (Farmer 1996; Dunlap and Stettler 1996; Howe et al. 2003). Furthermore, differentiation among populations is generally much higher for adaptive traits than for neutral genetic markers (Merila and Crnokrak 2001; McKay and Latta 2002; Howe et al. 2003; Savolainen et al. 2007), which suggests that divergent selection has played a dominant role in shaping adaptive genetic variation. Finally, and most importantly, genecological studies have revealed strong and repeatable correspondence between clinal genetic variation for adaptive traits, and climatic and geographic factors are believed to be important agents of natural selection (Morgenstern 1996; St Clair et al. 2005; Aitken et al. 2008). For example, genetic variation for phenological traits occurs along latitudinal gradients in multiple *Populus* species (Farmer 1996; Böhlenius et al. 2006; Hall et al. 2007; Friedman et al. 2008), suggesting that tradeoffs between length of growing season and risk of frost damage have driven patterns of adaptive differentiation in *Populus* populations. All of this suggests that *Populus* populations are likely to be locally adapted (i.e., genotypes originating from a given habitat tend to have higher fitness in that habitat than genotypes originating from other habitats; Kawecki and Ebert 2004). This has serious implications for potential responses of *Populus* species to rapid anthropogenic climate change (Savolainen et al. 2007; Aitken et al. 2008), although the situation for *Populus* is arguably better than that for other, less vagile forest trees.

## 1.5 Domestication and Silviculture

*Populus* cultivation began in Europe, Asia, and the Middle East over 300 years ago (Evelyn 1670). Breeding programs have focused on the production of interspecific hybrids since the early days of tree improvement in this genus, capitalizing on the hybrid vigor of the $F_1$ generation. These first-generation genetic gains can be readily captured for silvicultural purposes because of the ease with which many *Populus* hybrids can be vegetatively propagated (Stanton and Villar 1996; Stettler et al. 1996b). In Europe, spontaneous hybrids of the imported *P. deltoides* and the European native black poplar, *P. nigra*, were discovered around 1775. Initially named black Italian poplars, they were later renamed *P. x canadensis* by Mönch in 1795 (McNabb 1997). Although naturally-occurring poplar hybrids were common in 18th and 19th century Europe, systematic poplar breeding began only in the early 20th century at the Kew Botanical Gardens, England. The first recorded attempt of crossing poplars was made on flowers of *P. deltoides* with pollen of *P. trichocarpa*, with the resultant hybrid named as *P. x generosa*. Subsequent poplar breeding programs began around 1920 in Canada, USA, Denmark, and France (McNabb 1997).

A major consolidation of global poplar research was initiated through the establishment of the International Poplar Commission (IPC) in 1947, which is one of the technical statutory bodies of the Food and Agriculture Organization (FAO) of the United Nations. Their mandate is to promote cultivation, conservation and utilization of poplars and willows (*www. fao.org/forestry/site/ipc/en*). The IPC assists researchers in the direction and coordination of scientific efforts and promotes conservation and exchange of poplar germplasm among the member nations. The IPC also supports research and management activities that explore issues of concern to its 37 member countries through six working parties: harvesting and utilization;

diseases; insect pests; genetics, conservation and improvement; production systems; and environmental applications. The IPC also promotes the economic and ecological benefits of poplars and willows in developed and developing countries.

Although poplar cultivation has been traditionally integrated into many temperate and sub-tropical agricultural systems, the use of poplars and willows as biomass for renewable energy is a novel application (Dinus et al. 2001; Perlack et al. 2005). Research specific to Short Rotation Woody Crops (SRWCs), such as hybrid poplar, has been taking place around the US for the last 25 years, with particularly active programs in the northwest and the upper midwest. SRWCs are seen as a way to grow more wood fiber on less land in order to reduce potential conflicts between increasing demand for wood fiber and other forest land uses. Researchers felt that optimum tree growth could be realized by borrowing the intensive cultivation techniques common in other forms of agriculture, including the use of hybrids. Most genetic and silvicultural research in SRWCs involved hardwoods, particularly poplars, because of their rapid growth. Interest in SRWCs intensified in 1977 when the oil embargo provided an impetus for additional funding from the US Department of Energy (DOE) to consider wood as an alternative to fossil fuels. There has been a recent resurgence in interest since the spike in oil prices during 2007 and 2008, with a concomitant increase in federal funding for improvement of short rotation biomass crops, including *Populus* (Rubin 2008).

High intensity cultivation of woody biomass crops has great economic potential, and *Populus* is one of the premiere candidates for implementation of intensive, biotechnology-driven forestry (Balatinecz and Kretschmann 2001; Sedjo 2001; Strauss 2003). Over 70 million ha combined of natural poplar stands existed in 2004 in the 21 countries belonging to the IUFRO poplar commission (*http://www.fao.org/forestry/ipc2004*). The three countries with the largest areas of poplar forest were Canada (28.3 million ha), the Russian Federation (21.9 million ha) and the US (17.7 million ha). The next six countries containing significant areas of natural poplar forests are China (2.1 million ha), Germany (100,000 ha), Finland (67,000 ha), France (39,800 ha), India (10,000 ha) and Italy (7,200 ha). The global area of poplar plantations in 2004 was 6.7 million ha, of which 3.8 million ha (57%) was planted primarily for wood production and 2.9 million ha (43%) for environmental purposes. China reported the most planted poplar overall (4.9 million ha, or 73% of the global total), followed by India with 1.0 million ha. The other countries with significant areas of planted poplar included France with 236,000 ha, Turkey with 130,000 ha, Italy with 118,800 ha, Argentina with 63,500 ha and Chile with 15,000 ha. Only two countries reported significant annual removals of wood from natural stands of poplars: the Russian Federation (100 million m$^3$) and Canada (16 million m$^3$). Other countries that reported annual removals of more than 1 million m³ of poplar wood from planted forests were Turkey (3.8 million m³), China (1.85 million m³), France (1.8 million m³), Italy (1.4 million m³), and India (1.2 million m³) (*http://www.fao.org/forestry/ipc2004*).

Poplars provide a wide range of wood products, including industrial roundwood and poles, pulp and paper, reconstituted boards, plywood, veneer, sawn timber, packing crates, pallets, and furniture. Non-wood products from poplar include fodder, fuelwood, and bioenergy. Finally, poplar provides valuable services such as shelter, shade, conservation and protection of soil, water, crops, livestock and dwellings (Balatinecz and Kretschmann 2001; Stanton et al. 2002). Pulp, paper and cardboard have traditionally been one of the most favored end uses in Europe, North America, China and Argentina, but recent drops in global pulp prices have driven these countries toward using *Populus* for solid wood products, including plywood, oriented strand board, and laminated veneer (Stanton et al. 2002). Packaging (pallets, boxes and crates) is another favored end use in Europe, the Republic of Korea, the Russian Federation, Canada, China and India. Poplar is also used for matches in Chile, the Russian Federation, India, the Republic of Korea and Sweden. The use of poplar as an energy feedstock is most advanced in Sweden, the United Kingdom and Turkey, but this use is expected to grow in coming years (Stanton et al. 2002; Perlack et al. 2005). Finally the principal use of poplar and willow resources in China, the Republic of Korea, Serbia, Montenegro and Sweden is for environmental conservation, thus providing valuable services rather than forest products.

## 1.6 Concluding Remarks

Members of the genus *Populus* form an exceptional group of plants that has attracted the interest of researchers, funding agencies, and land managers for many years. In the genomics era *Populus* has now been catapulted into the forefront of basic and applied research in tree biology and tree improvement (Wullschleger et al. 2002; Bhalerao et al. 2003; Brunner et al. 2004; Jansson and Douglas 2007). We have summarized some of the salient biological characteristics that have driven this interest. Other, extensive reviews are available for those who would like more details about this fascinating genus (Stettler et al. 1996a; Dickmann et al. 2001; Dickmann and Kuzovkina 2008).

## Acknowledgements

This research was supported by funding from the BioEnergy Science Center, a U.S. DOE Bioenergy Research Center (Office of Biological and Environmental Research in the DOE Office of Science), the DOE EPSCOR

Laboratory Partnership Program, the NSF FIBR program, the USDA/DOE Plant Feedstock Genomics for Bioenergy Program , and USDA CSREES grant 2008-01105.

# References

Aitken SN, Yeaman S, Holliday JA, Wang T, Curtis-McLane S (2008) Adaptation, migration or extirpation: climate change outcomes for tree populations. Evol Appl 1: 95–111.

Ally D, Ritland K, Otto SP 2008. Can clone size serve as a proxy for clone age? An exploration using microsatellite divergence in *Populus tremuloides*. Molecular Ecology 17: 4897–4911.

Angiosperm Phylogeny Group (2003) An update of the Angiosperm Phylogeny Group classification for the orders and families of flowering plants: APG II. Bot J o Linn Soc 141: 399–436.

Arens P, Coops H, Jansen J, Vosman B (1998) Molecular genetic analysis of black poplar (*Populus nigra* L.) along Dutch rivers. Mol Ecol 7: 11–18.

Bailey, JK, Deckert R, Schweitzer JA, Rehill BJ, Lindroth RL, Gehring C, Whitham TG (2005) Host plant genetics affect hidden ecological players: links among *Populus*, condensed tannins, and fungal endophyte infection. Can J Bot 83: 356–361.

Balatinecz JJ, Kretschmann DE (2001) Properties and utilization of poplar wood. In: D Dickmann, JG Isebrands , JE Eckenwalder , J Richardson (eds) Poplar Culture in North America. NRC Press, Ottawa, ON, Canada, pp 277–290.

Balloux F, Lehmann L, de Meeûs T (2003) The population genetics of clonal and partially clonal diploids. Genetics 164: 1635–1644.

Barnes BV (1967) Indications of possible mid-Cenozoic hybridization in aspens of the Columbia Plateau. Rhodora 69: 70–81.

Barnes BV (1975) Phenotypic variation of trembling aspen in western North America. For Sci 21: 319–328.

Barsoum N, Muller E, Skot L (2004) Variations in levels of clonality among *Populus nigra* L. stands of different ages. Evol Ecol 18: 601–624.

Bhalerao R, Nilsson O, Sandberg G (2003) Out of the woods: forest biotechnology enters the genomic era. Curr Opin Biotechnol 14: 206–213.

Boes TK, Strauss SH (1994) Floral phenology and morphology of black cottonwood, *Populus trichocarpa* (Salicaceae). Am J Bot 81: 562–567.

Böhlenius H, Huang T, Charbonnel-Campaa L, Brunner AM, Jansson S, Strauss SH, Nilsson O (2006) CO/FT regulatory module controls timing of flowering and seasonal growth cessation in trees. Science 312: 1040–1043.

Braatne JH, Rood SB, Heilman PE (1996) Life history, ecology, and reproduction of riparian cottonwoods in North America. In: RF Stettler, HD Bradshaw Jr, PE Heilman, TM Hinckley (eds) Biology of *Populus* and Its Implications for Management and Conservation. NRC Research Press, Ottawa, ON, Canada, pp 57–85.

Bradshaw HD, Strauss SH (2001) Breeding strategies for the 21st century: Domestication of poplar. In: DI Dickmann , JG Isebrands , JE Eckenwalder, J Richardson (eds) Poplar Culture in North America. NRC Research Press, Ottawa, ON, Canada, pp 383–394.

Brunner AM, Busov VB, Strauss SH (2004) Poplar genome sequence: functional genomics in an ecologically dominant plant species. Trends Plant Sci 9: 49–56.

Campbell JS, Mahoney JM, Rood SB (1993) A lack of heterosis in natural poplar hybrids in southern Alberta. Can J Bot 71: 37–42.

Cervera MT, Storme V, Soto A, Ivens B, Van Montagu M, Rajora OP, Boerjan W (2005) Intraspecific and interspecific genetic and phylogenetic relationships in the genus *Populus* based on AFLP markers. Theor Appl Genet 111: 1440–1456.

Chase MW, Zmarzty S, Lledo MD, Wurdack KJ, Swensen SM, and Fay MF (2002) When in doubt put it in Flacourtiaceae: a molecular phylogenetic analysis based on plastid rbcL DNA sequences. Kew Bull 57: 141–181.

Cheliak WM, Dancik BP (1982) Genic diversity of natural populations of a clone-forming tree *Populus tremuloides*. Can J Genet Cytol 24: 611–616.

Chen SL, Lia JK, Fritz E, Wang SS, Huttermann A (2002) Sodium and chloride distribution in roots and transport in three poplar genotypes under increasing NaCl stress. For Ecol Manag 168: 217–230.

Chu Y, Su X, Huang Q, Zhang X (2009) Patterns of DNA sequence variation at candidate gene loci in black poplar (*Populus nigra* L.) as revealed by single nucleotide polymorphisms. Genetica (Dordrecht) 137: 141–150.

Cole CT (2005) Allelic and population variation of microsatellite loci in aspen (*Populus tremuloides*). New Phytol 167: 155–164.

Cronk QCB (2005) Plant eco-devo: the potential of poplar as a model organism. New Phytol 166: 39–48.

de Meeûs T, Prugnolle F, Agnew P (2007) Asexual reproduction: Genetics and evolutionary aspects. Cell Mol Life Sci 64: 1355–1372.

DeBell DS (1990) *Populus trichocarpa* Torr. & Gray, Black Cottonwood. In: RM Burns, BH Honkala (eds) Silvics of North America, vol. 2. Hardwoods. USDA For. Serv. Agric. Handbook 654. USDA Forest Service. Washington D.C. pp 570–576.

Dewit L, Reid DM (1992) Branch abscission in balsam poplar (*Populus balsamifera*): Characterization of the phenomenon and the influence of wind. Int J Plant Sci 153: 556–564.

DeWoody J, Rowe CA, Hipkins VD, Mock KE (2008) "Pando" lives: molecular genetic evidence of a giant aspen clone in central Utah. WN Am Naturalist 68: 493–497.

Dickmann DI (2001) An overview of the genus *Populus*. In: DI Dickman, JG Isebrands, JE Eckenwalder , J Richardson (eds) Poplar Culture in North America, part A. NRC Research Press, Ottawa, ON, Canada, pp 1–42.

Dickmann DI, Kuzovkina J (2008) Poplars and willows of the world, with emphasis on silvicuturally important species. In: Poplar and Willows in the World. FAO, Rome, Italy.

Dickmann DI, Baer J, Bowersox T, Drew A, Monroe M, Ostry M, Rousseau RJ, Solomon J, Wright LL (1987) Super trees or prima donnas? The truth about poplars. Am Nurseryman 109–117.

Dickmann DI, Isebrands JG, Eckenwalder JE, Richardson J (2001) Poplar Culture in North America. NRC Research Press. Ottawa, ON, Canada.

Dinus RJ, Payne P, Sewell MM, Chiang VL, Tuskan GA (2001) Genetic modification of short rotation poplar wood: properties for ethanol fuel and fiber productions. Crit Rev Plant Sci 20: 51–69.

Dixon MD (2003) Effects of flow pattern on riparian seedling recruitment on sandbars in the Wisconsin River, Wisconsin, USA. Wetlands 23: 125–139.

Dixon MD, Turner MG (2006) Simulated recruitment of riparian trees and shrubs under natural and regulated flow regimes on the Wisconsin River, USA. River Res Appl 22: 1057–1083.

Dunlap JM, Stettler RF (1996) Genetic variation and productivity of *Populus trichocarpa* and its hybrids9. Phenology and Melampsora rust incidence of native black cottonwood clones from four river valleys in Washington. For Ecol Manag 87: 233–256.

Eckenwalder JE (1984a) Natural intersectional hybridization between North American species of *Populus* (Salicaceae) in sections *Aigeiros* and *Tacamahaca*. II. Taxonomy. Can J Bot 62: 325–335.

Eckenwalder JE (1984b) Natural intersectional hybridization between North American species of *Populus* (Salicaceae) in sections *Aigeiros* and *Tacamahaca*. III. Paleobotany and Evolution. Can J Bot 62: 336–342.

Eckenwalder JE (1984c) Natural intersectional hybridization between North American species of *Populus* (Salicaceae) in sections *Aigeiros* and *Tacamahaca*. I. Population studies of *P. x parryi*. Can J Bot 62: 317–324.

Eckenwalder JE (1996) Systematics and evolution of *Populus*. In: RF Stettler ,HD Bradshaw Jr, PE Heilman, TM Hinckley (eds) Biology of *Populus* and its Implications for Management and Conservation, NRC Research Press, Ottawa, ON, Canada, pp 7–32.

Evelyn J (1670) Sylva, or A discourse of Forest-trees, and the Propagation of Timber in His Majesties Dominions. John Martyn, London, UK.

Ewen KR, Bahlo M, Treloar RF, Levinson DF, Mowry B, Barlow JW, Foote SJ (2000) Identification and analysis of error types in high-throughput genotyping. Am J Hum Genet 67: 727–736.

Farmer RE Jr (1996) The genecology of *Populus*. In: RF Stettler , HD Bradshaw Jr, PE Heilman, TM Hinckley (eds) Biology of *Populus* and its Implications for Management and Conservation, NRC Research Press, Ottawa, ON, Canada, pp 33–55.

Farmer RE, Cheliak WM, Perry DJ, Knowles P, Barrett J, Pitel JA (1988) Isozyme Variation in balsam poplar along a latitudinal transect in northwestern Ontario. Can J For Res 18: 1078–1081.

Floate KD (2004) Extent and patterns of hybridization among the three species of *Populus* that constitute the riparian forest of southern Alberta, Canada. Can J Bot 82: 253–264.

Friedman JM, Roelle JE, Gaskin JF, Pepper AE, Manhart JR (2008) Latitudinal variation in cold hardiness in introduced *Tamarix* and native *Populus*. Evol Appl 1: 598–607.

Galloway G, Worrall J (1979) Cladoptosis: a reproductive strategy in black cottonwood? Can J For Res 9: 122–125.

Gilchrist EJ, Haughn GW, Ying CC, Otto SP, Zhuang J, Cheung D, Hamberger B, Aboutorabi F, Kalynyak T, Johnson L, Bohlmann J, Ellis BE, Douglas CJ, Cronk QCB (2006) Use of Ecotilling as an efficient SNP discovery tool to survey genetic variation in wild populations of *Populus trichocarpa*. Mol Ecol 15: 1367–1378.

Gom LA, Rood SB (1999a) Fire induces clonal sprouting of riparian cottonwoods. Can J Bot 77: 1604–1616.

Gom LA, Rood SB (1999b) Patterns of clonal occurrence in a mature cottonwood grove along the Oldman River, Alberta. Can J Bot 77: 1095–1105.

González-Martínez SC, Ersoz E, Brown GR, Wheeler NC, Neale DB (2006a) DNA sequence variation and selection of tag single-nucleotide polymorphisms at candidate genes for drought-stress response in *Pinus taeda* L. Genetics 172: 1915–1926.

González-Martínez SC, Krutovsky KV, Neale DB (2006b) Forest-tree population genomics and adaptive evolution. New Phytol 170: 227–238.

Grant MC, Mitton JB (1979) Elevational gradients in adult sex ratios and sexual differentiation in vegetative growth rates of *Populus tremuloides* Michx. Evolution 33: 914–918.

Halkett F, Simon JC, Balloux F (2005) Tackling the population genetics of clonal and partially clonal organisms. Trends Ecol Evol 20: 194–201.

Hall D, Luquez V, Garcia VM, St Onge KR, Jansson S, Ingvarsson PK (2007) Adaptive population differentiation in phenology across a latitudinal gradient in European aspen (*Populus tremula*, L.): A comparison of neutral markers, candidate genes and phenotypic traits. Evolution 61: 2849–2860.

Hamrick JL, Godt MJW, Sherman-Broyles SL (1992) Factors influencing levels of genetic diversity in woody plant species. New For 6: 95–124.

Hamzeh M, Dayanandan S (2004) Phylogeny of *Populus* (Salicaceae) based on nucleotide sequences of chloroplast TRNT-TRNF region and nuclear rDNA. Am J Bot 91: 1398–1408.

Hamzeh M, Perinet P, Dayanandan S (2006) Genetic relationships among species of *Populus* (Salicaceae) based on nuclear genomic data. J Torrey Bot Soc 133: 519–527.

Hamzeh M, Sawchyn C, Perinet P, Dayanandan S (2007) Asymmetrical natural hybridization between *Populus deltoides* and *P. balsamifera* (Salicaceae). Can J Bot 85: 1227–1232.

Hedrick PW (2005) Genetics of Populations. 3rd edn. Jones and Bartlett, Sudbury, Massachusetts, USA.

Heuertz M, De Paoli E, Kallman T, Larsson H, Jurman I, Morgante M, Lascoux M, Gyllenstrand N (2006) Multilocus patterns of nucleotide diversity, linkage disequilibrium and demographic history of Norway spruce [*Picea abies* (L.) Karst]. Genetics 174: 2095–2105.

Hewitt N (1998) Seed size and shade-tolerance: a comparative analysis of North American temperate trees. Oecologia 114: 432–440.

Howe GT, Aitken SN, Neale DB, Jermstad KD, Wheeler NC, Chen THH (2003) From genotype to phenotype: unraveling the complexities of cold adaptation in forest trees. Can J Bot 81: 1247–1266.

Howe WH, Knopf FL (1991) On the imminent decline of Rio Grande cottonwoods in central New Mexico USA. SW Naturalist 36: 218–224.

Hukin D, Cochard H, Dreyer E, Le Thiec D, Bogeat-Triboulot MB (2005) Cavitation vulnerability in roots and shoots: does *Populus euphratica* Oliv., a poplar from arid areas of Central Asia, differ from other poplar species? J Exp Bot 56: 2003–2010.

Hultine KR, Bush SE, West AG, Ehleringer JR (2007) Population structure, physiology and ecohydrological impacts of dioecious riparian tree species of western North America. Oecologia 154: 85–93.

Hyun JO, Rajora OP, Zsuffa L (1987) Genetic-Variation in Trembling Aspen in Ontario Based on Isozyme Studies. Can J For Res 17: 1134–1138.

Imbert E, Lefèvre F (2003) Dispersal and gene flow of *Populus nigra* (Salicaceae) along a dynamic river system. J Ecol 91: 447–456.

Ingvarsson PK (2005a) Molecular population genetics of herbivore-induced protease inhibitor genes in European aspen (*Populus tremula* L., Salicaceae). Mol Biol Evol 22: 1802–1812.

Ingvarsson PK (2005b) Nucleotide polymorphism and linkage disequilbrium within and among natural populations of European aspen (*Populus tremula* L., Salicaceae). Genetics 169: 945–953.

Ingvarsson PK (2008) Multilocus patterns of nucleotide polymorphism and the demographic history of *Populus tremula*. Genetics 180: 329–340.

Ingvarsson PK, Garcia MV, Hall D, Luquez V, Jansson S (2006) Clinal variation in phyB2, a candidate gene for day-length-induced growth cessation and bud set, across a latitudinal gradient in European aspen (*Populus tremula*). Genetics 172: 1845–1853.

Jansson S, Douglas CJ (2007) *Populus*: A model system for plant biology. Annu Rev Plant Biol 58: 435–458.

Jelinski DE, Cheliak WM (1992) Genetic diversity and spatial subdivision of *Populus tremuloides* (Salicaceae) in a heterogeneous landscape. Am J Bot 79: 728–736.

Johnson RL (1990) *Populus heterophylla* L., Swamp Cottonwood. In: RM Burns ,BH Honkala (eds) Silvics of North America, vol. 2: Hardwoods. USDA For Serv Agri Handbook 654. USDA Forest Service Washington DC, USA, pp 551–554.

Johnson WC (1994) Woodland expansion in the Platte River, Nebraska: Patterns and causes. Ecol Monogr 64: 45–84.

Karrenberg S, Edwards PJ, Kollmann J (2002) The life history of Salicaceae living in the active zone of floodplains. Freshwater Biol 47: 733–748.

Kawecki TJ, Ebert D (2004) Conceptual issues in local adaptation. Ecol Lett 7: 1225–1241.

Keim P, Paige KN, Whitham TG, Lark KG (1989) Genetic analysis of an interspecific hybrid swarm of *Populus*: occurrence of unidirectional introgression. Genetics 123: 557–565.

Keller SR, Olson MS, Silim S, Schroeder W, Tiffin P (2010) Genomic diversity, population structure, and migration following rapid range expansion in the Balsam Poplar, *Populus balsamifera*. Mol Ecol 19: 1212–1226.

Kemperman JA, Barnes BV (1976) Clone size in American aspens. Can J Bot 54: 2603–2607.

Kim S, Plagnol V, Hu TT, Toomajian C, Clark RM, Ossowski S, Ecker JR, Weigel D, Nordborg M (2007) Recombination and linkage disequilibrium in *Arabidopsis thaliana*. Nat Genet 39: 1151–1155.

Krieger MJB, Keller L (1998) Estimation of the proportion of triploids in populations with diploid and triploid individuals. J Hered 89: 275–279.

Krutovsky KV, Neale DB (2005) Nucleotide diversity and linkage disequilibrium in cold-hardiness- and wood quality-related candidate genes in Douglas fir. Genetics 171: 2029–2041.

Laidly PR (1990) *Populus grandidentata* Michx., Bigtooth Aspen. In: RM Burns, BH Honkala (eds) Silvics of North America, vol. 2. Hardwoods. USDA For. Serv. Agric. Handbook 654, USDA Forest Service, Washington DC, pp 544–550.

Legionnet A, Lefèvre F (1996) Genetic variation of the riparian pioneer tree species *Populus nigra* L. I. Study of population structure based on isozymes. Heredity 77: 629–637.

Legionnet A, Faivre-Rampant P, Villar M, Lefèvre F (1997) Sexual and asexual reproduction in natural stands of *Populus nigra*. Bot Acta 110: 257–263.

Leskinen E, Alstrm-Rapaport C (1999) Molecular phylogeny of Salicaceae and closely related Flacourtiaceae: evidence from 5.8 S, ITS 1 and ITS 2 of the rDNA. Plant Syst Evol 215: 209–227.

Lexer C, van Loo M (2006) Contact zones: Natural labs for studying evolutionary transitions. Curr Biol 16: R407–R409.

Lexer C, Fay MF, Joseph JA, Nica MS, Heinze B (2005) Barrier to gene flow between two ecologically divergent *Populus* species, *P. alba* (white poplar) and *P. tremula* (European aspen): the role of ecology and life history in gene introgression. Mol Ecol 14: 1045–1057.

Lexer C, Buerkle CA, Joseph JA, Heinze B, Fay MF (2007) Admixture in European *Populus* hybrid zones makes feasible the mapping of loci that contribute to reproductive isolation and trait differences. Heredity 98: 74–84.

Liu Z, Furnier GR (1993) Comparison of allozyme, RFLP, and RAPD markers for revealing genetic variation within and between trembling aspen and bigtooth aspen. Theor Appl Genet 87: 97–105.

Lund ST, Furnier GR, Mohn CA (1992) Isozyme variation in quaking aspen in Minnesota. Can J For Res 22: 521–529.

Ma HC, Fung L, Wang SS, Altman A, Hutterman A (1997) Photosynthetic response of *Populus euphratica* to salt stress. For Ecol Manag 93: 55–61.

Martinsen GD, Whitham TG, Turek RJ, Keim P (2001) Hybrid populations selectively filter gene introgression between species. Evolution 55: 1325–1335.

Marty, T. L. (1984) Population Variability and Genetic Diversity of Eastern Cottonwood (*Populus deltoides* Bartr.). Ph.D. Thesis, Univ of Wisconsin, Madison, WI, USA.

Mckay JK, Latta RG (2002) Adaptive population divergence: markers, QTL and traits. Trends Ecol Evol 17: 285–291.

McNabb HS Jr (1997) Sentinels of the Prairie speak (by *Populus* species). In: NB Klopfenstein, YB Chun , MS Kim, MR Ahuja (eds) Micropropagation, Genetic Engineering, and Molecular Biology of *Populus*. USDA Forest Service, Fort Collins, CO. pp 1–3.

Merilä J, Crnokrak P (2001) Comparison of genetic differentiation at marker loci and quantitative traits. J Evol Biol 14: 892–903.

Mitton JB, Grant MC (1996) Genetic variation and the natural history of quaking aspen. BioScience 46: 25–31.

Mock KE, Rowe CA, Hooten MB, DeWoody J, Hipkins VD (2008) Clonal dynamics in western North American aspen (*Populus tremuloides*). Mol Ecol 17: 4827–4844.

Morgenstern EK (1996) Geographic Variation in Forest Trees: Genetic Basis and Application of Knowledge in Silviculture. UBC Press, Univ of British Columbia, Vancouver BC, Canada.

Namroud MC, Park A, Tremblay F, Bergeron Y (2005) Clonal and spatial genetic structures of aspen (*Populus tremuloides* Michx.). Mol Ecol 14: 2969–2980.

Neale DB, Savolainen O (2004) Association genetics of complex traits in conifers. Trends Plant Sci 9: 325–330.

Nei M (1973) Analysis of gene diversity in subdivided populations. Proc Natl Acad Sci USA 70: 3321–3323.

Olson MS, Robertson AL, Takebayashi N, Silim S, Schroeder WR, Tiffin P (2010) Nucleotide diversity and linkage disequilibrium in balsam poplar (*Populus balsamifera*). New Phytol 186: 526–536.

Perala D (1990) A. *Populus tremuloides* Michx., Quaking Aspen. In: RM Burns, BH Honkala. (eds) Silvics of North America, vol. 2. Hardwoods. Agriculture Handbook 654. U.S. Department of Agriculture, Forest Service, Washington DC, pp 1082–1115.

Perlack RD, Wright LL, Turhollow AF, Graham RL, Stokes BJ, Erbach DC (2005) Biomass as Feedstock for a Bioenergy and Bioproducts Industry: The Technical Feasibility of a Billion-Ton Annual Supply. Oak Ridge National Laboratory, Oak Ridge, TN.

Pospíšková M, Šálková I (2006) Population structure and parentage analysis of black poplar along the Morava River. Can J For Res 36: 1067–1076.

Rajora OP, Zsuffa L, Dancik BP (1991) Allozyme and leaf morphological variation of eastern cottonwood at the northern limits of its range in Ontario. For Sci 37: 688–702.

Ridout MS (2000) Improved estimation of the proportion of triploids in populations with diploid and triploid individuals. J Hered 91: 57–60.

Romme WH, Turner MG, Gardner RH, Hargrove WW, Tuskan GA, Despain DG, Renkin RA (1997) A rare episode of sexual reproduction in aspen (*Populus tremuloides* Michx) following the 1988 Yellowstone fires. Nat Areas J 17: 17–25.

Romme WH, Turner MG, Tuskan GA, Reed RA (2005) Establishment, persistence, and growth of aspen (*Populus tremuloides*) seedlings in Yellowstone National Park. Ecology 86: 404–418.

Rood SB, Campbell JS, Despins T (1986) Natural poplar hybrids from southern Alberta. I. Continuous variation for foliar characteristics. Can J Bot 64: 1382–1388.

Rood SB, Hillman C, Sanche T, Mahoney JM (1994) Clonal reproduction of riparian cottonwoods in southern Alberta. Can J Bot 72: 1766–1774.

Rood SB, Kalischuk AR, Polzin ML, Braatne JH (2003) Branch propagation, not cladoptosis, permits dispersive, clonal reproduction of riparian cottonwoods. Extensive pollen flow in two ecologically contrasting populations of *Populus trichocarpa*. For Ecol Manag 186: 227–242.

Rood SB, Goater LA, Mahoney JM, Pearce CM, Smith DG (2007) Flood, fire, and ice: disturbance ecology of riparian cottonwoods. Can J Bot 85: 1019–1032.

Rottenberg A, Nevo E, Zohary D (2000) Genetic variability in sexually dimorphic and monomorphic populations of *Populus euphratica* (Salicaceae). Can J For Res 30: 482–486.

Rowland DL, Beals L, Chaudhry AA, Evans AS, Grodeska LS (2001) Physiological, morphological, and environmental variation among geographically isolated cottonwood (*Populus deltoides*) populations in New Mexico. WN Am Naturalist 61: 452–462.

Rowland DL, Garner R, Jespersen M (2002) A rare occurrence of seed formation on male branches of the dioecious tree, *Populus deltoides*. Am Midland Naturalist 147: 185–187.

Rubin EM (2008) Genomics of cellulosic biofuels. Nature 454: 841–845.

Sakai AK, Burris TA (1985) Growth in Male and Female Aspen Clones - A 25-Year Longitudinal-Study. Ecology 66: 1921–1927.

Savolainen O, Pyhäjärvi T (2007) Genomic diversity in forest trees. Curr Opin Plant Biol 10: 162–167.

Savolainen O, Pyhäjärvi T, Knurr T (2007) Gene flow and local adaptation in trees. Annu Rev Ecol Evol Syst 38: 595–619.

Schweitzer JA, Martinsen GD, Whitham TG. (2002) Cottonwood hybrids gain fitness traits of both parents: A mechanism for their long-term persistence? Am J Bot 89: 981–990.

Schweitzer JA, Bailey JK, Hart SC, Wimp GM, Chapman SK, Whitham TG (2005) The interaction of plant genotype and herbivory decelerate leaf litter decomposition and alter nutrient dynamics. Oikos 110: 133–145.

Sedjo R (2001) The economic contribution of biotechnology and forest plantations in global wood supply and forest conservation. In: SH Strauss, HD Bradshaw Jr (eds) Proceedings of the First International Symposium on Ecological and Societal Aspects of Transgenic Plantations. Corvallis, OR: College of Forestry, Oregon State University, pp 29–46.

Slavov GT, Leonardi S, Burczyk J, Adams WT, Strauss SH, DiFazio SP (2009) Extensive pollen flow in two ecologically contrasting populations of *Populus trichocarpa*. Mol Ecol 18: 357–373.

Smith RL, Sytsma KJ (1990) Evolution of *Populus nigra* (Sect *Aigeiros*)—Introgressive hybridization and the chloroplast contribution of *Populus alba* (Sect *Populus*). Am J Bot 77: 1176–1187.

Smulders MJM, Cottrell JE, Lefèvre F, Van Der Schoot J, Arens P, Vosman B, Tabbener HE, Grassi F, Fossati T, Castiglione S, Krystufek V, Fluch S, Burg K, Vornam B, Pohl A, Gebhardt K, Alba N, Agundez D, Maestro C, Notivol E, Volosyanchuk R, Pospiskova M, Bordacs S, Bovenschen J, van Dam BC, Koelewijn HP, Halfmaerten D, Ivens B, Van Slycken J, Broeck AV, Storme V, Boerjan W. 2008. Structure of the genetic diversity in black poplar (*Populus nigra* L.) populations across European river systems: Consequences for conservation and restoration. For Ecol Manag 255: 1388–1399.

St Clair JB, Mandel NL, Vance-Boland KW (2005) Genecology of Douglas fir in western Oregon and Washington. Ann Bot 96: 1199–1214.

Stanton BJ, Eaton J, Johnson J, Rice D, Schuette W, Moser, B (2002) Hybrid poplar in the Pacific Northwest—The effects of market-driven management. J For 100: 28–33.

Stanton BJ, Villar M (1996) Controlled reproduction of *Populus*. In: RF Stettler, HD Bradshaw Jr, PE Heilman, TM Hinckley (eds) Biology of *Populus* and its Implications for Management and Conservation, NRC Research Press, Ottawa, ON, Canada, pp 113–138.

Stella JC, Battles JJ, Orr BK, McBride JR (2006) Synchrony of seed dispersal, hydrology and local climate in a semi-arid river reach in California. Ecosystems 9: 1200–1214.

Stettler R 1971. Variation in sex expression of black cottonwood and related hybrids. Silvae Genet 20: 42–46.

Stettler RF, Bradshaw HD Jr, Heilman PE, Hinckley TM (1996a) Biology of *Populus* and its implications for management and conservation. NRC Research Press, Ottawa, ON, Canada.

Stettler RF, Zsuffa L, Wu R (1996b) The role of hybridization in the genetic manipulation of *Populus*. In: RF Stettler, HD Bradshaw Jr, PE Heilman, TM Hinckley (eds) Biology of *Populus* and its Implications for Management and Conservation, NRC Research Press, Ottawa, ON, Canada.

Strauss SH (2003) Genetic technologies—Genomics, genetic engineering, and domestication of crops. Science 300: 61–62.

Suvanto LI Latva-Karjanmaa TB (2005) Clone identification and clonal structure of the European aspen (*Populus tremula*). Molecular Ecology 14: 2851–2860.

Tabbener HE, Cottrell J. E (2003) The use of PCR based DNA markers to study the paternity of poplar seedlings. For Ecol Manag 179: 363–376.

Tuskan GA, DiFazio SP, Jansson S, Bohlmann J, Grigoriev I, Hellsten U, Putnam N, Ralph S, Rombauts S, Salamov A, Schein J, Sterck L, Aerts A, Bhalerao RR, Bhalerao RP, Blaudez D, Boerjan W, Brun A, Brunner AM, Busov V, Campbell M, Carlson J, Chalot M, Chapman J, Chen GL, Cooper D, Coutinho PM, Couturier J, Covert S, Cronk Q, Cunningham R, Davis J, Degroeve S, Déjardin A, dePamphilis C, Detter J, Dirks B, Dubchak I, Duplessis S, Ehlting J, Ellis B, Gendler K, Goodstein D, Gribskov M, Grimwood J, Groover A, Gunter L, Hamberger B, Heinze B, Helariutta Y, Henrissat B, Holligan D, Holt R, Huang W, Islam-Faridi N, Jones S, Jones-Rhoades M, Jorgensen R, Joshi C, Kangasjärvi J, Karlsson J, Kelleher C, Kirkpatrick R, Kirst M, Kohler A, Kalluri U, Larimer F, Leebens-Mack J, Leplé JC, Locascio P, Lou Y, Lucas S, Martin F, Montanini B, Napoli C, Nelson DR, Nelson C, Nieminen K, Nilsson O, Pereda V, Peter G, Philippe R, Pilate G, Poliakov A, Razumovskaya J, Richardson P, Rinaldi C, Ritland K, Rouzé P, Ryaboy D, Schmutz J, Schrader J, Segerman B, Shin H, Siddiqui A, Sterky F, Terry A, Tsai CJ, Uberbacher E, Unneberg P, Vahala J, Wall K, Wessler S, Yang G, Yin TM, Douglas C, Marra M, Sandberg

G, Van de Peer Y, and Rokhsar D (2006) The genome of black cottonwood, *Populus trichocarpa* (Torr. & Gray). Science 313: 1596–1604.

van Loo M, Joseph JA, Heinze B, Fay MF, Lexer C (2008) Clonality and spatial genetic structure in *Populus x canescens* and its sympatric backcross parent *P. alba* in a Central European hybrid zone. New Phytol 177: 506–516.

Vanden Broeck A, Cottrell J, Quataert P, Breyne P, Storme V, Boerjan W, Van Slycken J (2006) Paternity analysis of *Populus nigra* L. offspring in a Belgian plantation of native and exotic poplars. Ann For Sci 63: 783–790.

Wang XZ, Curtis PS (2001) Gender-specific responses of *Populus tremuloides* to atmospheric $CO_2$ enrichment. New Phytol 150: 675–684.

Weber JC, Stettler RF (1981) Isoenzyme variation among ten [riparian] populations of *Populus trichocarpa* Torr. et Gray in the Pacific Northwest. Silvae Genet 30: 82–87.

Whitham TG, Bailey JK, Schweitzer JA, Shuster SM, Bangert RK, LeRoy CJ, Lonsdorf EV, Allan GJ, DiFazio SP, Potts BM, Fischer DG, Gehring CA, Lindroth RL, Marks JC, Hart SC, Wimp GN, Wooley SC (2006) A framework for community and ecosystem genetics: from genes to ecosystems. Nat Rev Genet 7: 510–523.

Whitham TG, DiFazio SP, Schweitzer JA, Shuster SM, Allan GJ, Bailey JK, Woolbright SA (2008) Extending genomics to natural communities and ecosystems. Science 320: 492–495.

Willing RR, Pryor LD (1976) Interspecific Hybridization in Poplar. Theor Appl Genet 47: 141–151.

Wimp GM, Martinsen GD, Floate KD, Bangert RK, Whitham TG (2005) Plant genetic determinants of arthropod community structure and diversity. Evolution 59: 61–69.

Wright S (1965) The interpretation of population structure by F-statistics with special regard to systems of mating. Evolution 19: 295–420.

Wullschleger SD, Janssson S, Taylor G (2002) Genomics and forest biology: *Populus* emerges as the perennial favorite. Plant Cell 14: 2651–2655.

Xu X, Peng GQ, Wu CC, Korpelainen H, Li CY (2008a) Drought inhibits photosynthetic capacity more in females than in males of *Populus cathayana*. Tree Physiol 28: 1751–1759.

Xu X, Yang F, Xiao XW, Zhang S, Korpelainen H, Li CY (2008b) Sex-specific responses of *Populus cathayana* to drought and elevated temperatures. Plant Cell Environ 31: 850–860.

Yeh FC, Chong DK, Yang R-C (1995) RAPD Variation within and among natural populations of trembling aspen (*Populus tremuloides* Michx.) from Alberta. J Hered 86: 454–460.

Yin TM, DiFazio SP, Gunter LE, Zhang X, Sewell MM, Woolbright SA, Allan GJ, Kelleher CT, Douglas CJ, Wang M, Tuskan GA (2008) Genome structure and emerging evidence of an incipient sex chromosome in *Populus*. Genome Res 18: 422–430.

Zasada JC, Phipps HM (1990) *Populus balsamifera* L., Balsam Poplar. In: RM Burns, BH Honkala (eds) Silvics of North America, vol. 2. Hardwoods. Agriculture Handbook 654. Department of Agriculture, USDA Forest Service, Washington DC, pp 1019–1043.

Zhu Z, Kang X, Zhang Z, Zhu ZT, Kang XY, Zhang ZY (1998) Studies on selection of natural triploids of *Populus tomentosa*. Sci Silvae Sin 34: 22–31.

# 2

# *Populus* Genomic Resources

*Derek R. Drost,[1] Catherine I. Benedict,[2,a] Evandro Novaes[2,b]*
*and Matias Kirst[2,c,*]*

## ABSTRACT

With advances in modern genomic technology, genome-level resources
for genus *Populus* have evolved rapidly since the initial whole genome
sequencing project of P. trichocarpa. These resources have been derived
from a substantial number of structural and functional genomic studies,
resulting in a composite picture of the genomes and transcriptomes of
several Populus species. Insights into genome structure and function
have been developed from an extensive suite of molecular markers to
develop genetic maps and characterize variation in natural populations.
Similarly, transcriptomic studies based on modern EST sequencing and
microarray platforms have led to sound initial understanding of the
role of transcript-level variation in the phenotypic diversity of Populus.
Thanks to extensive curation in public online data repositories, Populus
genomic data is generally easily accessible via tools for comparative
and functional analyses. Application of these resources has already
yielded a number of fundamental insights into genome structure
and evolution, the molecular control of developmental transitions,
and transcriptional responses to biotic and abiotic stresses. Despite
these early successes, substantial effort is still needed to improve the
genome assembly and annotation, sequence of additional genotypes
in each of the major sections of the genus, and develop databases and
technology tools to compile and unite multiple levels of "omics" data

[1]Graduate Program in Plant Molecular and Cellular Biology, School of Forest Resources and
Conservation, Genetics Institute, University of Florida, PO Box 110410, Gainesville, FL 32611,
USA; e-mail: *ddrost@ufl.edu*
[2]School of Forest Resources and Conservation, Genetics Institute, University of Florida, PO
Box 110410, Gainesville, FL 32611, USA;
[a]e-mail: *cibenedict@ufl.edu*
[b]e-mail: *evandro@ufl.edu*
[c]e-mail: *mkirst@ufl.edu*
*Corresponding author

into functional analytical frameworks. With strategic investments in these areas, Populus should continue to develop as one of the primary model organisms for plant research, providing major insights into tree biology and plant evolutionary biology.

**Keywords:** Genomics, transcriptome, microarray, *Populus*, EST, mapping, QTL

## 2.1 Introduction

Poplars have recently become the model woody plant for study of secondary cell-wall formation and perennial growth habit. Several *Populus* species were adopted initially due to the ease of vegetative propagation, transformability, and the availability of saturated genetic maps in several well-established pedigrees (Bradshaw et al. 2000; Taylor 2002). The research interest increased significantly with the availability of the genome sequence of the *P. trichocarpa* genotype Nisqually-1, released publicly in 2006 (Tuskan et al. 2006). *P. trichocarpa* was only the third plant species, and the first woody perennial to be sequenced, following the release of the *Arabidopsis thaliana* genome in 2000, and the rice genome in 2002 (Goff et al. 2002; Yu et al. 2002).

Since the sequencing of the *P. trichocarpa* genome, significant advances have been made in the development of gene models and annotation of the genome sequence (Tuskan et al. 2006), construction of bacterial artificial chromosome (BAC) and full-length cDNA libraries (Nanjo et al. 2004; Tuskan et al. 2006; Kelleher et al. 2007; Ralph et al. 2008), and design of whole-genome microarrays. These advances have created a wealth of resources that are fueling research on the molecular basis of woodiness and perenniality, and comparative genomic studies. As a consequence, the number of publications referencing the *Populus* genome sequence has been increasing rapidly since the release of the sequence in September 2006 (Fig. 2-1). However, these resources still lag significantly behind those developed for the main plant model *Arabidopsis* and some agricultural crops, particularly rice and maize.

The primary purpose of this chapter is to review the existing genomic resources for poplar research, and outline those that are missing and most urgently needed. The chapter is separated into three sections. Initially we describe a summary of the genomic resources that have been derived directly from sequencing the *P. trichocarpa* genome, including DNA sequence databases, polymorphism information, and other DNA-related resources. In the second section, we review the existing transcriptome studies that provide an atlas of transcription regulation under a variety of conditions, as well as existing resources for surveying the *Populus* transcriptome. Finally, the third section is dedicated to a discussion about new resources and tools that are necessary to sustain the current surge in *Populus* genomic research in the medium- and long-term. The development of these resources should solidify the relevance of *Populus* as a model woody species, particularly with the expected sequencing of other competing woody models in the next few years. Conversely, should these resources remain lacking, it may lead to a shift of focus in the research community towards new models.

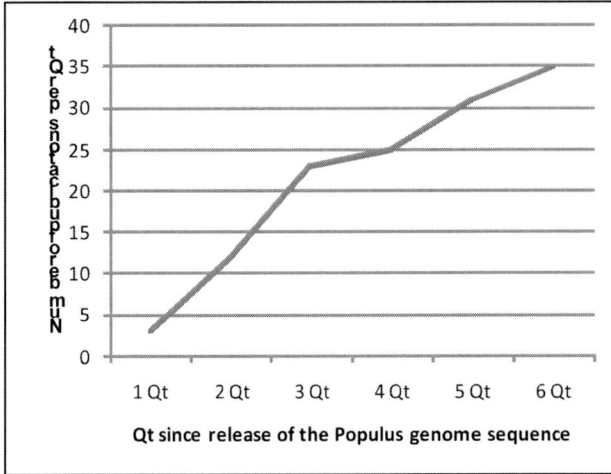

**Figure 2-1** Number of publications referencing the *Populus* genome sequence publication (Tuskan et al. 2006) every quarter (Qt), since its release in September 2006 (*Source:* ISI Web of Knowledge).

## 2.2 Populus Genome Resources

Although the groundwork for poplar's emergence as a model woody plant began long ago (Bradshaw et al. 2000; Taylor 2002; Wullschleger et al. 2002; Bhalerao et al. 2003a; Brunner et al. 2004), the use of *Populus* increased rapidly with the public release of the draft genome sequence of the *P. trichocarpa* clone "Nisqually-1" (Tuskan et al. 2006). Here we introduce an overview of the information and resources that were derived from the genome sequence that will be useful for reference in this and subsequent chapters.

### 2.2.1 Genome Sequence

The *P. trichocarpa* genome sequencing, conducted by the Joint Genome Institute of the Department of Energy (JGI-DOE), was a landmark for genetic, molecular, biochemical and evolutionary studies in poplar. Whole-genome shotgun sequencing and reassembly resulted in over 470 Mbp

(≈ 95% of the estimated genome size) of sequence distributed in 22,136 scaffolds with a 7.5× average sequencing depth (Tuskan et al. 2006). The sequences are accessible at the JGI *Populus* genome (http://genome.jgi-psf. org/Poptr1_1/), NCBI (http://www.ncbi.nlm.nih.gov/sites/entrez?Db= genomeprj&Cmd=Retrieve&list_uids=17973) and EMBL-EBI (*http://www. ebi.ac.uk/ebisearch/search.ebi?db=nucleotideSequences&t=populus*) websites. JGI and NCBI websites offer genome browsers for *Populus*, where sequence gaps, annotated gene models and sequenced expressed sequence tags (ESTs) are depicted at each genomic location in a user-friendly interface. The NCBI genome browser does not depict unanchored scaffolds; i.e. sequences that were not addressed to a specific genetic linkage group. The JGI genome browser, on the other hand, contains the 19 linkage group-level scaffold assemblies, and also the unmapped scaffolds. In addition, the JGI browser displays BLAST alignments with protein and non-redundant nucleotide sequences from other organisms, as well as VISTA track to demonstrate homology levels with the *Arabidopsis* and *Physcomitrella* genomes. The NCBI browser, on the other hand, depicts genetic markers utilized to anchor sequencing scaffolds into genetic linkage maps. These tools facilitate research aiming to link genetic and phenotypic variation as well as population and evolutionary studies of *Populus*.

Although whole-genome sequencing focused on *P. trichocarpa*, other poplar species are equally important for economic and biological purposes. *P. deltoides* and *P. nigra*, for example, are extensively used in pure or interspecific hybrid cultivars planted throughout the US and Europe (Arens et al. 1998; Cervera et al. 2001). Moreover, many *Populus* species have ecological importance given their wide natural distribution across the Northern Hemisphere (Brunner et al. 2004) and their major role as foundational species of ecosystems (Wimp et al. 2007; Schweitzer et al. 2008). Despite their importance, genomic sequence availability for species other than *P. trichocarpa* is largely limited to EST collections and BAC libraries. The DNA sequence information available for other poplar species in public repositories is mostly restricted to targeted regions containing genes of interest (Kawai et al. 1993; Tsai et al. 1995; Negi et al. 2002; Lescot et al. 2004), microsatellites (Dayanandan et al. 1998; Smulders et al. 2001), or regions of interest for population genetics studies (Ingvarsson 2005). In addition, the complete chloroplast DNA sequence for *P. trichocarpa* (Tuskan et al. 2006) and *P. alba* (Okumura et al. 2006), are publicly available. Lack of genomic sequences for additional *Populus* species prohibits rapid identification of genetic elements influencing the extensive morphological, physiological and ecological variation found within the genus (Keim et al. 1989; Wu et al. 1998; Brunner et al. 2004). As has been demonstrated in other plant species (Clark et al. 2006; Chen et al. 2007; Cong et al. 2008), genetic diversity in regulatory regions plays a significant role in quantitative phenotypic variation. Such diversity can only be assessed in large scale through genome sequencing and subsequent comparative studies. Therefore, it is crucial that future sequencing efforts include other *Populus* species for comparative genomics studies, as discussed later in this chapter.

New sequencing technologies (Holt et al. 2008; Ralph et al. 2008) could extend genome sequencing among *Populus* species rapidly. Moreover, these breakthrough sequencing platforms, combined with a 9.4 × BAC physical map available for *P. trichocarpa* (Kelleher et al. 2007), establish the necessary resources to close gaps and to address previously unanchored scaffolds in the "Nisqually-1" genome sequence. A BAC physical map is also being established for a *Populus deltoides* genotype (clone D124) that was used for quantitative trait loci (QTL) analysis in an interspecific pseudo-backcross (G. Tuskan, pers. comm.). The availability of several physical maps may serve as a temporary alternative for the lack of genome sequences from different genotypes or species, as it would create the framework required for characterization of species diversity in targeted regions.

## 2.2.2 Resources for Comparative Genomics

As the third plant genome to be sequenced after *Arabidopsis* and rice (Arabidopsis Genome Initiative 2000; Goff et al. 2002; Yu et al. 2002), *Populus* provided a novel resource for comparative genomic studies and understanding the evolution of flowering plants. From the 45,555 gene models identified in *P. trichocarpa*, 88% (40,307) had detectable similarity to gene sequences in *Arabidopsis* (*E* value ≤ 1 × 10⁻³). A primary set of 13,019 orthologs was initially identified and characterized. Whole-genome pairwise alignments between poplar and both *Arabidopsis* and *Oryza* indicated significant similarity between 58 and 38% of the coding regions of the genomes, respectively. Among the 5,248 *P. trichocarpa* genes with no similarity to *Arabidopsis*, more than 35% (1,883) have expression evidence in manually curated EST databases (Tuskan et al. 2006). These genes may be novel "tree-specific" genes important for the distinction of poplar from non-woody dicots, though further investigation is necessary to elucidate their potential roles.

Comparative genomic analysis of metabolic and regulatory pathways can identify novel genes that encode for enzymes in these pathways, in different genera. Gene-family expansion/reduction in one genome relative to another may define pathways or processes that differentiate plant species. For example, the comparative analysis of several cell-wall related gene families revealed a substantial expansion in *Populus* relative to *Arabidopsis* (Tuskan et al. 2006). *Populus* has 93 cellulose-synthesis related genes, a 20% expansion relative to *Arabidopsis* that is driven largely by the presence of almost twice as many cellulose synthase *CesA* genes (Tuskan et al. 2006). A similar expansion was observed for disease resistance gene families in poplar (Tuskan et al. 2006). Microsyntenic sequence blocks between *Arabidopsis* and poplar also provide a powerful tool to identify orthologous genes, as described for the invertase (Bocock et al. 2008) and expansin (Sampedro et al. 2006) gene families, and the poplar *MONOPTEROS* genes (Johnson and Douglas 2007).

Comparing poplar and *Salix* has also provided interesting clues about the evolution of the modern poplar genome. The *Populus* and *Salix* lineages share a whole-genome duplication event (Tuskan et al. 2006) and have largely colinear genetic maps (Hanley et al. 2007). Models produced based on the nucleotide substitution rates observed in the Brassicaceae suggest that the duplication occurred very recently, 8–13 million years ago (MYA; Sterck et al. 2005). However, fossil records indicate that *Populus* and *Salix* split 60–65 MYA (Collinson 1992); thus, the rate of sequence evolution must be considerably slower (~ 6-fold) in *Populus* than *Arabidopsis*. As a consequence, the *Populus* genome may bear substantial resemblance to the genome of the ancient Eurosid-era ancestor, albeit more complex owing to at least two rounds of whole-genome duplication. Utilizing comparisons between *Arabidopsis* and poplar, the common Eurosid ancestor was determined to contain at least 11,666 genes—in good agreement with a four-fold increase in gene content in *Populus* since the inception of modern speciation from this ancestor (Tuskan et al. 2006).

## 2.2.3 DNA Marker Resources

Reliable methods to genotype DNA sequence polymorphisms are essential for population and quantitative genetics, and evolutionary and comparative genomics studies. In poplar, several types of genetic markers are available and have been utilized to characterize genetic variation. Early population genetic studies in *P. fremontii* and *P. angustifolia* utilized restriction fragment length polymorphism (RFLP) (Keim et al. 1989). However, RFLPs were largely abandoned after the development of PCR-based markers due to the low multiplex ratio of each assay, and the high cost and labor required per genotypic data point. Development of PCR-based markers such as random amplified polymorphic DNA (RAPD) and amplified fragment length polymorphism (AFLP) made possible the development of the first high-coverage genomic surveys of polymorphisms in *Populus* (Yin et al. 2001, 2004). However, dominant inheritance and low transferability among genotypes and species made them less suitable for adoption by the scientific community. Anchoring these markers to the genome sequence can also be challenging (Gupta and Rustgi 2004), although sequence-tagged markers were shown to be crucial for assembling the *P. trichocarpa* Nisqually-1 genome sequence (Tuskan et al. 2006). Co-dominant markers such as microsatellites (or simple sequence repeats—SSRs) and single nucleotide polymorphism (SNPs) have become more widely used in animal and plant genetics. The main advantages of SSRs are their multi-allelic nature and high levels of heterozygosity, while SNPs are the most abundant type of DNA sequence variation. Since probes or primers used to assay SSRs or SNPs are designed for each locus specifically, they are frequently transferable within and (sometimes) among species. As a consequence, these two categories of genetic markers are particularly useful to the scientific community. Below is a summary of the SSR and SNP resources available for poplars as well as some of their applications.

## 2.2.3.1 Microsatellite (SSR) Resources

The International *Populus* Genome Consortium (IPGC) provides a large repository of 4,166 microsatellites. Primer sequences flanking these markers can be found in the IPGC website (*http://www.ornl.gov/sci/ipgc/ssr_resource.htm*). Most of the SSRs (84%) available for poplar were identified from the genome and BAC-end sequences of Nisqually-1 (Tuskan et al. 2006; Kelleher et al. 2007). The remaining were generated from several other studies (Dayanandan et al. 1998; Rahman et al. 2000; van der Schoot et al. 2000; Smulders et al. 2001; Tuskan et al. 2004). One-third of the primers (1,395) have been tested in at least one *Populus* pedigree, where 553 of them (40%) were successfully mapped. In addition, a set of 92 of the SSR markers were tested for their transferability in 23 species, including five from the *Salix* genus. The SSR resource is being extensively utilized for genetic mapping and analysis of QTLs. SSRs are also being used to estimate population structure (Wyman et al. 2003; Cole 2005; Hall et al. 2007; Ingvarsson et al. 2008), clone spatial distribution in natural areas (Namroud et al. 2005), genetic diversity in species threatened with extinction (Storme et al. 2004), and introgression in hybrid zones (Fossati et al. 2003; Woolbright et al. 2008). From a practical standpoint, SSRs have been applied in breeding programs for fingerprinting and ancestry determination of cultivars (Rahman et al. 2002; Rajora and Rahman 2003). Recently, SSR mapping data supported the indication of *Populus'* chromosome XIX as an incipient sex chromosome (Yin et al. 2008).

## 2.2.3.2 Single Nucleotide Polymorphisms (SNPs): Current Uses and Future Perspectives

While SSRs have been implemented frequently in poplar research, the use of SNPs is only at its infancy. The largest SNP discovery for the genus came as a by-product of the *Populus trichocarpa* genome sequence (Tuskan et al. 2006). Since poplars are obligate outcrossing species, the sequencing and assembly of a single heterozygous genotype (Nisqually-1) resulted in identification of over one million SNPs and 162 thousand single nucleotide insertion/deletion (indel) polymorphisms, with 20% of all polymorphisms affecting coding sequence in genes. Table 2-1 depicts the distribution of polymorphisms in each major sequencing scaffold as well as estimations of genetic diversity found in Nisqually-1. As expected given the action of purifying selection, diversity of SNPs was lower (1.4×) in genes when compared to the diversity found in the whole genome. Selection is apparently even stronger against indels within gene sequences—1.8× lower compared to the whole-genome. SNP diversity is variable between scaffolds of Nisqually-1, with maximum diversity found on LG_V (average of 1 SNP every 330 bp) and a minimum on LG_XVI (average of 1 SNP every 484 bp). Diversity found in gene sequences ranged from 591 bp per SNP on LG_XII to 433 bp per SNP on LG_XIII. However, variation in SNP diversity between scaffolds should be analyzed with caution as differences in sequencing depth were not considered in these estimations. In another study (Unneberg et al. 2005) 1,635 and 610 SNPs were identified in *P. tremula* and *P. trichocarpa*, respectively, by aligning approximately 70,000 ESTs. Aside from these two genome level studies (Unneberg et al. 2005; Tuskan et al. 2006), additional poplar SNPs have been discovered and characterized in only a few targeted genes. These genes were sequenced in a sample of genotypes in studies aimed at estimating population genetic parameters (Ingvarsson 2005; Gilchrist et al. 2006; Joseph and Lexer 2008), searching for departures from neutral evolution (Ingvarsson 2005; Keurentjes et al. 2007), or contributing to candidate gene association studies (Ingvarsson et al. 2006; Hall et al. 2007; Ingvarsson et al. 2008).

The main limitation of SNPs identified in these studies lies in the fact that forest species generally have an excess of low frequency alleles (Ingvarsson 2005; Krutovsky et al. 2005; Heuertz et al. 2006). As a result, many SNPs identified in a given study may not be suitable as molecular markers in different genetic backgrounds. However, as sequencing technologies advance (Ralph et al. 2008), polymorphisms will be discovered at markedly increased rates. As demonstrated in *Eucalyptus grandis* (Novaes et al. 2008), novel sequencing approaches offer a high-throughput platform for SNP discovery even in uncharacterized genomes. The availability of the *P. trichocarpa* genome sequence offers a framework for alignment of short sequence reads (30–35 bp) generated with the highest-throughput sequencing methods, opening new doors through which to characterize genetic variation in *Populus*.

**Table 2-1** DNA sequence polymorphisms derived from the diploid genome sequence of *P. trichocarpa* Nisqually-1.

| Scaffolds | Genome | | | | Genes[a] | | | |
|---|---|---|---|---|---|---|---|---|
| | Number of polymorphisms | | Diversity[b] (bp/polymorphism) | | Number of polymorphisms | | Diversity (bp/polymorphism) | |
| | Indel | SNPs | Indel | SNPs | Indel | SNPs | Indel | SNPs |
| LG_I | 13,344 | 85,410 | 2,410.01 | 376.53 | 2,132 | 16,772 | 4,409.58 | 560.53 |
| LG_II | 11,811 | 70,007 | 1,984.01 | 334.73 | 1,759 | 13,769 | 4,151.55 | 530.36 |
| LG_III | 7,767 | 46,293 | 2,246.28 | 376.88 | 1,340 | 9,721 | 3,990.46 | 550.07 |
| LG_IV | 5,798 | 39,659 | 2,600.94 | 380.25 | 953 | 7,180 | 4,244.67 | 563.39 |
| LG_IX | 5,589 | 33,874 | 2,220.73 | 366.41 | 1,005 | 7,980 | 4,321.07 | 544.19 |
| LG_V | 8,116 | 51,571 | 2,099.02 | 330.33 | 1,060 | 8,383 | 4,501.44 | 569.19 |
| LG_VI | 7,623 | 46,963 | 2,316.31 | 375.98 | 1,217 | 9,816 | 4,676.68 | 579.82 |
| LG_VII | 5,175 | 32,691 | 2,300.35 | 364.15 | 764 | 6,730 | 4,788.73 | 543.62 |
| LG_VIII | 6,940 | 42,758 | 2,223.37 | 360.87 | 1,184 | 10,362 | 4,678.92 | 534.63 |
| LG_X | 7,773 | 47,329 | 2,471.87 | 405.96 | 1,375 | 11,533 | 4,908.94 | 585.26 |
| LG_XI | 4,792 | 32,942 | 2,748.75 | 399.85 | 854 | 6,502 | 4,217.58 | 553.95 |
| LG_XII | 4,875 | 31,368 | 2,671.80 | 415.23 | 766 | 5,963 | 4,599.58 | 590.86 |
| LG_XIII | 4,979 | 32,977 | 2,308.85 | 348.60 | 921 | 8,179 | 3,844.72 | 432.94 |
| LG_XIV | 5,574 | 34,953 | 2,454.74 | 391.46 | 914 | 7,437 | 4,559.25 | 560.33 |
| LG_XIX | 3,644 | 24,744 | 2,806.48 | 413.30 | 636 | 5,832 | 4,378.72 | 477.51 |
| LG_XV | 4,059 | 25,327 | 2,510.66 | 402.37 | 653 | 5,418 | 4,808.18 | 579.50 |
| LG_XVI | 4,458 | 26,454 | 2,875.34 | 484.55 | 842 | 6,465 | 4,219.58 | 549.56 |
| LG_XVII | 2,238 | 16,291 | 2,432.56 | 334.18 | 344 | 2,808 | 3,855.58 | 472.34 |
| LG_XVIII | 4,568 | 29,995 | 2,722.86 | 414.67 | 752 | 5,976 | 4,638.20 | 583.66 |
| Unmapped scaffolds | 42,767 | 327,755 | 3,343.42 | 436.27 | 6,951 | 64,742 | 4,653.55 | 499.63 |
| Total or Average | 161,890 | 1,079,361 | 2,487.42 | 385.63 | 26,422 | 221,568 | 4,422.35 | 543.07 |

[a] Genes are comprised of 45,555 currently annotated gene models plus 10,288 additional less supported models.
[b] Sequence gaps of known size (symbolized in the genome sequence with 'Ns') were excluded from the calculation of diversity.

## 2.3 *Populus* Transcriptome Resources

Identifying patterns of genomic DNA sequence conservation and variation represents an important first step toward understanding the basic biology of *Populus* at the genetic level. However, it is only through analyses of gene expression within the genomic context that we can begin to interpret how differences in gene content, copy number, and coding sequences lead to phenotypic differences in individual trees. Here, we introduce transcriptome-level resources that have been produced for *Populus*, including EST sequence collections and microarray platforms. We also briefly detail genome-wide transcriptomics experiments carried out for *Populus* to date, which serve as a resource for gene expression data mining.

### 2.3.1 EST Sequence Collections and Databases

Prior to sequencing, assembling and annotating the *P. trichocarpa* genome (Tuskan et al. 2006), hundreds of thousands of ESTs had already been generated by groups in Canada, France, Sweden, and US (Sterky et al. 1998; Bhalerao et al. 2003b; Kohler et al. 2003; Dejardin et al. 2004; Ranjan et al. 2004; Sterky et al. 2004), strengthening the case for using *Populus* as the model tree species to be sequenced. Since the first *Populus* genome release in September of 2006, progress by other groups (Brosche et al. 2005; Lee et al. 2005; Nanjo et al. 2007; Pavy et al. 2007; Yu et al. 2007; Ralph et al. 2008) has resulted in the public availability of at least 421,350 *Populus* EST sequences (as of August 31, 2010) at NCBI. Approximately 40% of the total number of ESTs at NCBI comes from either North American *Populus trichocarpa* (black cottonwood) or European *Populus tremula x Populus tremuloides* (hybrid aspen). However, the list of other species represented includes *P. canadensis, P. deltoides, P. euphratica, P. nigra, P. tremula, P. tremula x P. alba*, and many hybrids among these (Table 2-2).

More recently, large-scale sequencing efforts have focused on characterizing the genomic population of small RNAs in *Populus*. These small RNAs include microRNAs (miRNAs), as well as the trans-acting (tasiRNAs) and natural antisense (nat-siRNAs) classes of small interfering RNAs (siRNAs). These small RNA classes are differentiated on the basis of their biogenesis (imperfect foldbacks of single-stranded RNA in the case of miRNAs, or perfectly complementary double-stranded RNA), but are not distinguishable on the basis of sequence or size alone—all range between 20–24 nucleotides (Barakat et al. 2007). Genome and database sequence information can often be used in post-sequencing steps to sort the 20–24 nt reads into their respective classes (Lu et al. 2005; Barakat et al. 2007; Lindow et al. 2007).

**Table 2-2** Populus EST sequence contributions by species (as of August 31, 2008).

| Species | #ESTs |
| --- | --- |
| *Populus trichocarpa* | 89,943 |
| *Populus tremula x Populus tremuloides* | 76,160 |
| *Populus trichocarpa x Populus deltoides* | 53,208 |
| *Populus nigra* | 51,361 |
| *Populus tremula* | 37,313 |
| *Populus trichocarpa x Populus nigra* | 20,130 |
| *Populus tremula x Populus alba* | 15,392 |
| *Populus deltoides* | 14,661 |
| *Populus euphratica* | 13,979 |
| *Populus tremuloides* | 12,813 |
| *Populus x canadensis* | 10,157 |
| *Populus alba x Populus tremula var. glandulosa* | 10,005 |
| *Populus simonii x Populus nigra* | 9,168 |
| *Populus fremontii x Populus angustifolia* | 5,410 |
| *Populus tomentiglandulosa* | 1,650 |
| **Total** | 421,350 |

While NCBI (*http://www.ncbi.nlm.nih.gov/*) provides the most comprehensive collection of *Populus* EST sequences, complementary EST repositories have also been created at the aforementioned PopulusDB (*http://www.populus.db.umu.se/*), MycorWeb (*http://mycor.nancy.inra.fr/PoplarDB/*), AspenDB (*http://aspendb.mtu.edu/*), Arborea (*http://www.arborea.ulaval.ca/en/*), Treenomix (*http://www.treenomix.ca/Home/ResearchActivities/Databases.aspx*) and ForestTreeDB (http://foresttree.org/ftdb). The results from small RNA sequencing projects (Lu et al. 2005; Barakat et al. 2007) can be found at miRBase (*http://microrna.sanger.ac.uk/sequences/*).

### 2.3.2 Poplar Microarray Platforms

Several platforms are currently available for whole-transcriptome analysis of *Populus*. These platforms differ in the probe type (cDNA or oligonucleotide) and synthesis methods. Microarray platforms can be produced by spotting cDNA or oligonucleotide probes onto slides, or by in situ synthesis of oligonucleotides using either masked (Affymetrix) or "maskless" (e.g., Roche Nimblegen and Agilent) probe synthesis technologies. Several academic groups have produced spotted cDNA arrays for *Populus* (Andersson et al. 2004; Moreau et al. 2005; Ralph et al. 2006; Zhang et al. 2007) based on EST sequencing resources. Spotted cDNA microarrays for *Populus* are also commercially available (e.g., *www.picme.at*) and details about the microarray design can be found in NCBI's Gene Expression Omnibus (GEO) Platform database (GPL4874). Due to the methods used to generate these microarrays, individual probe sizes vary and fewer design controls for cross-hybridization between highly similar paralogs exist.

In contrast, the advantage of the microarrays produced by in situ synthesis is that probes are designed computationally, maximizing specificity and minimizing self-complementarity, while maintaining constant annealing temperatures relative to other probes in the chip. Affymetrix commercializes a poplar genome microarray with 61,251 probe sets, representing 56,055 transcripts (GEO Platform GPL4359). The transcripts are derived from the assembly of genomic, EST and mRNA sequences from *Populus* species, deposited in GenBank, as well as the first predicted gene set from the *P. trichocarpa* genome sequence. Each probe set is composed of 11 25-mer perfect match probes, and the same number of probes containing a base mismatch in the 13th position (mismatch probe). The mismatch probes are used to quantify background and cross-hybridization signal in the Affymetrix array design. In contrast, Agilent and Roche Nimblegen (Rinaldi et al. 2007; Quesada et al. 2008) microarrays rely on probes that are typically 50–60 bases long to represent individual transcripts. These longer probes are reported to give a better signal to noise ratio, sensitivity and specificity (Hughes et al. 2001). The "maskless" technologies used to produce these microarrays are also more flexible and less costly—as a consequence, the cost of designing and producing a new microarray is relatively small. In addition, the resulting quantitations are thought to be more robust in the face of target-to-probe mismatches due to extensive polymorphisms present among different *Populus* species and genotypes assayed. While neither Agilent nor Roche Nimblegen offer catalog microarrays for *Populus*, both companies allow custom microarray design with spot densities capable of encompassing the predicted genomic complement of 45,555 protein coding gene transcripts in *Populus trichocarpa*. A microarray design that includes three probes representing all 45,555 predicted gene models form the *P. trichocarpa* genome, as well as approximately 10,000 less reliable models and a similar number of aspen ESTs has been developed recently and is deposited at NCBI (GEO Platform GPL2618).

The choice of microarray platform depends largely on the type of experiment to be performed. While short oligonucleotide microarrays (e.g., Affymetrix) are suited for transcriptome analysis in the model tree *P. trichocarpa*, gene expression surveys in hybrid trees or pedigrees may require the development of new designs, or the use of long oligonucleotide-based microarrays that are less prone to confounding effects of sequence variation. Because some have shown that cross-platform integration of data for meta-analyses is problematic (Jarvinen et al. 2004), it is also recommended to consider the eventual comparisons with respect to previously generated datasets.

### 2.3.3 Poplar Microarray Databases

While common gene expression databases are already well-established for *Arabidopsis* (as reviewed in Rensink et al. 2005), publicly available common collections of poplar transcriptomic experiments are currently spread into collections at UPSC-BASE (Sjodin et al. 2006), GEO (Edgar et al. 2002), ArrayExpress (Parkinson et al. 2005), and PlexDB (Wise et al. 2007). Of these, UPSC-BASE has implemented the most tools for cross-comparisons of array datasets, and so far, contain the largest number of publicly available experiments.

### 2.3.4 An Overview of Poplar Transcriptomics Experiments

Much of the recent interest in poplar transcriptomics has been focused on understanding tree-specific developmental processes and transitions, or environmental responses of trees to biotic and abiotic stresses. These experiments have largely relied on the use of microarrays to assess and identify differences in gene expression. Alternatively, "digital" or "electronic Northern" strategies have also been used in poplar transcriptomics experiments. In this strategy, the number of times ESTs representing a given gene are detected in two or more cDNA libraries is compared—statistically significant differences point to genes that are over- or under-expressed in the tissue from which the cDNA library was created. For example, digital Northerns that use the database of EST sequences from cDNA libraries deposited in PopulusDB (Sterky et al. 2004; Sjodin et al. 2006) have been successfully used to quantify mRNA and miRNA abundances and tissue specificity for genes of interest in aspen (Benedict et al. 2006; Geisler-Lee et al. 2006; Matsubara et al. 2006; Street et al. 2006; Barakat et al. 2007; Segerman et al. 2007). In the following subsections we highlight the main transcriptome resources for *Populus*, which serve as references to researchers who wish to investigate gene expression changes during development or as a response to stress. These studies are summarized in Table 2-3.

#### 2.3.4.1 The Poplar Transcriptome

EST sequencing projects provided an initial overview of the *Populus* transcriptome. ESTs have been generated from tissue-specific libraries (Sterky et al. 1998; Bhalerao et al. 2003b; Kohler et al. 2003; Dejardin 2004; Ranjan et al. 2004; Sterky et al. 2004; Lee et al. 2005; Ralph et al. 2006; Nanjo et al. 2007), resulting in datasets which inform on the expressed gene complement from a broad spectrum of vegetative and reproductive organs, and developmental stages in *Populus* (Table 2-3). Other transcriptome studies based on microarrays identified 22,616 predicted transcriptional

**Table 2-3** Populus EST libraries categorized according to growth condition and treatments, stage of development, tissue or organ from which mRNA was obtained, and the construction method (A = amplified, FL = full-length, N = normalized, S = standard).

| Treatment | Category | Organ/Tissue | Construction Method | References |
|---|---|---|---|---|
| **Untreated** | Vegetative tissue | *Young leaves* | S, FL | Bhalerao et al. 2003b; Sterky et al. 2004; Ralph et al. 2006 |
| | | *Leaves* | S, FL, N, A* | Ranjan et al. 2004; Zhang et al. 2007; Ralph et al. 2006; Barakat et al. 2007* |
| | | *Senescing leaves* | S | Bhalerao et al. 2003b; Sterky et al. 2004 |
| | | *Petiole* | S | Sterky et al. 2004 |
| | | *Imbibed seeds* | S | Sterky et al. 2004 |
| | | *Bark* | S | Sterky et al. 2004 |
| | | *Apical shoot* | S, N** | Sterky et al. 2004; Ranjan et al. 2004 |
| | | *Shoot meristem* | A | Sterky et al. 2004 |
| | | *Buds* | S, FL, N, A* | Ralph et al. 2006; Nanjo et al. 2007; Barakat et al. 2007* |
| | | *Stem* | FL | Ranjan et al. 2004; Nanjo et al. 2007; Pavy et al.2007 |
| | | *Cambial zone* | S, FL | Sterky et al. 1998; Sterky et al. 2004; Dejardin et al. 2004; Ralph et al. 2006 |
| | | *Active cambium* | A | Sterky et al. 2004 |
| | | *Tension wood* | S, FL | Sterky et al. 2004; Dejardin et al. 2004; Pavy et al. 2007 |
| | | *Opposite wood* | S, FL | Dejardin et al. 2004; Pavy et al.2007 |
| | | *Differentiating xylem* | S, FL | Sterky et al. 1998; Dejardin et al. 2004; Ralph et al. 2006; Pavy et al. 2007; Lu et al. 2005* |
| | | *Mature xylem* | S, FL | Sterky et al. 2004; Dejardin et al. 2004; Pavy et al.2007 |
| | | *Roots* | S, FL | Kohler et al. 2003; Sterky et al. 2004; Nanjo et al. 2007; Pavy et al. 2007 |
| | | *Root tips* | N** | Ranjan et al. 2004 |

| | | | | |
|---|---|---|---|---|
| | Dormant | *Dormant buds* | *A* | Sterky et al. 2004 |
| | | *Dormant cambium* | *A* | Sterky et al. 2004 |
| | Reproductive | *Floral bud* | *S* | Sterky et al. 2004 |
| | | *Male catkins* | *S* | Sterky et al. 2004 |
| | | *Female catkins* | *S* | Sterky et al. 2004 |
| | Undifferentiated | *Cell culture* | *S* | Ralph et al. 2006; Lee et al. 2005 |
| **Abiotic stress** | Drought | *Leaves, Roots* | *A, FL, N\*\** | Kohler et al. 2003; Nanjo et al. 2004; Brosché et al. 2005 |
| | Cold | *Leaves, Roots* | *S, FL, N\*\** | Sterky et al. 2004; Nanjo et al. 2004; Brosché et al. 2005 |
| | Salt | *Leaves, Roots* | *FL, N\*\** | Nanjo et al. 2004; Brosché et al. 2005 |
| | Heat | *Leaves* | *FL* | Nanjo et al. 2004 |
| | Flood | *Leaves, Roots* | *N\*\** | Brosché et al. 2005 |
| | H2O2 | *Leaves* | *FL* | Nanjo et al. 2004 |
| | Ozone | *Leaves, Roots* | *N\*\** | Brosché et al. 2005 |
| | Cadmium | *Leaves, Roots* | *N\*\** | Brosché et al. 2005 |
| | Safener | | | |
| | Nitrogen deprived | *Roots* | *S* | Ralph et al. 2006 |
| | Systemically wounded | *Leaves* | *S* | Christopher et al. 2004 |
| | Girdled/nitrogen fertilized | *Leaves, Roots, Stems* | *S* | Morency et al. (Arborea Genbank submission 2004) |
| **Biotic stress** | Virally infected | *Leaves* | *S* | Sterky et al. 2004 |
| | Fungally infected | *Leaves* | *S* | Sterky et al. 2004; Zhang et al. 2007 |
| | Locally FTC-herbivorized | *Leaves* | *S, FL* | Ralph et al. 2006 |
| | Systemically FTC-herbivorized | *Leaves* | *S, FL* | Ralph et al. 2006 |

*Table 2-3 contd....*

*Table 2-3 contd....*

| Treatment | Category | Organ/Tissue | Construction Method | References |
|---|---|---|---|---|
| | Willow weevil herbivorized | *Bark* | *S, N* | Ralph et al. 2006 |
| | *Pollacia radiosa* extract | *Cell culture* | *S* | Ralph et al. 2006 |
| | Chitosan (chemical elicitor) | *Cell culture* | *S* | Ralph et al. 2006 |
| **Hormonal** | ABA | *Leaves* | *FL* | Nanjo et al. 2004 |
| | SA | *Cell culture* | *S* | Ralph et al. 2006 |
| | MeJA | *Cell culture* | *S* | Ralph et al. 2006 |
| | Benzathiadiazole (SA mimic) | *Cell culture* | *S* | Ralph et al. 2006 |

units expressed above background in vegetative organs, with approximately 1/3rd of these displaying tissue-specific regulation (Tuskan et al. 2006; Quesada et al. 2008). Expansion of this study to qualitatively compare and contrast gene expression in the *P. tremula x P. alba* hybrid "INRA 717-1B4" with "Nisqually-1" demonstrated that there is a broad qualitative conservation of gene expression in many organs (Quesada et al. 2008).

### 2.3.4.2 Developmental Transitions and Responses

Undoubtedly, the primary focus of transcriptome studies in tree development has been at the level of wood formation—a process that clearly distinguishes poplar from other plant models, such as *Arabidopsis*. An initial insight into the transcriptomics of wood development came from Sterky and colleagues (Sterky et al. 1998) who developed EST libraries for the cambial zone of *P. tremula x P. tremuloides*, and the maturing xylem region of *P. trichocarpa*. This study laid the groundwork for future transcriptomics analysis of xylem and vascular cambium regions by demonstrating the power of isolating distinct zones of developing xylem for transcript quantification. A marked increase in the expression of genes related to secondary wall formation was noted in the xylem-specific library, whereas the cambial zone library (including phloem and xylem, as well as the cambial zone) sampled a higher overall proportion of transcripts. This stem sectioning approach was subsequently applied by several groups, who provided detailed "transcriptional roadmaps" to xylem differentiation (Hertzberg et al. 2001; Dejardin 2004), mature xylogenesis (Israelsson et al. 2003; Moreau et al. 2005), and cambial meristem identity (Schrader et al. 2004a) by isolation of specific cell layers within the cambial zone. Transcriptomics of wood development has also been studied through tension wood—a specialized tissue formed in response to a gravitational or mechanical stimulus. In these studies, specialized cDNA libraries were used to identify transcriptional and metabolic signatures for the tissue (Dejardin 2004; Sterky et al. 2004; Andersson-Gunneras et al. 2006) and individual genes that are responsive to this stress (Wu et al. 2000; Andersson-Gunneras et al. 2003).

Another primary focus of developmental transcriptomics research in poplar is the gene expression transitions which occur in response to seasonal shifts—particularly seasonal dormancy. Temperate, long-lived perennial species, such as poplars, require seasonal adjustments to survive unfavorable conditions in the winter while maintaining the ability to resume active growth in the spring. Therefore, studies have focused on gene expression and metabolomic responses associated with seasonal dormancy in vegetative tissues (Rohde et al. 2007; Ruttink et al. 2007) and cambial meristems (Schrader et al. 2004b; Druart et al. 2007; Park et al. 2008). These works paint a comprehensive picture of growth-to-dormancy seasonal transitions in poplar, whereby major transcriptional shifts in the two meristems largely mirror each other and receive inputs from analogous environmental and hormone-perception pathways. Likewise, shifts in gene expression in leaf tissue transitioning from active photosynthesis to autumn senescence are dramatic and affect large suites of metabolic genes (Bhalerao et al. 2003b; Andersson et al. 2004). The emerging paradigm is that a small number of shared transcriptional underpinnings could regulate these seasonal transitions and that epigenetic modifications may play a key role in acquisition and release from dormancy (Druart et al. 2007; Ruttink et al. 2007).

### 2.3.4.3 Environmental Responses

Because they are long-lived, trees must adapt and respond to changes in environmental conditions that occur over the course of a lifetime. One of the primary foci of research toward understanding the interaction of trees with their environments has been drought stress. Through EST- and microarray-based approaches (Brosche et al. 2005; Bogeat-Triboulot et al. 2007) transcriptional and metabolomic profiles of *P. euphratica*, a poplar species endemic to arid and semi-arid regions, have been produced and analyzed. These studies yielded a list of stress-responsive genes in poplar and identified genes that are differentially regulated in response to water stress acclimation in several tissues. A complimentary effort identified genetic and gene expression differences between *P. trichocarpa* and *P. deltoides* that may contribute to species-level differences in adaptation to drought stress (Street et al. 2006).

A second important topic of recent research is the response of poplar to changing atmospheric conditions. Taylor et al. (2005) have identified genes differentially regulated in response to free-air $CO_2$ enrichment (Poplar FACE or POPFACE) over the course of several years (Miglietta et al. 2001). The genes identified by this study provide initial clues to molecular mechanisms that trees may use to adjust to elevated atmospheric carbon levels in the midst of global climate change. Similarly, the transcriptome of susceptible and tolerant trees in the presence of elevated ozone has been characterized (Rizzo et al. 2007).

Finally, a large body of work has been invested in understanding transcriptome-level responses to biotic pests and pathogens. Transcript profiles in response to leaf rust inoculation (Miranda et al. 2007; Rinaldi et al. 2007), forest tent caterpillar feeding (Major et al. 2006; Ralph et al. 2006; Ralph et al. 2008), poplar mosaic virus inoculation (Smith et al. 2004), black spot disease (Zhang et al. 2007), systemic wounding (Christopher et al. 2004), willow weevil herbivory (Ralph et al. 2006), and biotic stress-inducible hormones (Ralph et al. 2006) have been undertaken. All studies mentioned here are summarized in Table 2-3.

## 2.4 Needed Genomic Resources and Tools for *Populus* Research

In the next section we outline some of the main resource needs for the advancement of *Populus* genomics research. The availability of these resources is crucial for a wider adoption of *Populus* as a model species by the scientific community. For plant species such as *Arabidopsis*, high-value resources like the genome sequence and large mutation line collections have driven research to higher levels and consolidated their roles as optimal model organisms. Articles referencing the use of *Arabidopsis thaliana* in scientific research have skyrocketed in the past decade, increasing approximately 200% from 1998 to 2008 (Fig. 2-2), coinciding with the sequencing of the genome and other resource milestones. Part of the growing interest in *Arabidopsis* research has also been fueled by targeted funding (e.g., Arabidopsis 2010 Program from the National Science Foundation in the US) leading to the development of novel, advanced resources which in turn propelled research

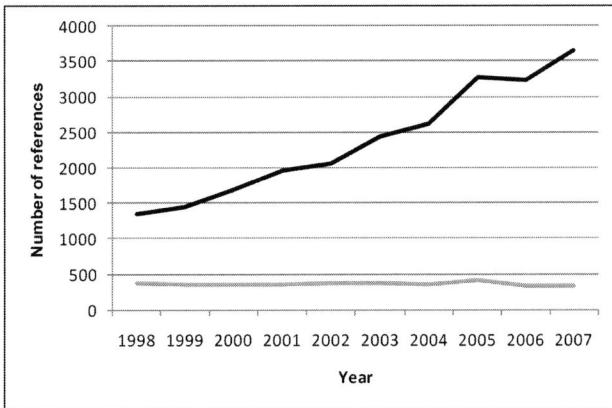

**Figure 2-2** Yearly number of publications referencing the use of *Arabidopsis thaliana* (black line) and *Nicotiana tabacum* (gray line) in the last decade (*Source:* ISI Web of Knowledge).

in the species even further. In contrast, the use of previously popular model species that have not seen such a rapid development in genomic resources has dwindled. As an example, references to *Nicotiana tabacum*, a previously popular plant model, have remained constant over the years or even declined (Fig. 2-2).

Support for the development of genomic resources for poplar is likely to continue in the near future. However, it will occur under the increasing competition from other plant species—particularly agricultural crops—that are being sequenced and may be more strongly supported by stakeholders. Therefore, it is crucial that poplar resources be developed rapidly to benefit from the unique advantages *Populus* currently has, including the availability of the genome sequence and its potential as a bioenergy crop. A very broad and extensive description of desired genomic resources for *Populus* has been outlined with the contribution of the scientific community in the booklet *"The Populus Genome Science Plan 2004–2009: From Draft Sequence to a Catalogue of All Genes Through the Advancement of Genomics Tools"* (*www. ornl.gov/sci/ipgc/the_populus_genome_science_plan.pdf*). In the next section we focus our discussion on the most urgently needed tools and resources to rapidly advance *Populus* genomics research, as well as improvements to those that already exist.

### 2.4.1 Improve Genome Assembly and Re-annotate the P. trichocarpa *Nisqualy-1 Genome*

The current release of the *Populus trichocarpa* genome assembly comprises 410 Mbp, or about 85% of the expected total genome size (485 ± 10 Mbp). Gene models were tentatively predicted based on a combination of ab initio, homology, and expressed sequence tag-based algorithms, which identified 45,555 transcriptional units (Tuskan et al. 2006). While the first draft of the *P. trichocarpa* DNA sequence provided the foundation for genomic research in the species, significant effort is still needed to achieve a near-finished quality by establishing resources that are reliable and complete. This need will become increasingly critical as the Nisqually-1 sequence is used as the reference for the genome assembly of additional *P. trichocarpa* genotypes and other *Populus* species. The resources necessary to fill most existing gaps, such as high coverage BAC libraries (Kelleher et al. 2007), have been developed and should now be utilized to create a sequence that is comprehensive and complete.

Refining the annotation of the genome is also a critical need. Transcriptional evidence for many gene models still needs to be verified. On the other hand, transcripts were detected for at least another ~ 4,000 potential transcriptional units. These are not currently annotated because of limited support from the gene-finding algorithms used in the genome analysis (Tuskan et al. 2006). Similarly, a recent survey of 4,487 full-length cDNAs identified 173 clones (4%) with no similarity to previously predicted poplar models, suggesting that they have been missed in the current annotation approach (Ralph et al. 2008). The structure of these genes may be such that they are not easily recognized by traditional gene-finding algorithms, requiring further manual annotation.

Finally, gene structure (e.g., UTRs, start and stop codons, intron-exon boundaries) needs to be refined for several gene models. Approaches such as the development of TILLING microarrays for verification of gene structure annotation were previously used in *Arabidopsis* (Kawai et al. 1993) with great success. We recently surveyed gene expression in several vegetative organs of a *P. trichocarpa* × *P. deltoides* hybrid, with a rudimentary tiling microarray (Drost et al. 2009). In this microarray, seven 60-mer non-overlapping probes were developed for each gene-model from the genome sequence, covering 420 bp of their predicted exosome. After hybridization with mRNA from a variety of vegetative tissues we identified probes for a large number of gene models with signal at or below background, in sharp contrast with other probes representing those genes (Drost et al. 2009). This suggests alternative spicing or that a predicted exon may not be correctly annotated. At the same time, several genes and gene families are currently being re-annotated by independent laboratories with gene-targeted research interest (Yu et al. 2002; Tuskan et al. 2006; Yang et al. 2006; Arnaud et al. 2007; Hamberger et al. 2007; Johnson and Douglas 2007; Kalluri et al. 2007; Oakley et al. 2007; Rajinikanth et al. 2007; Sasaki et al. 2007; Zhang et al. 2007; Bocock et al. 2008; Kohler et al. 2008; Ramirez-Carvajal et al. 2008; Zhuang et al. 2008). It will be essential for this information to be assembled and curated as a single, unified resource for the scientific community to make full use of it.

## 2.4.2 *Sequence Multiple* P. trichocarpa *Genotypes*

Sequencing and assembly of the *P. trichocarpa* genotype Nisqually-1 represented a new challenge in plant genome research, as it required creating contigs and scaffolds from sequences derived from two haplotypes of a highly heterozygous species. Now, future efforts should move toward incrementing these resources by sequencing multiple haplotypes.

In *A. thaliana*, microarray-based sequencing of 20 accessions created the first genomic survey of sequence polymorphisms in a broad spectrum of haplotypes, providing clues about the evolutionary forces that contribute to species variation while establishing the foundation for genome-wide linkage disequilibrium (LD) mapping (Clark et al. 2007; Kim et al. 2007). Plans for sequencing 1,001 accessions of *Arabidopsis* are now well under way (Ledford 2008), following similar plans for humans and *Drosophila*. We anticipate that in the relatively near future, entire breeding populations and large samples of wild natural populations will be fully sequenced for whole-genome population and quantitative genetic studies. In the meantime, production of draft-quality sequences of ~ 20–30 haplotypes that encompass the range of natural genetic diversity of the main *Populus* species will be critical for advancing genomics studies in these species. Sequencing of a few genotypes from *P. trichocarpa* is being completed by the Joint Genome Institute/DOE (G. Tuskan, pers. comm.) and will provide some guidance about the best approaches to carry out this work.

### 2.4.3 Genome Sequencing of **Populus** Species of Phylogenetic, Ecological and Economic Value

The genus *Populus* is commonly separated in six sections (*Abaso, Aigeiros, Leucoides, Populus, Tacamahaca, Turanga*) that comprise 29 species (Eckenwalder 1996). Poplars occur naturally in a wide spectrum of habitats, ranging from semi-arid areas of the Middle East and Asia (*P. euphratica*) to moist, cool temperate zones of North America (e.g., *P. tremuloides*). Sequencing the genome of multiple species will provide a unique opportunity for comparative studies aimed at understanding adaptation and species phenotypic and developmental diversity, much like the impact of sequencing the *Arabidopsis lyrata* had for the *Arabidopsis* community. Considering that poplar breeding heavily exploits the vigor of hybrids, genomic characterization of the most commercially relevant species will also be particularly important for understanding the molecular basis of heterosis and unique properties of some *Populus* species. Lower sequencing costs and the availability of the reference genome sequence of *P. trichocarpa* will provide the tools necessary for genome assembly and should significantly facilitate the development of these resources. Efforts should focus initially on sequencing one or more ecologically or commercially relevant genomes of each of the phylogenetic sections of the genus *Populus*.

### 2.4.4 Integrated Genome, Transcriptome, Proteome, Metabolome and Phenome Databases

*Populus* "-omics" resources are currently distributed across research institutions in North America and Europe. As the scale, complexity and breadth of these resources increases, so will the barriers against effective combination of this multilayered data. Systems Biology is expected to make it increasingly feasible to integrate information at the level of DNA, mRNA, protein and phenotype at multiple scales, largely through new cyberinfrastructures and bioinformatics tools. For that to become a reality, data and vocabularies will have to be standardized to create the desired interoperability in metadata repositories. Several efforts are underway to develop such integrated genomics resources in humans, plants and animals. The most advanced effort is being developed through the cancer Biomedical Informatics Grid (caBIG), which aims at enabling data and analytical tools created by public and private participating cancer research centers to be easily shared. caBIG is expected to provide the necessary cyberinfrastructure support for large genomic projects such as the *The Cancer Genome Atlas* (TCGA), through which all mutations related to cancer are being catalogued. The magnitude of these efforts in human genomics is much larger, and affects a broader and more diverse community. The development of these projects will create some of the frameworks by which other species, including poplar, will be studied. While some of the existing *Populus* genomic information and resources are already organized in public databases, most of the information is not yet easily accessible or well integrated. A significant effort should be directed towards this goal in the next decade.

## 2.4.5 Development and Repository for Advanced Germplasm and Molecular Resources

Organized, reliable and prompt availability of germplasm materials and molecular resources have propelled plant research in some major crop and model species. An outstanding example of the value of a germplasm and molecular "bank" comes from the Arabidopsis Biological Resource Center (ABRC) at Ohio State University (*www.biosci.ohio-state.edu/pcmb/Facilities/abrc/abrchome.htm*), which provides a repository for several thousand natural accessions (ecotypes), T-DNA insertion lines for the majority of *Arabidopsis* genes, and a broad range of recombinant inbred line (RIL) mapping populations. The ABRC also distributes DNA stocks that include genomic and cDNA libraries, as well as partial and full-length cDNAs, and BAC libraries.

Extensive germplasm resources have been generated in past decades for *Populus* species. Germplasm collections have been made from natural populations across the range of some species, capturing the genetic and phenotypic diversity that is the foundation for population and quantitative genetic studies. For *P. deltoides*, a collection that includes 815 unrelated genotypes has been obtained from 13 states in the central, southern and eastern US with the support of the Forest Service and other institutions. The *P. deltoides* collection is currently maintained in clone banks and seed orchards at Mississippi State University. More recently, over one thousand unrelated *P. trichocarpa* genotypes were collected in the US Pacific Northwest, as part of an effort funded by the Department of Energy to the Bioenergy Science Center (*http://bioenergycenter.org/*). This *P. trichocarpa* association population is now being established in three locations in northern California, western Oregon, and southern British Columbia for association genetics studies (S. DiFazio, pers. comm.). Smaller natural collections have also been made and established in common gardens for species such as *P. tremula* (Ingvarsson et al. 2006) and *P. fremontii*. Another valuable germplasm resource for *Populus* genomics research are the numerous intra- and interspecific pedigrees that have been generated in the past decades, for which detailed genetic maps are available (Bradshaw et al. 1994; Wu et al. 1997; Wu et al. 2000; Yin et al. 2004; Street et al. 2006).

In summary, germplasm resources are relatively well advanced regarding natural collections and family pedigrees, although a centralized location for their distribution would be highly valuable. However, particularly lacking are mutant collections for functional genomics studies, such as those available for the main model and agricultural crops. In particular, insertional mutagenesis using transposons or T-DNA has become an extremely valuable tool for functional genomic research in *Arabidopsis* and maize (Martienssen 1998; Krysan et al. 1999; Raizada et al. 2001), but the development of this tool has trailed in *Populus*. Small insertional mutagenesis collections have been developed for cottonwood (*P. trichocarpa*) and aspen (Busov and Strauss 2007) and it is clear from the high rate of insertions showing observable phenotypic changes that their occurrence is biased towards coding regions. Therefore, insertional mutagenesis could become an invaluable tool for functional genomics. However, a major effort by the poplar community is needed for functional characterization of the insertional lines.

A related issue concerns the public availability and distribution of poplar molecular resources, such as germplasm DNA, BAC and cDNA libraries or specific constructs. New molecular resources are being increasingly developed for *Populus* species. BAC libraries now exist for several species, including the *P. trichocarpa*, *P. deltoides*, *P. balsamifera* and interspecific hybrids. Similarly, numerous cDNA libraries have been created, targeting different species, tissues, developmental stages and growth conditions. While some of these resources may have been developed to address a specific biological hypothesis, their availability in standardized conditions and in a centralized facility would greatly facilitate molecular and genomic research in *Populus*.

## 2.5 Conclusions

Sequencing of the *P. trichocarpa* genome had the formidable effect of elevating the species to the status of a model plant, while organizing, expanding and uniting the poplar research community towards the development of common resource goals. Most of these goals were outlined in *The Populus Genome Science Plan* drafted in 2004 (*www.ornl.gov/sci/ipgc/*). The stimulus for *Populus* research resulted in significant advancement towards development of these resources—the production and annotation of the genome sequence draft and development of whole-genome microarrays are some of the objectives that have become a reality. However, most of the resources planned have yet to be developed (e.g., a genus-wide consensus genetic map) or are at early stages of development (e.g., genome browser) and funding opportunities to support their development are obviously necessary.

The future availability of funds for development, maintenance and distribution of existing and novel genomic resource for *Populus* is unpredictable. However, funding expectations in the US may be framed in the context of the National Plant Genome Initiative (NPGI), which was established in 1998. This interagency group has coordinated plant genome research among federal agencies. The National Research Council (NRC) recently released a report on the progress made in plant genome research since the establishment of the NPGI, and made recommendations for its future direction. While these recommendations may not be fully or even partially implemented, they provide a scholarly consensus within the plant scientific community concerning what direction the plant genome research should follow in the next decade. In its report, the NRC suggests an expansion of funding to a few species that have been adopted as models by the scientific community because of unique properties and biological advantages that distinguish them. Despite the lack of many important genomics resources for *Populus* research, it is unquestionable that it remains the best existing model for research on perenniality and wood formation. New funding programs, framed in a similar context as the National Science Foundation's Arabidopsis 2010, could aim to characterize all the "unique" poplar genes, or genes that may be specifically associated with wood formation and perenniality.

The value of developing genomic resources for *Populus* is not restricted to the worth of facilitating research on the fundamental physiological and molecular mechanisms that are unique to trees, but could include the significant practical impact of using *Populus* as a bioenergy crop. In the temperate zones of Europe and North America, poplars are the principal short rotation woody crop for providing clean, renewable and sustainable fuels. The genus *Populus* includes the fastest growing tree species within these climate zones. Rapid and significant gains can be made in biomass productivity and composition because poplars are at a very early stage of domestication. However, for these gains to occur a better understanding is needed of the molecular basis of natural variation in biomass yield and chemical composition, as well as the capacity of various genotypes to adapt to different environments. This gap in knowledge is the main barrier for efficient molecular breeding and selection of superior genotypes (high biomass productivity and conversion efficiency) that are well adapted to marginal sites. The availability of the *P. trichocarpa* genome sequence and the wealth of resources that have been generated as a result are unquestionably going to continue raising the status of *Populus* as a model species. However, new resources will be needed to turn these initial efforts into a valuable tool for discovery of genes and alleles associated with adaptive and quantitative variation in *Populus*.

# References

Arabidopsis Genome Initiative. Analysis of the genome sequence of the flowering plant *Arabidopsis thaliana*. Nature 408: 796–815.

Andersson A, Keskitalo J, Sjodin A, Bhalerao R, Sterky F, Wissel K, Tandre K, Aspeborg H, Moyle R, Ohmiya Y, Bhalerao R, Brunner A, Gustafsson P, Karlsson J, Lundeberg J, Nilsson O, Sandberg G, Strauss S, Sundberg B, Uhlen M, Jansson S, Nilsson P (2004) A transcriptional timetable of autumn senescence. Genome Biol 5: R24.

Andersson-Gunneras S, Hellgren JM, Bjorklund S, Regan S, Moritz T, Sundberg B (2003) Asymmetric expression of a poplar ACC oxidase controls ethylene production during gravitational induction of tension wood. Plant J 34: 339–49.

Andersson-Gunneras S, Mellerowicz EJ, Love J, Segerman B, Ohmiya Y, Coutinho PM, Nilsson P, . Henrissat B, Moritz T, Sundberg B (2006) Biosynthesis of cellulose-enriched tension wood in *Populus*: global analysis of transcripts and metabolites identifies biochemical and developmental regulators in secondary wall biosynthesis. Plant J 45: 144–65.

Arabidopsis Genome Initiative (2000) Analysis of the genome sequence of the flowering plant *Arabidopsis thaliana*. Nature 408: 796–815.

Arens P, Coops H, Jansen J, Vosman B (1998) Molecular genetic analysis of black poplar (*Populus nigra* L.) along Dutch rivers. Mol Ecol 7: 11–18.

Arnaud D, Dejardin A,. Leple JC, Lesage-Descauses MC, Pilate G (2007) Genome-wide analysis of LIM gene family in *Populus trichocarpa*, *Arabidopsis thaliana*, and *Oryza sativa*. DNA Res 14: 103–116.

Barakat A, Wall KP, Diloretto S, Depamphilis CW, Carlson JE (2007) Conservation and divergence of microRNAs in *Populus*. BMC Genom 8: 481.

Benedict C, Skinner JS, Meng R, Chang YJ, Bhalerao R, Huner NPA, Finn CE, Chen THH, Hurry V (2006) The CBF1-dependent low temperature signalling pathway, regulon and increase in freeze tolerance are conserved in *Populus* spp. Plant Cell Environ 29: 1259–1272.

Bhalerao R, Keskitalo J, Sterky F, Erlandsson R, Bjorkbacka H, Birve SJ, Karlsson J, Gardestrom P, Gustafsson P, Lundeberg J, Jansson S (2003b) Gene expression in autumn leaves. Plant Physiol 131: 430–42.

Bhalerao R, Nilsson O, Sandberg G (2003a) Out of the woods: forest biotechnology enters the genomic era. Curr Opin Biotechnol 14: 206–13.

Bocock PN, Morse AM, Dervinis C, Davis JM (2008) Evolution and diversity of invertase genes in *Populus trichocarpa*. Planta 227: 565–576.

Bogeat-Triboulot MB, Brosche M, Renaut J, Jouve L, Le Thiec D, Fayyaz P, Vinocur B, Witters E, Laukens K, Teichmann T, Altman A, Hausman JF, Polle A, Kangasjarvi J, Dreyer E (2007) Gradual soil water depletion results in reversible changes of gene expression, protein profiles, ecophysiology, and growth performance in *Populus euphratica*, a poplar growing in arid regions. Plant Physiol 143: 876–92.

Bradshaw HD, Villar M, Watson BD, Otto KG, Stewart S, Stettler RF (1994) Molecular genetics of growth and development in *Populus* .3. A genetic linkage map of a hybrid poplar composed of RFLP, STS, and RAPD markers. Theor Appl Genet 89: 167–178.

Bradshaw HD, Ceulemans R, Davis J, Stettler R (2000) Emerging model systems in plant biology: Poplar (*Populus*) as a model forest tree. J Plant Growth Regul 19: 306–313.

Brosche M, Vinocur B, Alatalo ER, Lamminmaki A, Teichmann T, Ottow EA, Djilianov D, Afif D, Bogeat-Triboulot MB, Altman A, Polle A, Dreyer E, Rudd S, Paulin L, Auvinen P, Kangasjarvi J (2005) Gene expression and metabolite profiling of *Populus euphratica* growing in the Negev desert. Genome Biol 6: R101.

Brunner AM, Busov VB, Strauss SH (2004) Poplar genome sequence: functional genomics in an ecologically dominant plant species. Trends Plant Sci 9: 49–56.

Busov V, Strauss SH (2007) Gene discovery in *Populus* using activation tagging. In vitro Cell Dev Biol Anim 43: S22–S22.

Busov V, Fladung M, Groover A, Strauss S (2005) Insertional mutagenesis in *Populus*: relevance and feasibility. Tree Genet Genom 1: 135–142.

Cervera MT, Storme V, Ivens B, Gusmao J, Liu BH, Hostyn V, Van Slycken J, Van Montagu M, Boerjan W (2001) Dense genetic linkage maps of three *Populus* species (*Populus deltoides*, *P. nigra* and *P. trichocarpa*) based on AFLP and microsatellite markers. Genetics 158: 787–809.

Chen KY, Cong B, Wing R, Vrebalov J, Tanksley SD (2007) Changes in regulation of a transcription factor lead to autogamy in cultivated tomatoes. Science 318: 643–5.

Christopher ME, Miranda M, Major IT, Constabel CP (2004) Gene expression profiling of systemically wound-induced defenses in hybrid poplar. Planta 219: 936–947.

Clark RM, Schweikert G, Toomajian C, Ossowski S, Zeller G, Shinn P, Warthmann N, Hu TT, Fu G, Hinds DA, Chen HM, Frazer KA, Huson DH, Schoelkopf B, Nordborg M, Raetsch G, Ecker JR, Weigel D (2007) Common sequence polymorphisms shaping genetic diversity in *Arabidopsis thaliana*. Science 317: 338–342.

Clark RM, Wagler TN, Quijada P, Doebley J (2006) A distant upstream enhancer at the maize domestication gene tb1 has pleiotropic effects on plant and inflorescent architecture. Nat Genet 38: 594–597.

Cole CT (2005) Allelic and population variation of microsatellite loci in aspen (*Populus tremuloides*). New Phytol 167: 155–164.

Collinson ME (1992) The early fossil history of *Salicaceae*: a brief review. Proc Roy Soc 98B: 155–167.

Cong B, Barrero LS, Tanksley SD (2008) Regulatory change in YABBY-like transcription factor led to evolution of extreme fruit size during tomato domestication. Nat Genet 40: 800–804.

Dayanandan S, Rajora OP, Bawa KS (1998) Isolation and characterization of microsatellites in trembling aspen (*Populus tremuloides*). Theor Appl Genet 96: 950–956.

Dejardin A, Leple JC, Lesage-Descauses MC, Costa G, Pilate G (2004) Expressed sequence tags from poplar wood tissues—A comparative analysis from multiple libraries. Plant Biol 6: 55–64.

Drost DR, Novaes E, Boaventura-Novaes C, Benedict CI, Brown RS, Yin TM, Tuskan GA, Kirst M (2009) A microarray-based genotyping and genetic mapping approach for highly heterozygous outcrossing species enables localization of a large fraction of the unassembled Populus trichocarpa genome sequence. Plant J. 58: 1054–1067.

Druart N, Johansson A, Baba K, Schrader J, Sjodin A, Bhalerao RR Resman L, Trygg J, Moritz T, Bhalerao RP (2007) Environmental and hormonal regulation of the activity-dormancy cycle in the cambial meristem involves stage-specific modulation of transcriptional and metabolic networks. Plant J 50: 557–573.

Eckenwalder JE (1996) Systematics and evolution of *Populus*. In: RF Stettler, HD Bradshaw, PE Heilman, TM Hinckley (eds) Biology of *Populus* and its implications for management and conservation. NRC Research Press, Ottawa, ON, Canada, pp 7–32.

Edgar R, Domrachev M, Lash AE (2002) Gene Expression Omnibus: NCBI gene expression and hybridization array data repository. Nucl Acids Res 30: 207–210.

Fossati T, Grassi F, Sala F, Castiglione S (2003) Molecular analysis of natural populations of *Populus nigra* L. intermingled with cultivated hybrids. Mol Ecol 12: 2033–2043.

Geisler-Lee J, Geisler M, Coutinho PM, Segerman B, Nishikubo N, Takahashi J, Aspeborg H, Djerbi S, Master E, Andersson-Gunneras S, Sundberg B, Karpinski S, Teeri TT, Kleczkowski LA, Henrissat B, Mellerowicz EJ (2006) Poplar carbohydrate-active enzymes. Gene identification and expression analyses. Plant Physiol 140: 946–962.

Gilchrist EJ, Haughn GW, Ying CC, Otto SP, Zhuang J, Cheung D, Hamberger B, Aboutorabi F, Kalynyak T, Johnson L, Bohlmann J, Ellis BE, Douglas CJ, Cronk QCB (2006) Use of Ecotilling as an efficient SNP discovery tool to survey genetic variation in wild populations of *Populus trichocarpa*. Mol Ecol 15: 1367–1378.

Goff SA, Ricke D, Lan TH, Presting G, Wang RL, Dunn M, Glazebrook J, Sessions A, Oeller P, Varma H Hadley D, Hutchinson D, Martin C, Katagiri F, Lange BM, Moughamer T, Xia Y, Budworth P, Zhong JP, Miguel T, Paszkowski U, Zhang SP, Colbert M, Sun WL, Chen LL, Cooper B, Park S, Wood TC, Mao L, Quail P, Wing R, Dean R, Yu YS, Zharkikh A,

Shen R, Sahasrabudhe S, Thomas A, Cannings R, Gutin A, Pruss D, Reid J, Tavtigian S, Mitchell J, Eldredge G, Scholl T, Miller RM, Bhatnagar S, Adey N, Rubano T, Tusneem N, Robinson R, Feldhaus J, Macalma T, Oliphant A, Briggs S (2002) A draft sequence of the rice genome (*Oryza sativa* L. ssp japonica). Science 296: 92–100.

Gupta P K, Rustgi S (2004) Molecular markers from the transcribed/expressed region of the genome in higher plants. Funct Integr Genom 4: 139–162.

Hall D, Luquez V, Garcia VM, St Onge KR, Jansson S, Ingvarsson PK (2007) Adaptive population differentiation in phenology across a latitudinal gradient in European aspen (*Populus tremula* L.): a comparison of neutral markers, candidate genes and phenotypic traits. Evolution 61: 2849–2860.

Hamberger B, Ellis M, Friedmann M, Souza CDA, Barbazuk B, Douglas CJ (2007) Genome-wide analyses of phenylpropanoid-related genes in *Populus trichocarpa, Arabidopsis thaliana*, and *Oryza sativa*: the *Populus* lignin toolbox and conservation and diversification of angiosperm gene families. Can J Bot 85: 1182–1201.

Hanley SJ, Mallott MD, Karp A (2007) Alignment of a Salix linkage map to the *Populus* genomic sequence reveals macrosynteny between willow and poplar genomes. Tree Genet Genomes 3: 35–48.

Hertzberg M, Aspeborg H, Schrader J, Andersson A, Erlandsson R, Blomqvist K, Bhalerao R, Uhlen M, Teeri TT, Lundeberg J, Sundberg B, Nilsson P, Sandberg G (2001) A transcriptional roadmap to wood formation. Proc Nl Acad Sci USA 98: 14732–14737.

Heuertz M, De Paoli E, Kallman T, Larsson H, Jurman I, Morgante M, Lascoux M, Gyllenstrand N (2006) Multilocus patterns of nucleotide diversity, linkage disequilibrium and demographic history of Norway spruce [*Picea abies* (L.) Karst]. Genetics 174: 2095–2105.

Holt RA, Jones SJ (2008) The new paradigm of flow cell sequencing. Genome Res 18: 839–846.

Hughes TR, Mao M, Jones AR, Burchard J, Marton MJ, Shannon KW, Lefkowitz SM, Ziman M,. Schelter JM, Meyer MR, Kobayashi S, Davis C, Dai HY, He YDD, Stephaniants SB, Cavet G, Walker WL, West A, Coffey E, Shoemaker DD, Stoughton R, Blanchard AP, Friend SH, Linsley PS (2001) Expression profiling using microarrays fabricated by an ink-jet oligonucleotide synthesizer. Nat Biotechnol 19: 342–347.

Ingvarsson PK (2005) Nucleotide polymorphism and linkage disequilbrium within and among natural populations of European Aspen (*Populus tremula* L., *Salicaceae*). Genetics 169: 945–953.

Ingvarsson PK, Garcia MV, Hall D, Luquez V, Jansson S (2006) Clinal variation in phyB2, a candidate gene for day-length-induced growth cessation and bud set, across a latitudinal gradient in European aspen (*Populus tremula*). Genetics 172: 1845–1853.

Ingvarsson PK, Garcia MV, Luquez V, Hall D, Jansson S (2008) Nucleotide polymorphism and phenotypic associations within and around the phytochrome B2 Locus in European Aspen (*Populus tremula* L., *Salicaceae*). Genetics 178: 2217–2226.

Israelsson M, Eriksson ME, Hertzberg M, Aspeborg H, Nilsson P, Moritz T (2003) Changes in gene expression in the wood-forming tissue of transgenic hybrid aspen with increased secondary growth. Plant Mol Biol 52: 893–903.

Jarvinen AK, Hautaniemi S, Edgren H, Auvinen P, Saarela J, Kallioniemi OP, Monni O (2004) Are data from different gene expression microarray platforms comparable? Genomics 83: 1164–1168.

Johnson LA, Douglas CJ (2007) Populus trichocarpa MONOPTEROS/AUXIN RESPONSE FACTOR5 (ARF5) genes: comparative structure, sub-functionalization, and *Populus* -*Arabidopsis* microsynteny. Can J Bot 85: 1058–1070.

Joseph JA, Lexer C (2008) A set of novel DNA polymorphisms within candidate genes potentially involved in ecological divergence between *Populus alba* and *P. tremula*, two hybridizing European forest trees. Mol Ecol Resour 8: 188–192.

Kalluri UC, DiFazio SP, Brunner AM, Tuskan GA (2007) Genome-wide analysis of Aux/IAA and ARF gene families in *Populus trichocarpa*. BMC Plant Biol 7: 59.

Kawai S, Matsumoto Y, Kajita S, Yamada K, Katayama Y, Morohoshi N (1993) Nucleotide sequence for the genomic DNA encoding an anionic peroxidase gene from a hybrid poplar, Populus kitakamiensis. Biosci, Biotechnol Biochem 57: 131–133.

Keim P, Paige KN, Whitham TG, Lark KG (1989) Genetic analysis of an interspecific hybrid swarm of *Populus*—Occurrence of unidirectional introgression. Genetics 123: 557–565.

Kelleher CT, Chiu R, Shin H, Bosdet IE, Krzywinski MI, Fjell CD, Wilkin J, Yin T, DiFazio SP, Ali J, Asano JK, Chan S, Cloutier A, Girn N, Leach S, Lee D, Mathewson CA, Olson T, O'Connor K, Prabhu AL, Smailus DE, Stott JM, Tsai M, Wye NH, Yang GS, Zhuang J, Holt RA, Putnam NH, Vrebalov J, Giovannoni JJ, Grimwood J, Schmutz J, Rokhsar D, Jones SJ, Marra MA, Tuskan GA, Bohlmann J, Ellis BE, Ritland K, Douglas CJ, Schein JE (2007) A physical map of the highly heterozygous *Populus* genome: integration with the genome sequence and genetic map and analysis of haplotype variation. Plant J 50: 1063–1078.

Keurentjes JJB, Fu JY,. Terpstra IR, Garcia JM, van den Ackerveken G, Snoek LB, Peeters AJM, Vreugdenhil D, Koornneef M, Jansen RC (2007) Regulatory network construction in *Arabidopsis* by using genome-wide gene expression quantitative trait loci. Proc Natl Acad Sci USA 104: 1708–1713.

Kim S, Plagnol V, Hu TT, Toomajian C, Clark RM, Ossowski S, Ecker JR, Weigel D, Nordborg M (2007) Recombination and linkage disequilibrium in *Arabidopsis thaliana*. Nat Genet 39: 1151–1155.

Kohler A, Delaruelle C, Martin D, Encelot N, Martin F (2003) The poplar root transcriptome: analysis of 7000 expressed sequence tags. FEBS Lett 542: 37–41.

Kohler A, Rinaldi C, Duplessis S, Baucher M, Geelen D, Duchaussoy F, Meyers BC, Boerjan W, Martin F (2008) Genome-wide identification of NBS resistance genes in *Populus trichocarpa*. Plant Mol Biol 66: 619–636.

Krutovsky KV, Neale DB (2005) Nucleotide diversity and linkage disequilibrium in cold-hardiness- and wood quality-related candidate genes in Douglas fir. Genetics 171: 2029–2041.

Krysan PJ, Young JC, Sussman MR (1999) T-DNA as an insertional mutagen in *Arabidopsis*. Plant Cell 11: 2283–2290.

Lee H, Lee JS, Noh EW, Bae EK, Choi YI, Han (2005) Generation and analysis of expressed sequence tags from poplar (*Populus alba* x *P. tremulla* var. glandulosa) suspension cells. Plant Sci 169: 1118–1124.

Lescot M, Rombauts S, Zhang J, Aubourg S, Mathe C, Jansson S, Rouze P, Boerjan W (2004) Annotation of a 95-kb *Populus deltoides* genomic sequence reveals a disease resistance gene cluster and novel class I and class II transposable elements. Theor Appl Genet 109: 10–22.

Lindow M, Jacobsen A, Nygaard S, Mang Y, Krogh A (2007) Intragenomic matching reveals a huge potential for miRNA-mediated regulation in plants. PLoS Comput Biol 3: e238.

Lu SF, Sun YH, Shi R, Clark C, Li LG, Chiang VL (2005) Novel and mechanical stress-responsive microRNAs in *Populus trichocarpa* that are absent from *Arabidopsis*. Plant Cell 17: 2186–2203.

Major IT, Constabel CP (2006) Molecular analysis of poplar defense against herbivory: comparison of wound- and insect elicitor-induced gene expression. New Phytol 172: 617–635.

Martienssen RA (1998) Functional genomics: Probing plant gene function and expression with transposons. Proc Natl Acad Sci USA 95: 2021–2026.

Matsubara S, Hurry V, Druart N, Benedict C, Janzik I, Chavarria-Krauser A, Walter A, Schurr U (2006) Nocturnal changes in leaf growth of *Populus deltoides* are controlled by cytoplasmic growth. Planta 223: 1315–1328.

Miglietta F, Hoosbeek MR, Foot J, Gigon F, Hassinen A, Heijmans M, Peressotti A, Saarinen T, Van Breemen N, Wallen B (2001) Spatial and temporal performance of the miniface (free air CO2 enrichment) system on Bog Ecosystems in northern and Central Europe. Environ Monitor Assess 66: 107–127.

Miranda M, Ralph SG, Mellway R, White R, Heath MC, Bohlmann J, Constabel CP (2007) The transcriptional response of hybrid poplar (*Populus trichocarpa* x *P. deltoides*) to infection by *Melampsora medusae* leaf rust involves induction of flavonoid pathway genes leading to the accumulation of proanthocyanidins. Mol Plant-Microb Interact 20: 816–831.

Moreau C, Aksenov N, Lorenzo MG, Segerman B, Funk C, Nilsson P, Jansson S, Tuominen H (2005) A genomic approach to investigate developmental cell death in woody tissues of *Populus* trees. Genome Biol 6: R34.

Namroud MC, Park A, Tremblay F, Bergeron Y (2005) Clonal and spatial genetic structures of aspen (*Populus tremuloides* Michx.). Mol Ecol 14: 2969–2980.

Nanjo T, Futamura N, Nishiguchi M, Igasaki T, Shinozaki K, Shinohara K (2004) Characterization of full-length enriched expressed sequence tags of stress-treated poplar leaves. Plant Cell Physiol 45: 1738–1748.

Nanjo T, Sakurai T, Totoki Y, Toyoda A, Nishiguchi M, Kado T, Igasaki T, Futamura N, Seki M, Sakaki Y, Shinozaki K, Shinohara K (2007) Functional annotation of 19,841 *Populus nigra* full-length enriched cDNA clones. BMC Genom 8: 448.

Negi MS, Rajagopal J, Chauhan N, Cronn R, Lakshmikumaran M (2002) Length and sequence heterogeneity in 5S rDNA of *Populus deltoides*. Genome 45: 1181–1188.

Novaes E, Drost DR, Farmerie WG, Pappas Jr GJ, Grattapaglia D, Sederoff RR, Kirst M (2008) High-throughput gene and SNP discovery in *Eucalyptus grandis*, an uncharacterized genome. BMC Genom 30: 312.

Oakley RV, . Wang YS, Ramakrishna W, Harding SA, Tsai CJ (2007) Differential expansion and expression of alpha- and beta-tubulin gene families in *Populus*. Plant Physiol 145: 961–973.

Okumura S, Sawada M, Park YM, Hayashi T, Shimamura M, Takase H, Tomizawa K (2006) Transformation of poplar (*Populus alba*) plastids and expression of foreign proteins in tree chloroplasts. Transgen Res 15: 637–646.

Park S, . Keathley DE, Han KN (2008) Transcriptional profiles of the annual growth cycle in *Populus deltoides*. Tree Physiol 28: 321–329.

Parkinson H, Sarkans U, Shojatalab M, Abeygunawardena N, Contrino S, Coulson R, Farne A, . Lara GG, Holloway E, Kapushesky M, Lilja P, Mukherjee G, Oezcimen A, Rayner T, Rocca-Serra P, Sharma A, Sansone S, Brazma A (2005) ArrayExpress—a public repository for microarray gene expression data at the EBI. Nucl Acids Res 33: D553–D555.

Pavy N, Johnson JJ, Crow JA, Paule C, Kunau T, MacKay J, Retzel EF (2007) ForestTreeDB: a database dedicated to the mining of tree transcriptomes. Nucl Acids Res 35: D888–D894.

Quesada T, Li Z, Dervinis C, Li Y, Bocock PN, Tuskan GA, Casella G, Davis JM, Kirst M (2008) An atlas of the vegetative transcriptome of *Populus trichocarpa* (Torr. & Gray ex Brayshaw) and its comparison to *Arabidopsis*. New Phytol 180: 408–420.

Rahman MH, Rajora OP (2002) Microsatellite DNA fingerprinting, differentiation, and genetic relationships of clones, cultivars, and varieties of six poplar species from three sections of the genus *Populus*. Genome 45: 1083–1094.

Rahman MH, Dayanandan S, Rajora OP (2000) Microsatellite DNA markers in *Populus tremuloides*. Genome 43: 293–297.

Raizada MN, Nan GL, Walbot V (2001) Somatic and germinal mobility of the RescueMu transposon in transgenic maize. Plant Cell 13: 1587–1608.

Rajinikanth M, Harding SA, Tsai CJ (2007) The glycine decarboxylase complex multienzyme family in *Populus*. J Exp Bot 58: 1761–1770.

Rajora P, Rahman H (2003) Microsatellite DNA and RAPD fingerprinting, identification and genetic relationships of hybrid poplar (*Populus* x *canadensis*) cultivars. Theor Appl Genet 106: 470–477.

Ralph S, Oddy C, Cooper D, Yueh H, Jancsik S, Kolosova N, Philippe RN, Aeschliman D, White R, Huber D, Ritland CE, Benoit F, Rigby T, Nantel A, Butterfield YS, Kirkpatrick R, Chun, E Liu J, Palmquist D, Wynhoven B, Stott J, Yang G, Barber S, Holt RA, Siddiqui A, Jones SJ, Marra MA, Ellis BE, Douglas CJ, Ritland K, Bohlmann J (2006) Genomics of hybrid poplar (*Populus trichocarpa* x *deltoides*) interacting with forest tent caterpillars (*Malacosoma disstria*): normalized and full-length cDNA libraries, expressed sequence tags, and a cDNA microarray for the study of insect-induced defences in poplar. Mol Ecol 15: 1275–1297.

Ralph SG, . Chun HJ, Cooper D, Kirkpatrick R, Kolosova N, Gunter L, Tuskan GA, Douglas CJ, Holt RA, Jones SJ, Marra MA, Bohlmann J (2008) Analysis of 4,664 high-quality sequence-finished poplar full-length cDNA clones and their utility for the discovery of genes responding to insect feeding. BMC Genom 9: 57.

Ramirez-Carvajal GA, Morse AM, Davis JM (2008) Transcript profiles of the cytokinin response regulator gene family in Populus imply diverse roles in plant development. New Phytol 177: 77–89.

Ranjan P, Kao YY, Jiang HY, Joshi CP, Harding SA, Tsai CJ (2004) Suppression subtractive hybridization-mediated transcriptome analysis from multiple tissues of aspen (*Populus tremuloides*) altered in phenylpropanoid metabolism. Planta 219: 694–704.

Rensink WA, Buell CR (2005) Microarray expression profiling resources for plant genomics. Trends Plant Sci 10: 603–609.

Rinaldi C, Kohler A, Frey P, Duchaussoy F, Ningre N, Couloux A, Wincker P, Le Thiec D, Fluch S, Martin F, Duplessis S (2007) Transcript profiling of poplar leaves upon infection with compatible and incompatible strains of the foliar rust *Melampsora larici-populina*. Plant Physioly 144: 347–366.

Rizzo M, Bernardi R, Salvini M, Nali C, Lorenzini G, Durante M (2007) Identification of differentially expressed genes induced by ozone stress in sensitive and tolerant poplar hybrids. J Plant Physiol 164: 945–949.

Rohde A, Ruttink T, Hostyn V, Sterck L, Van Driessche K, Boerjan W (2007) Gene expression during the induction, maintenance, and release of dormancy in apical buds of poplar. J Exp Bot 58: 4047–4060.

Ruttink T, Arend M, Morreel K, Storme V, Rombauts S, Fromm J, Bhalerao RP, Boerjan W, Rohde A (2007) A molecular timetable for apical bud formation and dormancy induction in poplar. Plant Cell 19: 2370–2390.

Sampedro J, Carey RE, Cosgrove DJ (2006) Genome histories clarify evolution of the expansin superfamily: new insights from the poplar genome and pine ESTs. J Plant Res 119: 11–21.

Sasaki S, Shimizu M, Wariishi H, Tsutsumi Y, Kondo R (2007) Transcriptional and translational analyses of poplar anionic peroxidase isoenzymes. J Wood Sci 53: 427–435.

Schrader J, Moyle R, Bhalerao R, Hertzberg M, Lundeberg J, Nilsson P, Bhalerao RP (2004a) Cambial meristem dormancy in trees involves extensive remodeling of the transcriptome. Plant J 40: 173–187.

Schrader J, Nilsson J, Mellerowicz E, Berglund A, Nilsson P, Hertzberg M, Sandberg G (2004b) A high-resolution transcript profile across the wood-forming meristem of poplar identifies potential regulators of cambial stem cell identity. Plant Cell 16: 2278–2292.

Schweitzer JA, Bailey JK, Fischer DG, LeRoy CJ, Lonsdorf EV, Whitham TG, Hart SC (2008) Plant-soil microorganism interactions: heritable relationship between plant genotype and associated soil microorganisms. Ecology 89: 773–781.

Segerman B, Jansson S, Karlsson J (2007) Characterization of genes with tissue-specific differential expression patterns in *Populus*. Tree Genet and Genomes 3: 351–362.

Sjodin A, Bylesjo M, Skogstrom O, Eriksson D, Nilsson P, Ryden P, Jansson S, Karlsson J. (2006) UPSC-BASE—*Populus* transcriptomics online. Plant J 48: 806–817.

Smith CM, Campbell MM (2004) Complete nucleotide sequence of the genomic RNA of poplar mosaic virus (*Genus carlavirus*). Arch Virol 149: 1831–1841.

Smulders MJM, Van Der Schoot J, Arens P, Vosman B (2001) Trinucleotide repeat microsatellite markers for black poplar (*Populus nigra* L.). Molecular Ecology Notes 1: 188–190.

Sterck L, Rombauts S, Jansson S, Sterky F, Rouze P, Van de Peer Y (2005) EST data suggest that poplar is an ancient polyploid. New Phytol 167: 165–170.

Sterky F, Regan S, Karlsson J, Hertzberg M, Rohde A, Holmberg A, Amini B, Bhalerao R, Larsson M, Villarroel R, Van Montagu M, Sandberg G, Olsson O, Teeri TT, Boerjan W,

Gustafsson P, Uhlen M, Sundberg B, Lundeberg J (1998) Gene discovery in the wood-forming tissues of poplar: analysis of 5, 692 expressed sequence tags. Proc Natl Acad Sci USA 95: 13330–13335.

Sterky F, . Bhalerao RR, . Unneberg P, Segerman B, Nilsson P, Brunner AM, Charbonnel-Campaa L, Lindvall JJ, Tandre K, Strauss SH, Sundberg B, Gustafsson P, Uhlen M, Bhalerao RP, Nilsson O, Sandberg G, Karlsson J, Lundeberg J, Jansson S (2004) A *Populus* EST resource for plant functional genomics. Proc Natl Acad Sci USA 101: 13951–13956.

Storme V, Vanden Broeck A, Ivens B, Halfmaerten D, Van Slycken J, Castiglione S, Grassi F, Fossati T, Cottrell JE, Tabbener HE, Lefevre F, Saintagne C, Fluch S, Krystufek V, Burg K, Bordacs S, Borovics A, Gebhardt K, Vornam B, Pohl A, Alba N, Agundez D, Maestro C, Notivol E, Bovenschen J, van Dam BC, van der Schoot J, Vosman B, Boerjan W, Smulders MJ (2004) Ex-situ conservation of Black poplar in Europe: genetic diversity in nine gene bank collections and their value for nature development. Theor Appl Genet 108: 969–981.

Street NR, Skogstrom O, Sjodin A, Tucker J, Rodriguez-Acosta M, Nilsson P, Jansson S, Taylor G (2006) The genetics and genomics of the drought response in *Populus*. Plant J 48: 321–341.

Taylor G (2002) Populus: Arabidopsis for forestry. Do we need a model tree? Ann. Bot. 90: 681–689.

Taylor G, Street NR, Tricker PJ, Sjödin A, Graham LE, Skogström O, Calfapietra C, Scarascia-Mugnozza G and Jansson S (2005) The transcriptome of *Populus* in elevated CO2. New Phytol 167: 143–154.

Tsai CJ, Podila GK, Chiang VL (1995) Nucleotide sequence of a *Populus tremuloides* gene encoding bispecific caffeic acid/5-hydroxyferulic acid O-methyltransferase. Plant Physiol 107: 1459.

Tuskan GA, Gunter LE, Yang ZMK, Yin TM, Sewell MM, DiFazio SP (2004) Characterization of microsatellites revealed by genomic sequencing of *Populus trichocarpa*. Can J For Res 34: 85–93.

Tuskan GA, Difazio S, Jansson A, Schein J, Sterck L, Aerts A, Bhalerao RR, Bhalerao RP, Blaudez D, Boerjan W, Brun A, Brunner A, Busov V, Campbell M, Carlson J, Chalot M, Chapman J, Chen GL, Cooper D, Coutinho PM, Couturier J, Covert S, Cronk Q, Cunningham R, Davis J, Degroeve S, Dejardin A, Depamphilis C, Detter J, Dirks B, Dubchak I, Duplessis S, Ehlting J, Ellis B, Gendler K, Goodstein D, Gribskov M, Grimwood J, Groover A, Gunter L, Hamberger B, Heinze B, Helariutta Y, Henrissat B, Holligan D, Holt R, Huang W, Islam-Faridi N, Jones S, Jones-Rhoades M, Jorgensen R, Joshi C, Kangasjarvi J, Karlsson J Kelleher C, Kirkpatrick R, Kirst M, Kohler A, Kalluri U, Larimer F, Leebens-Mack, J Leple JC, Locascio P, Lou Y, Lucas S, Martin F, Montanini B, Napoli C, Nelson DR, Nelson C, Nieminen K, Nilsson O, Pereda V, Peterr G, Philippe R, Pilate G, Poliakov A, Razumovskaya J, Richardson P, Rinaldi C, Ritland K, Rouze P, Ryaboy D, Schmutz J, Schrader J, Segerman B, Shin H, Siddiqui A, Sterky F, Terry A, Tsai CJ, Uberbacher E, Unneberg P, Vahala J, Wall K, Wessler S, Yang G, Yin T, Douglas C, Marra M, Sandberg G, Van de Peer Y, Rokhsar D (2006) The genome of black cottonwood, Populus trichocarpa (Torr. & Gray). Science 313: 1596–1604.

Unneberg P, Stromberg M, Lundeberg J, Jansson S, Sterky F (2005) Analysis of 70,000 EST sequences to study divergence between two closely related *Populus* species. Tree Genet Genomes 1: 109–115.

van der Schoot J, Pospíšková M, Vosman B, Smulders MJM (2000) Development and characterization of microsatellite markers in black poplar (*Populus nigra* L.). Theor Appl Genet 101: 317–322.

Wimp GM, Wooley S, Bangert RK, Young WP, Martinsen GD, Keim P, Rehill B, Lindroth RL, Whitham TG (2007) Plant genetics predicts intra-annual variation in phytochemistry and arthropod community structure. Mol Ecol 16: 5057–5069.

Wise RP, Caldo RA, Hong L, Shen L, Cannon E, Dickerson JA (2007) BarleyBase/PLEXdb: A Unified Expression Profiling Database for Plants and Plant Pathogens. Meth Mol Biol 406: 347–364.

Woolbright SA, Difazio SP, Yin T, Martinsen GD, Zhang X, Allan GJ, Whitham TG, Keim P (2008) A dense linkage map of hybrid cottonwood (*Populus fremontii* x *P. angustifolia*) contributes to long-term ecological research and comparison mapping in a model forest tree. Heredity 100: 59–70.

Wu L, Joshi CP, Chiang VL (2000) A xylem-specific cellulose synthase gene from aspen (*Populus tremuloides*) is responsive to mechanical stress. Plant J 22: 495–502.

Wu R, Bradshaw HD, Stettler RF (1997) Molecular genetics of growth and development in *Populus* (*Salicaceae*) .5. Mapping quantitative trait loci affecting leaf variation. Am J Bot 84: 143–153.

Wu R, Bradshaw HD, Stettler RF (1998) Developmental quantitative genetics of growth in *Populus*. Theor Appl Genet 97: 1110–1119.

Wu RL, Han YF, Hu JJ, Fang JJ, Li L, Li ML, Zeng ZB (2000) An integrated genetic map of *Populus deltoides* based on amplified fragment length polymorphisms. Theor Appl Genet 100: 1249–1256.

Wullschleger SD, Jansson S, Taylor G (2002) Genomics and forest biology: *Populus* emerges as the perennial favorite. Plant Cell 14: 2651–2655.

Wyman J, Bruneau A, Tremblay MF (2003) Microsatellite analysis of genetic diversity in four populations of *Populus tremuloides* in Quebec. Canadian Journal of Forest Research 81: 360–367.

Yang XH, Tuskan GA, Cheng ZM (2006) Divergence of the Dof gene families in poplar, *Arabidopsis*, and rice suggests multiple modes of gene evolution after duplication. Plant Physiol 142: 820–830.

Yin T, Huang M, Wang M, Zhu LH, Zeng ZB, Wu R (2001) Preliminary interspecific genetic maps of the populus genome constructed from RAPD markers. Genome 44: 602–609.

Yin T, Difazio SP, Gunter LE, Zhang X, Sewell MM, Woolbright SA, Allan GJ, Kelleher CT, Douglas CJ, Wang M, Tuskan GA (2008) Genome structure and emerging evidence of an incipient sex chromosome in *Populus*. Genome Res 18: 422–430.

Yin TM, DiFazio SP, Gunter LE, Riemenschneider D, Tuskan GA (2004) Large-scale heterospecific segregation distortion in *Populus* revealed by a dense genetic map. Theor Appl Genet 109: 451–463.

Yu J, Hu SN, Wang J, Wong GKS, Li SG, Liu B, Deng YJ, Dai L, Zhou Y, Zhang XQ, Cao ML, Liu J, Sun JD, Tang JB, Chen YJ, Huang XB, Lin W, Ye C, Tong W, Cong LJ, Geng JN, Han YJ, Li L, Li W, Hu GQ, Huang XG, Li WJ, Li J, Liu ZW, Li L, Liu JP, Qi QH, Liu JS, Li L, Li T, Wang XG, Lu H, Wu TT, Zhu M, Ni PX, Han H, Dong W, Ren XY, Feng XL, Cui P, Li XR, Wang H, Xu X, Zhai WX, Xu Z, Zhang JS, He SJ, Zhang JG, Xu JC, Zhang KL, Zheng XW, Dong JH, Zeng WY, Tao L, Ye J, Tan J, Ren XD, Chen, XW He J, Liu DF, Tian W, Tian CG, Xia HG, Bao QY, Li G, Gao H, Cao T, Wang J, Zhao WM Li P, Chen W, Wang XD, Zhang Y, Hu JF, Wang J, Liu S, Yang J, Zhang GY, Xiong YQ, Li ZJ, Mao L, Zhou CS, Zhu Z, Chen RS, Hao BL, Zheng WM, Chen SY, Guo W, Li GJ, Liu SQ, Tao M, Wang J, Zhu LH, Yuan LP, Yang HM (2002) A draft sequence of the rice genome (*Oryza sativa* L. ssp indica). Science 296: 79–92.

Yu YH, Xia XL, Yin WL, Zhang HC (2007) Comparative genomic analysis of CIPK gene family in *Arabidopsis* and *Populus*. Plant Growth Regul 52: 101–110.

Zhang Y, Zhang X, Chen Y, Wang Q, Wang M, Huang M (2007) Function and chromosomal localization of differentially expressed genes induced by *Marssonina brunnea* f. sp. *multigermtubi* in *Populus deltoides*. J Genet Genom 34: 641–648.

Zhuang J, Cai B, Peng RH, Zhu B, Jin XF, Xue Y, Gao F, Fu XY, Tian YS, Zhao W, Qiao YS, Zhang Z, Xiong AS, Yao QH (2008) Genome-wide analysis of the AP2/ERF gene family in *Populus trichocarpa*. Biochem Biophys Res Comm 371: 468–474.

# 3

# Haplotype Analysis of Complex Traits in Outcrossing Tree Species: Allele Discovery of Quantitative Trait Loci

*Tian Liu,[1] Chunfa Tong,[2] Jiangtao Luo,[2]Jiasheng Wu,[3] Bo Zhang,[4] Yuehua Cui,[5] Yao Li,[6] Yanru Zeng[3] and Rongling Wu[2,]\**

## ABSTRACT

Genetic mapping of quantitative trait loci (QTLs) based on a controlled cross has been widely useful for studying the genetic basis of a phenotypic trait. However, it is less likely that this approach provides a high-resolution mapping of genes that control complex quantitative traits. In particular, this approach is unable to characterize the linkage phase between alleles at the markers and QTL in a full-sib family derived from two heterozygous parents, thus limiting its application to marker-assisted selection in a plant breeding program. Currently, an emerging strategy for tracing haplotype segregation and transmission shows tremendous potential for characterizing QTL alleles for a quantitative trait and the molecular cloning of a QTL. This strategy is especially useful for QTL mapping in outcrossing species, because it can precisely estimate the allelic structure and organization of a QTL that is not possible to know from traditional QTL mapping. In this

[1]Human Genetics Group, Genome Institute of Singapore, 60 Biopolis Street, Singapore 138672; e-mail: *liut2@gis.a-star.edu.sg*
[2]Center for Statistical Genetics, Pennsylvania State University, Hershey, PA 17033, USA.
[3]School of Forestry and Biotechnology, Zhejiang Agricultural and Forestry University, Lin'an, Zhejiang 311300, China; e-mail: *zengyr@hotmail.com* and *yrzeng@zjfc.edu.cn*
[4]The Poplar Research Institute, Key Laboratory of Forest Genetics & Biotechnology, Nanjing Forestry University, Nanjing, Jiangsu 210037, China.
[5]Department of Statistics and Probability, Michigan State University, East Lansing, MI 48824, USA.
[6]Department of Statistics, West Virginia University, Morgantown, WV 26506, USA.
*Corresponding author: *rwu@hes.hmc.psu.edu*

chapter, we review statistical models and algorithms for haplotyping a complex trait in outcrossing species. With the increasing availability of polymorphic markers in a diverse range of species, haplotyping approaches will help to model a complex network of genetic regulation that includes the interactions between different haplotypes and between haplotypes and environments.

**Key words:** EM algorithm, Haplotype, Outcrossing species, Quantitative trait loci

## 3.1 Introduction

Plant height, branch structure, leaf display, wood production and wood gravity are important traits for the selection of superior tree types in forest improvement programs. These traits are inherently complex, determined by many genetic and environmental factors. The genetic factors, called quantitative trait loci (QTLs), are usually studied in an experimental cross derived from two parental lines through molecular mapping approaches. For many crops and model plant species, two phenotypically contrasting homozygous lines are crossed to generate an $F_1$ progeny from which advanced generations, such as backcross, $F_2$, or advanced intercross lines, are created for genetic mapping. The advantages of such a crossing procedure include the production of maximum disequilibria (i.e., non-random associations) between different loci and the precise characterization of a QTL allele from the marker alleles due to a known marker-QTL linkage phase. Because of these advantages, QTL mapping with an experimental cross has been instrumental for the identification of genes in a variety of plants from annuals to perennials (Grattapaglia et al. 1996; Wu et al. 1998; Boerjan 2005; Paterson 2006; Li et al. 2006), and will continue to play a role in unraveling the genetic architecture of plant traits.

Despite its practical usefulness, QTL mapping has a few significant limitations. First, its resolution for the characterization of genes that underlie a complex trait is low, especially when the number of progeny (and therefore meioses) in a cross is not sufficiently large. It is unlikely that the DNA polymorphism underlying a QTL can be identified with the markers linked with the QTL. Second, for outcrossing species like forest trees, QTL mapping is often based on a single $F_1$ cross between two heterozygous lines, providing inadequate information about the linkage phase of a QTL and its linked markers. Since the allele of a QTL cannot be predicated from observed marker genotypes, it is difficult to incorporate marker-assisted selection into a practical tree improvement program. Also, the number of alleles at a segregating locus in a single cross ranges from 2 to 4, making it challenging to determine the number of QTL alleles and, therefore, the genetic effects of a QTL on a quantitative trait.

Recently, it has been suggested that haplotype analysis with molecular markers may be useful for fine mapping of complex traits and the discovery of alleles at a functional QTL. A haplotype is the combination of alleles at a set of linked loci on the same chromosomal region. Several examples for successful haplotype analysis include the detection of haplotypes associated with biomedical traits (Park et al. 2003; Wiltshire et al. 2003; Wang et al. 2004; Payseur and Place 2007). Liu et al. (2004) developed a statistical model for identifying haplotypes that encode a quantitative trait from phase-unknown marker data collected in a natural population. This model is based on the population genetic properties of gene segregation at the haplotype level. Hou et al. (2007) extended Liu et al.'s idea to haplotype a complex trait in an experimental cross and make a conceptual link between the recombination fraction and linkage disequilibrium. These haplotyping models provide a general means for estimating the effect of haplotypes that encode a phenotypic trait and discovering the favorable alleles of a QTL for the trait value.

In this chapter, we will review the general principle and statistical model for haplotyping a complex trait. We will describe the population genetic theory that connects linkage and linkage disequilibrium in an experimental cross. The relationship between linkage analysis and linkage disequilibrium analysis will be explored in different types of crosses. We will show how the linkage phase between different markers can be determined in an outbred cross.

## 3.2 Haplotype and Diplotype

A haplotype represents a linear arrangement of alleles at different markers on a single chromosome, or part of a chromosome. The pair of haplotypes is called a diplotype. The observed phenotype of a diplotype is called a genotype. A diplotype is always constructed by two haplotypes, one from the maternal parent and the other from the paternal parent. Suppose there are two different markers on the same genomic region, one with two alleles *A* and *a* and the other with two alleles *B* and *b*, respectively. Allele *A* from marker 1 and allele *B* from marker 2 are located on the first homologous chromosome, whereas allele a from marker 1 and allele *b* from marker 2 located on the second homologous chromosome. Thus, *AB* is one haplotype and ab is a second haplotype, and both constitute a diplotype *AB* | *ab*, where the vertical line denotes separate haplotypes (Fig. 3-1).

In a practical genetic analysis, we can only observe the genotype expressed as *Aa/Bb*, where the slash is used to separate the genotypes at single markers when the phase is undetermined. However, the double heterozygote may be one (and only one) of two possible diplotypes *AB* | *ab* and *Ab* | *aB*. But these two diplotypes cannot be directly observed and should be inferred from marker genotype data (Fig. 3-2). In practice, it is important to estimate haplotype effects on a quantitative trait based on the diplotypes and therefore genotypes. For example, if a poplar tree carries haplotype *AB*, it will grow better than the trees that carry any other haplotypes, *Ab*, *aB*, and ab. For this reason, the same genotype *Aa/Bb* may perform differently, depending on what diplotype it carries. If this genotype is diplotype *AB | ab*, then it will have better growth. If the tree is diplotype *Ab | aB*, its growth will be poorer. The statistical model being developed will

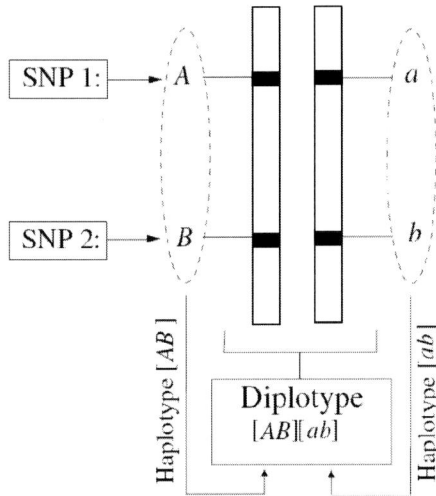

**Figure 3-1** Haplotype configuration of a diplotype for two markers.

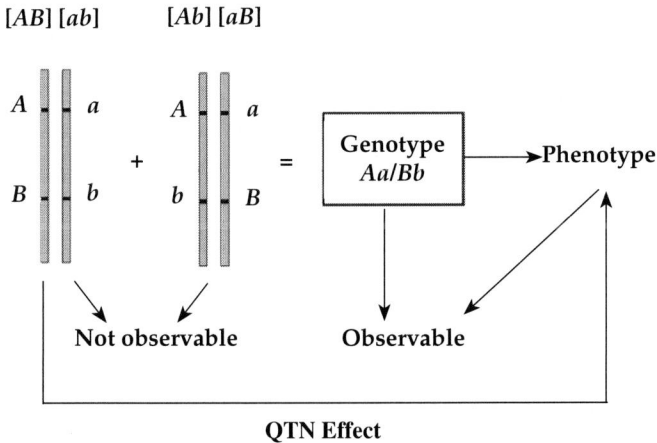

**Figure 3-2** Diplotype configuration of a genotype for two markers.

be used to determine which diplotype is associated with better growth in experimental crosses.

## 3.3 Marker Types in a Full-sib Family

Different from the backcross or $F_2$ derived from two inbred lines, a full-sib family derived from two parents of an outcrossing species may include up to four marker alleles, besides a null allele, at a single locus. Also, in such a full-sib family, the number of alleles may vary over loci. Each of the marker alleles, symbolized by $a$, $b$, $c$ and $d$, is dominant to the null allele, symbolized with $o$. Assume that all markers follow a Mendelian segregation without distortion. Depending on how different alleles are combined between the two parents used for the cross, there exists a total of 18 possible cross types for a segregating marker locus (Table 3-1). Based on unique information about linkage analysis provided by both parental and offspring marker band patterns, these cross types can be classified into seven groups:

**Table 3-1** Possible marker genotype cross combinations and observed marker band patterns for parents and their offspring in a full-sib family derived from two heterozygous lines.

| Cross type | | | Parent | | | Offspring | |
|---|---|---|---|---|---|---|---|
| | | Cross | Observed band | Symmetry | Observed bands | Segre-gation |
| A | 1 | $ab \times cd$ | $ab \times cd$ | No | $ac, ad, bc, bd$ | 1:1:1:1 |
| | 2 | $ab \times ac$ | $ab \times ac$ | No | $aa, ac, ba, bc$ | 1:1:1:1 |
| | 3 | $ab \times co$ | $ab \times c$ | No | $ac, a, bc, b$ | 1:1:1:1 |
| | 4 | $ao \times bo$ | $a \times b$ | No | $ab, a, b, o$ | 1:1:1:1 |
| B  B$_1$ | 5 | $ab \times ao$ | $ab \times a$ | No | $ab, 2a, b$ | 1:2:1 |
| B$_2$ | 6 | $ao \times ab$ | $a \times ab$ | No | $ab, 2a, b$ | 1:2:1 |
| B$_3$ | 7 | $ab \times ab$ | $ab \times ab$ | Yes | $a, 2ab, b$ | 1:2:1 |
| C | 8 | $ao \times ao$ | $a \times a$ | Yes | $3a, o$ | 3:1 |
| D  D$_1$ | 9 | $ab \times cc$ | $ab \times c$ | No | $ac, bc$ | 1:1 |
| | 10 | $ab \times aa$ | $ab \times a$ | No | $aa, ab$ | 1:1 |
| | 11 | $ab \times oo$ | $ab \times o$ | No | $a, b$ | 1:1 |
| | 12 | $bo \times aa$ | $b \times a$ | No | $ab, a$ | 1:1 |
| | 13 | $ao \times oo$ | $a \times o$ | No | $a, o$ | 1:1 |
| D$_2$ | 14 | $cc \times ab$ | $c \times ab$ | No | $ac, bc$ | 1:1 |
| | 15 | $aa \times ab$ | $a \times ab$ | No | $aa, ab$ | 1:1 |
| | 16 | $oo \times ab$ | $o \times ab$ | No | $a, b$ | 1:1 |
| | 17 | $aa \times bo$ | $a \times b$ | No | $ab, a$ | 1:1 |
| | 18 | $oo \times ao$ | $o \times a$ | No | $a, o$ | 1:1 |

Symmetry refers to whether two parents have the same marker genotype.

A. Fully informative markers that are heterozygous in both parents and segregate in a 1:1:1:1 ratio, including four alleles ab × cd, three non-null alleles *ab × ac*, three non-null alleles and a null allele ab × co, and two null alleles and two non-null alleles *ao × bo*;

B. Intercross markers that are heterozygous in both parents and segregate in a 1:2:1 ratio, which include three groups:

$B_1$. Three alleles form a non-symmetrical cross type between the two parents. Of the three alleles, one is a null-allele in one parent, e.g., *ab × ao*;

$B_2$. The reciprocal of $B_1$;

$B_3$. Two alleles form a symmetrical type between the two parents, i.e., *ab × ab*;

C. Dominant markers that are heterozygous in both parents and segregate in a 3:1 ratio, i.e., *ao × ao*;

D. Testcross markers that are segregating in one parent but not in the other parent and segregate in a 1:1 ratio, which include two groups:

$D_1$. Heterozygous in one parent and homozygous in the other, including three alleles *ab × cc*, two alleles *ab × aa*, *ab × oo* and *bo × aa*, and one allele (with three null alleles) *ao × oo*;

$D_2$. The reciprocals of $D_1$.

Some or all of these marker types occur in an outcrossing genome. A general statistical algorithm for two- and three-point linkage analysis with these marker types has been developed (Wu et al. 2002; Lu et al. 2004). In outcrossing species, testcross markers (D) are also called the pseudotestcross markers. The pseudotestcross strategy is widely used to construct linkage maps in forest trees based on the segregating patterns of the two parents (Grattapaglia and Sederoff 1994). Figure 3-3 gives part of linkage maps for interspecific hybrids between *P. deltoides* cv. "I-69" (D) and *P. euramericana* cv. "I-45" (E) based on testcross markers. Other marker types that are segregating in both parents can be used to align two parent-specific linkage maps and also bridge them with the published reference map through orthologous markers (Zhang et al. 2008). Haplotype analysis of a quantitative trait for an outcrossing population will be based on genetic linkage maps constructed from different types of markers.

## 3.4 Haplotyping with Testcross and Fully Informative Markers

### 3.4.1 Linkage Analysis for Testcross Markers

Suppose there is a chromosomal interval, bracketed by a pair of markers with a recombination fraction *r*, in which a QTL exists to affect a quantitative trait. A full-sib family is made with two heterozygous parents to detect the favorable allele of this QTL. Consider one parent, which produces segregating gametes during meiosis. Let 1 and 2 denote the two alleles at each of the two markers. The genotype of this double heterozygous parent is denoted as 12/12. This parent may have two possible diplotypes, 11 | 22 and 12 | 21. Let $p$ denote the probability with which this parent has diplotype 11 | 22, thus 1–$p$ denote the probability of the parent to be diplotype 12 | 21. Assuming Mendelian segregation, the haplotypes produced by this parent and their frequencies in terms of the recombination fraction r are expressed as

| Heterozygous Parent | | Haplotype | | | |
|---|---|---|---|---|---|
| Diplotype | Probability | 11 | 12 | 21 | 22 |
| 11 \| 22 | $p$ | $\frac{1}{2}(1-r)$ | $\frac{1}{2}r$ | $\frac{1}{2}r$ | $\frac{1}{2}(1-r)$ |
| 12 \| 21 | $1-p$ | $\frac{1}{2}r$ | $\frac{1}{2}(1-r)$ | $\frac{1}{2}(1-r)$ | $\frac{1}{2}r$ |

If these two markers are of the testcross type, they will produce four genotypes or diplotypes (11 | 22, 12 | 22, 21 | 22, and 22 | 22) in the full-sib family. Let $n_{11}$, $n_{12}$, $n_{21}$, and $n_{22}$ denote sample sizes for these genotypes, respectively. The Expectation-Maximization (EM) algorithm has been derived to estimate the recombination fraction ($r$) and parental diplotype probability ($p$) (Wu et al. 2002; Lu et al. 2004). In the Expectation step, the posterior probabilities with which a given diplotype contains recombinant (or parental) haplotypes are calculated by

$$\phi_1 = \frac{(1-p)r}{p(1-r)+(1-p)r}, \text{ for 12 | 21 diplotypes, and}$$

$$\phi_2 = \frac{pr}{pr+(1-p)(1-r)}, \text{ for 11 | 22 diplotypes.}$$

In the Maximization step, the MLEs of $r$ and $p$ are obtained by

$$r = \frac{\phi_1(n_{11}+n_{22})+\phi_2(n_{12}+n_{21})}{n_{11}+n_{12}+n_{21}+n_{22}},$$

$$p = \frac{(1-\phi_1)(n_{11}+n_{22})+\phi_2(n_{12}+n_{21})}{n_{11}+n_{12}+n_{21}+n_{22}}$$

The E and M steps are iterated until the estimates of the parameters converge to a stable value.

### 3.4.2 *Analysis of Variance*

To investigate how diplotypes are associated with a complex trait, an analysis of variance can be performed in which the sources of variance are due to the differences among and within diplotypes (Table 3-2). The $F$-value is calculated to test the significance of the difference among the four diplotypes by comparing this value against the critical $F_{(3,n-4;0.05)}$ value (with 3 vs. $n-4$ degrees of freedom at the $\alpha = 0.05$ significance level). If a significant difference is detected, this indicates that diplotypes (i.e., haplotypes in this case) trigger significant effects on phenotypic variation. The proportion of the total variance explained by diplotype difference is calculated as

$$R = \frac{\sigma_h^2}{\sigma_h^2 + \sigma_e^2} = \frac{MS_2 - MS_1}{MS_2 + (k-1)MS_1},$$

where $\sigma_h^2$ is the genetic variance due to the differences among diplotypes, $\sigma_e^2$ is the residual variance, and $k$ is the harmonic mean (Table 3-2).

**Table 3-2** ANOVA for haplotype analysis for testcross markers in a full-sib family.

| Source | df | Square | Mean Square | F-value |
|---|---|---|---|---|
| Among Diplotypes | 3 | MS1 | $\sigma_e^2 + k\sigma_h^2$ | MS1/MS2 |
| Within Diplotypes | $n-4$ | MS2 | $\sigma_e^2$ | |

$k$ is the harmonic mean expressed as $\dfrac{n_{11} + n_{12} + n_{21} + n_{22}}{\dfrac{1}{n_{11}} + \dfrac{1}{n_{12}} + \dfrac{1}{n_{21}} + \dfrac{1}{n_{22}}}$

### 3.4.3 t-Test

A specific risk haplotype for a trait can be detected by assuming that any one of the haplotypes, 11, 12, 21, and 22, is the risk haplotype. If haplotype 11 is assumed as the risk haplotype (labeled as $A$) and the other three assumed as the non-risk haplotype (labeled as $\overline{A}$), then these risk and non-risk haplotypes form two possible composite diplotypes in the backcross (Table 3-3). Then, a t-test approach is used to detect whether haplotype 11 exerts a significant effect on the phenotypic value of a trait. The $t$-test statistic is calculated as

$$t_{11} = \frac{\mu_{11} - \frac{1}{3}(\mu_{12} + \mu_{21} + \mu_{22})}{s\sqrt{\frac{1}{n_{11}} + \frac{1}{9n_{12}} + \frac{1}{9n_{21}} + \frac{1}{9n_{22}}}}$$

where

$$s = \frac{(n_{11}-1)s_{11}^2 + (n_{12}-1)s_{12}^2 + (n_{21}-1)s_{21}^2 + (n_{22}-1)s_{22}^2}{n_{11} + n_{12} + n_{21} + n_{22} - 4}$$

**Table 3-3** Basic statistics of two testcross markers in a full-sib family.

| Diplotype | Sample Size | Mean | Sampling Variance | Composite Risk Diplotype 11 | 12 | 21 | 22 |
|---|---|---|---|---|---|---|---|
| 11 \| 22 | $n_{11}$ | $\mu_{11}$ | $s_{11}^2$ | $\overline{AA}$ | $\overline{AA}$ | $\overline{AA}$ | $\overline{AA}$ |
| 12 \| 22 | $n_{12}$ | $\mu_{12}$ | $s_{12}^2$ | $\overline{AA}$ | $\overline{AA}$ | $\overline{AA}$ | $\overline{AA}$ |
| 21 \| 22 | $n_{21}$ | $\mu_{21}$ | $s_{21}^2$ | $\overline{AA}$ | $\overline{AA}$ | $\overline{AA}$ | $\overline{AA}$ |
| 22 \| 22 | $n_{22}$ | $\mu_{22}$ | $s_{22}^2$ | $\overline{AA}$ | $\overline{AA}$ | $\overline{AA}$ | $AA$ |
| Total | $n$ | | t-value | $n_{11}$ | $n_{12}$ | $n_{21}$ | $n_{22}$ |

If the calculated $t$-value is greater than the critical $t$ threshold with $n_{11} + n_{12} + n_{21} + n_{22} - 4$ degrees of freedom at the 5% significance level, this means that 11 is a significant risk haplotype. Otherwise, it is not a risk haplotype. An optimal risk haplotype can be selected on the basis of $t$-values calculated when different risk haplotypes are assumed, i.e.,

$$t_{12} = \frac{\mu_{12} - \frac{1}{3}\left(\mu_{11} + \mu_{21} + \mu_{22}\right)}{s\sqrt{\frac{1}{n_{12}} + \frac{1}{9n_{11}} + \frac{1}{9n_{21}} + \frac{1}{9n_{22}}}}, \quad t_{21} = \frac{\mu_{21} - \frac{1}{3}\left(\mu_{11} + \mu_{12} + \mu_{22}\right)}{s\sqrt{\frac{1}{n_{21}} + \frac{1}{9n_{11}} + \frac{1}{9n_{12}} + \frac{1}{9n_{22}}}},$$

$$t_{22} = \frac{\mu_{22} - \frac{1}{3}\left(\mu_{11} + \mu_{12} + \mu_{21}\right)}{s\sqrt{\frac{1}{n_{22}} + \frac{1}{9n_{11}} + \frac{1}{9n_{12}} + \frac{1}{9n_{21}}}}$$

for haplotypes 12, 21, and 22, respectively.

In practice, two-point analysis based on the $t$-test can be used to scan for every two adjacent markers over a linkage map, from which $t$-values are then plotted against map position. Genomic regions that carry significant haplotypes can be detected from the $t$ profile.

### 3.4.4 Example

Suppose there are two linked testcross markers on a chromosome. The heterozygous parent has the diplotype 11 | 22, which generates four possible groups of genotypes (and therefore diplotypes) at the two markers. Observations, means and sampling variances for each genotype group in the full-sib family were tabulated as follows:

| Offspring Diplotype | Observation | Sample Mean | Sample Variance |
|---|---|---|---|
| 11 \| 22 | 49 | 30 | 4 |
| 12 \| 22 | 3 | 28 | 6 |
| 21 \| 22 | 2 | 43 | 5 |
| 22 \| 22 | 51 | 29 | 5 |

Offspring diplotypes 12|22 and 21|22 should be derived from the recombinant haplotype of the diplotype parent, whereas diplotypes 11|22 and 22|22, derived from the non-recombinant haplotype. Thus, with the data given above, the recombination fraction between the two markers is estimated as

$$\hat{r} = \frac{3+2}{49+3+2+51} = 0.05.$$

The pooled sampling variance is estimated as

$$s^2 = \frac{(49-1) \times 4 + (3-1) \times 6 + (2-1) \times 5 + (51-1) \times 5}{49+3+2+51-4} = 4.54.$$

By assuming that 11 is a risk haplotype, the t-value is calculated as

$$t_{11} = \frac{30 - \frac{1}{3}(28+43+29)}{\sqrt{4.54} \times \sqrt{\frac{1}{49} + \frac{1}{9\times3} + \frac{1}{9\times2} + \frac{1}{9\times51}}} = -6.40.$$

Similarly, we calculate the *t*-values as $t_{12} = -4.49$, $t_{21} = 8.93$, and $t_{22} = -6.47$, when the risk haplotype is assumed to be 12, 21, and 22, respectively. By comparison, we conclude that haplotype 21 is an optimal risk haplotype. Haplotype 21 as a risk haplotype is statistically significant because its t-value (8.93) is greater than the critical value $t_{df=49+3+2+51-4,0.05} = 1.66$.

For testcross markers, models for linkage analysis can be performed at the gametic level because gametes provide the same information as zygotes do. For fully informative markers, both parents for a full-sib family are heterozygous and also have different alleles. Linkage analysis for fully informative markers should be carried out with zygotic data. Because the information of fully informative markers is just double that of testcross markers, the methods for estimating the recombination fraction and parental diplotype probability and testing haplotype effects on a quantitative trait with testcross markers can be simply extended to deal with fully informative markers.

## 3.5 Haplotyping with Symmetrical Intercross Markers

### 3.5.1 Haplotype, Diplotype, and Genotype Frequencies

A symmetrical intercross marker is segregating in a pattern similar to an $F_2$ marker, except that the linkage phases of two heterozygous parents are unknown. For intercross markers, a simple *t*-test is not adequate for haplotyping because there exists a double heterozygote that may carry any one of two possible diplotypes (Fig. 3-3). Suppose a panel of symmetrical intercross markers are typed, each of which is segregating in a 1:2:1

Mendelian ratio in the full-sib family derived from two parents P and Q. As above, we use 1 and 2 to denote alternative alleles at each marker. Consider two linked markers with the recombination fraction of $r$. Assume that the probability with which P is diplotype 11 | 22 is $p$, and then the probability with which this parent is diplotype 12 | 21 is $1-p$. Similar notation for the diplotype probability of Q is $q$ and $1-q$, respectively. These unknown parameters are arrayed in $\Omega = (r, p, q)$. We present an approach that uses the EM algorithm to obtain maximum likelihood estimates (MLEs) of haplotype

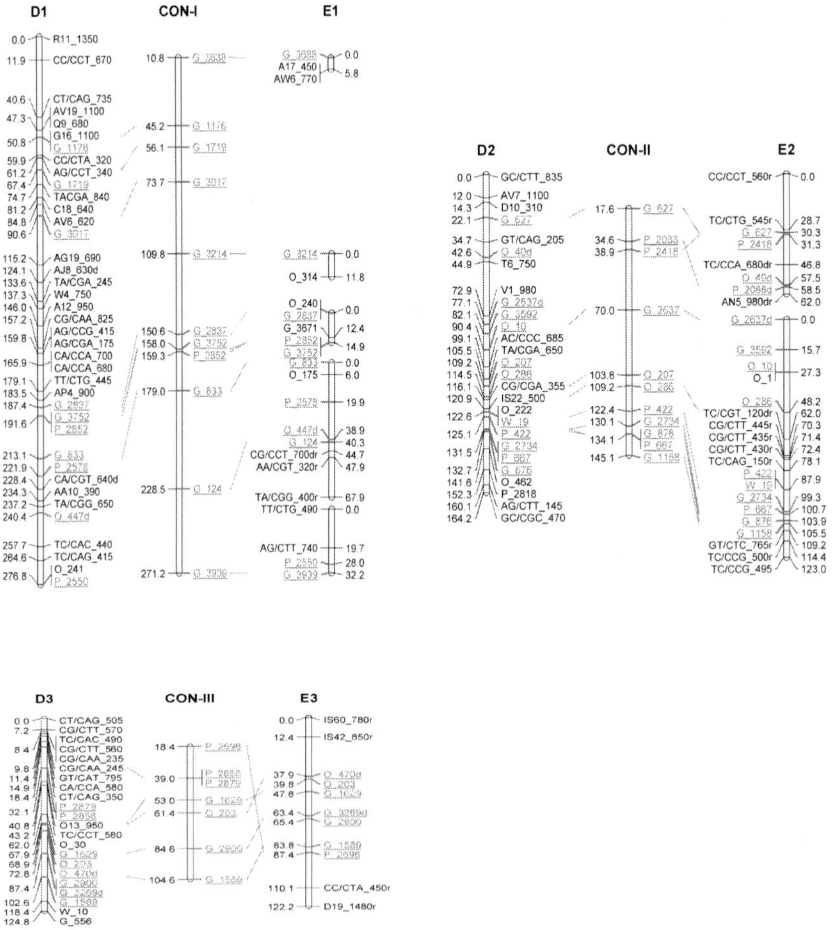

**Figure 3-3** Part of two linkage maps for interspecific hybrids between *P. deltoides* cv. "I-69" (D) and *P. euramericana* cv. "I-45" (E). The D and E maps are aligned to the reference linkage map of *Populus* (CON). Markers in red are homologous loci among the three maps and red lines link the homologous locations of the three genomes.

*Color image of this figure appears in the color plate section at the end of the book.*

frequencies, which are then used in conjunction with phenotypic data to derive MLEs of haplotype effects.

Considering the two possible diplotypes for each parent, haplotype frequencies produced by different parents P and Q are expressed as

$$p_{11}^P = \tfrac{1}{2}(1-r)p + \tfrac{1}{2}r(1-p),\ p_{12}^P = \tfrac{1}{2}rp + \tfrac{1}{2}(1-r)(1-p),$$

$$p_{21}^P = \tfrac{1}{2}rp + \tfrac{1}{2}(1-r)(1-p),\ p_{22}^P = \tfrac{1}{2}(1-r)p + \tfrac{1}{2}r(1-p),$$

$$p_{11}^Q = \tfrac{1}{2}(1-r)q + \tfrac{1}{2}r(1-q),\ p_{12}^Q = \tfrac{1}{2}rq + \tfrac{1}{2}(1-r)(1-q),$$

$$p_{21}^Q = \tfrac{1}{2}rq + \tfrac{1}{2}(1-r)(1-q),\ p_{22}^Q = \tfrac{1}{2}(1-r)q + \tfrac{1}{2}r(1-q),$$

which are used to calculate the diplotype and genotype frequencies given in Table 3-4.

**Table 3-4** Diplotypes and their frequencies for each of nine genotypes at two symmetrical intercross markers, haplotype composition frequencies for each genotype, and composite risk diplotypes for four possible risk haplotypes.

| | Diplotype | | Risk Haplotype | | | |
|---|---|---|---|---|---|---|
| Genotype | P\|Q | Frequency | 11 | 12 | 21 | 22 |
| 11/11 | 11\|11 | $p_{11}^P p_{11}^Q$ | $AA$ | $\overline{AA}$ | $\overline{AA}$ | $\overline{AA}$ |
| 11/12 | $\begin{cases}11\|12\\12\|11\end{cases}$ | $\begin{cases}p_{11}^P p_{12}^Q\\p_{12}^P p_{11}^Q\end{cases}$ | $\begin{cases}A\overline{A}\\A\overline{A}\end{cases}$ | $\begin{cases}A\overline{A}\\A\overline{A}\end{cases}$ | $\begin{cases}\overline{AA}\\\overline{AA}\end{cases}$ | $\begin{cases}\overline{AA}\\\overline{AA}\end{cases}$ |
| 11/22 | 12\|12 | $p_{12}^P p_{12}^Q$ | $\overline{AA}$ | $AA$ | $\overline{AA}$ | $\overline{AA}$ |
| 12/11 | $\begin{cases}11\|21\\21\|11\end{cases}$ | $\begin{cases}p_{11}^P p_{21}^Q\\p_{21}^P p_{11}^Q\end{cases}$ | $\begin{cases}A\overline{A}\\A\overline{A}\end{cases}$ | $\begin{cases}\overline{AA}\\\overline{AA}\end{cases}$ | $\begin{cases}A\overline{A}\\A\overline{A}\end{cases}$ | $\begin{cases}\overline{AA}\\\overline{AA}\end{cases}$ |
| 12/12 | $\begin{cases}11\|22\\22\|11\\12\|21\\21\|12\end{cases}$ | $\begin{cases}p_{11}^P p_{22}^Q\\p_{22}^P p_{11}^Q\\p_{12}^P p_{21}^Q\\p_{21}^P p_{12}^Q\end{cases}$ | $\begin{cases}A\overline{A}\\A A\\\overline{AA}\\\overline{AA}\end{cases}$ | $\begin{cases}\overline{AA}\\\overline{AA}\\A\overline{A}\\A\overline{A}\end{cases}$ | $\begin{cases}\overline{AA}\\\overline{AA}\\A\overline{A}\\A\overline{A}\end{cases}$ | $\begin{cases}\overline{AA}\\\overline{AA}\\\overline{AA}\\\overline{AA}\end{cases}$ |
| 12/22 | $\begin{cases}12\|22\\22\|12\end{cases}$ | $\begin{cases}p_{12}^P p_{22}^Q\\p_{22}^P p_{12}^Q\end{cases}$ | $\begin{cases}\overline{AA}\\\overline{AA}\end{cases}$ | $\begin{cases}A\overline{A}\\A\overline{A}\end{cases}$ | $\begin{cases}\overline{AA}\\\overline{AA}\end{cases}$ | $\begin{cases}A\overline{A}\\A\overline{A}\end{cases}$ |
| 22/11 | 21\|21 | $p_{21}^P p_{21}^Q$ | $\overline{AA}$ | $\overline{AA}$ | $AA$ | $\overline{AA}$ |
| 22/12 | $\begin{cases}21\|22\\22\|21\end{cases}$ | $\begin{cases}p_{21}^P p_{22}^Q\\p_{22}^P p_{21}^Q\end{cases}$ | $\begin{cases}\overline{AA}\\\overline{AA}\end{cases}$ | $\begin{cases}\overline{AA}\\\overline{AA}\end{cases}$ | $\begin{cases}A\overline{A}\\A\overline{A}\end{cases}$ | $\begin{cases}A\overline{A}\\A\overline{A}\end{cases}$ |
| 22/22 | 22\|22 | $p_{22}^P p_{22}^Q$ | $\overline{AA}$ | $\overline{AA}$ | $\overline{AA}$ | $AA$ |
| Likelihood | | | $L_{11}$ | $L_{12}$ | $L_{21}$ | $L_{22}$ |

### *3.5.2 Likelihood and Estimation for Frequency Parameters*

Based on observed genotypes ($n_{../.}$) and diplotype frequencies of the progeny, a multinomial likelihood of marker data (**S**) is constructed as

$$\log L(\Omega \,|\, \mathbf{S}) = \text{constant} + n_{11/11} \log( p_{11}^P p_{11}^Q ) + n_{11/12} \log( p_{11}^P p_{12}^Q + p_{12}^P p_{11}^Q )$$

$$+ n_{11/22} \log( p_{12}^P p_{12}^Q ) + n_{12/11} \log( p_{11}^P p_{21}^Q + p_{21}^P p_{11}^Q ) + n_{12/12} \log( p_{11}^P p_{22}^Q + p_{22}^P p_{11}^Q + p_{12}^P p_{21}^Q + p_{21}^P p_{12}^Q )$$

$$+ n_{12/22} \log( p_{12}^P p_{22}^Q + p_{22}^P p_{12}^Q ) + n_{22/11} \log( p_{21}^P p_{21}^Q ) + n_{22/12} \log( p_{21}^P p_{22}^Q + p_{21}^P p_{22}^Q )$$

$$+ n_{22/22} \log( p_{22}^P p_{22}^Q ).$$

The EM algorithm can be derived to obtain the MLEs of the recombination fraction ($r$) and parental diplotype probabilities ($p$ and $q$). In the E step, calculate the expected numbers of recombinants within each genotype by

$$\phi_1 = \frac{ r(1-r)[(1-p)q + p(1-q)] + 2r^2(1-p)(1-q) }{ \omega_1^P \omega_1^Q },$$

$$\phi_2 = \frac{ 2r(1-r)[pq + (1-p)(1-q)] + 2r^2[(1-p)q + p(1-q)] }{ \omega_1^P \omega_2^Q + \omega_2^P \omega_1^Q },$$

$$\phi_3 = \frac{ r(1-r)[p(1-q) + (1-p)q] + 2r^2 pq }{ \omega_2^P \omega_2^Q },$$

$$\phi_4 = \frac{ 2r(1-r)[p(1-q) + (1-p)q] + 2r^2[pq + (1-p)(1-q)] }{ \omega_1^P \omega_1^Q + \omega_2^P \omega_2^Q },$$

$$\phi_5 = \frac{ 2r(1-r)[pq + (1-p)(1-q)] + 2r^2[p(1-q) + (1-p)q] }{ \omega_2^P \omega_1^Q + \omega_1^P \omega_2^Q },$$

the expected numbers of *nonrecombinants for parent P* within each genotype by

$$\psi_1^P = \frac{ (1-r)p }{ \omega_1^P },$$

$$\psi_2^P = \frac{ (1-r)p\omega_2^Q + rp\omega_1^Q }{ \omega_1^P \omega_2^Q + \omega_2^P \omega_1^Q },$$

$$\psi_3^P = \frac{ rp }{ \omega_2^P },$$

$$\psi_4^P = \frac{ (1-r)p\omega_1^Q + rp\omega_2^Q }{ \omega_1^P \omega_1^Q + \omega_2^P \omega_2^Q },$$

$$\psi_5^P = \frac{rp\omega_1^Q + (1-r)p\omega_2^Q}{\omega_2^P\omega_1^Q + \omega_1^P\omega_2^Q},$$

and the expected numbers of *nonrecombinants for parent Q* within each genotype by

$$\psi_1^Q = \frac{(1-r)q}{\omega_1^Q},$$

$$\psi_2^Q = \frac{(1-r)q\omega_2^P + rq\omega_1^P}{\omega_1^P\omega_2^Q + \omega_2^P\omega_1^Q},$$

$$\psi_3^Q = \frac{rq}{\omega_2^Q},$$

$$\psi_4^Q = \frac{(1-r)q\omega_1^P + rq\omega_2^P}{\omega_1^P\omega_1^Q + \omega_2^P\omega_2^Q},$$

$$\psi_5^Q = \frac{rq\omega_1^P + (1-r)q\omega_2^P}{\omega_2^P\omega_1^Q + \omega_1^P\omega_2^Q},$$

where

$$\omega_1^P = (1-r)p + r(1-p),$$
$$\omega_1^Q = (1-r)q + r(1-q),$$
$$\omega_2^P = (1-r)(1-p) + rp),$$
$$\omega_2^Q = (1-r)(1-q) + rq.$$

In the M step, estimate the recombination fraction and diplotype probabilities with the numbers estimated from the E step by

$$r = \frac{1}{2n}[\phi_1(n_{11/11} + n_{22/22}) + \phi_2(n_{11/12} + n_{12/11}) + \phi_3(n_{11/22} + n_{22/11}) + \phi_4 n_{12/12} + \phi_5(n_{12/22} + n_{22/12})],$$

$$p = \frac{1}{2n}[\psi_1^P(n_{11/11} + n_{22/22}) + \psi_2^P(n_{11/12} + n_{12/11}) + \psi_3^P(n_{11/22} + n_{22/11}) + \psi_4^P n_{12/12} + \psi_5^P(n_{12/22} + n_{22/12})],$$

and

$$q = \frac{1}{2n}[\psi_1^Q(n_{11/11} + n_{22/22}) + \psi_2^Q(n_{11/12} + n_{12/11}) + \psi_3^Q(n_{11/22} + n_{22/11}) + \psi_4^Q n_{12/12} + \psi_5^Q(n_{12/22} + n_{22/12})].$$

Iterations between the E and M steps are repeated until the estimates of parameters converge. The estimates at the convergence are the MLEs of $r$, $p$, and $q$.

### 3.5.3 Likelihood and Estimation for Effect Parameters

To estimate haplotype effects on a quantitative trait, we need to assume that one haplotype is the risk haplotype (designated as $A$), and thus all the others are the non-risk haplotypes (designated as $\overline{A}$ ). The risk and non-risk haplotypes will form three composite diplotypes AA (**2**), $A\overline{A}$ (**1**), and $\overline{A}\overline{A}$ (**0**), whose genotypic values are expressed as $\mu_2$, $\mu_1$, and $\mu_0$, respectively. The genotypic values of different composite diplotypes and residual variance are arrayed in $\Theta = (\mu_2, \mu_1, \mu_0, \sigma^2)$.

The mixture-based likelihood of phenotypic (y) and marker data (**S**) and the estimated frequency parameters is constructed as

$$\log L\!\left(\Theta \mid y, S, \hat{\Omega}_p\right) = \sum_{i=1}^{n_{11/11}} \log f_2(y_i) + \sum_{i=1}^{\dot{n}} \log f_1(y_i) + \sum_{i=1}^{\ddot{n}} \log f_0(y_i) + \sum_{i=1}^{n_{12/12}} \log[\phi f_1(y_i) + (1-\phi) f_0(y_i)],$$

where

$f_j(y_i)$ is the normal distribution density function of the phenotypic trait with mean, $\mu_j$ (j = 2, 1, 0), and variance, $\sigma^2$.

$$\phi = \frac{\hat{p}_{11}^P \hat{p}_{22}^Q + \hat{p}_{22}^P \hat{p}_{11}^Q}{\hat{p}_{11}^P \hat{p}_{22}^Q + \hat{p}_{22}^P \hat{p}_{11}^Q + \hat{p}_{12}^P \hat{p}_{21}^Q + \hat{p}_{21}^P \hat{p}_{12}^Q}$$

$$\dot{n} = n_{11/12} + n_{12/11},$$

$$\ddot{n} = n_{11/22} + n_{12/22} + n_{21/21} + n_{21/22} + n_{22/22}.$$

The EM algorithm is used to estimate the haplotype effect parameters. In the E step, calculate the posterior probability with which a double heterozygote i carries diplotype 11 | 22 by

$$\Phi_{1|22i} = \frac{\phi f_1(y_i)}{\phi f_1(y_i) + (1-\phi) f_0(y_i)}$$

In the M step, estimate the effect parameters and variance by

$$\mu_2 = \frac{\sum_{i=1}^{n_{11/11}} y_i}{n_{11/11}}, \quad \mu_1 = \frac{\sum_{i=1}^{\dot{n}} y_i + \sum_{i=1}^{n_{12/12}} \Phi_{1|2i} y_i}{\dot{n} + \sum_{i=1}^{n_{12/12}} \Phi_{1|22i}}, \quad \mu_0 = \frac{\sum_{i=1}^{\ddot{n}} y_i + \sum_{i=1}^{n_{12/12}} (1 - \Phi_{1|22i}) y_i}{\ddot{n} + \sum_{i=1}^{n_{12/12}} (1 - \Phi_{1|22i})},$$

$$\sigma^2 = \frac{1}{n}\left\{ \sum_{i=1}^{n_{11|11}}(y_i - \mu_2)^2 + \sum_{i=1}^{\hat{n}}(y_i - \mu_1)^2 + \sum_{i=1}^{\bar{n}}(y_i - \mu_0)^2 + \sum_{i=1}^{n_{12|12}}\left[\Phi_{1|22i}(y_i - \mu_1)^2 + (1 - \Phi_{1|22i})(y_i - \mu_0)^2\right]\right\}.$$

The E and M steps are iterated until the estimates converge to stable values. An optimal risk haplotype is chosen on the basis of the likelihoods calculated by assuming all possible risk haplotypes. Table 3-4 provides the composite diplotypes of two markers when different haplotypes are assumed as a risk haplotype. Under each assumption, the likelihood is calculated. The haplotype that gives the largest likelihood is regarded as an optimal risk haplotype. After an optimal risk haplotype is determined, hypothesis tests can be performed about the existence of significant risk haplotypes as well as the additive and dominance genetic effects on the trait conferred by these haplotypes.

## 3.6 Haplotyping with Nonsymmetrical Intercross Markers

In Table 3-1, there is a unique type of marker $B_1$ or $B_2$, called the nonsymmetrical intercross marker, which contains a null allele (denoted by 0) for one of the parents. Suppose there is a nonsymmetrical intercross marker $B_1$ in which parent P has genotype 12/12 (producing four haplotypes, 11, 12, 21, and 22, in a frequency of $p_{11}^P$, $p_{12}^P$, $p_{21}^P$, $p_{22}^P$ and, respectively), whereas parent Q has genotype 10/10 (producing four haplotypes 11, 10, 01, and 00, in a frequency of $p_{11}^Q$, $p_{10}^Q$, $p_{01}^Q$, $p_{00}^Q$ and, respectively). Although four haplotypes from each parent combine randomly to generate 16 diplotypes, only nine genotypes can be observed. Table 3-5 gives all possible genotypes and their diplotypes and diplotype frequencies.

The diplotype of the two parents is unknown. Let $p$ be the probability with which parent P carries diplotype 11 | 22 and, thus, $1-p$ be the probability of diplotype 12 | 21. We use $q$ to denote the probability with which parent Q carries diplotype 11 | 00 and $1-q$ to denote the probability of diplotype 10 | 01. The two markers are linked with a recombination fraction of $r$. Haplotype frequencies produced by the two parents can be expressed as a mixture of haplotype frequencies under each possible parental diplotype, i.e.,

$$p_{11}^P = \tfrac{1}{2}(1-r)p + \tfrac{1}{2}r(1-p), \; p_{12}^P = \tfrac{1}{2}rp + \tfrac{1}{2}(1-r)(1-p),$$
$$p_{21}^P = \tfrac{1}{2}rp + \tfrac{1}{2}(1-r)(1-p), \; p_{22}^P = \tfrac{1}{2}(1-r)p + \tfrac{1}{2}r(1-p),$$

for parent P, and

$$p_{11}^Q = \tfrac{1}{2}(1-r)q + \tfrac{1}{2}r(1-q), \; p_{10}^Q = \tfrac{1}{2}rq + \tfrac{1}{2}(1-r)(1-q),$$
$$p_{01}^Q = \tfrac{1}{2}rq + \tfrac{1}{2}(1-r)(1-q), \; p_{00}^Q = \tfrac{1}{2}(1-r)q + \tfrac{1}{2}r(1-q),$$

for parent Q.

**Table 3-5** Diplotypes and their frequencies for each of nine genotypes at two nonsymmetrical intercross markers, haplotype composition frequencies for each genotype, and composite risk diplotypes for seven possible risk haplotypes.

| Genotype | Diplotype | | Risk Haplotype | | | | | | |
|---|---|---|---|---|---|---|---|---|---|
| | P\|Q | Frequency | 11 | 12 | 21 | 22 | 10 | 01 | 00 |
| 1_/1_ | 11\|11 | $p_{11}^P p_{11}^Q$ | $AA$ | $\bar{A}\bar{A}$ | $\bar{A}\bar{A}$ | $\bar{A}\bar{A}$ | $\bar{A}\bar{A}$ | $\bar{A}\bar{A}$ | $\bar{A}\bar{A}$ |
| | 11\|10 | $p_{11}^P p_{10}^Q$ | $A\bar{A}$ | $\bar{A}\bar{A}$ | $\bar{A}\bar{A}$ | $\bar{A}\bar{A}$ | $\bar{A}A$ | $\bar{A}\bar{A}$ | $\bar{A}\bar{A}$ |
| | 11\|01 | $p_{11}^P p_{01}^Q$ | $A\bar{A}$ | $\bar{A}\bar{A}$ | $\bar{A}\bar{A}$ | $\bar{A}\bar{A}$ | $\bar{A}\bar{A}$ | $\bar{A}A$ | $\bar{A}\bar{A}$ |
| | 11\|00 | $p_{11}^P p_{00}^Q$ | $A\bar{A}$ | $\bar{A}\bar{A}$ | $\bar{A}\bar{A}$ | $\bar{A}\bar{A}$ | $\bar{A}\bar{A}$ | $\bar{A}\bar{A}$ | $\bar{A}A$ |
| 1_/12 | 12\|11 | $p_{12}^P p_{11}^Q$ | $\bar{A}A$ | $A\bar{A}$ | $\bar{A}\bar{A}$ | $\bar{A}\bar{A}$ | $\bar{A}\bar{A}$ | $\bar{A}\bar{A}$ | $\bar{A}\bar{A}$ |
| | 12\|01 | $p_{12}^P p_{01}^Q$ | $\bar{A}\bar{A}$ | $A\bar{A}$ | $\bar{A}\bar{A}$ | $\bar{A}\bar{A}$ | $\bar{A}\bar{A}$ | $\bar{A}A$ | $\bar{A}\bar{A}$ |
| 1_/20 | 12\|10 | $p_{12}^P p_{10}^Q$ | $\bar{A}\bar{A}$ | $A\bar{A}$ | $\bar{A}\bar{A}$ | $\bar{A}\bar{A}$ | $\bar{A}A$ | $\bar{A}\bar{A}$ | $\bar{A}\bar{A}$ |
| | 12\|00 | $p_{12}^P p_{00}^Q$ | $\bar{A}\bar{A}$ | $A\bar{A}$ | $\bar{A}\bar{A}$ | $\bar{A}\bar{A}$ | $\bar{A}\bar{A}$ | $\bar{A}\bar{A}$ | $\bar{A}A$ |
| 12/1_ | 21\|11 | $p_{21}^P p_{11}^Q$ | $\bar{A}A$ | $\bar{A}\bar{A}$ | $A\bar{A}$ | $\bar{A}\bar{A}$ | $\bar{A}\bar{A}$ | $\bar{A}\bar{A}$ | $\bar{A}\bar{A}$ |
| | 21\|10 | $p_{21}^P p_{10}^Q$ | $\bar{A}\bar{A}$ | $\bar{A}\bar{A}$ | $A\bar{A}$ | $\bar{A}\bar{A}$ | $\bar{A}A$ | $\bar{A}\bar{A}$ | $\bar{A}\bar{A}$ |
| 12/12 | 22\|11 | $p_{22}^P p_{11}^Q$ | $\bar{A}A$ | $\bar{A}\bar{A}$ | $\bar{A}\bar{A}$ | $A\bar{A}$ | $\bar{A}\bar{A}$ | $\bar{A}\bar{A}$ | $\bar{A}\bar{A}$ |
| 12/20 | 22\|10 | $p_{22}^P p_{10}^Q$ | $\bar{A}\bar{A}$ | $\bar{A}\bar{A}$ | $\bar{A}\bar{A}$ | $A\bar{A}$ | $\bar{A}A$ | $\bar{A}\bar{A}$ | $\bar{A}\bar{A}$ |
| 20/1_ | 21\|01 | $p_{21}^P p_{01}^Q$ | $\bar{A}\bar{A}$ | $\bar{A}\bar{A}$ | $A\bar{A}$ | $\bar{A}\bar{A}$ | $\bar{A}\bar{A}$ | $\bar{A}A$ | $\bar{A}\bar{A}$ |
| | 21\|00 | $p_{21}^P p_{00}^Q$ | $\bar{A}\bar{A}$ | $\bar{A}\bar{A}$ | $A\bar{A}$ | $\bar{A}\bar{A}$ | $\bar{A}\bar{A}$ | $\bar{A}\bar{A}$ | $\bar{A}A$ |
| 20/12 | 22\|01 | $p_{22}^P p_{01}^Q$ | $\bar{A}\bar{A}$ | $\bar{A}\bar{A}$ | $\bar{A}\bar{A}$ | $A\bar{A}$ | $\bar{A}\bar{A}$ | $\bar{A}A$ | $\bar{A}\bar{A}$ |
| 20/20 | 22\|00 | $p_{22}^P p_{00}^Q$ | $\bar{A}\bar{A}$ | $\bar{A}\bar{A}$ | $\bar{A}\bar{A}$ | $A\bar{A}$ | $\bar{A}\bar{A}$ | $\bar{A}\bar{A}$ | $\bar{A}A$ |
| Likelihood | | | $L_{11}$ | $L_{12}$ | $L_{21}$ | $L_{22}$ | $L_{10}$ | $L_{01}$ | $L_{00}$ |

These haplotype frequencies are used to express diplotype and genotype frequencies as shown in Table 3-4. A similar EM procedure described for linkage analysis with symmetrical intercross markers can be used to estimate haplotype frequencies, parental diplotype probabilities, the recombination fraction, and haplotype effects on a quantitative trait. For nonsymmetrical intercross markers, the choice of an optimal risk haplotype will be made among haplotypes 11, 12, 21, 22, 10, 01, and 00 (Table 3-5). Under each possible risk haplotype, genotypic values of composite diplotypes are estimated and then likelihoods are calculated. The maximum likelihood is due to an optimal risk haplotype.

## 3.7 Haplotyping with Dominant Markers

Unlike humans and several model systems, such as the mouse and *Arabidopsis*, genetic mapping of many underrepresented species, like forest trees, still heavily relies on simple and cheap dominant marker techniques. It is therefore necessary to develop a statistical model for haplotyping a complex trait with dominant markers. Consider a pair of parents P and Q, typed for two dominant markers with a recombination fraction of $r$. Let 1 and 0 denote the dominant and recessive alleles at each marker, respectively. For a heterozygous parent with genotype 10/10, it should carry one of two possible diplotypes 11|00 (with a probability of $p$) and 10|01 (with a probability of $1-p$). The haplotype frequencies produced by the parents are expressed as

$$p_{11}^P = \tfrac{1}{2}(1-r)p + \tfrac{1}{2}r(1-p), \, p_{10}^P = \tfrac{1}{2}rp + \tfrac{1}{2}(1-r)(1-p),$$
$$p_{01}^P = \tfrac{1}{2}rp + \tfrac{1}{2}(1-r)(1-p), \, p_{00}^P = \tfrac{1}{2}(1-r)p + \tfrac{1}{2}r(1-p),$$

and

$$p_{11}^Q = \tfrac{1}{2}(1-r)q + \tfrac{1}{2}r(1-q), \, p_{10}^Q = \tfrac{1}{2}rq + \tfrac{1}{2}(1-r)(1-q),$$
$$p_{01}^Q = \tfrac{1}{2}rq + \tfrac{1}{2}(1-r)(1-q), \, p_{00}^Q = \tfrac{1}{2}(1-r)q + \tfrac{1}{2}r(1-q).$$

These haplotype frequencies are used to calculate diplotype frequencies and therefore genotype frequencies. As shown in Table 3-6, different diplotypes may be collapsed into the same genotype based on the dominant inheritance of the markers. The EM algorithm is then used to estimate haplotype frequencies, parental diplotype probabilities, the recombination fraction, and haplotype effects on a quantitative trait. The best risk haplotype is chosen among all possible haplotypes, 11, 10, 01, and 00, based on the likelihoods calculated under different assumptions of risk haplotypes (Table 3-6).

## 3.8 Haplotyping with SNP Markers

The availability of efficient sequencing methods has made it possible to develop a new class of polymorphic markers at single nucleotide sites, called single nucleotide polymorphisms (SNPs), in functional genes. These techniques have been developed in outcrossing species and display tremendous potential for natural resource management, genetic conservation, and forest tree breeding. For a given outcrossing line, the genotype for each SNP can be either homozygous (11 or 22), or heterozygous (12). For any pair of SNPs, the genotype can be 11/11 (1), 11/12 (2), 11/22 (3), 12/11 (4), 12/12 (5), 12/22 (6), 22/11 (7), 22/12 (8), or 22/22 (9). If both outcrossing lines (P and Q) are homozygous for both loci, their cross will not produce any segregation. Thus, such a cross type will not be useful for haplotyping analysis. Risk haplotypes can be discerned when parent cross types P × Q = (1,2,3,4,5,6,7,8,9) × (2,4,5,6,8) or (2,4,5,6,8) × (1,2,3,4,5,6,7,8,9).

**Table 3-6** Diplotypes and their frequencies for each of four genotypes at two dominant markers, haplotype composition frequencies for each genotype, and composite diplotypes for four possible risk haplotypes.

| | Diplotype | | Risk Haplotype | | | |
|---|---|---|---|---|---|---|
| Genotype | P\|Q | Frequency | 11 | 10 | 01 | 00 |
| 1_/1_ | 11\|11 | $p_{11}^P p_{11}^Q$ | $AA$ | $\overline{A}\overline{A}$ | $\overline{A}\overline{A}$ | $\overline{A}\overline{A}$ |
| | 11\|10 | $p_{11}^P p_{10}^Q$ | $A\overline{A}$ | $A\overline{A}$ | $\overline{A}\overline{A}$ | $\overline{A}\overline{A}$ |
| | 11\|01 | $p_{11}^P p_{01}^Q$ | $A\overline{A}$ | $\overline{A}\overline{A}$ | $A\overline{A}$ | $\overline{A}\overline{A}$ |
| | 11\|00 | $p_{11}^P p_{00}^Q$ | $A\overline{A}$ | $\overline{A}\overline{A}$ | $\overline{A}\overline{A}$ | $A\overline{A}$ |
| | 10\|11 | $p_{10}^P p_{11}^Q$ | $A\overline{A}$ | $A\overline{A}$ | $\overline{A}\overline{A}$ | $\overline{A}\overline{A}$ |
| | 10\|01 | $p_{10}^P p_{01}^Q$ | $\overline{A}\overline{A}$ | $A\overline{A}$ | $A\overline{A}$ | $\overline{A}\overline{A}$ |
| | 01\|11 | $p_{01}^P p_{11}^Q$ | $A\overline{A}$ | $\overline{A}\overline{A}$ | $A\overline{A}$ | $\overline{A}\overline{A}$ |
| | 01\|10 | $p_{01}^P p_{10}^Q$ | $\overline{A}\overline{A}$ | $A\overline{A}$ | $A\overline{A}$ | $\overline{A}\overline{A}$ |
| | 00\|11 | $p_{00}^P p_{11}^Q$ | $A\overline{A}$ | $\overline{A}\overline{A}$ | $\overline{A}\overline{A}$ | $A\overline{A}$ |
| 1_/00 | 10\|10 | $p_{10}^P p_{10}^Q$ | $\overline{A}\overline{A}$ | $AA$ | $\overline{A}\overline{A}$ | $\overline{A}\overline{A}$ |
| | 10\|00 | $p_{10}^P p_{00}^Q$ | $\overline{A}\overline{A}$ | $A\overline{A}$ | $\overline{A}\overline{A}$ | $A\overline{A}$ |
| | 00\|10 | $p_{00}^P p_{10}^Q$ | $\overline{A}\overline{A}$ | $A\overline{A}$ | $\overline{A}\overline{A}$ | $A\overline{A}$ |
| 00/1_ | 01\|01 | $p_{01}^P p_{01}^Q$ | $\overline{A}\overline{A}$ | $\overline{A}\overline{A}$ | $AA$ | $\overline{A}\overline{A}$ |
| | 01\|00 | $p_{01}^P p_{00}^Q$ | $\overline{A}\overline{A}$ | $\overline{A}\overline{A}$ | $A\overline{A}$ | $A\overline{A}$ |
| | 00\|01 | $p_{00}^P p_{01}^Q$ | $\overline{A}\overline{A}$ | $\overline{A}\overline{A}$ | $A\overline{A}$ | $A\overline{A}$ |
| 00/00 | 00\|00 | $p_{00}^P p_{00}^Q$ | $\overline{A}\overline{A}$ | $\overline{A}\overline{A}$ | $\overline{A}\overline{A}$ | $AA$ |
| Likelihood | | | $L_{11}$ | $L_{10}$ | $L_{01}$ | $L_{00}$ |

When a double heterozygote (5) is crossed with any genotype (2,4,5,6,8), the EM algorithm is needed to estimate the recombination fraction, diplotype probabilities, and quantitative genetic parameters. In the progeny of these cross types, there is a double heterozygote that has diplotypes 11 | 22 or 12 | 21. For cross types (1,2,3,4,6,7,8,9) × (2,4,6,8) or (2,4,6,8) × (1,2,3,4,6,7,8,9), a simple *t*-test will be sufficient to detect a risk haplotype because no double heterozygote is formed from these cross types.

It can be seen that genetic haplotyping is more informative than genetic mapping for a full-sib family. First, QTL mapping in a single full-sib family cannot estimate the linkage phase between markers and a QTL bracketed by the markers, although the genetic effect of the QTL can be estimated. Second, QTL mapping can use a very limited number of markers segregating in a full-sib family. For example, cross types (1,2,3,4,6,7,8,9) × (2,4,6,8) or (2,4,6,8) × (1,2,3,4,6,7,8,9) do not provide any information for linkage analysis. Of course, genetic haplotyping largely relies on the availability of a high-density SNP-based linkage map, so in practice the two approaches work in conjunction.

## 3.9 Three-Point Analysis

The procedure for analyzing two markers with a *t*-test can be extended to include three or more markers. Multi-point analysis may not provide more information about the test of haplotype effects compared with two-point analysis (Hou et al. 2007), but the former helps to improve linkage estimation through considering interference at adjacent intervals. A three-point analysis can test the significance of two types of interference during meioses.

For three linked markers, a triple heterozygous parent produces eight different haplotypes, i.e., 111, 112, 121, 122, 211, 212, 221, and 222. Let $D_{12}$, $D_{23}$, and $D_{13}$ be the linkage disequilibria between the first and second markers, between the second and third markers, and between the first and third markers, and $D_{123}$ be the linkage disequilibrium among the three markers. Since allele frequencies at each marker are $\frac{1}{2}$, assuming no segregation distortion, the haplotype frequencies in a gametic population are expressed as

$$P_{111} = \tfrac{1}{8} + \tfrac{1}{2}D_{12} + \tfrac{1}{2}D_{13} + \tfrac{1}{2}D_{23} + D_{123}$$

$$P_{112} = \tfrac{1}{8} + \tfrac{1}{2}D_{12} - \tfrac{1}{2}D_{13} - \tfrac{1}{2}D_{23} - D_{123}$$

$$P_{121} = \tfrac{1}{8} - \tfrac{1}{2}D_{12} + \tfrac{1}{2}D_{13} - \tfrac{1}{2}D_{23} - D_{123}$$

$$P_{122} = \tfrac{1}{8} - \tfrac{1}{2}D_{12} - \tfrac{1}{2}D_{13} + \tfrac{1}{2}D_{23} + D_{123}$$

$$P_{211} = \tfrac{1}{8} - \tfrac{1}{2}D_{12} - \tfrac{1}{2}D_{13} + \tfrac{1}{2}D_{23} - D_{123}$$

$$P_{212} = \tfrac{1}{8} - \tfrac{1}{2}D_{12} + \tfrac{1}{2}D_{13} - \tfrac{1}{2}D_{23} + D_{123}$$

$$P_{221} = \tfrac{1}{8} + \tfrac{1}{2}D_{12} - \tfrac{1}{2}D_{13} - \tfrac{1}{2}D_{23} + D_{123}$$

$$P_{222} = \tfrac{1}{8} + \tfrac{1}{2}D_{12} + \tfrac{1}{2}D_{13} + \tfrac{1}{2}D_{23} - D_{123}$$

By solving the above equations, the four disequilibrium coefficients can be estimated as

$$D_{12} = \tfrac{1}{4}[(P_{111} + P_{112} + P_{221} + P_{222}) - (P_{121} + P_{122} + P_{211} + P_{212})],$$

$$D_{23} = \tfrac{1}{4}[(P_{111} + P_{211} + P_{122} + P_{222}) - (P_{112} + P_{212} + P_{121} + P_{221})],$$

$$D_{13} = \tfrac{1}{4}[(P_{111} + P_{121} + P_{212} + P_{222}) - (P_{112} + P_{122} + P_{211} + P_{221})],$$

$$D_{123} = \tfrac{1}{8}[(P_{111} + P_{122} + P_{212} + P_{221}) - (P_{112} + P_{121} + P_{211} + P_{222})].$$

The first three first-order linkage disequilibria can be used to describe the linkage between different markers and crossover interference, whereas the last second-order linkage disequilibrium is thought to be associated with chromatid interference (Wu and Lin 2006).

## 3.10 Prospects

Quantitative trait locus (QTL) mapping aims to identify narrow chromosomal segments for a quantitative trait by using a statistical method, and has proven its value to study the genetic architecture of the trait in a variety of species (Frary et al. 2000; Mackay 2001; Li et al. 2006). The limitations of this technique lie in its inability to characterize the structure and organization of DNA sequences and statistical difficulty in deriving the distribution of test statistics under the null hypothesis of no QTL (Lander and Schork 1994). At least partly for these reasons, despite thousands of QTL reported for different traits and populations, a very small portion of them have been cloned (Flint et al. 2005). With the completion of genome sequencing projects for several important organisms, a new line of thought in the post genomic era has begun to emerge for the identification of specific combinations of nucleotides or haplotypes that contribute to a complex quantitative trait (Liu et al. 2004; Lin and Wu 2006).

The pseudotestcross strategy has been widely used for map construction and QTL identification in outcrossing species, including many forest trees (Grattapaglia and Sederoff 1994). This strategy uses testcross markers, i.e., those that are heterozygous in one parent but homozygous in the other, to construct two linkage maps, each for a different parent. This strategy is effective for estimating the recombination fraction between two markers, the linkage phase of the heterozygous parent, and the genetic effect of the QTL. However, this strategy cannot be used to estimate the linkage phase between the markers and QTL. Haplotype analysis has the power to estimate the marker-QTL linkage phase from phase-unknown marker data and, thus, identify the favorable allele of a QTL for a quantitative trait. The information provided by haplotype analysis will have immediate implications for marker-assisted breeding programs.

The haplotyping model offers a powerful tool for positional cloning of QTLs that affect a complex trait. Flint et al. (2005) reviewed the potential of currently available cloning strategies, such as probabilistic ancestral haplotype reconstruction, Yin-Yang crosses, and in silico analysis of sequence variants, to identify genes that underlie QTL in rodents. Haplotyping models, in conjunction with these strategies, may open a new gateway for the illustration of a detailed picture of the genetic architecture for a complex trait.

# Acknowledgments

We thank Dr. Stephen DiFazio for constructive comments on the chapter. The preparation of this manuscript is partially supported by NSF grant Joint DMS/NIGMS-0540745 and NNSF of China grant 30771752.

# References

Boerjan W (2005) Biotechnology and the domestication of forest trees. Curr Opin Biotechnol 16: 159–166.

Flint J, Valdar W, Shifman S, Mott R, et al. (2005) Strategies for mapping and cloning quantitative trait genes in rodents. Nat Rev Genet 6: 271–286.

Frary A, Nesbitt TC, Frary A, Grandillo S, Knaap E, Cong B, Liu J, Meller J, Elber R, Alpert KB, et al. (2000) fw2. 2: A quantitative trait locus key to the evolution of tomato fruit size. Science 289: 85–88.

Grattapaglia D, Sederoff RR (1994) Genetic linkage maps of Eucalyptus grandis and *Eucalyptus urophylla* using a pseudo-testcross: mapping strategy and RAPD markers. *Genetics* 137: 1121–1137.

Grattapaglia D, Bertolucci FL, Penchel R, Sederoff RR(1996) Genetic mapping of quantitative trait loci controlling growth and wood quality traits in *Eucalyptus grandis* using a maternal half-sib family and RAPD markers. Genetics 144: 1205–1214.

Hou W, Yap JSF, Wu S, Liu T, Cheverud JM, Wu RL (2007) Haplotyping a quantitative trait with a high-density map in experimental crosses. PLoS ONE 2: e732.

Lander ES, Botstein D (1989) Mapping Mendelian factors underlying quantitative traits using RFLP linkage maps. Genetics 121: 185–199.

Lander ES, Schork NJ (1994) Genetic dissection of complex traits. Science 265: 2037–2048.

Li, Zhou A, Sang T (2006) Rice domestication by reducing shattering. Science 311: 1936–1939.

Lin M, Wu RL (2006) Detecting sequence-sequence interactions for complex diseases. Curr Genom 7: 59–72.

Liu T, Johnson JA, Casella G, Wu RL (2004) Sequencing complex diseases with HapMap. Genetics 168: 503–511.

Lu Q, Cui YH, Wu RL (2004) A multilocus likelihood approach to joint modeling of linkage, parental diplotype and gene order in a full-sib family. BMC Genet 5: 20.

Mackay TFC (2001) Quantitative trait loci in Drosophila. Nat Rev Genet 2: 11–20.

Park YG, Clifford R, Buetow KH, Hunter KW (2003) Multiple cross and inbred strain haplotype mapping of complex-trait candidate genes. Genome Res 13: 118–121.

Paterson AH (2006) Leafing through the genomes of our major crop plants: strategies for capturing unique information. Nat Rev Genet 7: 174–184.

Payseur BA, Place M (2007) Prospects for association mapping in classical inbred mouse strains. Genetics 175: 1999–2008.

Wang X, Korstanje R, Higgins D, Paig en B (2004) Haplotype analysis in multiple crosses to identify a QTL gene. Genome Res 14: 1767–1772.

Wiltshire T, Pletcher MT, Batalov S, Barnes SW, Tarantino LM, Cooke MP, Wu H, Smylie K, Santrosyan A, Copeland NG, et al. (2003) Genome-wide single-nucleotide polymorphism analysis defines haplotype patterns in mouse. Proc Natl Acad Sci USA 100: 3380–3385.

Wu RL, Lin M (2006) Functional mapping – How to study the genetic architecture of dynamic complex traits. Nat Rev Genet 7: 229–237.

Wu RL, Bradshaw HD, Stettler RF (1998) Developmental quantitative genetics of growth in *Populus*. Theor Appl Genet 97: 1110–1119.

Wu RL, Zeng Z-B, McKend SE, O'Malley DM (2000) The case for molecular mapping in forest tree breeding. Plant Breed Rev 19: 41–68.

Wu RL, Ma C-X , Painter I, Zeng Z-B (2002) Simultaneous maximum likelihood estimation of linkage and linkage phases in outcrossing species. Theor Popul Biol 61: 349–363.

Zhang B, Tong CF, Yin TM, Zhang XY, Zhuge Q, Huang MR, Wang MX, Wu RL (2008) Detection of quantitative trait loci influencing growth trajectories of adventitious roots in Populus using functional mapping. Tree Genet Genom (in press).

# 4

# The *Populus* Genome Sequence

*Stephen P. DiFazio,*[1,*] *Xiaohan Yang*[2,a] and *Gerald A. Tuskan*[2,b]

## ABSTRACT

The *Populus trichocarpa* genome sequence has been a major stimulus for tree genomics research, enabling genome-scale studies on multiple aspects of tree structure and function. The project was a gigantic undertaking, bringing together over one hundred researchers from eight countries. The initial genome assembly was derived from 7.5 million paired end sequences from libraries with 4 insert sizes. These were assembled into more than 20,000 scaffolds, which were then anchored to a genetic map using SSR markers. The final outcome was a reference sequence containing the majority of the euchromatic genome organized in chromosome-sized scaffolds. Analysis of the genome sequence has revealed a strong but variable relationship between genetic and physical distance, evidence of whole genome duplications and rearrangements, and identification of a large number of sequences and scaffolds that appear to be of symbiotic origin. There are approximately 40,000 predicted genes in the *Populus* genome, and comparison with grape and *Arabidopsis* has revealed a greater degree of conservation with grape, despite greater phylogenetic distance, suggesting slower evolutionary rates in the two perennial taxa compared to Arabidopsis. Comparison of categories of genes over-represented in each taxon yields some general insights into woody habit and divergent evolution in metabolic networks. This initial analysis indicates the power of comparative genomics to provide insights into plant biology, a promise that will be fulfilled in coming years as genome sequences become commonplace across the world.

**Keywords:** *Populus*, *Vitis*, sequence, repeats, genome duplication, gene content, symbionts

[1]Department of Biology, West Virginia University, Morgantown, West Virginia 26506-6057;
e-mail: *spdifazio@mail.wvu.edu*
[2]Biosciences Division, Oak Ridge National Laboratory, Oak Ridge, Tennessee 37831.
[a]e-mail: *yangx@ornl.gov*
[b]e-mail: *tuskanga@ornl.gov*
*Corresponding author

## 4.1 Overview of the *Populus* Genome Sequencing Project

The genome sequence of *Populus trichocarpa* was a major breakthrough in plant biology and, in particular, in forest genetics. This resource has propelled *Populus* to the forefront of model organisms for forestry, and has generated substantial interest and attention in the genomics community as a whole. At the time of its selection in 2003, *Populus* was already well-established as a model organism because of its experimental tractability, potential economic importance, and its central role in many ecosystems (Taylor 2002; Jansson and Douglas 2007). The rationale for selecting *Populus* was based in part on its importance, and in part on its tractability as a model organism for genomics research. The *Populus* whole genome sequencing project was initially proposed by Roger Dahlman, program manager of the US Department of Energy's (DOE) Terrestrial Carbon program, in January 2001. The DOE had made major infrastructure and personnel investments for the human genome project, which was nearing completion in 2000. The available sequencing capacity was therefore available for organisms with special relevance to the DOE research missions related to energy production and its consequences. *Populus* was well-positioned in this regard because there was a long history of DOE-funded research on *Populus* as a potential bioenergy crop (Dinus et al. 2001), and a surge of interest in *Populus* as a possible solution for carbon sequestration to counter anthropogenic climate change (Tuskan and Walsh 2001). Following substantial grass-roots lobbying and the production of a series of white papers, the sequencing project was approved by DOE's Office of Science in October 2001. The *Populus* genomics community quickly organized around this project, and the International *Populus* Genome Consortium (*http://www.ornl.gov/sci/ipgc/*) was formed in 2002 to spearhead the project.

The *Populus* genome sequencing project was largely the result of a joint effort involving 40 laboratories in eight countries, with 108 researchers directly contributing to the sequencing and subsequent analysis. The sequencing itself took place primarily at DOE's Joint Genome Institute (JGI), and this certainly represented the bulk of the expense of the project. However, major contributions were made by cooperating institutions around the world. Some of this work was funded by the US National Science Foundation Plant Genome Research Program, which supported efforts to create a genome portal and to enhance *Populus* bioinformatics tools, led by Oak Ridge National Laboratory, which also coordinated the sequencing effort. The University of British Columbia and Genome BC contributed full-length cDNAs (Ralph et al. 2008), a bacterial artificial chromosome (BAC) physical map and BAC end sequences to enhance assembly (Kelleher et al. 2007). The Umea Plant Sciences center spearheaded the collection and analysis of approximately 350,000 expressed sequence tag (EST) sequences provided by laboratories around the world (Unneberg et al. 2005). Finally, the University of Ghent customized software for gene prediction and annotation (Lescot et al. 2004). Countless other individual researchers around the world provided in-kind contributions to the project, in the form of physical resources like template DNA, ESTs or BAC libraries, or analytical expertise related to some aspect of the genome analysis.

## 4.2 Tools in Place at the Time of Selection

The genome was ultimately sequenced using a hybrid strategy. A physical map was produced by fingerprinting and end-sequencing a *Populus* BAC library (Kelleher et al. 2007), while the JGI performed a whole genome shotgun for this same genotype. The end sequences were used to integrate the physical map with the sequence, and gaps and problematic areas of the assembled sequence were filled by shotgun sequencing of selected BAC clones. The whole-genome shotgun assembly was based on paired clone-end sequencing of random genomic fragments in independently-prepared libraries of three main sizes: 3 kb inserts, 8 kb inserts in standard plasmid libraries, and 40 kb inserts in fosmid libraries. The 3 kb libraries were intended to minimize cloning bias, while the larger libraries enhanced contiguity of scaffolds by bridging small repetitive regions of the genome.

Many of the tools and resources required for a successful whole-genome sequence project were already in place for *Populus* prior to the initiation of sequencing. First, it was clear that *Populus* had a relatively tractable genome based on genetic maps and cytogenetic analyses. Genetic maps were available for multiple species (Bradshaw and Stettler 1995; Cervera et al. 2001; Yin et al. 2001), and flow cytometry indicated that the haploid genome size of the major *Populus* species was approximately 550 Mb (Bradshaw and Stettler 1993), about four times larger than *Arabidopsis*, but a fraction of the size of most conifers. This was one of the primary factors in the choice of *Populus* as the first sequenced tree over more commercially important species like loblolly pine (*Pinus taeda*) and Douglas-fir (*Pseudotsuga menziesii*), which had received the bulk of research attention prior to the genomics era. Another major factor was the phylogenetic similarity between *Populus* and *Arabidopsis*. Prior to sequencing, approximately 150,000 expressed sequence tags were available for *Populus*, and these showed high sequence conservation with known *Arabidopsis* genes (Sterky et al. 1998, 2004), thereby facilitating provisional annotation of predicted genes based on similarity to intensively-studied model annual species. Finally, BAC libraries were already available for *Populus*, and shotgun sequencing of some of these BACs had revealed microsynteny between *Arabidopsis* and *Populus*, and a moderate density of repeat elements in *Populus* (Lescot et al. 2004; Stirling et al. 2003). This initial sequencing suggested that assembly of a whole-genome shotgun sequence was feasible, and that existing algorithms for gene prediction in Angiosperms would function reasonably well for *Populus* (Lescot et al. 2004).

## 4.3 Overview of the Genome Sequence

### 4.3.1 Sequencing Strategy

At the initiation of the project there was substantial debate within the sequencing consortium about the overall approach to the project. The primary question was whether the genome should be approached as a true whole-genome shotgun project, or whether a BAC-by-BAC approach would be more prudent. The whole-genome shotgun approach, in which the genome would be randomly sheared into small pieces, which would then be sequenced separately and subsequently assembled computationally, had previously been used successfully for a number of large whole-genome sequencing projects at JGI and elsewhere (Aparicio et al. 2002; Dehal et al. 2002). The alternative BAC-by-BAC approach requires a physical map of tiled BACs, which is produced with restriction fragment length polymorphism fingerprinting, followed by shotgun sequencing of individual BACs of known position in the physical map (Marra et al. 1997). A similar debate had taken place within the Human Genome Sequencing Projects, with the public sequencing consortium initially attempting BAC-by-BAC approach (Lander et al. 2001), and a private effort led by Celera and J. Craig Venter attempting a whole-genome shotgun approach (Venter et al. 2001). The primary advantage of the whole-genome shotgun approach is speed, and when it became clear that the private sequencing effort would be completed well ahead of the public effort, the public human genome sequencing project was forced to partially switch to the shotgun strategy as well (Green 2001). In the case of *Populus*, it was unclear if a whole-genome shotgun would be feasible for two reasons: 1.) the repeat composition and high heterozygosity of the genome could prevent coalescence of sequence contigs into coherent scaffolds, and 2.) there was some evidence of a recent genome duplication in *Populus* (Bradshaw et al. 1994), and it was feared that this would complicate the assembly.

### 4.3.2 The Sequenced Genotype

The selection of the genotype to be sequenced was another crucial decision. As described in Chapter 1, the *Populus* genus is quite diverse, and different sections of the genus are important in different regions. There was a strong argument for selecting an aspen (*Populus tremula*, *P. tremuloides*, or *P. alba*), because most of the model transformation clones were derived from this section, and the majority of ESTs were also from that section of the genus. However, the cottonwoods are much more important commercially in the United States, and cottonwood hybrids were the leading candidates for high-yield plantations for bioenergy and carbon sequestration (Tuskan 1998). Furthermore, most genetic maps and pedigrees were for cottonwoods (Bradshaw et al. 1994; Cervera et al. 2001; Yin et al. 2004b, 2008), and the existing BAC libraries were also from cottonwoods (Stirling et al. 2001; Lescot et al. 2004), so the most relevant resources for genome sequencing and assembly were already in hand. Therefore, for strategic reasons, and to accelerate the production of template for the sequencing pipeline, it was decided to sequence a black cottonwood tree (*Populus trichocarpa* Torr. & Gray). The selected genotype, clone number 383–2499, was originally collected along the Nisqually River in Washington State. This clone, commonly called "Nisqually-1," was the maternal parent for the largest pedigree produced for *Populus*, a cross with *P. deltoides* clone ILL-101 to produce a family of 2,028 $F_1$ progeny. The purpose of this pedigree was to isolate a major gene conferring resistance to a hybrid leaf pathogen, *Melamspora x columbiana* 3, which was segregating in a 1:1 ratio in this pedigree (Stirling et al. 2001).

For this purpose, a 9.5× BAC library was also prepared for this pedigree by partially digesting high molecular weight genomic DNA with *Hind*III (Stirling et al. 2001). The existence of the large pedigree and the BAC library, coupled with the availability of abundant material in clone banks at the University of Washington and elsewhere, was enough to tip the balance in favor of this genotype. There are several ironies about Nisqually-1:1.) the disease resistance gene that originally piqued interest in this genotype still has not been isolated, due in part to high complexity of this genomic region, which may be linked to suppression of recombination in the large pedigree, thereby making map-based cloning nearly impossible (Stirling et al. 2001; Yin et al. 2004a); 2.) the original ortet has since been destroyed by flooding in its native habitat; and 3.) Nisqually-1 has proven to be somewhat difficult to handle in tissue culture. Even though transformation protocols have been successfully developed (Ma et al. 2004; Song et al. 2006), this clone is unlikely to supplant aspen hybrids as the model of choice for functional genomics in *Populus* (Busov et al. 2005). However, there is enough sequence conservation on a genome-wide scale between the model aspens and *P. trichocarpa* that for most purposes genomic resources from one species can be used informatively for other species in the genus (Sterky et al. 2004).

## 4.3.3 Shotgun Sequencing and Assembly

The initial sequencing template was prepared from surface-sterilized leaves of Nisqually-1 using a CTAB-based protocol. This template was used to construct the 3 kb and 8 kb libraries that form the basis for most of the sequence data (Table 4-1). A second set of templates was also prepared from root tissue grown in hydroponics and tissue culture. The DNA extraction protocol involved a nucleii isolation step using a sucrose gradient followed by a cesium chloride gradient centrifugation step. This DNA was expected to be virtually free of plastid contamination, and was used to construct the fosmid libraries.

**Table 4-1** Description of sequencing libraries generated for the *Populus trichocarpa* genome sequencing project by JGI as of January 2004. The difference between theoretical and actual coverage of the genome is based on the cumulative length of sequence actually assembled into contigs (i.e., excluding the unassembled reads and organellar scaffolds).

| Insert Size (kb) | Libraries | Sequences | Theoretical Sequence Coverage[a] | Assembled Sequence Coverage | Clone Coverage[b] |
|---|---|---|---|---|---|
| 3 | 4 | 4,427,983 | 5.48 | 3.57 | 13.69 |
| 8 | 4 | 2,570,799 | 3.18 | 2.14 | 21.20 |
| 36 | 3 | 651,211 | 0.81 | 0.62 | 24.17 |
| 100 | 1 | 81,904 | 0.11 | 0.10 | 9.49 |
| Total | 11 | 7,649,993 | 9.46 | 6.33 | 6 |

[a]Sequence coverage is calculated based on the total amount of sequence in each library divided by the estimated genome size of 485 Mb.
[b]Clone coverage is the total insert size of the clones (assuming the averages given in the insert size column) divided by the estimated genome size of 485 Mb.

A total of 7.65 million sequence reads were generated from these libraries, with 4.4 million reads coming from 3 kb libraries and 2.5 million reads from 8 kb libraries, and 650,000 reads from fosmid libraries (Table 4-1). In addition, 81,904 end sequences were obtained from BAC clones that averaged 100 kb in size (Kelleher et al. 2007). This resulted in a theoretical sequence coverage of the genome of nearly 10× (i.e., an average of 10 sequences representing each nucleotide position), and an expected clone coverage of nearly 70 × (i.e., the average number of clone inserts covering each position in the genome, though only the ends of the clones are represented by actual sequence). Therefore, a highly contiguous assembly was expected.

The shotgun sequences were initially assembled based on homology and paired end read information using the JAZZ assembler (Aparicio et al. 2002). The assembly process began with identification and masking of reads derived from repetitive regions of the genome. This was accomplished by counting the number of occurrences of all 16-mer "words" in the entire set of 7.65 million sequences, and then masking of 16-mers that occurred more than 32 times, numbers which were empirically determined to yield the best tradeoff between contiguity and completeness of assembly of repetitive, polymorphic genomes. The initial assembly utilized 4.8 million of the sequence reads to form approximately 45,970 sequence contigs of at least 1 kb in length, resulting in approximately 427 Mb of assembled genome sequence contigs, excluding gaps. These contigs were grouped together using paired clone end information into 22,136 sequence scaffolds that covered 464 Mb of assembled sequence and "captured" sequence gaps (the size of which was estimated based on average clone insert size). Half of this scaffold sequence was contained in 62 major scaffolds, each of which was at least 2 Mb in size. The maximum contig size was 1.7 Mb, and the maximum scaffold size was nearly 12.5 Mb.

### 4.3.4 Organellar, Microbial and Contaminant Sequences

Approximately 25% of the original sequence reads could not be assembled into meaningful sequence contigs in the whole-genome shotgun assembly. Approximately 750,000 sequences were simply low-quality or chimeric reads that were excluded by the assembler. Another 2.1 million reads were of high quality, and their failure to assemble was apparently due to a variety of causes. First, two sets of sequences with uniform sequence depth (954 × and 60 ×, respectively) were assembled into chloroplast and mitochondrial genomes, respectively. These accounted for approximately 300,000 of the unassembled sequences. Another 613,000 sequence reads matched *Populus* repeat elements, as determined by high 16-mer composition and comparison to *Populus* repeat libraries. The remaining 1.1 million unassembled sequences were compared to the NCBI non-redundant nucleotide database using WU-BLASTN searches. Approximately 600,000 of these sequences showed no homology to known sequences, and are therefore of unknown origin. An additional 482,199 had significant hits to known sequences from organisms other than *Populus*. Of these, the vast majority had hits (E < 1e$^{-10}$) to other plants, and likely represent unassembled portions of the *Populus* genome. However, nearly 25,000 of the remaining sequences had matches to fungi, bacteria, and viruses that were likely endophytic or pathogenic contaminants of the sequencing template, despite the fact that the leaves and roots were surface-sterilized prior to extraction (Table 4-2). Similar trends were seen for small scaffolds from the sequencing dataset, where nearly 300 of the scaffolds < 10 kb in size were apparently of microbial origin. This provides a potentially-interesting window into the invisible and largely unknown microbial associates of *Populus* (Germaine et al. 2004). These apparent contaminants have mostly been screened and removed from subsequent sequence assemblies.

**Table 4-2** Kingdoms represented among unassembled sequence reads and small scaffolds from the *Populus* shotgun sequence dataset, based on WU-BLASTN searches versus the NCBI non-redundant nucleotide database.

| Kingdom | Unassembled Reads | | Small Scaffolds (< 10 Kb) | |
| --- | --- | --- | --- | --- |
| | Taxa | Sequences | Taxa | Sequences |
| Fungi | 78 | 540 | 1 | 1 |
| Metazoa | 175 | 10,638 | 6 | 45 |
| Archaea | 9 | 54 | 0 | 0 |
| Bacteria | 291 | 13,656 | 40 | 231 |
| Eukaryota | 40 | 477 | 2 | 2 |
| Viruses | 27 | 407 | 0 | 0 |
| Vector | 67 | 1,996 | 5 | 7 |
| Viridiplantae | 723 | 577,511 | 35 | 2,900 |
| Total | 1,410 | 605,279 | 89 | 3,186 |

### 4.3.5 BAC Physical Map

A physical map was constructed using a 9.5× library derived from Nisqually-1. The library was constructed from high molecular weight DNA extracted from leaf tissue and partially digested with *Hind*III (Stirling et al. 2001). Restriction enzyme fingerprinting using *Hind*III was performed on 46,025 clones from this library with an average insert size of 100 kb, providing 9.4-fold coverage of the physical map (Kelleher et al. 2007). This resulted in production of 3471 contigs containing an average of 11 BAC clones each. The relative lack of contiguity in this library appears to be the result of complex haplotype structure in the *Populus* genome. In particular, there were over 8,000 heterozygous polymorphisms at *Hind*III sites, which prevented haplotypes from converging in the assembly, leading to complex forking patterns in the tiling path (Kelleher et al. 2007). Another complicating factor is the apparent existence of *Hind*III "deserts" in the *Populus* genome: large regions entirely lacking *Hind*III sites. These regions would not be represented in this BAC library, since it was constructed with a single restriction enzyme. Such complexity greatly mitigates the advantages of a BAC-by-BAC sequencing approach in a highly heterozygous organism like *Populus*, and tips the balance strongly in favor of the more efficient whole-genome shotgun approach. Nevertheless, BAC end sequences and the BAC physical map were extremely useful in enhancing the contiguity of the shotgun assembly, so BAC fingerprinting and mapping still plays a vital role in genome sequencing projects.

### 4.3.6 Map-Based Assembly

The genome sequence was consolidated into segments approximating the 19 *Populus* chromosomes using genetic mapping with sequence-tagged markers, principally microsatellites (Yin et al. 2004b, 2008). Primer sequences from 356 loci were identified in the assembled genome using BLASTN. Anchored loci had primers that matched the genome sequence in inverse orientation at a distance that was consistent with the known size of the amplified SSR product. This resulted in linking 155 major sequence scaffolds and 335 Mb of sequence into chromosomal linkage groups (LGs) (Fig. 4-1).

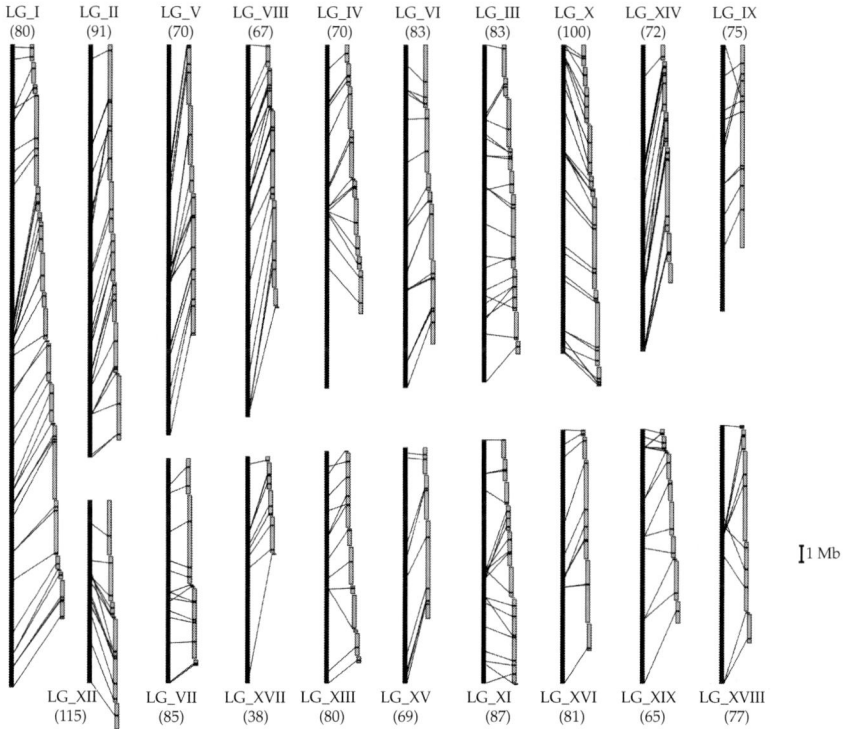

**Figure 4-1** Representation of a recent map-based assembly of the *Populus* genome. Black bars represent chromosomal linkage groups derived from genetic mapping, and gray bars represent sequence scaffolds from the sequence-based assembly. Positions of microsatellite markers are indicated by lines connecting the linkage groups to the sequence scaffolds. Number in parentheses below linkage group is estimated sequence coverage of the chromosome, based on average physical to genetic distance ratios across the genome.

The map-based assembly greatly enhanced the contiguity and utility of the genome sequence by allowing chromosome-scale analyses and facilitating integration with quantitative trait loci (QTL) studies. However, there are some important caveats about this initial assembly. First, the vast majority of the markers used for genome assembly were mapped with only 44 progeny, so the resolution of the map is quite low, and small sequence scaffolds are not always positioned or oriented accurately. This is also a problem for scaffolds that are only mapped with one marker, which is true for 75 out of 155 scaffolds. The assembly has since been improved with single nucleotide polymorphism (SNP) markers and additional SSR loci (Drost et al. 2009), but much uncertainty still remains. Second, scaffolds representing different haplotypes are sometimes assembled consecutively rather than being merged and assembled to the same position. This type of misassembly can easily be misinterpreted as large-scale tandem or segmental duplications. It is extremely difficult to distinguish between misassemblies of this type and true duplication events. Examples of problematic regions of the assembly are the peritelomeric (top) portions of LG_X and LG_XIX (Fig. 4-1). LG_XIX has been investigated in some detail, and demonstrated by intensive mapping to have very strong haplotypic divergence in this peritelomeric region (Yin et al. 2008). This, coupled with strong suppression of recombination and the mapping of sex determination to this region (Markussen et al. 2007; Gaudet et al. 2008), has led to speculation that this chromosome might be in the early stages of evolving into a sex chromosome (Yin et al. 2008).

Integration of the physical and genetic maps provided the opportunity to examine the ratio of physical to genetic distance in *Populus*. The median ratio of physical to genetic distance was 118.5 kb/centiMorgan, based on 54 "framework" SSR markers (mapped with at least 150 progeny) located on the same sequence scaffold. There is, as expected, a substantial amount of variation in these estimates (Fig. 4-2), reflecting real differences in recombination frequency across the genome, as well as errors in estimation of physical genome size caused by sequence gaps, imprecision in mapping positions, and actual differences in genome composition between Nisqually-1 and the clone used in the mapping pedigree, 93–968 (Yin et al. 2008). Efforts are now underway to use Illumina BeadArray technology to map 2000 SNP in a pedigree derived from Nisqually-1, as well as another 4000 loci genotyped for over 800 progeny in another family derived from clone 93–968.

## 4.4 *Populus* Genome Characteristics

### 4.4.1 Gene content

Gene content prediction was carried out using four different approaches, and the results were merged to provide a consensus list of gene models. The four main gene-calling algorithms were ab initio FgenesH, homology-based FgenesH (which uses EST evidence) (Solovyev et al. 2006), Genewise, GrailExp6, and EuGène (Foissac et al. 2008), all of which were trained with a set of over 4,664 full-length cDNA sequences (Ralph et al. 2008). These gene predictions were carried out by three independent groups

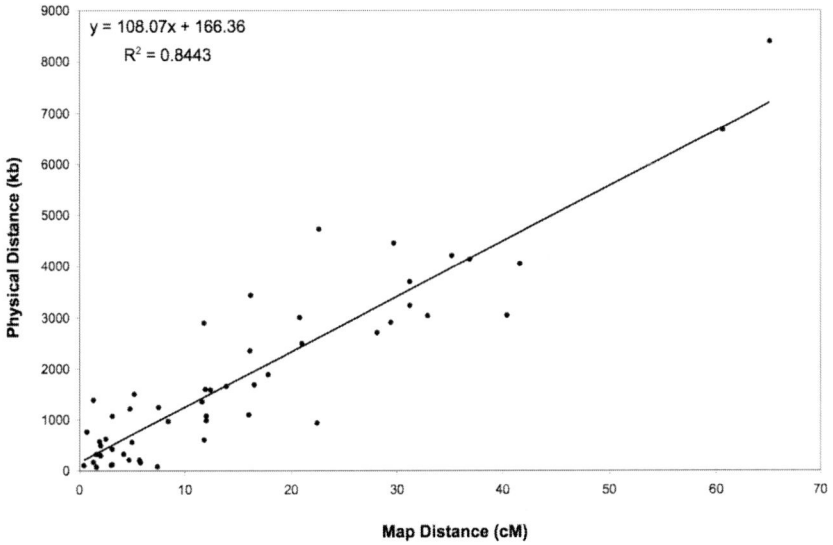

Figure 4-2 Relationship between genetic distance and physical distance for *Populus* sequence scaffolds for *Populus* framework markers, which were mapped with over 150 progeny (Yin et al. 2004b).

(JGI, ORNL, and the University of Ghent), and then merged by the JGI to produce consensus predictions. Most gene prediction programs provide markedly different results, and each has its own strengths and weaknesses, with major tradeoffs between specificity and sensitivity, depending on the weight given to different evidence sources (ESTs, full-length cDNAs, alignments to other genomes) in the training and analysis phases (Foissac et al. 2008). As expected, the gene finding algorithms produced quite different results for *Populus* as well (Fig. 4-3), and it was challenging to identify the best model at each locus, and derive a consensus set of predicted genes that are likely to be true, protein coding genes. The initial consensus set of contained 58,036 putative genes, only 25% of which were predicted by two or more algorithms. This set was quickly discovered to contain many pseudogenes and transposable elements, and was gradually reduced to the publicly-released set of 45,555. However, this set still contains approximately 370 genes with strong hits to known transposable elements. Furthermore, it appears that many bona fide genes from the initial set are not included in the final set of released genes. BLASTP searches revealed that 4,520 of these excluded genes have significant hits to plant proteins in the NCBI nonredundant database. Furthermore, 3,675 of these excluded genes showed some evidence of expression in whole-genome microarray experiments using diverse *Populus* tissues (Rodgers-Melnick et al., submitted). Therefore, the gene content of *Populus* is still poorly determined, and efforts to improve the selection and annotation of gene models are continuing. Nevertheless, it seems clear that the final number of *Populus* genes will exceed 40,000, based on expression evidence and homology to known genes.

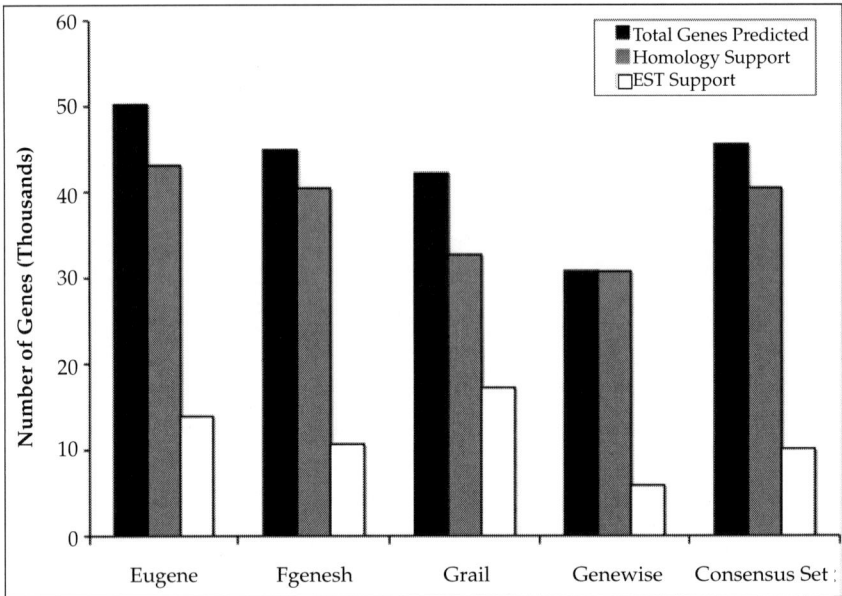

**Figure 4-3** Number of genes predicted by each set of gene prediction programs, as well as the consensus set of gene models released by the genome consortium in 2004. "Homology support" indicates the number of gene models with significant BLASTN (E < 1e$^{-10}$) alignments versus know proteins in the NCBI non-redundant database, and "EST Support" is the number of gene models with significant BLASTN alignments to one of the ~250,000 *Populus* ESTs that were available at the time of prediction.

### 4.4.2 Comparison of Gene Content with **Arabidopsis** and **Grape**

In one sense, the gene content of *Populus* is surprisingly similar to that of *Arabidopsis*, despite the considerable phylogenetic distance between these taxa and their obviously contrasting biological characteristics. *Populus* is a member of the Eurosid I subclass, while *Arabidopsis* is in the Eurosid II subclass. Furthermore, *Arabidopsis* is a diminutive herbaceous annual with perfect flowers and largely selfing mating system, while *Populus trichocarpa* is the tallest perennial angiosperm in the Northern Hemisphere, with a dioecious, completely outcrossing breeding system (Dickmann et al. 2001). Nevertheless, 72% of *Populus* genes had significant BLASTP hits to at least one *Arabidopsis* gene, with an average expectation score of 7.3 x 10$^{-13}$ and an average of 59% (+/–1.6%) amino acid identity over their aligned lengths. The comparison looks even more favorable in the opposite direction, with

87.4% of *Arabidopsis* genes showing significant hits to *Populus* proteins, with an average amino acid identity of 61% (+/–1.6%) over their aligned length, with 17% of gene models having 80% or greater amino acid identity (Fig. 4-4). The discrepancy in the reciprocal comparisons probably reflects a higher frequency of incorrectly annotated pseudogenes in *Populus*, since the *Arabidopsis* annotation has received substantially more attention and resources than the *Populus* annotation (Swarbreck et al. 2008).

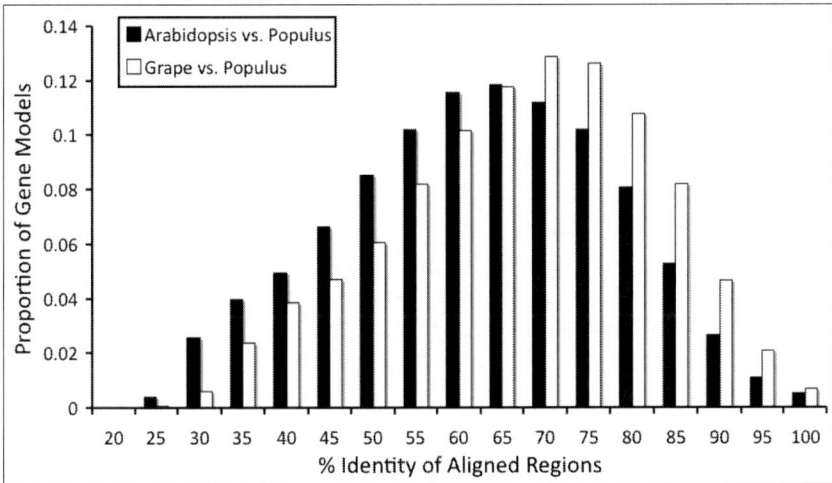

**Figure 4-4** Comparison of *Populus* predicted proteins to those in the *Arabidopsis* and Grape genomes. Comparison only includes genes with significant BLASTP alignments (E score < 1e-10). The % identity is the weighted average of all High Scoring Pairs from the BLASTP alignments. The *Arabidopsis* comparison is based on 26,994 significant alignments out of 30,900 proteins compared (87.4%), while the grape comparison is based on 46,288 significant alignments out of 55,990 proteins compared (82.7%).

The complete genome of the wine grape (*Vitis vinifera*) was also recently published (Jaillon et al. 2007), and this genome makes for some interesting comparisons with *Populus*. First, one might expect more similarity in coding sequences because grape is also a perennial woody plant. However, the phylogenetic position of grape relative to the rosids has been the subject of some controversy, and recent analyses seem to place the order Vitales as a sister group to the Rosids (Jansen et al. 2006). Therefore, based on phylogeny one might expect grape proteins to be more divergent from *Populus* than *Arabidopsis* proteins. Surprisingly, the percentage of proteins showing significant hits to *Populus* proteins was somewhat higher in grape (84.9% of 30,442 proteins) compared to *Arabidopsis*, using the criteria described above. Furthermore, the amino acid identity was 65.1% (+/–1.49%), with 26.5% of models showing 80% or greater amino acid identity to *Populus* genes

(Fig. 4-4). Therefore, grape genes have substantially higher identity to *Populus* genes than to *Arabidopsis* genes, despite the fact that *Arabidopsis* is closer evolutionarily (Velasco et al. 2007). This is probably due to the higher rate of evolution in *Arabidopsis* and other herbaceous annuals, which have many more generations per time interval than long-lived woody perennials like *Populus* and grape (Tuskan et al. 2006; Sémon and Wolfe 2007).

The relative abundance of genes in functional categories provides a more sensitive and informative measure than gross gene content differences between genomes. The Gene Ontology classification system provides a convenient means of doing this (Harris et al. 2006). Because GO classifications have not yet been performed for *Populus*, we assigned provisional GO classifications using *Arabidopsis* best hits, utilizing the simpler GO-Slim terms. There were significant differences between *Populus* and *Arabidopsis* for almost all GO-Slim categories, but only five categories were different between *Populus* and grape (Fig. 4-5). Interestingly, the classes that showed significant differences between grape and *Populus* were transcription factor activity and kinase activity, which were over-represented in *Populus*, and "response to stress" and "cell wall", which were over-represented in grape. Both of these latter classes were in turn strongly over-represented in *Populus* compared to *Arabidopsis*, so perhaps both of these classes are related to the woody, perennial habit. However, in the case of the cell wall class, a large portion of those over-represented in grape are related to flavonoid and polyphenol production, which is one of the aspects of grape chemistry that makes this species so desirable for wine production (Jaillon et al. 2007; Velasco et al. 2007).

### 4.4.3 Genome Structure

Assembly of the genome and corresponding gene content to linkage groups made it possible to investigate the gross structure of the genome at a chromosomal scale. This revealed the striking existence of two whole-genome duplication events. This was accomplished by making pairwise comparisons among all *Populus* genes using double-affine Smith-Waterman alignments. This revealed the presence of large syntenic blocks of genes on different linkage groups that had approximately concordant genetic distances (Fig. 4-6). Blocks of these syntenic genes were defined based on the existence of two or more genes aligning on different chromosomes, with fewer than 10 intervening, non-aligning genes. The genetic distance between these aligning genes was calculated based on the number of transversion substitutions at four-fold degenerate nucleotide sites (4DTV), which is a conservative estimate of genetic divergence that should be less susceptible to multiple substitutions than more commonly-occurring synonymous substitutions (Comeron 1995). Comparison of the size of the syntenic blocks versus the mean 4DTV value for those blocks revealed two clear groups of blocks that were of approximately uniform age (Fig. 4-7). The group of larger blocks centered at 4DTV = 0.068 represents the most recent whole genome duplication in *Populus*, while the group of smaller blocks at 4DTV = 0.31 represents a more ancient duplication event

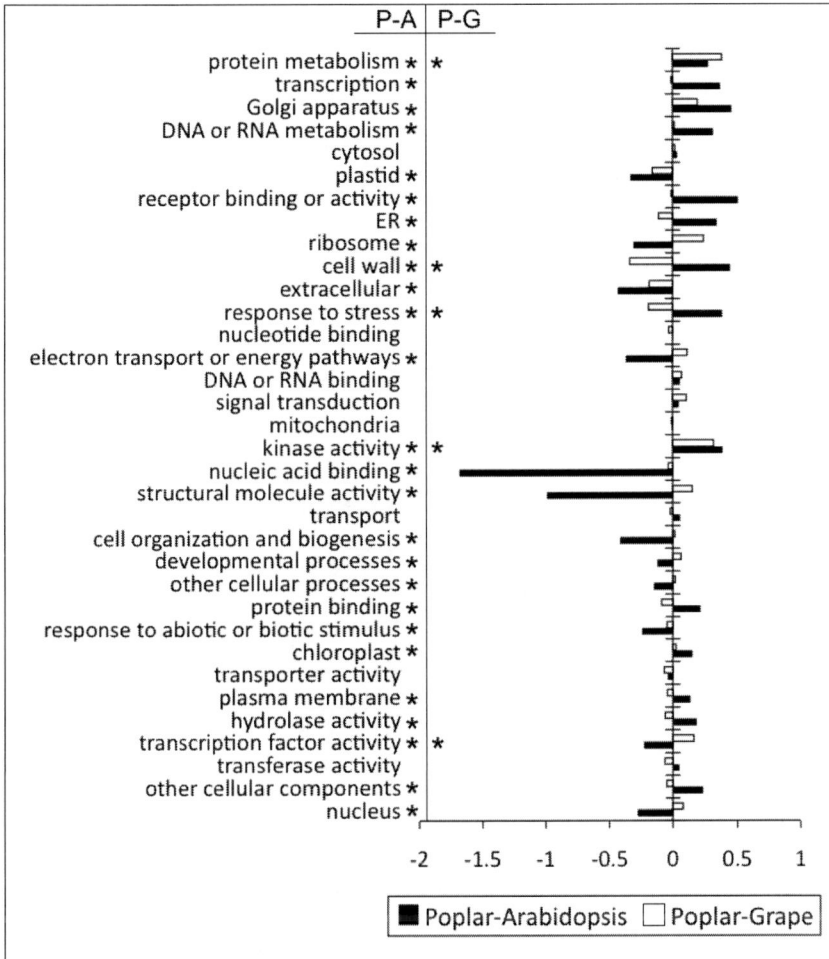

**Figure 4-5** Comparison of *Arabidopsis* and grape gene content to that of *Arabidopsis*, organized by GOslim categories. The relative difference in proportions of genes in each category is calculated as (P-X)/P, where P is the proportion of *Populus* genes in that category, and X is the proportion of *Arabidopsis* or grape genes in that category. Asterisks indicate significant deviations ($p < 0.001$) from expected proportions, based on a Chi-squared test, with *Populus-Arabidopsis* comparisons on top, and *Arabidopsis*-Grape comparisons on the bottom. GOslim groups are arranged from the most abundant (left) to the least abundant class in Arabidopsis.

# LG_VIII LG_X

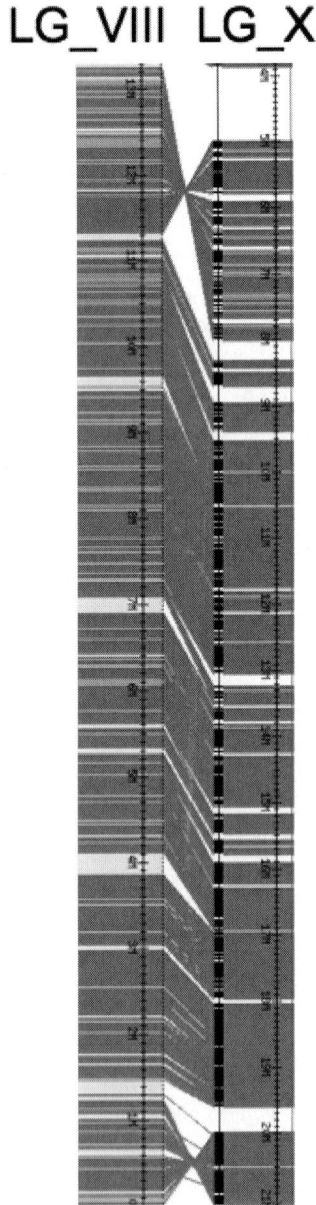

**Figure 4-6** Comparison of genes with significant alignments between two linkage groups, chromosome VIII and Chromosome X. Genes are gray with black lines connecting the scales representing the chromosomes. Position on the linkage group is given in Megabases to the side of each group. Genetic distances between genes in these large syntenic blocks are highly concordant, indicating that these syntenic chromosome blocks originated from whole-genome duplication events.

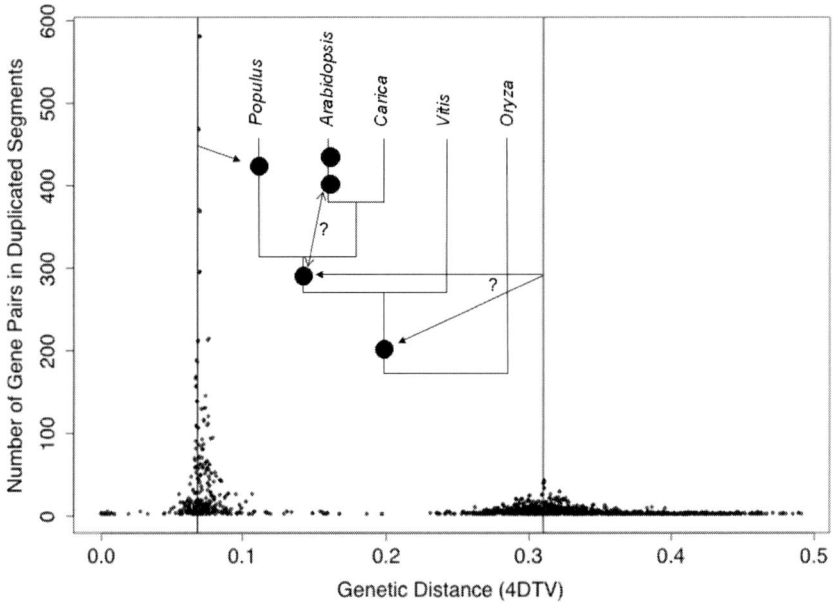

**Figure 4-7** Observed number of genes aligned in approximately syntenic blocks between chromosomes versus the genetic distance between genes, as measured by the rate of transversions at four-fold degenerate nucleotide sites (4DTV). Peaks corresponding to the putative whole-genome duplication events are indicated as peaks at the centers of clusters of similar 4DTV values, indicating the existence of large numbers of duplicated blocks of approximately the same age. The timing of the duplication events relative to species divergence is controversial, with some evidence pointing to the older event occurring prior to the split of *Arabidopsis* and *Populus,* and other evidence indicating that the older event dates back to the triplication event(s) apparently shared by all angiosperms.

(Sterck et al. 2005). Dating of these events is difficult, because the *Populus* genome is evolving considerably slower than genomes that have previously been used to calibrate the angiosperm molecular clock. Using the molecular clock as calibrated by fossil records for the Brassicaceae, for example, the most recent duplication dates to 8 million years ago (Sterck et al. 2005). However, the *Populus* genus has been in existence for at least 50 million years, and the genome duplication is shared by many species in the genus (Sterck et al. 2005), so the *Populus* genome is clearly evolving at a much slower rate than herbaceous angiosperms, which is to be expected based on generation time (Bell et al. 2005).

The *Arabidopsis* genome also shows evidence of at least two whole-genome duplication events (Blanc et al. 2003), but following these events the genome has become substantially rearranged, making it difficult to reconstruct the older events (Blanc et al. 2003). A similar rearrangement has occurred in *Populus*, but much less severe (see Fig. 2 in Tuskan et al. 2006). Extensive rearrangements following genome doubling is a common component of the diploidization process (Adams and Wendel 2005; Doyle et al. 2008). The structural complexity of these two genomes, coupled with the high rates of gene evolution in *Arabidopsis*, make it particularly difficult to establish orthology and determine whether the ancient duplication event in *Populus* is shared with *Arabidopsis*. The timing of the event appears to be similar to the timing of the split of the *Arabidopsis* and *Populus* lineages, as determined by pairwise comparisons of genetic distances between *Populus* duplicated genes, *Arabidopsis* duplicated genes, and between putative *Arabidopsis* and *Populus* orthologs (Tuskan et al. 2006). Given the close timing of these events, it is tempting to speculate that the genome duplication was a primary driver of the diversification of the rosids (Lynch and Conery 2003).

In contrast to *Populus* and *Arabidopsis*, the grape genome has been comparatively quiescent, with minimal rearrangements, and equivocal evidence of a single whole-genome duplication that could be shared with *Populus* and *Arabidopsis* (Jaillon et al. 2007; Velasco et al. 2007). Similarly, there is no evidence of a recent whole genome duplication event in the papaya (*Carica papaya*) genome (Ming et al. 2008). This structural simplicity has allowed reconstruction of the truly ancient whole-genome duplication event that is shared by all angiosperms. It appears that this event resulted in hexaploidy in the ancient angiosperm progenitor, as suggested by the presence of three syntenic blocks in rice, *Populus*, and *Arabidopsis* for every one block in grape and papaya (Jaillon et al. 2007; Tang et al. 2008b).

Comparison of syntenic regions between *Arabidopsis*, *Populus*, papaya, and grape, coupled with inferences about their relative phylogenetic positions, suggests that *Populus* and *Arabidopsis* do not share the putative Eurosid duplication event described above (Jaillon et al. 2007; Ming et al. 2008; Tang et al. 2008a). Based on these analyses, and distribution of synonymous substitutions for syntenic pairs, it appears that *Populus* only has the hexaploid duplication shared by all angiosperms and the recent Salicoid duplication. However, synonymous substitutions become saturated after a relatively short time period, so the ancient angiosperm duplication may be obscured by the Eurosid duplication event. There is a clearer separation when 4DTV distances are plotted (Fig. 4-7). If papaya is truly in the Brassicales, and lacks the Eurosid duplication event (Ming et al. 2008), then it is difficult to explain how *Populus* and *Arabidopsis* could share this duplication. These issues are likely to be resolved as more angiosperm genomes are sequenced over the next few years (Tang et al. 2008a).

## 4.5 Genome Portals

The *Populus* genome has been incorporated into numerous public web-based portals since the release of the sequence in 2004, facilitating a variety of computational analyses. The original portal is hosted by the DOE JGI (*http://genome.jgi-psf.org/Poptr1_1/Poptr1_1.home.html*), and remains one of the most heavily trafficked genome portals at JGI. The site provides various browse and search capabilities, as well as access to gene classifications and bulk data downloads. A new portal, the Populus Genome Integrative Explorer, PopGenIE (http://www.popgenie.org), provides much of the same functionality, in addition to a number of useful tools that integrate information about QTLs, gene expression patterns, and synteny within and among genomes to allow deeper inferences about the function and evolutionary history of genes and genome regions than has previously been possible (Sjodin et al. 2009).

The *Populus* genome has also been incorporated into several portals that are focused on comparative analysis of whole genome sequences from multiple organisms. The Department of Energy's Joint Genome Institute and the Center for Integrative Genomics developed Phytozome (*http:// www.phytozome.net/index.php*). This site provides access to 11 sequenced and annotated angiosperm genomes, including the most recent assembly and annotation of *Populus* (v 2.0, released January 2010), in addition to four monocots, and three lower plants. This site allows clustering of genes at 20 nodes that span the evolutionary history of green plants. The site also offers PFAM, KOG, KEGG and PANTHER assignments, facilitating identification of clusters of orthologous and paralogous genes that represent the modern descendents of ancestral gene sets, as well as clade specific genes and gene expansions.

*Populus* is also included in PlantGDB (*http://www.plantgdb.org/*) which provides annotated transcript assemblies for > 100 plant species, with transcripts mapped to their cognate genomic context integrated with a variety of sequence analysis tools and web services (Duvick et al. 2008). PlantGDB also hosts a plant genomics research outreach portal that facilitates access to a large number of training resources. Another plant comparative genomics database is Gramene (Liang et al. 2008), which contains genomic information for *Oryza sativa* var. *indica* and var. *japonica*, *O. glaberrima*, *O. rufipogon*, *Zea mays*, *Sorghum bicolor*, *Arabidopsis thaliana*, *Vitis vinifera* and *P. trichocarpa*, including genome assembly and annotations, cDNA/mRNA sequences, genetic and physical maps/markers, genes, quantitative trait loci, proteins, comparative ontologies.

Other, more specialized databases include particular aspects of *Populus* genomics. For example, multiple databases focus on particular EST and microarray data sets, including POPULUSDB (*http://www.populus.db.umu.se/*; Sterky et al. 2004), ASPENDB (*http://aspendb.uga.edu/*), and the poplar EFP browser (*http://bar.utoronto.ca/*; Wilkins et al. 2009). In addition, the RIKEN *Populus* database (*http://rpop.psc.riken.jp*) contains information covering 10 *Populus* species (*P. deltoides*, *P. euphratica*, *P. tremula*, *P. tremula* x *P. alba*,

*P. tremula* x *P. tremuloides*, *P. tremuloides*, *P. trichocarpa*, *P. trichocarpa* x *P. deltoides*, *P. trichocarpa* x *P. nigra* and *P.* x *canadensis*). PLEXdb (*http://plexdb. org/*), offers a unified web interface to support the functional interpretation of highly parallel microarray experiments integrated with traditional structural genomics and phenotypic data (Wise et al. 2006). PLEXdb contains information for 13 plant species such as *Arabidopsis*, barley, *Citrus*, cotton, *Vitis*, *Zea*, *Medicago*, *Populus*, *Oryza*, *Glycine*, sugarcane, tomato and wheat.

Other databases are devoted to particular classes of genes. The *Populus* transcription factor (TF) database (DPTF; *http://dptf.cbi.pku.edu.cn/*) contains 2,576 putative *Populus* TFs, distributed in 64 families (Zhu et al. 2007). It provides comprehensive information for *Populus* TFs such as sequence features, functional domains, GO assignment, expression evidence, phylogenetic tree of each family, and homologs in *Arabidopsis* and *Oryza*. A web-browsable database, RepPop (*http://csbl.bmb.uga.edu/~ffzhou/RepPop/*), offers resources for browsing repeat elements based on a genome-wide analysis of 9,623 repetitive elements in the *P. trichocarpa* genome (Zhou and Xu 2009). This database also provides various search capabilities and a Wiki system to facilitate functional annotation and curation of the repetitive elements. Finally, the ChromDB database (*http://www.chromdb.org*) displays chromatin-associated proteins, including RNAi-associated proteins, for a broad range of organisms such as *A. thaliana*, *O. sativa* ssp. *japonica*, *P. trichocarpa*, *Zea mays* and model animal and fungal species (Gendler et al. 2008). ChromDB contains three types of sequences: genomic-based (predominantly plant sequences); transcript-based (EST contigs or cDNAs for plants lacking a sequenced genome) and NCBI RefSeq sequences for a variety of model animal organisms.

## 4.6 Implications and Applications of the Genome Sequence

The genome sequence was truly a watershed event for the tree genetics community, and the impacts have reverberated throughout forest science and even into other parts of plant science. The genome sequence has provided a nearly complete catalog of all genes and regulatory elements in this model tree, thus opening up a whole realm of research that was not possible before the sequencing project. One index of the impact on the sequence is the number of citations of journal articles related to *Populus* has more than doubled since 2004, the year that the sequence was first publicly released, and the number of *Populus* publications has nearly doubled since 2000. The main article describing the genome sequence (Tuskan et al. 2006) has been cited over 250 times since it was published in September of 2006.

The genome sequence has already had extensive applications in applied science. For example, *Populus* is currently the focus of three major bioenergy projects, two in the US and one in Canada, with a total committed funding of more than US $20 million over the next few years. One of these projects, the DOE Bioenergy Science Center headed by Oak Ridge National Lab, is resequencing 18 *Populus* genotypes using next-generation sequencing technology. The project will characterize Single Nucleotide Polymorphisms across the genome for the purpose of genetic association studies to identify genes underlying cell wall biosynthesis, with the ultimate goal of reducing the recalcitrance of lignocellulosic feedstocks to cellulose extraction (Rubin 2008). The project will then use Illumina Bead Arrays to assay 10,000 SNPs for over 1,000 trees collected across the range of *Populus trichocarpa*. These trees will be established in three different common gardens encompassing most of the range of the species (California, Oregon, and British Columbia), and phenotyping will be performed for a large number of traits. This project, and others like it, will therefore propel *Populus* from the realm of comparative genomics, and almost complete reliance on gene homology to herbaceous models for functional annotation, to direct functional characterization of a large fraction of the genes in the genome. *Populus* will thus be solidly established as a premier model organism for functional genomics.

The impact in areas outside of genomics has been equally profound. The fields of Community Genetics and Ecological Genomics are flourishing, with *Populus* as one of the primary model organisms, driven by the availability of the genome sequence and the central importance of *Populus* in many ecosystems (Whitham et al. 2006; Whitham et al. 2008). The genome sequence is allowing exploration of diverse questions, such as exploration of the genetic architecture of species barriers, based on patterns of introgression across hybrid zones (Lexer et al. 2007), or the genetic basis of sexual selection (Yin et al. 2008). Furthermore, additional species associated with *Populus* have also been sequenced (Martin et al. 2004, 2008), and many more are in progress. We are truly on the threshold of a brave new era in which genome sequencing of entire communities will become entirely plausible, potentially allowing elucidation of fundamental truths about the mechanisms of the assemblage and persistence of ecological communities (Whitham et al. 2008). This is likely to fundamentally change the way we approach ecological and evolutionary research.

## Acknowledgements

The *Populus* Genome consortium, with leadership from Dan Rokhsar, Carl Douglas, Stefan Jansson, Goran Sandberg, and Yves Van de Peer, made all of the work reported in this chapter possible. In particular, we have directly co-opted analyses performed by Uffe Hellsten, Nik Putnam, and Igor

Grigoriev. This work was supported by the BioEnergy Science Center, a U.S. DOE Bioenergy Research Center (Office of Biological and Environmental Research in the DOE Office of Science), and the NSF Plant Genome Research Program.

# References

Adams KL, Wendel JF (2005) Polyploidy and genome evolution in plants. Curr Opin Plant Biol 8: 135–141.

Aparicio S, Chapman J, Stupka E, Putnam N, Chia J, Dehal P, Christoffels A, Rash S, Hoon S, Smit A, Gelpke MDS, Roach J, Oh T, Ho IY, Wong M, Detter C, Verhoef F, Predki P, Tay A, Lucas S, Richardson P, Smith SF, Clark MS, Edwards YJK, Doggett N, Zharkikh A, Tavtigian SV, Pruss D, Barnstead M, Evans C, Baden H, Powell J, Glusman G, Rowen L, Hood L, Tan YH, Elgar G, Hawkins T, Venkatesh B, Rokhsar D, Brenner S (2002) Whole-genome shotgun assembly and analysis of the genome of *Fugu rubripes*. Science 297: 1301–1310.

Bell CD, Soltis DE, Soltis PS (2005) The age of the angiosperms: A molecular timescale without a clock. Evolution 59: 1245–1258.

Blanc G, Hokamp K, Wolfe KH (2003) A recent polyploidy superimposed on older large-scale duplications in the *Arabidopsis* genome. Genome Res 13: 137–144.

Bradshaw HD, Jr, Stettler RF (1993) Molecular genetics of growth and development in Populus I. Triploidy in hybrid poplars. Theor Appl Genet 86: 301–307.

Bradshaw HD, Jr, Stettler RF (1995) Molecular Genetics of Growth and Development in Populus. IV. Mapping QTLs With Large Effects on Growth, Form, and Phenology Traits in a Forest Tree. Genetics 139: 963–973.

Bradshaw HD, Jr., Villar M, Watson BD, Otto KG, Stewart S, Stettler RF (1994) Molecular-genetics of growth and development in *Populus* 3. A genetic-linkage map of a hybrid poplar composed of RFLP, STS, and RAPD markers. Theor Appl Genet 89: 167–178.

Busov VB, Brunner AM, Meilan R, Filichkin S, Ganio L, Gandhi S, Strauss SH (2005) Genetic transformation: a powerful tool for dissection of adaptive traits in trees. New Phytol 167: 9–18.

Cervera MT, Storme V, Ivens B, Gusmao J, Liu BH, Hostyn V, Slycken Jv, Montagu M, Boerjan W (2001) Dense genetic linkage maps of three populus species (*Populus deltoides*, *P. nigra* and *P. trichocarpa*) based on AFLP and microsatellite markers. Genetics 158: 787–809.

Comeron JM (1995) A method for estimating the numbers of synonymous and nonsynonymous substitutions per site. J Mol Evol 41: 1152–1159.

Dehal P, Satou Y, Campbell RK, Chapman J, Degnan B, De Tomaso A, Davidson B, Di Gregorio A, Gelpke M, Goodstein DM, Harafuji N, Hastings KEM, Ho I, Hotta K, Huang W, Kawashima T, Lemaire P, Martinez D, Meinertzhagen IA, Necula S, Nonaka M, Putnam N, Rash S, Saiga H, Satake M, Terry A, Yamada L, Wang HG, Awazu S, Azumi K, Boore J, Branno M, Chin-bow S, DeSantis R, Doyle S, Francino P, Keys DN, Haga S, Hayashi H, Hino K, Imai KS, Inaba K, Kano S, Kobayashi K, Kobayashi M, Lee BI, Makabe KW, Manohar C, Matassi G, Medina M, Mochizuki Y, Mount S, Morishita T, Miura S, Nakayama A, Nishizaka S, Nomoto H, Ohta F, Oishi K, Rigoutsos I, Sano M, Sasaki A, Sasakura Y, Shoguchi E, Shin-i T, Spagnuolo A, Stainier D, Suzuki MM, Tassy O, Takatori N, Tokuoka M, Yagi K, Yoshizaki F, Wada S, Zhang C, Hyatt PD, Larimer F, Detter C, Doggett N, Glavina T, Hawkins T, Richardson P, Lucas S, Kohara Y, Levine M, Satoh N, Rokhsar DS (2002) The draft genome of Ciona intestinalis: Insights into chordate and vertebrate origins. Science 298: 2157–2167.

Dickmann DI, Isebrands JG, Eckenwalder JE, Richardson J (2001) Poplar Culture in North America. NRC Research Press ,Ottawa, ON, Canada.

Dinus RJ, Payne P, Sewell MM, Chiang VL, Tuskan GA (2001) Genetic modification of short rotation poplar wood: properties for ethanol fuel and fiber productions. Crit Rev Plant Sci 20: 51–69.

Doyle JJ, Flagel LE, Paterson AH, Rapp RA, Soltis DE, Soltis PS, Wendel JF (2008) Evolutionary genetics of genome merger and doubling in plants. Annual Review of Genetics 42: 443–461.

Drost DR, Novaes E, Boaventura-Novaes C, Benedict CI, Brown RS, Yin TM, Tuskan GA, Kirst M (2009) A microarray-based genotyping and genetic mapping approach for highly heterozygous outcrossing species enables localization of a large fraction of the unassembled *Populus trichocarpa* genome sequence. Plant J 58: 1054–1067.

Duvick J, Fu A, Muppirala U, Sabharwal M, Wilkerson MD, Lawrence CJ, Lushbough C, Brendel V (2008) PlantGDB: a resource for com-parative plant genomics. Nucleic Acids Res 36: D959–965.

Foissac S, Gouzy J, Rombauts S, Mathe C, Amselem J, Sterck L, Van de Peer Y, Rouze P, Schiex T (2008) Genome annotation in plants and fungi: EuGene as a model platform. Current Bioinformatics 3: 87–97.

Gaudet M, Jorge V, Paolucci I, Beritognolo I, Mugnozza GS, Sabatti M (2008) Genetic linkage maps of Populus nigra L. including AFLPs, SSRs, SNPs, and sex trait. Tree Genet Genomes 4: 25–36.

Gendler K, Paulsen T, Napoli C (2008) ChromDB: The chromatin database. Nucl Acids Res 36: D298–D302.

Germaine K, Keogh E, Garcia-Cabellos G, Borremans B, van der Lelie D, Barac T, Oeyen L, Vangronsveld J, Moore FP, Moore ERB, Campbell CD, Ryan D, Dowling DN (2004) Colonisation of poplar trees by gfp expressing bacterial endophytes. FEMA Microbiol Ecol 48: 109–118.

Green ED (2001) Strategies for the systematic sequencing of complex genomes. Nat Rev Genet 2: 573–583.

Harris MA, Clark JI, Ireland A, Lomax J, Ashburner M, Collins R, Eilbeck K, Lewis S, Mungall C, Richter J, Rubin GM, Shu SQ, Blake JA, Bult CJ, Diehl AD, Dolan ME, Drabkin HJ, Eppig JT, Hill DP, Ni L, Ringwald M, Balakrishnan R, Binkley G, Cherry JM, Christie KR, Costanzo MC, Dong Q, Engel SR, Fisk DG, Hirschman JE, Hitz BC, Hong EL, Lane C, Miyasato S, Nash R, Sethuraman A, Skrzypek M, Theesfeld CL, Weng SA, Botstein D, Dolinski K, Oughtred R, Berardini T, Mundodi S, Rhee SY, Apweiler R, Barrell D, Camon E, Dimmer E, Mulder N, Chisholm R, Fey P, Gaudet P, Kibbe W, Pilcher K, Bastiani CA, Kishore R, Schwarz EM, Sternberg P, Van Auken K, Gwinn M, Hannick L, Wortman J, Aslett M, Berriman M, Wood V, Bromberg S, Foote C, Jacob H, Pasko D, Petri V, Reilly D, Seiler K, Shimoyama M, Smith J, Twigger S, Jaiswal P, Seigfried T, Collmer C, Howe D, Westerfield M 2006. The Gene Ontology (GO) project in 2006. Nucleic Acids Res 34: D322–D326.

Jaillon O, Aury JM, Noel B, Policriti A, Clepet C, Casagrande A, Choisne N, Aubourg SÃ, Vitulo N, Jubin C, Vezzi A, Legeai F, Hugueney P, Dasilva C, Horner D, Mica E, Jublot D, Poulain J, BruyÃ¨re Cm, Billault A, Segurens BÃ, Gouyvenoux M, Ugarte E, Cattonaro F, Anthouard VÃ, Vico V, Del Fabbro C, Alaux Ml, Di Gaspero G, Dumas V, Felice N, Paillard S, Juman I, Moroldo M, Scalabrin S, Canaguier Al, Le Clainche I, Malacrida G, Durand Eo, Pesole G, Laucou Vr, Chatelet P, Merdinoglu D, Delledonne M, Pezzotti M, Lecharny A, Scarpelli C, Artiguenave Fo, PÃ¨ ME, Valle G, Morgante M, Caboche M, Adam-Blondon AFo, Weissenbach J, QuÃ©tier F, Wincker P (2007) The grapevine genome sequence suggests ancestral hexaploidization in major angiosperm phyla. Nature 449: 463–467.

Jansen RK, Kaittanis C, Saski C, Lee SB, Tomkins J, Alverson AJ, Daniell H (2006) Phylogenetic analyses of *Vitis* (Vitaceae) based on complete chloroplast genome sequences: effects of taxon sampling and phylogenetic methods on resolving relationships among rosids. BMC Evol Biol 6.

Jansson S, Douglas CJ (2007) Populus: A model system for plant biology. Annu Rev Plant Biol 58: 435–458.

Kelleher CT, Chiu R, Shin H, Bosdet IE, Krzywinski MI, Fjell CD, Wilkin J, Yin T, DiFazio SP, Ali J, Asano JK, Chan S, Cloutier A, Girn N, Leach S, Lee D, Mathewson CA, Olson T,

O'Connor K, Prabhu AL, Smailus DE, Stott JM, Tsai M, Wye NH, Yang GS, Zhuang J, Holt RA, Putnam NH, Vrebalov J, Giovannoni JJ, Grimwood J, Schmutz J, Rokhsar D, Jones SJ, Marra MA, Tuskan GA, Bohlmann J, Ellis BE, Ritland K, Douglas CJ, Schein JE (2007) A physical map of the highly heterozygous *Populus* genome: integration with the genome sequence and genetic map and analysis of haplotype variation. Plant J 50: 1063–1078.

Lander ES, Linton LM, Birren B, Nusbaum C, Zody MC, Baldwin J, Devon K, Dewar K, Doyle M, FitzHugh W, Funke R, Gage D, Harris K, Heaford A, Howland J, Kann L, Lehoczky J, LeVine R, McEwan P, McKernan K, Meldrim J, Mesirov JP, Miranda C, Morris W, Naylor J, Raymond C, Rosetti M, Santos R, Sheridan A, Sougnez C, Stange-Thomann N, Stojanovic N, Subramanian A, Wyman D, Rogers J, Sulston J, Ainscough R, Beck S, Bentley D, Burton J, Clee C, Carter N, Coulson A, Deadman R, Deloukas P, Dunham A, Dunham I, Durbin R, French L, Grafham D, Gregory S, Hubbard T, Humphray S, Hunt A, Jones M, Lloyd C, McMurray A, Matthews L, Mercer S, Milne S, Mullikin JC, Mungall A, Plumb R, Ross M, Shownkeen R, Sims S, Waterston RH, Wilson RK, Hillier LW, McPherson JD, Marra MA, Mardis ER, Fulton LA, Chinwalla AT, Pepin KH, Gish WR, Chissoe SL, Wendl MC, Delehaunty KD, Miner TL, Delehaunty A, Kramer JB, Cook LL, Fulton RS, Johnson DL, Minx PJ, Clifton SW, Hawkins T, Branscomb E, Predki P, Richardson P, Wenning S, Slezak T, Doggett N, Cheng JF, Olsen A, Lucas S, Elkin C, Uberbacher E, Frazier M, Gibbs RA, Muzny DM, Scherer SE, Bouck JB, Sodergren EJ, Worley KC, Rives CM, Gorrell JH, Metzker ML, Naylor SL, Kucherlapati RS, Nelson DL, Weinstock GM, Sakaki Y, Fujiyama A, Hattori M, Yada T, Toyoda A, Itoh T, Kawagoe C, Watanabe H, Totoki Y, Taylor T, Weissenbach J, Heilig R, Saurin W, Artiguenave F, Brottier P, Bruls T, Pelletier E, Robert C, Wincker P, Rosenthal A, Platzer M, Nyakatura G, Taudien S, Rump A, Yang HM, Yu J, Wang J, Huang GY, Gu J, Hood L, Rowen L, Madan A, Qin SZ, Davis RW, Federspiel NA, Abola AP, Proctor MJ, Myers RM, Schmutz J, Dickson M, Grimwood J, Cox DR, Olson MV, Kaul R, Raymond C, Shimizu N, Kawasaki K, Minoshima S, Evans GA, Athanasiou M, Schultz R, Roe BA, Chen F, Pan HQ, Ramser J, Lehrach H, Reinhardt R, McCombie WR, de la Bastide M, Dedhia N, Blocker H, Hornischer K, Nordsiek G, Agarwala R, Aravind L, Bailey JA, Bateman A, Batzoglou S, Birney E, Bork P, Brown DG, Burge CB, Cerutti L, Chen HC, Church D, Clamp M, Copley RR, Doerks T, Eddy SR, Eichler EE, Furey TS, Galagan J, Gilbert JGR, Harmon C, Hayashizaki Y, Haussler D, Hermjakob H, Hokamp K, Jang WH, Johnson LS, Jones TA, Kasif S, Kaspryzk A, Kennedy S, Kent WJ, Kitts P, Koonin EV, Korf I, Kulp D, Lancet D, Lowe TM, McLysaght A, Mikkelsen T, Moran JV, Mulder N, Pollara VJ, Ponting CP, Schuler G, Schultz JR, Slater G, Smit AFA, Stupka E, Szustakowki J, Thierry-Mieg D, Thierry-Mieg J, Wagner L, Wallis J, Wheeler R, Williams A, Wolf YI, Wolfe KH, Yang SP, Yeh RF, Collins F, Guyer MS, Peterson J, Felsenfeld A, Wetterstrand KA, Patrinos A, Morgan MJ (2001) Initial sequencing and analysis of the human genome. Nature 409: 860–921.

Lescot M, Rombauts S, Zhang J, Aubourg S, Mathe C, Jansson S, Rouze P, Boerjan W (2004) Annotation of a 95-kb Populus deltoides genomic sequence reveals a disease resistance gene cluster and novel class I and class II transposable elements. Theoret Appl Genetics 109: 10–22.

Lexer C, Buerkle CA, Joseph JA, Heinze B, Fay MF (2007) Admixture in European Populus hybrid zones makes feasible the mapping of loci that contribute to reproductive isolation and trait differences. Heredity 98: 74–84.

Liang CZ, Jaiswal P, Hebbard C, Avraham S, Buckler ES, Casstevens T, Hurwitz B, McCouch S, Ni JJ, Pujar A, Ravenscroft D, Ren L, Spooner W, Tecle I, Thomason J, Tung CW, Wei XH, Yap I, Youens-Clark K, Ware D, Stein L (2008) Gramene: a growing plant comparative genomics resource. Nucleic Acids Res 36: D947–D953.

Lynch M, Conery JS (2003) The evolutionary demography of duplicate genes. J Struct Funct Genom 3: 35–44.

Ma C, Strauss SH, Meilan R (2004) *Agrobacterium*-mediated transformation of the genome-sequenced poplar clone, Nisqually-1 (*Populus trichocarpa*). Plant Mol Biol Rep 22: 311–312.

Markussen T, Pakull B, Fladung M (2007) Positioning of sex-correlated markers for Populus in a AFLP- and SSR-Marker based genetic map of Populus tremula x tremuloides. Silvae Genet 56: 180–184.

Marra MA, Kucaba TA, Dietrich NL, Green ED, Brownstein B, Wilson RK, McDonald KM, Hillier LW, McPherson JD, Waterston RH (1997) High throughput fingerprint analysis of large-insert clones. Genome Res. 7: 1072–1084.

Martin F, Tuskan GA, DiFazio SP, Lammers P, Newcombe G, Podila GK (2004) Symbiotic sequencing for the Populus mesocosm. New Phytol 161: 330–335.

Martin F, Aerts A, Ahren D, Brun A, Danchin EGJ, Duchaussoy F, Gibon J, Kohler A, Lindquist E, Pereda V, Salamov A, Shapiro HJ, Wuyts J, Blaudez D, Buee M, Brokstein P, Canback B, Cohen D, Courty PE, Coutinho PM, Delaruelle C, Detter JC, Deveau A, DiFazio S, Duplessis S, Fraissinet-Tachet L, Lucic E, Frey-Klett P, Fourrey C, Feussner I, Gay G, Grimwood J, Hoegger PJ, Jain P, Kilaru S, Labbe J, Lin YC, Legue V, Le Tacon F, Marmeisse R, Melayah D, Montanini B, Muratet M, Nehls U, Niculita-Hirzel H, Secq MPO, Peter M, Quesneville H, Rajashekar B, Reich M, Rouhier N, Schmutz J, Yin T, Chalot M, Henrissat B, Kues U, Lucas S, Van de Peer Y, Podila GK, Polle A, Pukkila PJ, Richardson PM, Rouze P, Sanders IR, Stajich JE, Tunlid A, Tuskan G, Grigoriev IV (2008) The genome of Laccaria bicolor provides insights into mycorrhizal symbiosis. Nature 452: 88–92.

Ming R, Hou SB, Feng Y, Yu QY, Dionne-Laporte A, Saw JH, Senin P, Wang W, Ly BV, Lewis KLT, Salzberg SL, Feng L, Jones MR, Skelton RL, Murray JE, Chen CX, Qian WB, Shen JG, Du P, Eustice M, Tong E, Tang HB, Lyons E, Paull RE, Michael TP, Wall K, Rice DW, Albert H, Wang ML, Zhu YJ, Schatz M, Nagarajan N, Acob RA, Guan PZ, Blas A, Wai CM, Ackerman CM, Ren Y, Liu C, Wang JM, Wang JP, Na JK, Shakirov EV, Haas B, Thimmapuram J, Nelson D, Wang XY, Bowers JE, Gschwend AR, Delcher AL, Singh R, Suzuki JY, Tripathi S, Neupane K, Wei HR, Irikura B, Paidi M, Jiang N, Zhang WL, Presting G, Windsor A, Navajas-Perez R, Torres MJ, Feltus FA, Porter B, Li YJ, Burroughs AM, Luo MC, Liu L, Christopher DA, Mount SM, Moore PH, Sugimura T, Jiang JM, Schuler MA, Friedman V, Mitchell-Olds T, Shippen DE, dePamphilis CW, Palmer JD, Freeling M, Paterson AH, Gonsalves D, Wang L and Alam M (2008) The draft genome of the transgenic tropical fruit tree papaya (*Carica papaya* Linnaeus). Nature 452: 991–997.

Ralph SG, Chun HJ, Cooper D, Kirkpatrick R, Kolosova N, Gunter L, Tuskan GA, Douglas CJ, Holt RA, Jones SJ, Marra MA and Bohlmann J (2008) Analysis of 4,664 high-quality sequence-finished poplar full-length cDNA clones and their utility for the discovery of genes responding to insect feeding. BMC Genom 9: 57.

Rubin EM (2008) Genomics of cellulosic biofuels. Nature 454: 841–845.

Semon M, Wolfe KH (2007) Consequences of genome duplication. Curr Op Genet Dev 17: 505–512.

Sjodin A, Street NR, Sandberg G, Gustafsson P, Jansson S (2009) The Populus Genome Integrative Explorer (PopGenIE): a new resource for exploring the Populus genome. New Phytol 182: 1013–1025.

Solovyev V, Kosarev P, Seledsov I, Vorobyev D (2006) Automatic annotation of eukaryotic genes, pseudogenes and promoters. Genome Biol 7.

Song JY, Lu SF, Chen ZZ, Lourenco R, Chiang VL (2006) Genetic transformation of Populus trichocarpa genotype Nisqually-1: A functional genomic tool for woody plants. Plant Cell Physiol 47: 1582–1589.

Sterck L, Rombauts S, Jansson S, Sterky F, Rouze P, Van de Peer Y (2005) EST data suggest that poplar is an ancient polyploid. New Phytol 167: 165–170.

Sterky F, Regan S, Karlsson J, Hertzberg M, Rohde A, Holmberg A, Amini B, Bhalerao R, Larsson M, Villarroel R, Van Montagu M, Sandberg G, Olsson O, Teleri TT, Boerjan W, Gustafsson P, Uhlen M, Sundberg B, Lundeberg J (1998) Gene discovery in the wood-forming tissues of poplar: analysis of 5,692 expressed sequence tags. Proc Natl Acad Sci USA 95: 13330–13335.

Sterky F, Bhalerao RR, Unneberg P, Segerman B, Nilsson P, Brunner AM, Charbonnel-Campaa L, Lindvall JJ, Tandre K, Strauss SH, Sundberg B, Gustafsson P, Uhlen M, Bhalerao RP,

Nilsson O, Sandberg G, Karlsson J, Lundeberg J, Jansson S (2004) A *Populus* EST resource for plant functional genomics. Proc Nat Acad Sci USA 101: 13951–13956.

Stirling B, Newcombe G, Vrebalov J, Bosdet I, Bradshaw HD, Jr (2001) Suppressed recombination around the *MXC3* locus, a major gene for resistance to poplar leaf rust. Theor Appl Genet 103: 1129–1137.

Stirling B, Yang ZK, Gunter LE, Tuskan GA, Bradshaw HD, Jr (2003) Comparative sequence analysis between orthologous regions of the Arabidopsis and Populus genomes reveals substantial synteny and microcollinearity. Can J For Res 33: 2245–2251.

Swarbreck D, Wilks C, Lamesch P, Berardini TZ, Garcia-Hernandez M, Foerster H, Li D, Meyer T, Muller R, Ploetz L, Radenbaugh A, Singh S, Swing V, Tissier C, Zhang P, Huala E (2008) The Arabidopsis Information Resource (TAIR): gene structure and function annotation. Nucleic Acids Res 36: D1009–D1014.

Tang HB, Bowers JE, Wang XY, Ming R, Alam M, Paterson AH (2008a) Perspective—Synteny and collinearity in plant genomes. Science 320: 486–488.

Tang HB, Wang XY, Bowers JE, Ming R, Alam M, Paterson AH (2008b) Unraveling ancient hexaploidy through multiply-aligned angiosperm gene maps. Genome Res 18: 1944–1954.

Taylor G (2002) *Populus*: Arabidopsis for forestry. Do we need a model tree? Ann Bot 90: 681–689.

Tuskan GA (1998) Short-rotation woody crop supply systems in the United States: What do we know and what do we need to know? Biomass Bioenerg 14: 307–315.

Tuskan GA, Walsh ME (2001) Short-rotation woody crop systems, atmospheric carbon dioxide and carbon management: A US case study. For Chron 77: 259–264.

Tuskan GA, DiFazio S, Jansson S, Bohlmann J, Grigoriev I, Hellsten U, Putnam N, Ralph S, Rombauts S, Salamov A, Schein J, Sterck L, Aerts A, Bhalerao RR, Bhalerao RP, Blaudez D, Boerjan W, Brun A, Brunner A, Busov V, Campbell M, Carlson J, Chalot M, Chapman J, Chen GL, Cooper D, Coutinho PM, Couturier J, Covert S, Cronk Q, Cunningham R, Davis J, Degroeve S, Dejardin A, Depamphilis C, Detter J, Dirks B, Dubchak I, Duplessis S, Ehlting J, Ellis B, Gendler K, Goodstein D, Gribskov M, Grimwood J, Groover A, Gunter L, Hamberger B, Heinze B, Helariutta Y, Henrissat B, Holligan D, Holt R, Huang W, Islam-Faridi N, Jones S, Jones-Rhoades M, Jorgensen R, Joshi C, Kangasjarvi J, Karlsson J, Kelleher C, Kirkpatrick R, Kirst M, Kohler A, Kalluri U, Larimer F, Leebens-Mack J, Leple JC, Locascio P, Lou Y, Lucas S, Martin F, Montanini B, Napoli C, Nelson DR, Nelson C, Nieminen K, Nilsson O, Pereda V, Peter G, Philippe R, Pilate G, Poliakov A, Razumovskaya J, Richardson P, Rinaldi C, Ritland K, Rouze P, Ryaboy D, Schmutz J, Schrader J, Segerman B, Shin H, Siddiqui A, Sterky F, Terry A, Tsai CJ, Uberbacher E, Unneberg P, Vahala J, Wall K, Wessler S, Yang G, Yin T, Douglas C, Marra M, Sandberg G, Van de Peer Y, Rokhsar D (2006) The genome of black cottonwood, *Populus trichocarpa* (Torr. & Gray). Science 313: 1596–1604.

Unneberg P, Stromberg M, Lundeberg J, Jansson S, Sterky F (2005) Analysis of 70,000 EST sequences to study divergence between two closely related Populus species. Tree Genet Genomes 1: 109–115.

Velasco R, Zharkikh A, Troggio M, Cartwright DA, Cestaro A, Pruss D, Pindo M, Fitzgerald LM, Vezzulli S, Reid J, Malacarne G, Iliev D, Coppola G, Wardell B, Micheletti D, Macalma T, Facci M, Mitchell JT, Perazzolli M, Eldredge G, Gatto P, Oyzerski R, Moretto M, Gutin N, Stefanini M, Chen Y, Segala C, Davenport C, Dematte L, Mraz A, Battilana J, Stormo K, Costa F, Tao Q, Si-Ammour A, Harkins T, Lackey A, Perbost C, Taillon B, Stella A, Solovyev V, Fawcett JA, Sterck L, Vandepoele K, Grando SM, Toppo S, Moser C, Lanchbury J, Bogden R, Skolnick M, Sgaramella V, Bhatnagar SK, Fontana P, Gutin A, Van de Peer Y, Salamini F, Viola R (2007) A high quality draft consensus sequence of the genome of a heterozygous grapevine variety. PLoS ONE 2: e1326.

Venter JC, Adams MD, Myers EW, Li PW, Mural RJ, Sutton GG, Smith HO, Yandell M, Evans CA, Holt RA, Gocayne JD, Amanatides P, Ballew RM, Huson DH, Wortman JR, Zhang Q, Kodira CD, Zheng XH, Chen L, Skupski M, Subramanian G, Thomas PD, Zhang J,

Gabor Miklos GL, Nelson C, Broder S, Clark AG, Nadeau J, McKusick VA, Zinder N, Levine AJ, Roberts RJ, Simon M, Slayman C, Hunkapiller M, Bolanos R, Delcher A, Dew I, Fasulo D, Flanigan M, Florea L, Halpern A, Hannenhalli S, Kravitz S, Levy S, Mobarry C, Reinert K, Remington K, Abu-Threideh J, Beasley E, Biddick K, Bonazzi V, Brandon R, Cargill M, Chandramouliswaran I, Charlab R, Chaturvedi K, Deng Z, Di F, V, Dunn P, Eilbeck K, Evangelista C, Gabrielian AE, Gan W, Ge W, Gong F, Gu Z, Guan P, Heiman TJ, Higgins ME, Ji RR, Ke Z, Ketchum KA, Lai Z, Lei Y, Li Z, Li J, Liang Y, Lin X, Lu F, Merkulov GV, Milshina N, Moore HM, Naik AK, Narayan VA, Neelam B, Nusskern D, Rusch DB, Salzberg S, Shao W, Shue B, Sun J, Wang Z, Wang A, Wang X, Wang J, Wei M, Wides R, Xiao C, Yan C, Yao A, Ye J, Zhan M, Zhang W, Zhang H, Zhao Q, Zheng L, Zhong F, Zhong W, Zhu S, Zhao S, Gilbert D, Baumhueter S, Spier G, Carter C, Cravchik A, Woodage T, Ali F, An H, Awe A, Baldwin D, Baden H, Barnstead M, Barrow I, Beeson K, Busam D, Carver A, Center A, Cheng ML, Curry L, Danaher S, Davenport L, Desilets R, Dietz S, Dodson K, Doup L, Ferriera S, Garg N, Gluecksmann A, Hart B, Haynes J, Haynes C, Heiner C, Hladun S, Hostin D, Houck J, Howland T, Ibegwam C, Johnson J, Kalush F, Kline L, Koduru S, Love A, Mann F, May D, McCawley S, McIntosh T, McMullen I, Moy M, Moy L, Murphy B, Nelson K, Pfannkoch C, Pratts E, Puri V, Qureshi H, Reardon M, Rodriguez R, Rogers YH, Romblad D, Ruhfel B, Scott R, Sitter C, Smallwood M, Stewart E, Strong R, Suh E, Thomas R, Tint NN, Tse S, Vech C, Wang G, Wetter J, Williams S, Williams M, Windsor S, Winn-Deen E, Wolfe K, Zaveri J, Zaveri K, Abril JF, Guigo R, Campbell MJ, Sjolander KV, Karlak B, Kejariwal A, Mi H, Lazareva B, Hatton T, Narechania A, Diemer K, Muruganujan A, Guo N, Sato S, Bafna V, Istrail S, Lippert R, Schwartz R, Walenz B, Yooseph S, Allen D, Basu A, Baxendale J, Blick L, Caminha M, Carnes-Stine J, Caulk P, Chiang YH, Coyne M, Dahlke C, Mays A, Dombroski M, Donnelly M, Ely D, Esparham S, Fosler C, Gire H, Glanowski S, Glasser K, Glodek A, Gorokhov M, Graham K, Gropman B, Harris M, Heil J, Henderson S, Hoover J, Jennings D, Jordan C, Jordan J, Kasha J, Kagan L, Kraft C, Levitsky A, Lewis M, Liu X, Lopez J, Ma D, Majoros W, McDaniel J, Murphy S, Newman M, Nguyen T, Nguyen N, Nodell M (2001) The sequence of the human genome. Science 291: 1304–1351.

Whitham TG, Bailey JK, Schweitzer JA, Shuster SM, Bangert RK, LeRoy CJ, Lonsdorf EV, Allan GJ, DiFazio SP, Potts BM, Fischer DG, Gehring CA, Lindroth RL, Marks JC, Hart SC, Wimp GN, Wooley SC (2006) A framework for community and ecosystem genetics: from genes to ecosystems. Nat Rev Genet 7: 510–523.

Whitham TG, DiFazio SP, Schweitzer JA, Shuster SM, Allan GJ, Bailey JK, Woolbright SA (2008) Extending genomics to natural communities and ecosystems. Science 320: 492–495.

Wilkins O, Waldron L, Nahal H, Provart NJ, Campbell MM (2009) Genotype and time of day shape the *Populus* drought response. Plant J 60: 703–715.

Wise R, Caldo R, Hong L, Wu S, Cannon E and Dickerson J (2006) PLEXdb: A unified expression prolfiling database for plants and plant pathogens. Phytopathology 96: S161–S161.

Yin T, DiFazio SP, Gunter LE, Zhang X, Sewell MM, Woolbright SA, Allan GJ, Kelleher CT, Douglas CJ, Wang M, Tuskan GA (2008) Genome structure and emerging evidence of an incipient sex chromosome in *Populus*. Genome Res 18: 422–430.

Yin TM, Huang MR, Wang MX, Zhu LH, Zeng ZB, Wu RL (2001) Preliminary interspecific genetic maps of the Populus genome constructed from RAPD markers. Genome 44: 602–609.

Yin TM, DiFazio SP, Gunter LE, Jawdy SS, Boerjan W, Tuskan GA (2004a) Genetic and physical mapping of *Melampsora* rust resistance genes in *Populus* and characterization of linkage disequilibrium and flanking genomic sequence. New Phytol 164: 95–105.

Yin TM, DiFazio SP, Gunter LE, Riemenschneider D, Tuskan GA (2004b) Large-scale heterospecific segregation distortion in Populus revealed by a dense genetic map. Theor Appl Genet 109: 451–463.

Zhou FF, Xu Y (2009) RepPop: a database for repetitive elements in *Populus trichocarpa*. BMC Genom 10.

Zhu QH, Guo AY, Gao G, Zhong YF, Xu M, Huang MR, Luo JC (2007) DPTF: a database of poplar transcription factors. Bioinformatics 23: 1307–1308.

# 5

# Poplar Genome Microarrays

*Chung-Jui Tsai,[1,]\* Priya Ranjan,[2] Stephen P. DiFazio,[3]*
*Gerald A. Tuskan[4] and Virgil E. Johnson[5]*

## ABSTRACT

Microarray-based transcriptome profiling is arguably the most mature and widely practiced omics technology of the post-genomics era. It provides an intuitive measure of transcript abundance for comparative gene expression analysis and for construction of co-regulation networks. Whole-genome microarray analysis can also facilitate systems-level analysis and integration of data from other omics platforms that lack global coverage, such as proteome and metabolome analyses. Just as *Populus* expressed sequence tag collections have proven invaluable for gene discovery and functional characterization, the growing volume of whole-genome microarray data represents a rich resource that promises to support a wide range of scientific inquiries to advance *Populus* biology in years to come.

**Keywords:** Transcriptome; gene expression; POParray database; probe design

## 5.1 Introduction

The release of the *Populus trichocarpa* genome sequence and its initial annotation (Tuskan et al. 2006) has enabled development of powerful genomic tools for understanding tree biology and facilitating tree improvement. An important genome-enabled application is the development of high-density

[1]School Forestry and Natural Resources, and Department of Genetics, University of Georgia, 170 Green Street, Athens, Georgia 30602, USA; e-mail: *cjtsai@warnell.uga.edu*
[2]Department of Plant Sciences, University of Tennessee, Knoxville, Tennessee, 37996; and Biosciences Division, Oak Ridge National Laboratory, Oak Ridge, Tennessee, 37831, USA; e-mail: *ranjanp@ornl.gov*
[3]Department of Biology, West Virginia University, Morgantown, West Virginia 26506, USA; e-mail: *spdifazio@mail.wvu.edu*
[4]Biosciences Division, Oak Ridge National Laboratory, Oak Ridge, Tennessee 37831; and, Laboratory Science Program, Joint Genome Institute, Walnut Creek, California 94598, USA; e-mail: *tuskanga@ornl.gov*
[5]School of Forestry and Natural Resources, University of Georgia Athens, Georgia 30602, USA; e-mail: *vedjohns@uga.edu*
\*Corresponding author

oligo arrays for global transcriptome analysis. As is already the case for *Arabidopsis, Oryza* (rice) and other species with a sequenced or nearly sequenced genome, multiple whole-genome oligo array platforms are now available for *Populus,* including NimbleGen, Affymetrix and Agilent microArrays. These platforms differ from one another in many aspects, ranging from oligo fabrication technology and probe design to hybridization method and multiplexing capability. As the genus *Populus* is comprised of many species and hybrids that exhibit varying degrees of sequence divergence from the sequenced *P. trichocarpa* genome (Tuskan et al. 2004), probe design and properties (i.e., location and length of probes) strongly influence the specificity and/or cross-species utility of each microarray platform.

While the availability of various microarray platforms offers *Populus* researchers multiple options in gene expression analysis, cross-referencing between studies with different microarray systems, including previously developed expressed sequence tag (EST) microarrays, is difficult. The situation is further complicated by the constantly evolving nature of genome annotation. Computational annotation was accomplished using a variety of ab initio gene calling algorithms (Tuskan et al. 2006), and only a small number of gene models have been manually curated to date [Joint Genome Institute (JGI) *Populus trichocarpa* v1.1, *http://genome.jgi-psf.org/Poptr1_1*; and v2.0, *http://www.phytozome.net/poplar.php*]. Thus, multiple gene model predictions may be found at any given locus, and their inconsistent use in microarray probe design and annotation, as well as in the literature, represents a major challenge for comparative data analysis.

In this chapter, we provide an overview of the three poplar whole-genome microarray platforms and the recent development of the next-generation, ultra-high density microarrays for gene expression profiling analysis. Numerous *Populus* EST microarrays have been developed (Hertzberg et al. 2001; Kohler et al. 2003; Ranjan et al. 2004; Sterky et al. 2004; Ralph et al. 2006), and readers are referred to the corresponding papers for specifics. We also discuss development of a relational database, POParray, for facilitating cross-referencing among various *Populus* microarray platforms and JGI gene models.

## 5.2 Choosing a Suitable Microarray Platform

The choice of microarray platform for a given project is influenced by technical and biological factors, as well as cost. One common determinant is the microarray infrastructure support available at a given institution, although out-sourcing microarray hybridization to other core facilities is becoming a common practice. In general, in situ synthesized oligo arrays offer superior consistency relative to spotted (EST or oligo) arrays. However, hybridization performance achieved across different platforms has not been systematically investigated until recently. Below, we briefly review the MicroArray Quality Control (MAQC) project to draw on experience from the human microarray community, and discuss various factors for consideration when choosing a microarray platform.

## 5.2.1 What Can We Learn from the Human MAQC Project?

The MAQC project is spearheaded by the Food and Drug Administration (FDA) in a community-wide effort to establish QC metrics and thresholds for objectively assessing the performance of various microarray platforms, a critical first step toward applying microarray technology in clinical and regulatory settings (MAQC 2006; Patterson et al. 2006). The MAQC consortium involves over 100 participants across 51 academic, governmental and commercial institutions. In the Phase I project (MAQC-I), seven human microarray platforms were evaluated, including six commercial arrays (Applied Biosystems, Affymetrix, Agilent, Eppendorf, GE Healthcare, and Illumina) and one academic array produced at the National Cancer Institute. Using standardized protocols and a pair of high-quality, commercially available RNA reference samples of four titration pools, MAQC assessed intra- and inter-platform reproducibility based on results generated from five replicate assays at three test sites. Cross-platform data analysis was based on extensive probe mapping, and due to inherent probe design variability, was limited to a common set of 12,091 genes represented on all microarray platforms and to one probe per gene for each platform (MAQC 2006). For the high-density oligo arrays, the selected 12,091 probes represent only 22 to 36% of the total number of probes/probe sets present in each platform. Overall, the MAQC study showed high levels of intra- and inter-platform reproducibility. A few trends observed from the study are summarized as follows:

1. Intra-platform reproducibility: The median CV (coefficient of variation) values of signal intensities among replicate assays ranged from 5 to 15% across platforms, and from 10 to 20% when all replicates from all test sites were analyzed. Data obtained from the Affymetrix platform had the lowest replicate CVs, both within and across test sites.

2. Detection sensitivity: Evaluated on the basis of consistent qualitative calls (i.e., present or absent) according to platform-specific parameters, the Agilent and GE Healthcare arrays detected a higher number of genes, and with a higher level of consistency among replicates and across all test sites.

3. Inter-platform comparability: Cross-platform analysis was carried out indirectly by analyzing relative gene expression between the two reference RNA samples obtained on the same platform, followed by

comparing the list of differentially expressed genes from all test sites and platforms. A generally high degree of overlap between gene lists was reported, and in many cases exceeded 80%. When evaluated by Spearman's rank correlation of the log ratios, excellent comparability was also found between all pairwise comparisons ($r$ = 0.83 to 0.94). Of note, the Agilent platform detected more differentially expressed genes (44% of the 12,091 common gene list) than the other high-density platforms (34–38%) based on the cutoff of $P$ value < 0.001 and fold change > 2.

4. Relative accuracy: Relative gene expression between two reference RNA samples was also measured by quantitative RT-PCR and analyzed as described above. All platforms showed high correlations with the TaqMan assays (Applied Biosystems), with the Affymetrix, Agilent and Illumina arrays demonstrating an average correlation of greater than 0.9.

5. One-color versus two-color assays: No significant differences were found between data generated from one-color and two-color arrays, in terms of reproducibility, sensitivity, and concordance of gene lists identified as differentially expressed (Patterson et al. 2006). In general, data derived from one-color microarrays had slightly lower levels of inter- and intra-site variation than those from two-color platforms. In addition, in situ synthesized oligo arrays (e.g., Agilent) performed consistently better than spotted oligo arrays (e.g., CapitalBio and TeleChem).

The MAQC study provided important validation of microarray technology in general, and demonstrated the general comparability between microarray platforms and other alternative technologies. On the basis of equivalent data quality across platforms, therefore, the choice of a microarray platform depends greatly on specific research considerations and biological systems under investigation. A sampling of these factors is discussed below.

### 5.2.2 The Power of a Community Data Repository

Adoption of a given platform within the research community plays an important role in swaying the microarray choice of new users. For instance, the Affymetrix *Arabidopsis* ATH1 platform had more than 6,000 cataloged samples in the NCBI Gene Expression Omnibus database as of May 2010 (GEO accession number GPL198), presumably due to its head-start. The Agilent *Arabidopsis* platforms introduced 2–3 years later (GEO accession numbers GPL888, GPL2871 and GPL6177) had less than 200 affiliated samples in the GEO repository. Consequently, adopting the Affymetrix platform in new *Arabidopsis* experiments empowers researchers to perform comparative analyses across multiple organ/tissue types, treatment conditions and/or mutant backgrounds (e.g., Winter et al. 2007) in an unprecedented manner.

### 5.2.3 Probe length

The technical aspects of the microarray platform also influence the utility of the arrays. In general, shorter probe length affords better discrimination power among closely related transcripts than longer probes (Lockhart et al. 1996). However, short oligos suffer from higher background noise due to surface-binding (Guo et al. 1994). Conversely, long oligo arrays usually yield a superior signal-to-noise ratio and higher detection sensitivity, and are more tolerant of sequence mismatches during hybridization (Guo et al. 1994; Hughes et al. 2001).

### 5.2.4 Feature Size

Detection sensitivity is also influenced by the feature size of the array, with larger feature sizes affording better sensitivity. As shown in the MAQC (2006) study, the Agilent human array with a 125 µm feature size detected a higher number of "present" calls as well as differentially expressed genes than the Affymetrix platform with an 11 µm feature size.

### 5.2.5 Probe Design

The location and uniqueness of the probes significantly affect the hybridization performance of the array. Probes designed to target 3'-UTRs are usually more gene-specific, and hence can better discriminate between closely related gene family members. By the same token, however, arrays with 3'-UTR probes are less applicable for use with closely related, non-target species, due to potential sequence divergence. A second issue is the number of probes targeted per gene, which varies between platforms. Targeting multiple probes per gene (i.e., probe sets) helps maximize sequence representation, but uneven specificity among probes is a concern and can complicate data analysis.

### 5.2.6 Design Flexibility

With the explosive increase of genome sequences, the continuing improvement of microarray technology and constantly evolving gene annotation, platforms with design flexibility and user-friendly portals are gaining popularity. In this context, the Affymetrix system based on photolithographic fabrication techniques for array production is least flexible (or most costly) to accommodate probe design alterations. Both Agilent and NimbleGen platforms are more flexible with their maskless array fabrication techniques. Agilent's eArray web portal permits custom probe design and editing, thus representing a truly user-friendly and user-driven platform for iterative array design.

## 5.3 Poplar Whole-genome Microarrays

As mentioned, the three poplar genome arrays differ in their fabrication technologies and their design principles. A slightly different collection of target sequences was used in each case, in conjunction with proprietary algorithms for probe selection. Both Affymetrix and NimbleGen platforms contain probes that target, in addition to predicted *P. trichocarpa* nuclear gene models, EST sequences not present in the reference genome. Probes corresponding to chloroplast and mitochondrial genes are included on the Agilent and NimbleGen arrays. Key features of each array are summarized in Table 5-1, and described in further detail below.

**Table 5-1** Comparison of presently available poplar whole-genome microarray platforms.

| Platform | Affymetrix | Agilent | NimbleGen (1st generation) | NimbleGen (2nd generation) |
|---|---|---|---|---|
| Probe length | 25-mer | 60-mer | 60-mer | 60-mer |
| Probe number per gene model | 11 pairs | 1 | 3 | 7 |
| Feature density | > 1,350,000 | 4 × 44,000 (240,000) | 385,000 | 385,000 |
| Number of poplar probes or probe sets (see Table 5.2 for probe redundancy) | 61,251 | 43,795 | 65,965 | 52,225 |
| Multiplex capability | No | Yes | No | No |
| Feature size | 11 μm | 65 μm | 16 μm | 16 μm |
| Assay type | One-color | One- or two-color | One- or two-color | One- or two-color |

## *5.3.1 NimbleGen Poplar Genome Array*

The NimbleGen poplar array was first introduced in 2005. The first-generation array was originally designed to target all predicted genes in the *Populus trichocarpa* nuclear and organellar genomes, as well as divergent aspen (*Populus alba*, *P. tremula*, *P. tremuloides* and their hybrids) transcripts, and predicted miRNAs using 3 probes per target (Tuskan et al. 2006). The original design was based on NimbleGen's old 195K platform, with design ID 1482 and GEO accession number GPL2618 (Yang et al. 2008). The same design was upgraded (doubled) to the 385K platform, with NimbleGen design ID 7962, and GEO accession numbers GPL2699 and GPL7424. Its use has been reported in several published studies (Groover et al. 2006; Kalluri et al. 2007; Rinaldi et al. 2007; Kohler et al. 2008; Felten et al. 2009; Dharmawardhana et al. 2010). The array has since been redesigned (second generation) to target a manually-curated subset of the predicted *P. trichocarpa* genes using seven probes per gene (NimbleGen design ID 7399, design name 080401). This array targets 52,098 nuclear gene models predicted by the International *Populus* Genome Consortium, including thousands of manual curations. The array also has probes for chloroplast (69) and mitochondrial (58) genes that were predicted using the Oak Ridge National Laboratory prokaryotic pipeline (Badger and Olsen 1999, Delcher et al. 1999), followed by manual curation. In addition, 3,035 probes targeting random, noncoding portions of the *P. trichocarpa* genome were also included as negative controls.

Seven independent 60-mer oligos per transcript were designed according to a NimbleGen design algorithm that considers uniqueness, self-complementarity, and base composition. The seven best probes were initially selected for each target sequence when possible, resulting in a final set of 364,152 probes. For models that had non-unique probes, up to 100 bp of 3' genomic sequences (based on JGI v1.1) were appended to the predicted stop codon as putative untranslated regions to facilitate identification of unique probes. Many of these probes still have > 90% nucleotide identity due to the repetitive nature of the *Populus* genome, so cross-hybridization to non-target sequences is likely to be a problem for some loci.

A new, high-density platform for 2.1 million features (13 μm feature size) was introduced in 2009 that allows multiplexing 12 samples on one slide. The *Populus* version targets 48,146 nuclear genes as well as the organellar genes, using two to three probes per gene (design name 080925). This subset of genes was selected from among those included on the 1-plex array (design 080401) by eliminating gene models with strong similarity to known transposable elements, and which showed no evidence of expression in previous array studies. Furthermore, all probes with > 95% nucleotide identity to nontarget genes were also removed. This design will substantially reduce the cost of replicates in array experiments, thereby enhancing the robustness of experiments. Both the 1-plex and 12-plex poplar arrays are now available under NimbleGen's custom design catalog. Finally, a custom-modified version of the NimbleGen poplar array (GEO GPL7234), with one probe per transcript, has been developed by Matias Kirst at the University of Florida (Ramírez-Carvajal et al. 2009; Drost et al. 2010).

### 5.3.2 *Affymetrix Poplar GeneChip*

The Affymetrix Poplar GeneChip was introduced in 2005, through the Affymetrix GeneChip® Consortia Program. The poplar array is based on Affymetrix array format 49 with an 11-μm feature size, and contains 11 probe pairs per probe set. Probe design was based on JGI release v1.1 and the GenBank *Populus* mRNA and EST sequence collection up to spring 2005, including UniGene Build #6 and over 260,000 ESTs from 13 *Populus* species. As with the other Affymetrix expression array designs, probes were selected from the 3'-end of mRNA/EST or from predicted gene sequences for greater specificity. Whenever the consensus or exemplar mRNA/EST sequences overlap with predicted JGI gene models, the Affymetrix program preferentially targets EST/mRNA sequences for probe selection. Experience from the human, mouse and rat arrays suggested that the EST-based probe design could complement probes derived from the full-length mRNA/gene model collection, and will be more robust over time (Affymetrix, 2006). This is because ESTs often contain untranslated regions that may be absent in mRNA sequences or are not included in current gene model predictions. Therefore, specificity of probes derived from ESTs is less likely to be affected by the continuing growth of full-length mRNA collections in GenBank or in silico gene model predictions over time (see also discussion in the Agilent microarray section below).

The Affymetrix genome array contains 61,251 poplar probe sets, in addition to 62 reporter probe sets and 100 control probe sets. Of the 61,251 poplar probe sets, 47,835 were derived from JGI gene models (v1.0 and v1.1), 7,790 from the UniGene collection, 5,608 from ESTs and 18 from the Affymetrix proprietary database, based on the annotation version na26 (July 2008). Currently, about 63% of the probe sets are classified as Grade A, having nine or more probes from the probe set matching perfectly with a known mRNA or gene model sequence. Approximately 22% of the probe sets fall under Grades B and C, where the probe set's target or consensus sequence overlaps with the predicted transcript sequence, but the probes themselves do not match or overlap with the predicted transcript, presumably due to incomplete 3' UTR prediction. About 6 and 9% of the probe sets are derived from UniGene clusters (Grade E, for EST-only records) and individual ESTs (Grade R, for representative sequence-only records), respectively, and have no corresponding transcript prediction in the current version of poplar genome assembly and annotation. In other words, a total of 50,848 probe sets have transcript assignments corresponding to the JGI poplar genome v1.1, with the remaining 10,403 probes supported by UniGene/mRNA/EST evidence. In sum, the 50,848 probe sets with genome match correspond to 40,236 unique JGI v1.1 gene models.

Part of the probe set redundancy was due to the inclusion of alternative transcripts (especially of EST/mRNA origin) in the target sequence collection for probe design. Uniqueness of the probe set is indicated in the probe ID according to Affymetrix convention. For instance, probe sets that recognize alternative transcripts of the same gene are given the "_a" suffix, while those matching multiple transcripts from separate genes carry the "_s" suffix. A probe set that is annotated with the "_x" suffix may have unpredictable cross-hybridization potential, as it contains some probes that are identical or highly similar to other transcript(s). The Affymetrix poplar array contains a total of 2,868 "_a" probe sets, 9,008 "_s" probe sets, and 2,509 "_x" probe sets.

The Affymetrix poplar array is commercially available, with the annotation regularly updated via the NetAffx Annotation Center. Use of the Affymetrix array has been reported (Qin et al. 2008; Azaiez et al. 2009; Du et al. 2009; Wilkins et al. 2009; Yuan et al. 2009; Guo et al. 2010).

### 5.3.3 Agilent Poplar Array

The Agilent poplar array was introduced in January 2008. The design is based on Agilent's multiplex 4 × 44K platform with a per-array capacity of up to 43,803 probes and additional controls. Because the per-array density is slightly less than the number of JGI predicted nuclear gene models, and because there were known contaminating sequences in draft assembly v1.1, it was necessary to reduce the number of target genes for probe design. For this purpose, predicted poplar proteins that share high homology (BLASTP $E$-value $\leq 1e^{-10}$) with *Arabidopsis* proteins, based on TAIR (The Arabidopsis Information Resources) release 7, were assigned top priority, while those annotated as transposable elements were classified as group 6 (denoting a potentially lowest priority ranking). The rest of the models were searched against the NCBI nonredundant (nr) protein database, and assigned the following group numbers based on the taxonomic origins of the best BLAST hits: (1) plants, (2) fungi, (3) animals, (4) microbes, (5) no hits. As taxonomic data alone are not necessarily reliable indicators of non-poplar sequences, exclusion of gene models also took into account whether they are present in the 19 linkage groups (representing 19 chromosomes) or in the scaffolds (long contiguous sequences that have not been assembled into linkage groups due to sequence gaps). This is because draft assembly v1.1 contains over 22,000 scaffolds, a majority of which are less than 20 kb in length with many representing suspected contaminants or redundant sequences that failed to assemble with their cognate gene models due to sequence quality or other complications (Kelleher et al. 2007). As an additional measure, available expression data obtained from the NimbleGen (Dharmawardhana et al. 2010) and Affymetrix (Yuan et al. 2009) platforms were used to guide removal of "questionable" gene models in priority groups 2 to 6. These expression profiles were generated from a wide range of tissues and treatment conditions, originating from various *Populus* species and genotypes.

For the remaining nuclear gene models, up to 100 bp of 3'-sequences were appended to the predicted stop codon of each gene model to ensure uniform representation of (putative) 3'-UTRs. This was intended to minimize design artifacts (i.e., false "unique probe" selection) observed from a preliminary design where the JGI reference transcript sequences were used. Because of limited cDNA or EST support, averaging 22% of the predicted transcriptome at the time of the ab initio annotation (Tuskan et al. 2006), a majority of the JGI v1.1 reference transcript sequences did not contain 3'-UTRs. Consequently, a 3'-UTR probe might have met the design criteria as gene-specific simply because a 3'-UTR sequence was not present in the paralog sequence of the reference transcript set. Although most probe design algorithms favor 3'-UTRs (or 3'-end of input sequence) for uniqueness, 3'-UTRs are also among the most divergent among closely related *Populus* species and hybrids. Therefore, limiting the appended (putative) 3'-UTR to 100 bp would, in theory, force selection of probes closer to the 3'-ORF, maximizing the cross-species utility of the resultant array.

After the addition of plastidic and mitochondrial gene models, probe design was performed by Agilent and a total of 48,407 probes passed the selection criteria. The probe design algorithms consider probe uniqueness, base composition, and cross-hybridization potential with non-target sequences using a thermodynamic scan (Kronick 2004). The probe number was further reduced based on probe identity to non-target sequences ($\geq$ 95% over the entire 60-mer) in conjunction with scaffold size consideration as described above. Probes corresponding to nuclear gene models that match with organellar genes were removed. In addition, probes derived from Scaffold numbers > 3,500 (corresponding to < 2,402 bp in length) or Scaffold numbers > 1,000 in Groups 2–6 described above were also excluded. The final design for array production contains 43,801 custom features, representing 43,663 nuclear gene models (including two probes corresponding to putative splice variants), 85 chloroplastic gene models, 45 mitochondrial gene models and six commonly used reporter genes in transgenic research. Despite the extensive filtering, numerous spurious or partial gene predictions remained in the target dataset, which could be further excluded in future array design versions. In addition, approximately 20% of the final probe sets are flagged with "cross-hybridization potential" based on Agilent's probe design algorithms. As with the other systems, unequivocal expression comparisons of closely related gene members or paralogs will likely depend on alternative techniques, such as real-time RT-PCR, in post-array analysis.

This poplar array is publicly available to all researchers through the Agilent eArray's "Published Designs" portal. Two design IDs are associated with this array, 018004 and 021876; both have the same probe set, differing from one another only in the randomization of probe physical layout (due to eArray account changes). A new, high-density expression array platform introduced in 2010 accommodates 1 million features, with a 30 μm feature size. A number of multiplexing configurations will be available, including 2 × 400K, 4 × 180K and 8 × 60K, and should reduce the costs associated with microarray experiments.

## 5.4 POParray: A Database for Cross-reference of Multiple Microarray Platforms

As illustrated in the MAQC study, comparison of microarray data from different platforms can be convoluted due to differences in probe design and properties. The situation is further complicated in the case of *Populus* because of the draft nature of the genome annotation. In the *Populus* genome v1.1 release, a single locus may be represented by four, and sometimes as many as 12, gene predictions based on different ab initio gene calling algorithms. Thus, a major discrepancy of the target gene sets between different microarray platforms was due to the use of synonymous gene models (or in some cases, the corresponding ESTs or UniGene clusters) during probe design. Similar discrepancies are also found in the *Populus* literature where different gene models are used by different researchers, making data inference difficult. To facilitate cross-referencing among various *Populus* microarray platforms and JGI gene models, a relational database, POParray (*http://aspendb.uga.edu/poparray*), was developed to enable efficient extraction of probe/gene model information across platforms.

### 5.4.1 Probe Matching Process

A three-step process was employed to develop a master cross-reference table that links platform-specific probe (or probe set) identifiers to a non-redundant (reference) set of JGI v1.1 gene models. Platform-specific probe identifiers, probe sequences and the public identifiers (e.g., JGI v1.1 gene models or EST/UniGene identifiers) of their target genes were used, along with other available annotation information from each microarray system. Platform-specific probes designed for quality control purposes, as well as probes corresponding to reporter genes, and organellar genes (Agilent and NimbleGen) were excluded from this analysis.

1. Probe matching based on the JGI v1.1 gene model or UniGene identifier: In the first step of probe matching, a direct text search of the target gene identifiers among different platforms was performed. This is the simplest case of matching probes from one platform to another. For the

Affymetrix design, one gene model may be represented by more than one probe identifier, and one probe set may correspond to multiple gene models, due to the predicted cross-hybridization potential.

2. Probe matching by synonymous gene model mapping: Some probes could not be matched based on target gene identifiers due to the use of synonymous gene models between platforms. For instance, estExt_fgenesh4_pg.C_LG_II1020 (LG_II:8298968-8301040) and gw1.II.3483.1 (LG_II:8299090–8300967) are two computational predictions of the same locus. Both Affymetrix and NimbleGen used estExt_fgenesh4_pg.C_LG_II1020 in their design whereas Agilent used gw1.II.3483.1. To enable cross-referencing between these gene models, a synonymous gene model table was built based on the genomic location of the gene model predictions. The synonymous gene modeling mapping allowed additional probes to be matched across platforms. It also formed the basis for synonymous gene model findings in the POParray database. For each group of gene models mapped to the same genome locus, the JGI-promoted v1.1 gene model (in some cases, manually curated) is designated as the "preferred" gene model in the data output.

3. Probe-matching based on BLAST: For probes that were designed based on EST or UniGene sequences with no reference JGI v1.1 gene model information, the target sequences were subjected to BLASTN search against the JGI v1.1 reference gene sequences. The best BLAST hit (> 95% identity and < 1e-5) was then used for cross-platform probe matching. The lower E-value cutoff was necessary to match alternatively spliced transcripts (mostly of EST origin) to the cognate gene models, although in most cases, the E-value was well below the cutoff. For this reason, the cross-platform matching results should be used with caution.

Table 5-2 summarizes the results of the cross-platform probe matching, based on JGI v1.1 gene model mapping alone (steps 1 and 2 above). A total of 33,568 non-redundant gene models are represented in all three microarray platforms, and an additional 16,183 gene models are common between at least two platforms.

### 5.4.2 POParray Web Interface: Search Options and Data Access

The web interface of the POParray database provides two options for probe search and cross-reference between platforms. The first is to search with the probe ID of any microarray platform as query. The second option allows the use of gene models as queries to find matching probe IDs in the three microarray platforms, as well as the synonymous gene models. In addition, probe annotation information for all three platforms can be obtained using the probe ID as query. As newer (next-generation) microarray platforms arrive and the poplar genome annotation continues to improve, it becomes all the more important to maintain and regularly update the POParray database for cross-referencing between different identifiers.

**Table 5-2** Summary of probe cross-referencing among the three *Populus* microarrays based on JGI *Populus* genome v1.1 gene model match.

| Platform | Affymetrix | Agilent | NimbleGen[1] |
|---|---|---|---|
| Number of poplar probes or probe sets (including all nuclear and organellar genes) | 61,251 | 43,795 | 65,965 |
| Number of non-redundant, nuclear gene models represented[2] | 40,236 (50,848) | 43,663 (43,663) | 59,216 (59,216) |
| Probes present in all 3 platforms[3] | 45,709 (75%) | 33,568 (77%) | 33,568 (51%) |
| Probes uniquely present[3] | 94 (0.2%) | 79 (0.2%) | 9,769 (15%) |
| Probes common in only two platforms[3] | 5,045 (8%) | 10,016 (23%) | 15,879 (24%) |

[1]The analysis was performed using the first-generation, 385K NimbleGen *Populus* array.
[2]Numbers in parentheses indicate corresponding probe numbers.
[3]Numbers in parentheses indicate percentage of total poplar probes for each platform.

In closing, the *Populus* community will likely witness an upsurge of whole-genome oligo array data from a wide range of studies over the next few years. A grand challenge to *Populus* researchers is to reconcile different gene expression analysis platforms, including emerging 454-, Illumina- or SOLiD-based RNA-Seq approaches, toward building an expression omnibus to facilitate comparative transcriptomics analyses. Relative to *Arabidopsis*, such efforts are inherently more difficult with *Populus*, due to the greater degree of species divergence and variations in experimental conditions between studies. It should be noted that probe-matching alone does not necessarily imply corroborated expression data. Nevertheless, these transcriptomics resources will be of tremendous value in our quest of addressing the molecular mechanisms underlying the perennial growth habit and life cycle of woody species.

## Acknowledgments

The authors wish to thank Scott Harding for critical review and comments on an earlier version of the manuscript, and Amy Brunner for sharing data. The POParray database development was supported in part by a grant (DBI-0421756) from the National Science Foundation Plant Genome Program to CJT, and the first-generation NimbleGen array development was supported by grants from the US Department of Energy, Office of Science, Biological and Environmental Research (ERKP447) to GAT. Oak Ridge National Laboratory is managed by UT-Battelle, LLC, for the U.S. Department of Energy under contract DE-AC05-00OR22725.

# References

Affymetrix (2006) Transcript assignment for NetAffx™ annotations Affymetrix GeneChip® IVT Array Whitepaper Collection. Revision Version 2.3.

Azaiez A, Boyle B, Levée V, Séguin A (2009) Transcriptome profiling in hybrid poplar following interactions with *Melampsora* rust fungi. Mol Plant Microbe Interact 22: 190–200.

Badger JH, Olsen GJ (1999) CRITICA: coding region identification tool invoking comparative analysis. Mol Biol Evol 16: 512–524.

Delcher AL, Harmon D, Kasif S, White O, Salzberg SL (1999) Improved microbial gene identification with GLIMMER. Nucl Acids Res 27: 4636–4641.

Dharmawardhana P, Brunner AM, Strauss SH (2010) Genome-wide transcriptome analysis of the transition from primary to secondary stem development in *Populus trichocarpa*. BMC Genomics 11: 150.

Drost DR, Benedict CI, Berg A, Novaes E, Novaes CRDB, Yu Q, Dervinis C, Maia JM, Yap J, Miles B, Kirst M (2010) Diversification in the genetic architecture of gene expression and transcriptional networks in organ differentiation of *Populus*. Proc Natl Acad Sci USA 107: 8492–8497.

Du J, Mansfield SD, Groover AT (2009) The *Populus* homeobox gene ARBORKNOX2 regulates cell differentiation during secondary growth. Plant J 60: 1000–1014.

Felten J, Kohler A, Morin E, Bhalerao RP, Palme K, Martin F, Ditengou FA, Legué V (2009) The ectomycorrhizal fungus Laccaria bicolor stimulates lateral root formation in poplar and *Arabidopsis* through auxin transport and signaling. Plant Physiol 151: 1991–2005.

Groover A, Mansfield S, DiFazio S, Dupper G, Fontana J, Millar R, Wang Y (2006) The *Populus* homeobox gene ARBORKNOX1 reveals overlapping mechanisms regulating the shoot apical meristem and the vascular cambium. Plant Mol Biol 61: 917–932.

Gou J Strauss SH, Tsai CJ, Fang K, Chen Y, Jiang X, Busov VB (2010) Gibberellins regulate lateral root formation in *Populus* through interactions with auxin and other hormones. Plant Cell 22:623–639.

Guo Z, Guilfoyle RA, Thiel AJ, Wang R, Smith LM (1994) Direct fluorescence analysis of genetic polymorphisms by hybridization with oligonucleotide arrays on glass supports. Nucl Acids Res 22: 5456–5465.

Hertzberg M, Aspeborg H, Schrader J, Andersson A, Erlandsson R, Blomqvist K, Bhalerao R, Uhlen M, Teeri TT, Lundeberg J, Sundberg B, Nilsson P, Sandberg G (2001) A transcriptional roadmap to wood formation. Proc Natl Acad Sci USA 98: 14732–14737.

Hughes TR, Mao M, Jones AR, Burchard J, Marton MJ, Shannon KW, Lefkowitz SM, Ziman M, Schelter JM, Meyer MR, Kobayashi S, Davis C, Dai H, He YD, Stephaniants SB, Cavet G, Walker WL, West A, Coffey E, Shoemaker DD, Stoughton R, Blanchard AP, Friend SH, Linsley PS (2001) Expression profiling using microarrays fabricated by an ink-jet oligonucleotide synthesizer. Nat Biotechnol 19: 342–347.

Kelleher CT, Chiu R, Shin H, Bosdet IE, Krzywinski MI, Fjell CD, Wilkin J, Yin TM, DiFazio SP, Ali J, Asano JK, Chan S, Cloutier A, Girn N, Leach S, Lee D, Mathewson CA, Olson T, O'Connor K, Prabhu AL, Smailus DE, Stott JM, Tsai M, Wye NH, Yang GS, Zhuang J, Holt RA, Putnam NH, Vrebalov J, Giovannoni JJ, Grimwood J, Schmutz J, Rokhsar D, Jones SJM, Marra SM, Tuskan GA, Bohlmann J, Ellis BE, Ritland K, Douglas CJ, Schein JE (2007) A physical map of the highly heterozygous *Populus* genome: integration with the genome sequence and genetic map and analysis of haplotype variation. Plant J 50:1063–1078.

Kalluri U, DiFazio S, Brunner A, Tuskan G (2007) Genome-wide analysis of Aux/IAA and ARF gene families in *Populus trichocarpa*. BMC Plant Biol 7: 59.

Kohler A, Delaruelle C, Martin D, Encelot N, Martin F (2003) The poplar root transcriptome: analysis of 7000 expressed sequence tags. FEBS Lett 542: 37–41.

Kohler A, Rinaldi C, Duplessis S, Baucher M, Geelen D, Duchaussoy F, Meyers B, Boerjan W, Martin F (2008) Genome-wide identification of NBS resistance genes in *Populus trichocarpa*. Plant Mol Biol 66: 619–636.

Kronick MN (2004) Creation of the whole human genome microarray. Expert Rev Proteom 1: 19–28.

Lockhart DJ, Dong H, Byrne MC, Follettie MT, Gallo NV, Chee MS, Mittmann M, Wang C, Kobayashi M, Norton H, Brown EL (1996) Expression monitoring by hybridization to high-density oligonucleotide arrays. Nat Biotechnol 14: 1675–1680.

MAQC (2006) The MicroArray Quality Control (MAQC) project shows inter- and intraplatform reproducibility of gene expression measurements. Nat Biotechnol 24: 1151–1161.

Patterson TA, Lobenhofer EK, Fulmer-Smentek SB, Collins PJ, Chu TM, Bao W, Fang H, Kawasaki ES, Hager J, Tikhonova IR, Walker SJ, Zhang L, Hurban P, de Longueville F, Fuscoe JC, Tong W, Shi L, Wolfinger RD (2006) Performance comparison of one-color and two-color platforms within the MicroArray Quality Control (MAQC) project. Nat Biotechnol 24: 1140–1150.

Pertea G, Huang X, Liang F, Antonescu V, Sultana R, Karamycheva S, Lee Y, White J, Cheung F, Parvizi B, Tsai J, Quackenbush J (2003) TIGR Gene Indices clustering tools (TGICL): a software system for fast clustering of large EST datasets. Bioinformatics 19: 651–652.

Qin HZ, Feng T, Harding SA, Tsai CJ, Zhang SL (2008) An efficient method to identify differentially expressed genes in microarray experiments. Bioinformatics 24: 1583–1589.

Ralph S, Oddy C, Cooper D, Yueh H, Jancsik S, Kolosova N, Philippe RN, Aeschliman D, White R, Huber D, Ritland CE, Benoit F, Rigby T, Nantel A, Butterfield YSN, Kirkpatrick R, Chun E, Liu J, Palmquist D, Wynhoven B, Stott J, Yang G, Barber S, Holt RA, Siddiqui A, Jones SJM, Marra MA, Ellis BE, Douglas CJ, Ritland K, Bohlmann J (2006) Genomics of hybrid poplar (*Populus trichocarpa* x *deltoides*) interacting with forest tent caterpillars (*Malacosoma disstria*): normalized and full-length cDNA libraries, expressed sequence tags, and a cDNA microarray for the study of insect-induced defences in poplar. Mol Ecol 15: 1275–1297.

Ramírez-Carvajal GA, Morse AM, Dervinis C, Davis JM (2009) The cytokinin type-B response regulator PtRR13 is a negative regulator of adventitious root development in *Populus*. Plant Physiol 150: 759–771.

Ranjan P, Kao YY, Jiang HY, Joshi CP, Harding SA, Tsai CJ (2004) Suppression subtractive hybridization-mediated transcriptome analysis from multiple tissues of aspen (*Populus tremuloides*) altered in phenylpropanoid metabolism. Planta 219: 694–704.

Rinaldi C, Kohler A, Frey P, Duchaussoy F, Ningre N, Couloux A, Wincker P, Le Thiec D, Fluch S, Martin F, Duplessis S (2007) Transcript profiling of poplar leaves upon infection with compatible and incompatible strains of the foliar rust *Melampsora larici populina*. Plant Physiol 144: 347–366.

Sterky F, Bhalerao RR, Unneberg P, Segerman B, Nilsson P, Brunner AM, Charbonnel-Campaa L, Lindvall JJ, Tandre K, Strauss SH, Sundberg B, Gustafsson P, Uhlen M, Bhalerao RP, Nilsson O, Sandberg G, Karlsson J, Lundeberg J, Jansson S (2004) A *Populus* EST resource for plant functional genomics. Proc Natl Acad Sci USA 101: 13951–13956.

Tuskan GA, Gunter LE, Yang ZK, Yin T, Sewell MM, DiFazio SP (2004) Characterization of microsatellites revealed by genomic sequencing of *Populus trichocarpa*. Can J For Res 34: 85–93.

Tuskan GA, DiFazio S, Jansson S, Bohlmann J, Grigoriev I, Hellsten U, Putnam N, Ralph S, Rombauts S, Salamov A, Schein J, Sterck L, Aerts A, Bhalerao RR, Bhalerao RP, Blaudez D, Boerjan W, Brun A, Brunner A, Busov V, Campbell M, Carlson J, Chalot M, Chapman J, Chen GL, Cooper D, Coutinho PM, Couturier J, Covert S, Cronk Q, Cunningham R, Davis J, Degroeve S, Dejardin A, dePamphilis C, Detter J, Dirks B, Dubchak I, Duplessis S, Ehlting J, Ellis B, Gendler K, Goodstein D, Gribskov M, Grimwood J, Groover A, Gunter L, Hamberger B, Heinze B, Helariutta Y, Henrissat B, Holligan D, Holt R, Huang W, Islam-Faridi N, Jones S, Jones-Rhoades M, Jorgensen R, Joshi C, Kangasjarvi J, Karlsson J, Kelleher C, Kirkpatrick R, Kirst M, Kohler A, Kalluri U, Larimer F, Leebens-Mack J, Leple JC, Locascio P, Lou Y, Lucas S, Martin F, Montanini B, Napoli C, Nelson DR, Nelson C, Nieminen K, Nilsson O, Pereda V, Peter G, Philippe R, Pilate G, Poliakov A,

Razumovskaya J, Richardson P, Rinaldi C, Ritland K, Rouze P, Ryaboy D, Schmutz J, Schrader J, Segerman B, Shin H, Siddiqui A, Sterky F, Terry A, Tsai CJ, Uberbacher E, Unneberg P, Vahala J, Wall K, Wessler S, Yang G, Yin T, Douglas C, Marra M, Sandberg G, Van de Peer Y, Rokhsar D (2006) The Genome of Black Cottonwood, Populus trichocarpa (Torr. & Gray). Science 313: 1596–1604.

Yang X, Kalluri UC, Jawdy S, Gunter LE, Yin T, Tschaplinski TJ, Weston DJ, Ranjan P, Tuskan GA (2008) The F-box gene family is expanded in herbaceous annual plants relative to woody perennial plants. Plant Physiol 148: 1189–1200.

Yuan Y, Chung JD, Fu X, Johnson VE, Ranjan P, Booth SL, Harding SA, Tsai CJ (2009) Alternative splicing and gene duplication differentially shaped the regulation of isochorismate synthase in *Populus* and *Arabidopsis*. Proc Natl Acad Sci USA 106: 22020–22025.

Wilkins O, Nahal H, Foong J, Provart NJ, Campbell MM (2009) Expansion and diversification of the *Populus* R2R3-MYB family of transcription factors. Plant Physiol 149: 981–993.

Winter D, Vinegar B, Nahal H, Ammar R, Wilson GV, Provart NJ (2007) An "Electronic Fluorescent Pictograph" browser for exploring and analyzing large-scale biological data sets. PLoS ONE 2: e718.

# 6

# Poplar Proteomics: Update and Future Challenges

*Christophe Plomion,[1,a,]\* Delphine Vincent,[1,b] Frank Bedon,[1,c] Johann Joets,[2] Ludovic Bonhomme,[3] Domenico Morabito,[4] Sébastien Duplessis,[5] Robert Nilsson,[6,d] Gunnar Wingsle,[6,e] Christer Larsson,[7] Yves Jolivet,[8] Jenny Renaut,[9] Olga Pechanova[10] and Cetin Yuceer[11]*

## ABSTRACT

Poplar functional genomics is in its infancy. This statement particularly holds true for proteomics, the topic of this chapter. We will review the methodological developments that have taken place during the last decade and used by researchers to understand how forest trees develop and adapt to a changing environment over their entire life span. We will also present emerging issues of poplar proteomics and our vision toward the discovery about the cell's protein machinery that could yield important applications in forestry in terms of wood productivity and adaptability in the frame of global change.

**Key words**: proteomics; poplar; 2-DE; mass spectrometry; sub-proteome; abiotic stress, biotic stress; wood formation; database

## 6.1 Introduction

Integrated approaches that combine the systematic sequencing of expressed genes and the analysis of mRNA expression for a large number of genes are now considered strategies for tracking the genes of interest. However, while the poplar genome (*Populus* spp.; Tuskan et al. 2006) provides us with an overview of the genes in trees, it is becoming increasingly clear that this is only a very fragmentary beginning of understanding their role and function. The old paradigm of one gene for one protein is no longer valid, and in eukaryotic cells the situation is more like 6-8 proteins per gene. Thus, while there may be only 25,000 to 100,000 genes in a plant species, there

For affiliations see at the end of this chapter on page 165

are many more resultant proteins, including splice variants and essential post-transcriptional/translational modifications. Genome information is not sufficient to define all of these protein components. For example, studying mRNA quantities provides only a partial view of gene product expression due to: (i) large differences between mRNA and protein turnover (i.e., a protein can still be abundant while the mRNA is no longer detectable because its synthesis has stopped), (ii) post-translational modifications of proteins such as removal of signal peptides, phosphorylation and glycosylation, which play important roles in their activity and sub-cellular localization, and (iii) complex interactions with other proteins. These processes cannot be deduced from microarray or nucleic acid-based methodologies. The problem becomes even more complicated, because complex networks of proteins can be very divergent in different organs or developmental phases of the same organism, despite the same genomic information.

Given that the genetic information is only indicative of the cell's potential and does not reflect the actual state in a particular cell at a given time, the concept of "proteome" (protein complement expressed by a genome; Wilkins et al. 1996a, b), has emerged to provide complementary and critical information by revealing the regulation, activities, quantities and interaction of proteins in cells, as well as how their abundance responds to developmental and environmental signals. Consequently, proteomics is now considered a priority by many universities and research institutes and is starting to be widely applied to the model plants *Arabidopsis* and rice (Agrawal and Rakwal 2006), as well as to other important crop species (Jorrin et al. 2007; Carpentier et al. 2008). However, forest tree proteomics in general and more specifically poplar proteomics still remains largely embryonic, despite the fact that a large collection of expressed sequence tags (ESTs; Sterky et al. 2004; Ralph et al. 2006) and the genome sequence of poplar are available, which greatly simplify the identification of proteins from mass spectrometry data. The first large poplar proteomic analysis dates back to 2000 (van der Mijnsbrugge et al. 2000), but most of the papers have been published since 2006, showing clearly that this field is in its infancy compared with proteomic analyses of other plant species. Yet, the last three years have seen considerable progress in developing the field of poplar proteomics both in its technological and functional dimensions.

In this chapter we will review for the first time, the contribution of some of the major research groups involved in poplar proteomics. We will focus on published results. This chapter is divided into several sections, including methodology developments (from sample preparation to protein identification), sub-cellular and tissue-specific proteomes, organ development with emphasis on xylogenesis, and identification of abiotic and biotic stress responding-proteins. Then, we will summarize the main challenges of poplar proteomics and our vision toward the discovery about the cell's protein machinery in this model tree that could yield important applications in forestry. Although unpublished, it is worth acknowledging the contribution of Oak Ridge National Laboratory (ORNL, Oak Ridge, TN, USA), through a systems biology strategy, and Pacific Northwest National Laboratory (PNNL, Richland, WA, USA), using a more targeted approach, to poplar proteomics. We expect that the present contribution will provide molecular biologists a backbone for future poplar proteomics.

## 6.2 Technology Used For Poplar Proteomics

In this section, we will review the technologies favoured by the poplar proteomics community to recover, analyze, and identify soluble proteins expressed in different tissues/developmental stages or else responding to different abiotic and biotic stresses. A summary of all the specifics adopted in the proteomic studies presented here is provided in Table 6-1.

### 6.2.1 Protein Extraction

The extraction method greatly depends on the tissue considered and must be optimized with respect to the downstream analytical procedures. Most of the proteomic studies published so far on poplar have been carried out on leaves (Renaut et al. 2004, 2005; Bohler et al. 2007; Bogeat-Triboulot et al. 2007; Kieffer et al. 2008; He et al. 2008; Yuan et al. 2008). A significant body of data has also been obtained on wood forming tissues (van der Mijnsbrugge et al. 2000; Du et al. 2006). In this context, a global proteomic characterization of various poplar tissues (vegetative and reproductive) collected at different developmental stages was accomplished by Plomion et al. (2006). In this work, the same protocol was applied to all tissues, thereby enabling their comparison, to the detriment of optimal conditions for a specific tissue.

The most critical steps during the extraction of proteins include protein precipitation to remove non-protein contaminants and subsequently protein resuspension in an appropriate solution that is compatible with the follow-up protein resolving techniques. Both procedures cause protein losses, which must be minimized along with the interfering compounds.

### 6.2.2 Protein Precipitation

Proteins must first be released from cells for proper extraction. Plant cells, particularly in woody species, are protected by a highly resistant cell wall, which requires a strong mechanical action to be disrupted. Cell disruption in poplar was exclusively achieved by grinding plant material in liquid nitrogen using a mortar and pestle. Not only does this method ensure proper crushing of plant tissues, but the extremely low temperatures prevent protein modification by inhibiting hydrolytic enzyme activities.

**Table 6-1** Features of the analytical processes of the various publications on poplar proteomics based on 2-DE.

| 1st author | Bogeat-Triboulot | Bohler | Du | He | Kieffer | Plomion | Renaut | van der Mijnsbrugge | Yuan |
|---|---|---|---|---|---|---|---|---|---|
| year | 2007 | 2007 | 2006 | 2008 | 2008 | 2006 | 2004 | 2000 | 2008 |
| context | water deficit | ozone | SVS regeneration | drought, high temperature | cadmium | drought | chilling | wood formation | fungal diseases |
| organ | leaf | leaf | cambium | leaf | leaf | leaf, root, xylem, cambium, bud, inflorescence, | leaf | xylem | leaf |
| protein extraction | TCA/acetone | TCA/acetone | TCA/acetone | TCA/acetone | TCA/acetone | TCA/acetone | TCA/acetone | Phenol | TCA/acetone |
| sample homogenization | no | 50mM Tris-HCl, 25mM EDTA, 500mM thiourea, 0.5% DTT | 60mM Tris-HCl pH 6.8, 0.5% SDS, 10% glycerol, 5% 2-ME, 10% PVPP | no | 50mM Tris-HCl, 25mM EDTA, 500mM thiourea, 0.5% DTT | no | no | 5% PVPP, 0.7M sucrose, 40mM DTT, 50mM boric acid, 50mM ascorbic acid, pH9 | no |
| protein precipitation | cold 20% TCA/0.07% 2-ME/acetone | cold 20% TCA/0.1% DTT/acetone | cold acetone | TCA/acetone | cold 20% TCA/0.1% DTT/acetone | cold 10% TCA/0.07% 2-ME/acetone | cold 20% TCA/0.07% 2-ME/acetone | cold 0.1 M ammonium acetate/methanol | cold 10% TCA/0.07% DTT/acetone |
| pellet rinsing | cold 0.07% 2-ME/acetone | cold 0.1% DTT/acetone | cold acetone | N.P. | cold 0.1% DTT/acetone | cold 0.07% 2-ME/acetone | cold 0.07% 2-ME/acetone | cold methanol/acetone | cold 0.07% DTT/acetone |
| protein solubilization | 7M urea, 2M thiourea, 4% CHAPS, 30mM Tris | 7M urea, 2M thiourea, 4% CHAPS, 30mM tris | 8M urea, 4% CHAPS | N.P. | 7M urea, 2M thiourea, 4% CHAPS, 30mM tris | 7M urea, 2M thiourea, 0.4% Triton X-100, 4% CHAPS, 10mM DTT, 1% ampholytes | 8M urea, 2% CHAPS, 2% ampholytes | N.P. | 9M urea, 4% CHAPS, 1mM PMSF, 10mM DTT, 1% ampholytes |

*Table 6-1 contd....*

*Table 6-1 contd....*

| 1st author | Bogeat-Triboulot | Bohler | Du | He | Kieffer | Plomion | Renaut | van der Mijnsbrugge | Yuan |
|---|---|---|---|---|---|---|---|---|---|
| protein separation | 2-DE | 2-DE | 2-DE | 2-DE | 2-DE | 2-DE | 2-DE | 2-DE | 2-DE |
| protein labelling | CyDyes 2, 3, 5 | CyDyes 2, 3, 5 | no | no | CyDyes 2, 3, 5 | no | no | no | no |
| 1st dimension pH gradient | IPG-IEF 4-7 | IPG-IEF 4-7 | IPG-IEF 4-7 | IPG-IEF 4-7 | IPG-IEF 4-7 | IPG-IEF 4-7 | IPG-IEF 4-5, 4.5-5.5, 5-6 | CA-IEF 3-10 | IPG-IEF 4-7 |
| length (cm) | 24 | 24 | 18 | 18 | 24 | 24 | 24 | 12 | 17 |
| loading method | cup loading | cup loading | in-gel rehydration | in-gel rehydration | cup loading | in-gel rehydration | cup loading | cathodic end | in-gel rehydration |
| protein content | 3x50µg | 3x30µg | 1200µg | 2000µg | 3x30µg + 300µg | 300µg | 170µg | N.P. | 600µg |
| 2nd dimension % acrylamide | SDS-PAGE 12.5% | SDS-PAGE 8-14% | SDS-PAGE 15% | SDS-PAGE 15% | SDS-PAGE 8-14% | SDS-PAGE 12% | SDS-PAGE 12.5% | SDS-PAGE 12% | SDS-PAGE 12% |
| staining | no | no | CBB R-250 | CBB R-250 | Sypro Ruby | CBB G-250 | AgNO3 | CBB | CBB G-250 |
| acquisition | Typhoon Variable Mode Imager | Typhoon Variable Mode Imager | UMAX Powerlook 2100 | UMAX Powerlook 2100 | Typhoon Imager 9400 | M141 Image Scanner | Image Scanner II | N.P. | GS-800 Calibrated Densitometer |
| image analysis | Decyder | Decyder | Image Master 2D Elite | Image Master 2D Platinum | Decyder | Image Master 2D Elite | Image Master 2D Elite | N.P. | PDQuest |
| detected spots | N.P. | 1900 | | >1000 | ~1000 | Tissue-dependent | >1500 | N.P. | 550 |
| reproducible spots | N.P. | 529 | | N.P. | 717 | Tissue-dependent | 800 | N.P. | 500 |
| responsive proteins | 375 | 147 | 550 | 26 | 194 | | 60 | 40 | 40 |

| 1st author | Bogeat-Triboulot | Bohler | Du | He | Kieffer | Plomion | Renaut | van der Mijnsbrugge | Yuan |
|---|---|---|---|---|---|---|---|---|---|
| protein identification | MALDI-TOF-MS + MALDI-TOF-MS/MS | MALDI-TOF-MS + MALDI-TOF/TOF-MS/MS | MALDI-TOF-MS | MALDI-TOF/TOF-MS | MALDI-TOF/TOF-MS | nLC-ESI-MS/MS + MALDI-TOF-MS | MALDI-TOF-MS | Edman micro-sequencing | MALDI-TOF-MS |
| searched databases | SP + TrEMBL + nr viridiplantae | SP + TrEMBL + nr viridiplantae | nr + SP | nr viridiplantae | NCBI *Populus* ESTs | dbEST *Populus* ESTs + first draft of the JGI *Populus* genome assembly | SP + TrEMBL viridiplantae | *Populus trichocarpa* ESTs | nr viridiplantae |
| analyzed proteins (a) | 100 | 144 | 550 | 26 | 194 | 398 + 320 | 60 | 15+25 | 40 |
| proteins with hit (including unknown function) | N.P. | N.P. | 244 | 26 | N.P. | N.P. + 163 | N.P. | 1 | 40 |
| identified proteins (known function only) (b) | 39 | 71 | 199 | 24 | 118 | 363 + 83 | 26 | 14+5 | 37 |
| identification rate (b/a) | 39.0% | 49.3% | 39.8% | 92.3% | 60.8% | 91.2% + 43% | 43.3% | 93.3% + 20% | 92.5% |

Once the cell content is liberated, proteins must be isolated by precipitation while the other compounds remain in solution. It is worth mentioning that all of the poplar proteomic projects reviewed in this chapter were carried out under denaturing conditions. No native proteins, protein complexes, or polymers have been studied.

The majority of poplar studies have employed a trichloroacetic acid (TCA)/acetone extraction procedure, originally developed by Damerval et al. (1986), albeit with some modifications. This procedure is not only highly efficient on photosynthetically active organs and/or young tissues, but also rapid and easy to perform, hence enabling its broad use in proteomics. The process involves protein denaturation by the reducing reagent 2-mercaptoethanol (0.07% 2-ME) and precipitation by both the strong acid trichloroacetic acid (10% TCA) and the organic solvent acetone. The modifications brought to the initial protocol are outlined in Table 6-1. For example, the TCA concentration was either increased, and/or 2-ME was replaced with another reducing reagent dithiothreitol (DTT), which is less toxic and less potent (Renaut et al. 2004, 2005; Bohler et al. 2007; Bogeat-Triboulot et al. 2007; Kieffer et al. 2008; Yuan et al. 2008). The final modification consisted of homogenization of the ground material in Tris buffer prior to TCA/acetone precipitation (Bohler et al. 2007; Kieffer et al. 2008). A variation of this technique was introduced by Du et al (2006), which involves an initial solubilization of ground tissues in a different Tris buffer prior to cold acetone precipitation (without the addition of TCA or 2-ME).

An alternative to the TCA/acetone procedure is the phenol/ammonium acetate extraction protocol developed by Hurkman and Tanaka (1986), which proved highly successful on recalcitrant plant material. This more sophisticated, time-consuming, and toxic procedure is based on a phase-partition method during which proteins are first recovered in a phenol phase settling above an aqueous phase, and then precipitated by 0.1% ammonium acetate/methanol. van der Mijnsbrugge et al. (2000) modified the aqueous homogenization buffer to extract proteins from xylem tissues (Table 6-1). While TCA/acetone proved suitable for young white roots (Plomion et al. 2006), it failed to produce the expected protein patterns from mature roots (D. Vincent et al. unpublished). The phenol/ammonium acetate extraction method overcame this problem as can be seen in Fig. 6-1. The anionic detergent sodium deoxycholate (DOC) can also be added to the aqueous homogenization buffer to improve protein extraction, since it remains in the aqueous phase and does not interfere with the downstream isoelectric focusing (IEF).

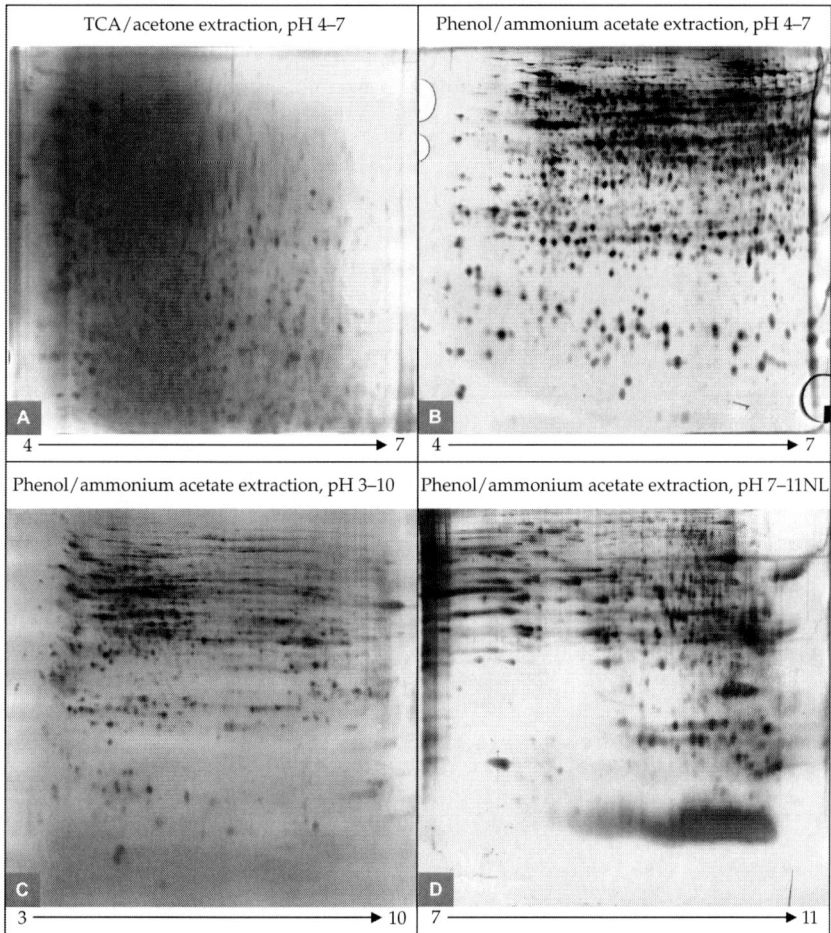

**Figure 6-1** Effect of protein extraction method and pH gradient on poplar mature roots.
A/ TCA/acetone extraction method. B-D/ phenol/ammonium acetate extraction method.
A-B/ 4-7 pH gradient. C/ 3–10 pH gradient. D/ 4-7NL pH gradient.

## 6.2.3 Protein Solubilization

Once pelleted, proteins are recovered by resuspension in a solution compatible with the subsequent separating tools. An ideal resuspension solution comprises chaotropes (urea, thiourea), reducing reagents (2-ME, DTT), detergents (3-[(3-Cholamidopropyl)dimethylammonio]-1-propanesulfonate (CHAPS), Triton X-100), and other chemicals such as anti-proteases (phenylmethanesulphonylfluoride, PMSF), broad spectrum cocktail), anti-oxidases (polyvinyl pyrrolidone, PVP), chelators (EDTA), buffering compounds (Tris), and ampholytes, which are all to ensure protein integrity and solubility, albeit with the loss of native conformation.

Various resuspension solutions have been used on different poplar tissues (Table 6-1). Some were quite simple, composed only of urea, CHAPS, and ampholytes (Renaut et al. 2004, 2005; Du et al. 2008), whereas others were slightly more complex, consisting of PMSF and DTT (Yuan et al. 2008). Urea was often combined with thiourea after Rabilloud (1998) demonstrated its advantages upon using IPG-IEF. A low concentration of Tris was also included in the solution to resolubilize leaf proteins (Bohler et al. 2007; Bogeat-Triboulot et al. 2007; Kieffer et al. 2008). On poplar roots, Vincent et al. (unpublished) combined 1% DTT and 1% 2-ME. Plomion et al. (2006) used a reagent developed by Gion et al. (2005), combining Triton X-100 with CHAPS to resuspend the proteins from different tissues. Using various detergents and/or reducing reagents appear to be particularly efficient for tissues from trees (Valcû and Schlink 2006a, b). Surprisingly, very few studies have reported the use of protease inhibitors. Yuan et al. (2008) used PMSF. On mature roots, 0.5%, proteinase inhibitor mix (GE Healthcare, Uppsala, Sweden) was used (D. Vincent et al. unpublished).

### 6.2.4 Protein Separation And Quantitation

Poplar proteomic studies published thus far have used gel-based strategies, particularly two-dimensional gel electrophoresis (2-DE) (O'Farrell 1975). Although many gel-free MS-based technologies with quantitative analyses have recently emerged, 2-DE remains the preferred approach for protein profiling, except that Renaut et al. (2005) used 1-DE (Laemmli 1970) in combination with Western blotting (Burnette 1981) to track the amount of three proteins during cold acclimation.

### 6.2.5 DIGE Labelling

Three studies on poplar leaves have adopted difference gel electrophoresis (DIGE) technology (Unlü et al. 1997) to follow protein expression under ozone treatment (Bohler et al. 2007), water deficit (Bogeat-Triboulot et al. 2007), and cadmium exposure (Kieffer et al. 2008). In this strategy, depending on the amount of material available, two to three protein extracts were separately labelled with fluorescent CyDyes™, and pooled together prior to classical 2-DE. A scanner amenable to fluorescence was then used to digitize gel images. The fact that gel staining is not needed saves time. The other benefit of DIGE labelling is high reproducibility of IEF and sodium dodecyl sulphate-polyacrylamide gel electrophoresis (SDS-PAGE) steps, greatly facilitating image comparisons.

## 6.2.6 Two-Dimensional Gel Electrophoresis

Most poplar proteomic results were obtained using Immobilized pH gradient-isoelectric focusing (IPG-IEF) during the first dimension of 2-DE, with the exception of van der Mijnsbrugge et al. (2000) who used Carrier-Ampholyte (CA)-IEF (12 cm) to separate xylem proteins. The majority of studies have focused on acidic proteins (pH 4-7), which were resolved on either 18 cm or 24 cm IPG strips (Du et al. 2006; Plomion et al. 2006; Bogeat-Triboulot et al. 2007; Bohler et al. 2007; He et al. 2008; Kieffer et al. 2008; Yuan et al. 2008). Basic IPG strips (pH 7-11NL) also successfully resolved alkaline proteins from poplar mature roots (D. Vincent et al. unpublished), and the gain in spot resolution upon using a narrow range gradient is illustrated in Fig. 6-1. This was also demonstrated by Renaut et al. (2004) who utilized overlapping narrow range IPG strips (24 cm, pH 4–5, 4.5–5.5 and 5–6) to separate poplar leaf proteins.

The second dimension of 2-DE was obtained with only slight variations from one study to another, generally using 12% (van der Mijnsbrugge et al. 2000; Renaut et al. 2004; Plomion et al. 2006; Bogeat-Triboulot et al. 2007; Yuan et al. 2008) or 15% (Du et al. 2006; He et al. 2008) acrylamide gels to resolve proteins whose molecular weights (MW) fall within the 10–150 kDa range. Alternatively, an 8–14% gradient (Bohler et al. 2007; Kieffer et al. 2008) was used to separate proteins with MW outside the 15–150 kDa range, which reduced the resolution. Likewise, the staining procedures were quite consistent across all the studies that used CBB implemented by either R-250 (Du et al. 2006; He et al. 2008) or G-250 (van der Mijnsbrugge et al. 2000; Plomion et al. 2006; Yuan et al. 2008). Silver nitrate was also used to a lesser degree (Renaut et al. 2004; D. Vincent., unpublished data; Fig. 6-1). Fluorescent dyes (Ruby Sypro, Bio-Rad) were only used in one instance by Kieffer et al. (2008).

## 6.2.7 Image Analysis

A fairly common feature in proteomics is the use of image analysis software to detect and quantify 2-D protein spots. Most groups have used Image Master 2D Elite/Platinum (Amersham Biosciences) to analyze 2-D images (Renaut et al. 2004; Du et al. 2006; Plomion et al. 2006; He et al. 2008), except that Yuan et al. (2008) used PDQuest (Bio-Rad), whereas Bohler et al. (2007), Bogeat-Triboulot et al. (2007), and Kieffer et al. (2008) used Decyder (GE Healthcare), which is amenable to fluorescent dyes. It would be interesting to estimate the impact of detection and quantitation algorithms by comparing the quantitative data obtained using the same set of images across several software packages.

The 2-D patterns and number of detected spots varied by organ, extraction method, resolving conditions, staining procedure, and image analysis software. In general, the use of IPG-IEF resolves at least 1,000 spots, especially when only acidic pH ranges are considered. Among all the detected spots, only the reproducible ones present on all of the biological/technical replicates are used for the subsequent statistical analyses. Due to technical difficulties, the number of reproducible spots can be quite different, as evidenced by Bohler et al. (2007) who detected 1,900 spots but retained only 529 spots for statistical analyses, despite the use of highly reproducible 2-D DIGE technology.

## 6.2.8 Protein Identification

Aside from van der Mijnsbrugge et al. (2000) who applied Edman degradation (Edman 1949) to micro-sequence proteins of interest, mass spectrometry (MS) has been the method of choice to identify poplar proteins thus far. Methods utilized for peptide mass fingerprinting (PMF) include matrix-assisted laser desorption ionization-time of flight-mass spectrometry (MALDI-TOF-MS) or tandem MS (MS/MS) through MALDI-TOF/TOF-MS or nanoscale liquid chromatography-electrospray ionization-mass sSpectrometry (nLC-ESI-MS/MS).

## 6.2.9 PMF

The sequencing of the *Populus trichocarpa* (Torr. & Gray) genome (Tuskan et al. 2006) has enabled adoption of a PMF strategy to identify proteins from poplar tissues. PMF generates a list of peptide mass values. Because such a sequence-free approach relies on the presence of protein sequences in the database searched, it is best exploited with sequenced genomes. PMF of the 60 cold-responsive leaf proteins using SwissProt (SP) and TrEMBL Viridiplantae databanks only allowed retrieval of 26 (43.3%) proteins with a known function (Renaut et al. 2004). Out of 550 differentially expressed proteins during secondary vascular system regeneration that were subjected to MALDI-TOF-MS analysis, 244 (44.4%) displayed PMFQ scores high enough (> 2) for further DB searching, and 199 (81.6%) presented a hit when non-redundant (nr) and SP databases were searched (Du et al. 2006). Yuan et al. (2008) were recently able to successfully identify 37 disease-induced leaf proteins out of 40 MS-analyzed (92.5%) using the non-redundant Viridiplantae database.

### *6.2.10 MS/MS*

Unlike PMF, MS/MS gives access to peptide sequence information, provided that the sequence exists in the query database. Alternatively, de novo sequencing can be attempted. Both He et al. (2008) and Kieffer et al. (2008) recently used MALDI-TOF/TOF-MS to analyze proteins of interest with high identification rate, 24/26 (92.3%) and 118/194 (60.8%), respectively. The lower rate obtained by Kieffer et al. (2008) arises from the fact that only NCBI *Populus* ESTs were used. On the other hand, He et al. (2008) searched non-redundant Viridiplantae, and indeed only two proteins (isoforms) matched a poplar sequence. However, all but three proteins displayed very high identification scores.

### *6.2.11 PMF vs. MS/MS*

Several groups have adopted both PMF and MS/MS to maximize protein identification success rate while minimizing costs (Bohler et al. 2007; Bogeat-Triboulot et al. 2007). However, only Plomion et al. (2006) drew a comparison. Surprisingly only 71/144 (49%) (Bolher et al. 2007) and 39/100 (Bogeat-Triboulot et al. 2007) poplar leaf proteins were successfully identified by searching SP, TrEMBL, and NCBI viridiplantae databases in both instances. Plomion et al. (2006) reported that the excellent identification rate (363/398, 91.2%) observed upon using an nLC-ESI-MS/MS device dramatically dropped (163/320, 51%) when MALDI-TOF-MS was employed.

### 6.3 Sub-Proteome Analysis: Subcellular Fractions

The complexity of the plant system should be considered in terms of not only interdependent organs but also numerous highly specialized tissues within each organ, as well as specialized organelles/compartments within individual cells. With the proteomic analytical tools used today, it is not possible to obtain information of proteins with low or even medium expression levels in a whole cell extract. However, sub-cellular fractionation with subsequent mass spectrometry analysis has been shown to be an efficient tool to localize proteins and track low abundant proteins (Haynes and Roberts 2007).

Recently, a sub-cellular approach was developed to highlight ongoing developmental processes taking place in poplar during regeneration in spring (R Nilsson, G. Wingsle, C. Larsson, unpublished data). To identify integral and peripheral proteins associated with the plasma membrane (PM), highly purified poplar PMs from the leaf and stem were analyzed by mass spectrometry (Alexandersson et al. 2004). During leafing in spring there is a rapid development of leaves and wood forming tissue near the bark in stem. The PM is probably the most diverse membrane of the cell, and its protein composition is expected to vary depending on the environment, developmental stage, and tissue. PMs were prepared using aqueous two-phase partitioning, and proteins were identified using electrospray ionization-quadrupole time of flight (ESI-QTOF) and MALDI-TOF. Plant material was harvested in early June from a natural stand of small poplar trees. PMs were extracted from leaves, cambium/phloem, and xylem. The samples were separated on SDS-PAGE, individual protein bands were digested, and the resulting peptide extracts were analyzed using mass spectrometry. Proteins were identified by a local version of the MASCOT search program (v2.1.04; Matrix Science Limited, *http://www. matrixscience.com*) and the Mascot Daemon application (V2.1.6) using the 45,555 translated poplar gene models downloaded from JGI (*http://genome. jgi-psf.org*). Identifications will be available in a peptide database (*http:// www.funcfiber.se*) as they are published.

There is still no definitive method to establish the purity of a sub-cellular fraction. In this study we used antibodies for different types of proteins localized in organelles such as chloroplasts and mitochondria to validate the purity of our PM fraction. Additionally, different procedures were performed to filter the resulting peptides for true membrane protein identification. For example, a cytosolic xylem peptide library produced in a similar way as described above was used. The study has so far resulted in more than a thousand identified proteins. The distributions of identified proteins containing transmembrane domains are illustrated in Figure 6-2.

Isolation of pure PM proteins, particularly from woody tissues, is a difficult task. Cytoskeleton proteins (tubulin and actin) are more frequently found in this tissue. Vesicles, organelles, and membranes are linked by the cytoskeleton to enable the transport of material required for the developing secondary cell walls in the stem. Proteins commonly regarded as soluble cytosolic proteins were repeatedly found in the PM fraction. These proteins lack TM-domains or other apparent domains (GPI anchor, myristoyl chain) that are known to attach to the membranes. Compartments and structures closely associated with the membrane, such as the cellulose synthase rosette might dock with these proteins/enzymes using the cytoskeleton. Additional subfractionation of the PM proteome was also performed. Isolation of so-called "lipid rafts" can enhance our understanding of the localization of proteins in the PM. The identified PM proteins will act as a backbone for more comprehensive developmental studies.

Using quantitative, label-free LC/MS strategies or isobaric multiplex tagging (iTRAQ) would enable monitoring of the identified proteins in detail. Additional research is being undertaken with material collected at

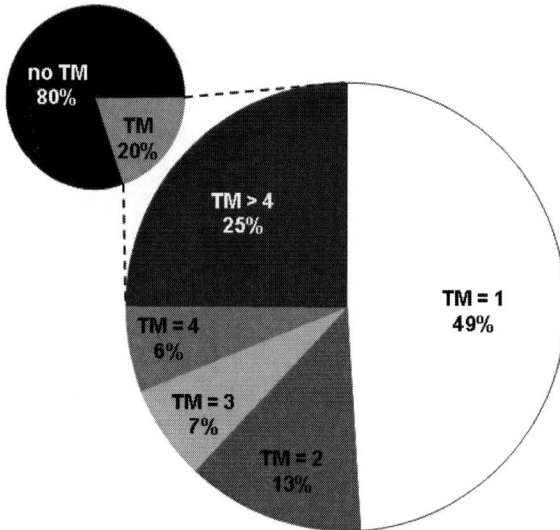

**Figure 6-2** Predicted transmembrane domains (TM) of more than 1000 proteins identified in *Populus* plasma membranes using ESI-qTOF and MALDI-TOF. Transmembrane domains were predicted by the Phobius web server (Käll et al. 2004) using 45,555 translated gene models from *Populus trichocarpa*.

different time points to further study wood development during the leafing period. A label-free LC/MS strategy has been developed and recently published (Bylesjö et al. 2009). In essence, samples can be of different origin, prefractionated as described for the PM sample above or used as a total digest of a more complex sample. Multivariate statistics will distinguish which peptides differ in the samples from a first MS run. These peptides will then be identified by MS/MS in a second run.

## 6.4 Tissue Specific Proteomes

Plomion et al. (2006) were the first to study the proteomes of different organs/tissues from poplar trees. They produced 2-DE reference maps (pH gradient ranging from 4 to 7) corresponding to the soluble proteome of eight tissues, including young and expanding leaves, dormant vegetative bud, reproductive male bud, young female inflorescence, root tips, as well as for the cambial zone and the developing secondary xylem of the wood forming tissues. The number of spots detected ranged from 567 for the cambial zone to 1,600 for the expanding leaf with an average of 1,015. The successful MS analysis of 363 out of 8,122 detected spots provided a snapshot of the most abundant proteins in each tissue. Overall 163 different functions were detected and classified into functional categories: 19, 13 and 11% falling into carbohydrate metabolism, defense, and energy categories, respectively. A significant fraction (4.7%) of the spots corresponded to proteins for which no function had yet been characterized, which is intriguing given their abundance. This first attempt to characterize the poplar proteome was completed by the construction of a database (see Section 5) where any spot can be searched by keywords and spot ID.

In trees, biomass accumulates as wood (secondary xylem), which originates from the activity of a secondary meristem: the vascular cambium. The large amount of wood forming tissues produced during the growing season enabled the study of molecular mechanisms involved in secondary xylogenesis at the proteome level in pine (Plomion et al. 2000; Gion et al. 2005), eucalyptus (Plomion et al. 2003; Fiorani Celedon et al. 2007) and poplar (van der Mijnsbrugge et al. 2000; Du et al. 2006). Based on the assumption that proteins directly involved in secondary cell wall formation should be up-regulated in differentiating xylem, van der Mijnsbrugge et al. (2000) compared the expression profile of phenol extracted proteins by 2-DE gels analysis from young and mature differentiating xylem versus the bark at different times of the year. They reported that xylem-preferentially expressed proteins were more abundant when harvested in summer compared to spring and winter, and over-expressed in young compared to mature differentiating xylem. They successfully assigned functions to 17 of the most-abundant xylem proteins using micro-sequencing, and observed that these proteins belonged to the phenylpropanoid pathway which leads to the biosynthesis of phenolic and lignin compounds.

In a very interesting study, Du et al. (2006) investigated the initiation and differentiation of cambium cells in Chinese white poplar by removing the outer- and inner-bark of 4 year-old trees and wrapping the exposed tissue before scraping the regenerated tissues from the trunk surface at different times (6 to 30 days) after girdling. Out of 550 differentially expressed proteins characterized by MS, they found that 244 spots presented significant scores, among which 199 had strong similarities with existing predicted protein sequences. In particular, 22 enzymes classified in the cell wall formation category were found predominantly regulated in the latest harvested sample undergoing lignification (i.e., 18 and 22 days after girdling).

The identification of key poplar proteins involved in wood formation will require further investigation. The anatomical, chemical, and physical properties of this tissue are extremely variable, depending on genetic, environmental, and ontogenic factors. Whether this variability can be attributed to changes in the proteome was tested in other tree species (Gion et al. 2005; Paiva et al. 2008) and seems a promising route to discover key molecular players involved in wood formation and underlying the variability of this tissue.

## 6.5 Identification Of Biotic And Abiotic Stress Responding Proteins

### *6.5.1 Introduction*

How plants respond and adapt to their environment remains one of the most important questions in plant biology. This question is particularly central in trees, since their growth and development are challenged by biotic and abiotic stresses throughout their entire lifetime. In addition, accelerated climatic changes and the biotic environmental challenges resulting from them suggest that these long-lived organisms will have to cope with increasing stresses within few if not a single generation. If these species are to survive in their current geographical distributions, they will need to respond via both phenotypic plasticity and selection-driven changes in allele frequencies. It is therefore of paramount importance to know whether the present structure and level of genetic diversity existing in forest tree populations is sufficient to allow adaptation to these future conditions. Addressing this question will help underpin actions to preserve adaptability and possibly avoid major losses, especially in managed forests.

In this section we will highlight how the toolkits of proteomics allow remarkable variation in protein expression to be discovered in poplar, revealing key molecular mechanisms underlying plant response and adaptation to the environment, thus revealing "expressional candidate genes" for many traits of ecological interest (e.g., cold, drought, heavy metal and atmospheric pollutant tolerance, disease resistance). Considering how and why this variation matters, perhaps affecting the function and fitness of individual trees in natural populations is a further step that requires investigating the relationships between the variability observed at the molecular level (allelic and/or expressional variations) and the variability of adaptive traits.

### *6.5.2 Effects of Biotic Stress*

Dissection of plant defence responses through transcriptome-assisted profiling has greatly helped in drawing a general molecular model for pathogen resistance in plants (Nimchuk et al. 2003). However, it is not known whether it applies to long-term adaptation of resistance mechanisms in perennial species like trees. Such long-lived organisms are more subjected to pathogen attacks before reproduction, and their extended generation time makes it difficult to cope with pathogens that can produce several generations every year. Analysis of the poplar genome content revealed a large expansion of NBS-leucine-rich repeat (LRR) resistance gene families (Tuskan et al. 2006; Kohler et al. 2008), indicating a possible adaptation to long-term exposure to pathogens. To date, our knowledge of defence mechanisms in poplar at the molecular level is based on transcriptomic studies during interactions with forest tent caterpillars (Major and Constabel 2006; Ralph et al. 2006), the poplar mosaic virus (Smith et al. 2004), and more recently fungal leaf-rust pathogens (Miranda et al. 2007; Rinaldi et al. 2007). Only a few studies addressed the role of proteins in pathogen-associated defence mechanisms in poplar (Rouhier et al. 2004). In recent years, several proteomic-based studies were conducted in model plant species to identify new molecular markers, allowing comparisons with previously described transcript profiles (Jorrin et al. 2007). These early studies were not only confirmatory for well-known resistance- and pathogenesis-related proteins but also unravelled proteins not previously thought to be involved in plant response to pathogen attacks (Thurston et al. 2006).

Proteomics was recently applied to study the effect of biotic stress in poplar upon infection with the fungal pathogen *Marssonina brunnea* f. sp. *Multigermtubi*. This pathogen causes the black spot disease, which is a major threat to poplar plantations, particularly in China (Han et al. 2000). Defence response was monitored in a poplar clone known to be highly resistant to the disease in a time-course experiment after inoculation of the fungal pathogen. Proteins were isolated from infected leaves of *P. euramericana* clone "NL895" at different time-points within 3 days of post-inoculation (PI). Of approximately 500 reproducible protein spots on 2-DE, 40 protein spots were subjected to MALDI-TOF MS (Yuan et al. 2008). Major changes in protein expression were observed after twenty-four hours and 48 hours PI. In this first attempt to analyze proteins expressed in a compatible poplar-fungus interaction, the authors identified few defence response-related proteins, and most functions were related to photosynthesis (e.g., RubisCO, RubisCO activase). Interestingly several spots corresponded to ascorbate peroxidases that all showed consistent expression profiles in the time-course infection. This function was previously identified among accumulated transcripts in poplar leaves infected by the fungus *Melampsora larici-populina* responsible for the rust disease, another major threat of poplar trees in plantations (Rinaldi et al. 2007). Proteomic profiles in poplar subjected to *M. brunea* f. sp. *Multigermtubi* also identified a couple of other proteins of interest, but as emphasized by the authors, interference with RubisCO abundance was a major problem to identify more defence-related functions.

A genetically-oriented approach that combined proteomics and transcriptomics was carried out in an interspecific poplar pedigree that showed a major resistance quantitative trait loci (QTL) to rust (Dowkiw and Bastien 2004) in order to derive reliable molecular markers for breeding selection based on partial resistance to the rust fungus *M. larici-populina* (Lalanne et al. 2007). The proteomic approach aimed at targeting differences in two groups of poplar genotypes with contrasting defence reactions to the rust fungus. After 2-DE separation of proteins and selection of appropriate candidates expressed in infected versus non-infected leaves at 2 and 4 days PI, about 50 protein spots were subjected to MS/MS characterization. Interestingly in this case, the proteins overexpressed in the more resistant genotypes (e.g., thaumatin-like proteins, glutathione S-transferases) correlated quite well with the transcriptomic data (S Duplessis, C Plomion and C Lalanne, unpublished data) and with rust-induced transcripts previously identified in poplar (Rinaldi et al. 2007).

A proteomic approach applied to poplar to isolate proteins differentially expressed under biotic stress is still in its infancy. Recent reports focusing on conifer-herbivore interactions have proven promising for the use of combined proteomic/transcriptomic approaches (Lippert et al. 2007) and, as highlighted by the previous examples, similar approaches will undoubtedly help our understanding of poplar defence mechanisms against pathogens.

### 6.5.3 Effects of Cold

Due to their inherent properties, temperate woody plants have to endure harsh climatic conditions during their life cycle. One of the most important global ecological issues today is global warming. The increase in average annual temperature and precipitation in many parts of the world is resulting in phenological modifications, such as earlier flowering for many temperate plants. But this phenomenon will not protect plants against unexpected frost events, damaging flower buds, ovaries and leaves during the growing season (Inouye 2001).

The ability of plants to cold acclimate is a quantitative trait involving the action of many genes with small additive effects (Thomashow 1999). An important factor in the cold acclimation process is the development of tolerance to cellular dehydration. Accumulation of compatible solutes (e.g., proline, glycine-betaine), sugars, and certain proteins is thought to protect cell structures during dehydration by binding water molecules (see review in Ingram and Bartels 1996; Beck et al. 2007). Antifreeze proteins, heat-shock proteins, dehydrins and pathogenesis-related proteins have been extensively studied to unravel their involvement in cold acclimation and/or freezing tolerance (Guy et al. 1994; Close 1997; Rinne et al. 1999; Wisniewski et al. 1999; Welling et al. 2002; Griffith and Yaish 2004; Lopez-Matas et al. 2004; Ukaji et al. 2004; Griffith et al. 2005; Renaut et al. 2006).

To protect themselves against unfavourable conditions, poplars have shown an ability to counteract a decrease in temperature by an array of metabolic and physiological modifications (Sauter et al. 1996, 1998, 1999; Hausman et al. 2000; Cox and Stushnoff 2001).

In a first experiment of cold acclimation, focusing on extremely narrow isoelectric points (pI 4–5; 4.5–5.5 and 5–6), several proteins were selected on 2D gels that were silver-stained to visualize proteins extracted from leaves of actively growing poplars (*Populus tremula x tremuloides*) (Renaut et al. 2004). Their abundance was either increased or decreased upon exposure to low temperature (4°C) over 2 weeks. Different classes of "more-abundant" proteins were identified: drought-responsive proteins, proteins of stress signalling or transduction, and enzymes involved in antioxidant pathways. Chaperones, heat-shock proteins, late embryogenesis abundant proteins, dehydrin or thioredoxin are some examples taken from these different classes. On the other hand, proteins involved in energy metabolic pathways were less abundant in stressed samples (ATP synthase and several enzymes involved in carbohydrate metabolism).

Recently, a 2-D DIGE experiment was carried out to compare the differential effects of heavy metal (20 $\mu$M CdSO$_4$), chilling temperature (4°C), and their combination on poplar leaves (S. Kieffer et al., unpublished data). It is important to specify that this experiment was conducted in hydroponic conditions. The exposure to low temperature was concomitant with the appearance of chlorosis and growth inhibition. A broader pH range was used for this analysis (24 cm IPG strips, pH 4–7). More than 300 spots (out of the 1,100 spots common to all gels) showed a variation in their abundance upon cold, cadmium, and the combination of cold and cadmium treatments. We will discuss here only the effect of cold on the abundance of identified proteins (i.e., 92 proteins out of 326). The effect of low temperature involved accumulation of 74 spots, including isoforms of well-known cold responsive proteins (e.g., proteolysis, protein folding, cold stress response proteins, regulation, phospholipid metabolism, carbohydrate metabolism such as starch and sucrose metabolism, glycolysis, pentose phosphate pathway, oxidative stress response, storage, ATP synthesis and RNA metabolism). An interesting observation was the accumulation of proteins involved in methionine metabolism (e.g., methionine synthase) as recently observed in chilled potato (Renaut et al. 2008) and in frozen onions (J. Renaut et al. unpublished data). Down-regulated proteins concerned mostly photosynthesis, but also ATP synthesis, nitrogen and arabinose metabolism, oxidative stress regulation and cell structure.

Cold, as other abiotic stress factors, is known to induce oxidative stress in cells (Lukatkin 2002) leading to the accumulation of reactive oxygen species (ROS). The damage that could be caused by these ROS can be reduced by the accumulation of detoxifying enzymes (e.g., glutathione peroxidase) and compatible solutes. Methionine synthase is involved in methionine synthesis, as well as in providing single carbons in plants (Hanson et al. 2000). Methionine synthase is also the starting point of glycinebetaine, methylated polyols, polyamine and ethylene biosyntheses via the production of S-adenosyl-L-methionine and its decarboxylated form (Eckermann et al. 2000). This increase in methionine synthase has already been mentioned at the mRNA level in response to several abiotic constraints and at the protein level when barley was exposed to salt stress (Narita et al. 2004). The presence of proteases could be involved in protein breakdown, mobilization of storage proteins, and recycling of damaged proteins (e.g., oxidized or denatured proteins) (Gruis et al. 2004; Smalle and Vierstra 2004; van der Hoorn and Jones 2004). The protection of membranes and macromolecules could be improved by the accumulation of HSPs, dehydrins, lipocalins, protein disulphide isomerases, but also via carbohydrate accumulation (Oliver et al. 1998). The changes in proteins linked to photosynthesis could also be related to accumulation of carbohydrates, and consequently to the cessation of growth and probably the onset of ecological dormancy. All of these events are pointing to an acute protection of plants regarding their survival under low temperature stress.

### 6.5.4 *Effects of Drought*

Water deficit is one of the most important abiotic stresses affecting forest ecosystems (Loustau et al. 2005; Bréda et al. 2006). It could directly impact biomass production by decreasing wood formation (Zahner 1968; Rozenberg et al. 2002) and by making trees more sensitive to other biotic or abiotic constraints (Desprez-Loustau et al. 2006; Mittler 2006). In the current global climatic change context (IPCC 2007), woody species with large water requirements may suffer from the expected increase of drought intensities and frequencies (Saxe et al. 2001). Poplars are the fastest growing forest trees from temperate latitudes. Their high productivity is associated with a tight dependency upon water availability (Tschaplinski and Blake 1989; Tschaplinski et al. 1994; Zsuffa et al. 1996). Although poplars are among the most susceptible woody plants to drought (Dreyer et al. 2004), they display a consistent genetic variation in their patterns of drought response (Ceulemans et al. 1978; Pallardy and Kozlowski 1981; Liu and Dickmann 1996; Marron et al. 2002, 2003; Monclus et al. 2006). This variation encompasses several physiological and morphological traits, such as: stomatal closure (Marron et al. 2002), osmotic adjustment (Gebre et al. 1994; Marron et al. 2002), leaf expansion, leaf abscission, root/shoot ratio (Liu and Dickmann 1992; Chen et al. 1997; Ibrahim et al. 1997; Tschaplinski et al. 1998; Marron et al. 2003) and productivity (Monclus et al. 2006; Bonhomme et al. 2008). However, little information has been reported on the molecular mechanisms involved in drought tolerance. The identification of genes and proteins, which may allow poplar to cope with drought-prone periods, will be a valuable tool for marker-assisted selection or genetic engineering.

In poplar as in herbaceous plants, the loss of water from a cell triggers a cellular signal transduction pathway (Chefdor et al. 2006), which may elicit a complex regulatory gene network leading to the accumulation of proteins involved in drought tolerance (Bray 1997). The first study concerning the effect of drought on poplar protein patterns has been performed using one-dimensional electrophoresis on *P. tremula* (Pelah et al. 1995). It was found that protein content of young shoots wilted to 80% of their fresh weight displayed an increase of a boiling stable protein. The authors suggested that its accumulation could contribute to membrane stability (Pelah et al. 1997). Nowadays, 2-DE is the dominant protein separation methodology used to study the effects of drought effect on the poplar proteome, but only a few studies have been published thus far. These studies were conducted from rooted cuttings belonging to three genotypes: *P. euphratica* Oliv., a phreatophyte and salt tolerant species able to grow in semi-arid areas (Bogeat-Triboulot et al. 2007) and two cultivated hybrid poplars i.e., *P. trichocarpa* x *P. deltoides* cv. " Beaupré " (Plomion et al. 2006) and *P. deltoides* x *P. nigra* cv. "74/76" (He et al. 2008). Proteomes were analyzed from roots and/or leaf samples separated across a 4–7 pH range.

A common physiological effect of drought in all studied genotypes was a decrease in stomatal conductance in order to minimize water loss. In *P. euphratica*, during the first stage of drought, the net $CO_2$ assimilation rate was maintained (Bogeat-Triboulot et al. 2007). This was accompanied by an increased abundance of photosynthesis-related proteins: an oxygen-evolving complex 33-kDa PSII, a RubisCO activase, a carbonic anhydrase, a chloroplast glyceraldehyde-3-P dehydrogenase and a phosphoglycerate kinase. These proteins may contribute to the maintenance of photosynthesis during the initial stages of water deficit. When drought becomes more pronounced, the net $CO_2$ assimilation rate decreases. In cv. "Beaupré", a decrease in abundance of proteins resulting from carbohydrate metabolism was observed: a malate dehydrogenase, a glucan endo-1,3-beta-glucosidase, an enolase, a 2,3-bisphosphoglycerate-independent phosphoglycerate mutase, an alpha-1,4-glucan-protein synthase and a triosephosphate isomerase. In cv. "74/76", the decrease of photosynthesis was associated with an up-regulation of degraded RubisCO isoforms (He et al. 2008).

All together, these results indicate that growth-related biochemical mechanisms were reduced. Plomion et al. (2006) suggested that protein degradation, such as RubisCO may result in the increased abundance of a bark storage protein in leaves, which may correspond to either an energy remobilization in order to face the constraint, or a temporary nitrogen storage under stress. This could enhance, for instance, root growth during drought by providing supplementary energy from leaves when photosynthesis collapses. When the environmental conditions return to optimal, it is assumed that stored nitrogen will be remobilized and will contribute to rapidly restore leaf growth. In addition, an up-regulation of a glutamine synthase in cv. "Beaupré" was proposed to contribute to the protection of photosynthesis. El-Khatib et al. (2004) supported this hypothesis using hybrid poplar lines (INRA 717 1-B4, *P. tremula* L. × *P. alba* L.) transformed to over-express a pine cytosolic glutamine synthetase (GS1) gene. In leaves of *P. euphratica* as in cv. "Beaupré", it was also observed that the abundance of HSP70 and chaperonins were decreased by water deficit.

Drought associated with high light intensity is able to promote ROS, thus inducing oxidative stress which can damage cellular structures and proteins (Smirnoff 1998). In order to counteract the deleterious effects of ROS, proteins involved in oxidative stress repair are up-regulated in response to drought (Plomion et al. 2006). In roots of cv. "Beaupré", a leucoanthocyanidin reductase involved in flavonoid production was newly observed in drought-stressed trees. In leaves, a ferredoxin-NADP reductase was also newly observed and two other proteins were strongly up-regulated: a [Cu–Zn] SOD and a polyphenol oxidase. In leaves of cv. "74/76", increased abundances of two detoxification enzymes were also reported: a methionine sulphoxide reductase A and a mitochondrial manganese superoxide dismutase. The increased accumulation of these proteins exemplifies that both enzymatic (Halli-well Asada pathway) and nonenzymatic (based on anti-oxidant molecules) systems were actively stimulated in drought-stressed poplars. However, in some cases, proteins could be damaged by ROS, aggregate and become toxic to the cells. In roots of cv. "Beaupré", a 26S protease regulatory subunit 6A, a component of the 26S proteasome involved in the ATP-dependent degradation of ubiquitinate proteins, has been newly observed in drought stressed plants whereas in leaves, an aminopeptidase 2 increases. All together, this suggests that protein breakdown and recycling via the ubiquitin-proteasome pathway or proteolytic enzymes are involved in the drought response (Plomion et al. 2006).

In poplar, proteomics has proven its value to assess drought-responsive proteins for a given genotype. However, since poplars display consistent genetic variation, genotype-dependent changes of expressed proteins in response to drought (genotype × environment interaction, G × E) could provide new insights into the molecular basis of drought tolerance. To obtain information on G × E interactions and on the genetic background of drought tolerance in poplars, responses need to be evaluated from studies combining different poplar genotypes and confirmed from mature trees. Two studies were conducted on poplar genotypes submitted to a water withholding period (Bonhomme et al., 2009a,b). The results support the conclusion that such a proteomic approach could be a reliable tool for the identification of proteins whose plasticity is genetically controlled. Although genotypic differences were overall observed regardless of the water regime, differential changes in response to drought were also reported and mainly involved photosynthesis related proteins.

In *P. euphratica*, as in cv. "Beaupré", the comprehensive regulation of drought-responsive proteins was completed from transcript fold-change studies using microarray analyses. Among the identified proteins responding to drought for which ESTs were present on the arrays, most of the transcripts did not display correlation with protein abundance (Plomion et al. 2006). As highlighted by Gygi et al. (1999), mRNAs are not always translated and, thus, transcript and protein abundances are not necessarily linked. Therefore, transcriptomics and proteomics constitute complementary approaches to investigate molecular drought responses.

## 6.5.5 Effects of Heavy Metals

Human activities have resulted in the release of toxic elements in the environment. Cadmium (Cd), a nonessential heavy metal, is one of the elements that contaminates soils and water, and accumulates in plants with drastic effects leading to chlorosis and reduction in plant development (Prasad 1995; Das et al. 1997; Sanità di Toppi and Gabbrielli 1999). To counteract cadmium accumulation in tissues, organisms have developed different strategies with undeniable consequences on the proteomic status of the cells. In mammal cells, chelation with small cysteine-rich proteins (metallothioneins) prevails, whereas in plants and many fungi, the cysteine-containing peptides glutathione and phytochelatins are preferentially synthesized (Cobbet and Goldsbrough 2002).

In a proteomic study carried out on leaves of young aspen trees subjected to 20 µM Cd for 14 days, glutamine synthetase and O-acetylserine(thiol) lyase, respectively, key enzymes in nitrogen assimilation and sulphate assimilation, increased in abundance (Kieffer et al. 2008). The higher activities of these enzymes may support a higher demand in glutathione and phytochelatin synthesis by the up-regulation of glutamate and sulphur amino acid biosynthesis. Other components of the GSH and phytochelatin biosynthetic pathways were also up-regulated in Cd-exposed herbaceous plants (Roth et al. 2006, Sarry et al. 2006; Aina et al. 2007), but the most striking effect occurred with yeasts subjected to high Cd exposure (Vido et al. 2001).

Higher levels of glutathione have also been obtained in poplar transgenic plants by over-expressing the gene encoding γ-glutamylcysteine synthetase, the rate-limiting enzyme in the biosynthesis of glutathione (Arisi et al. 2000; Koprivova et al. 2002; Bittsanszky et al. 2005). These transgenic plants were able to accumulate greater amount of cadmium or zinc but the improvement of heavy metal tolerance is still in debate.

Up-regulation of proteins involved in an oxidative stress response was also observed in poplar leaves exposed to Cd (Kieffer et al. 2008). In this case, higher amounts of peroxidase isoforms or aldehyde dehydrogenase were supposed to neutralize toxic compounds such as hydrogen peroxide and aldehyde while the up-regulation of a quinone oxidoreductase would be useful to limit the overproduction of ROS. However, a process of down-regulation for oxidative stress proteins seemed to prevail with lower amounts of Cu/Zn superoxide dismutase isoforms and thioredoxin reductase (Kieffer et al. 2008). For some of these enzymes, the competition between Cd and divalent cations, used as enzyme activators, may explain this down-regulation. At the same time, and linked to the development of an oxidative stress in the poplar leaf, a higher amount of several proteins belonging to the pathogenesis-related (PR) family or involved in protein degradation occurred.

Concerning carbon metabolism, a general depletion of proteins linked to carbon fixation, such as RubisCO activase and carbonic anhydrase isoforms, was noted (Kieffer et al. 2008). Conversely, enzymes involved in carbon remobilization of storage forms were up-regulated in leaves of poplar exposed to Cd. RubisCO subunit-binding isoforms, belonging to the HSP60 chaperonin family were also down-regulated. Generally, HSP proteins have been reported to be up-regulated in herbaceous plants exposed to cadmium or copper stress to prevent oxidative damage (Timperio et al. 2008).

## 6.5.6 Effects of Ozone

Ozone is an important atmospheric pollutant affecting herbaceous plants and forest trees (Sandermann et al. 1997; Ashmore 2005). In contrast with heavy metals, this reactive molecule is not accumulated in plant tissues where the internal concentration is quite low compared to external concentrations (Heath 2008). In fact, ozone can rapidly promote ROS formation in both apoplasm and symplasm compartments. In these conditions the level of proteins engaged in the redox status of the cells is altered. A proteomic study bearing on leaves of poplar subjected to a chronic ozone exposure showed that two enzymes, a cytosolic ascorbate peroxidase and an isoflavone reductase, were up-regulated (Bohler et al. 2007). Based on the expression of poplar ESTs, other anti-oxidant proteins such as SOD and glutathione S-transferase seems to be up-regulated in poplar leaves subjected to long-term ozone exposure (Gupta et al. 2005). Some of these enzymes were also up-regulated in a proteomic study on rice seedlings exposed to an acute ozone dose (Agrawal et al. 2002), as well as pathogenesis related proteins.

A larger number of defence proteins were induced by ozone, but their occurrence has been mainly observed in physiological or transcriptomic studies (Kangasjärvi et al. 1994; Rao and Davies 2001; Tamaoki et al. 2003; Li et al. 2006). Among these proteins, some are involved in the biosynthesis of protective molecules such as lignin, a compound which has been showed to increase and to be more condensed in leaves of poplar fumigated with ozone (Cabane et al. 2004). Finally, as previously mentioned for heavy metals, the up-regulation of proteins involved in protein folding also occurred upon ozone exposure (e.g., 2 HSP isoforms and 3 disulfide isomerase isoforms) (Bohler et al. 2007).

In agreement with previous studies based on the determination of enzyme activity, proteins involved in primary carbon metabolism were also regulated upon ozone exposure (Heath and Taylor 1997; Dizengremel et al. 2001; Heath 2008). For different herbaceous and woody species, the activity and amount of RubisCO decreased (see Dizengremel 2001 for a review). In a proteomic study on poplar leaves, it was the amount of RubisCO activase and Calvin cycle enzymes which decreased in response to chronic ozone exposure (Bohler et al. 2007). At the biochemical level, the activity of enzymes linked to carbohydrate breakdown were increased; in a context of ozone risk assessment it is quite relevant to consider the up-regulation of several enzymes delivering NADPH to sustain detoxification processes (Dizengremel et al. 2009).

Ozone stress also impacts N assimilation. Bohler et al. (2007) showed that the amount of cytosolic glutamine synthetase increased in poplar leaves exposed to ozone, allowing a shunt in the nitrogen assimilatory capacity from the chloroplast towards the cytosol. This result, added to the generation of other signals mentioned in physiological and transcriptomic studies, argues for an accelerated senescence triggered by long term ozone exposure (Miller et al. 1999; Gupta et al. 2005; Gielen et al. 2007).

## 6.6 Toward the Establishment of a Poplar Proteomic Database

Proteomics approaches involve various experimental techniques that produce large volumes of highly heterogeneous data. Typical projects include protein sample separation by 2-D gel electrophoresis, protein quantitation using 2-D gel image analysis, and protein identification by MS. Alternatively, relative protein abundance can also be obtained through MS analyses. The post-translational modification (PTM) status of proteins may also be investigated by MS or other biochemical approaches. All of these experiments provide distinct information about the same set of proteins: experimental setup, location on a gel, peptide sequence, function of proteins, PTMs, expression profile etc., all of which must be integrated into a single database environment to be easily queried and summarized.

The proteomics database put in place at INRA, is a first example of such database for poplar (*http://cbib1.cbib.u-bordeaux2.fr/Protic/public/index.*

*php*). It gives access to a set of annotated 2-D gel images from a panel of poplar tissues (Plomion et al. 2006). In addition to spot volume, protein identification and crosslinks to external sequence databases, users have access to individual genotype, experimental procedure, and complete MS data as required by proteomics publication guidelines (see Fig. 6-3). This web site is based on the PROTICdb freeware (Langella et al. 2007). Among other things, the latest version of PROTICdb, released in the fall of 2008, is fully compliant with major biological ontologies (Smith et al. 2007), and includes new tools for annotated mass spectrum visualization as well as data import/export in standard Portable Site Information format (Martens et al. 2008). Ongoing projects focus on new tools for statistical analysis, annotation and validation of proteomic data, as well as full integration into the world 2-D PAGE network of distributed gel-based proteomic databases (Hoogland et al. 2008).

Because poplar is a model tree, the establishment of a poplar proteomic reference database would benefit the whole forest tree community in several respects. The accumulation of results from a large set of experiments,

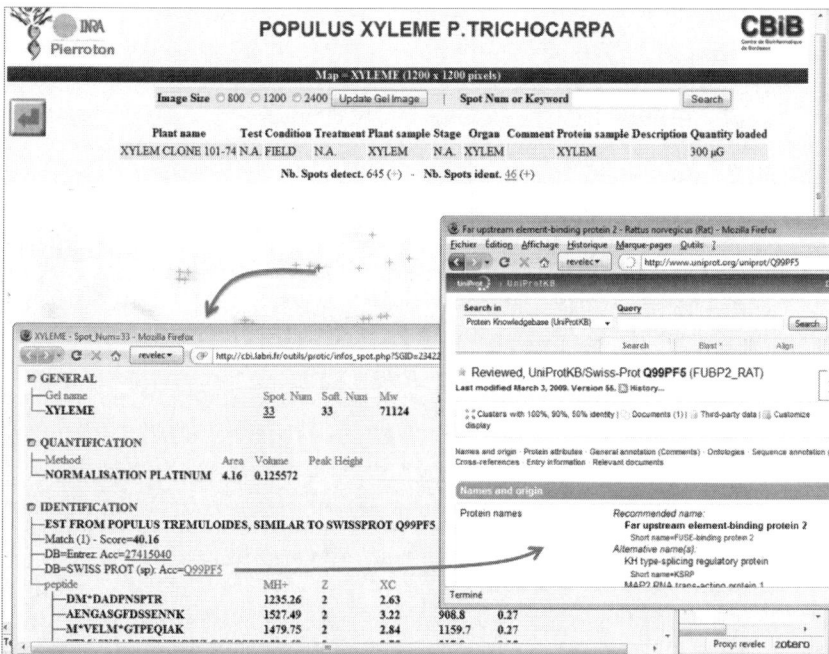

**Figure 6-3** The PROTICdb gel browser displays 2D-PAGE annotated images. Complete dataset for a spot including quantitative data and mass spectrometry identification are available as well as links to external sequence databases.

*Color image of this figure appears in the color plate section at the end of the book.*

therefore for many sub-proteomes, will offer broader views of proteome dynamics, and provide ways to unravel protein expression networks and identify key proteins for biological processes of interest. These data will also be essential for systems biology approaches that will require integration of proteomic data with other resources like mRNA and metabolite profiling databases in order to devise mathematical predictive models as well as genetic databases to identify key variations. The availability of MS data will provide translational evidence that will likely help to refine and validate poplar gene models, or even discover yet unpredicted protein-coding genes (Tanner et al. 2007). This will also be a highly valuable resource for projects such as the Plant Proteome Annotation Program (Schneider et al. 2005) and help increase the number of curated poplar protein entries in the Uniprot/ Swiss-Prot database (only 191 entries in 09/2008 release).

## 6.7 Challenges and Vision

Poplar proteomics has been making considerable progress but faces critical biological and technical challenges. Proteins are highly dynamic biological entities, readily responding to various internal and external stimuli. Protein variants may originate from extensive alternative splicing, single-point mutations, and numerous post-translational modifications, generating multiple protein isoforms from a single gene (Early et al. 1980; Rosenfield et al. 1982; Werneke et al. 1989). This heterogeneity ultimately leads to production of a large amount of different protein species that are diverse in their functions and physicochemical characteristics, varying in stability and longevity under different conditions, making the proteome immensely complex and dynamic. Moreover, the fact that proteins are not present in the cell in equal amounts adds to further complexity (Corthals et al. 2000; Gygi et al. 2000; Adkins et al. 2002). For example, while some proteins are highly abundant (housekeeping enzymes), others (regulatory proteins) are present at extremely low levels. This broad range in protein concentration (e.g., $10–10^{12}$ molecules per cell) poses a great challenge to poplar proteomic investigations and notably complicates the analyses of large-scale proteomics, since detection sensitivity becomes a major issue. In addition, poplar tissues are highly heterogenous in structure and development, influencing protein purification and purity.

Proteomics techniques have greatly advanced in recent years, owing to the introduction of improved protocols for sample preparation, integration of robust robotics, sensitive MS technologies, and powerful bioinformatics. However, MS-based proteome profiling or comparative quantitative proteomics still remains a challenging task and is influenced by multiple factors. The most critical step in proteomics is reproducible protein extraction and sample preparation. Thus, adequate, specific, and effective protein extraction procedures need to be used for every tissue/organelle to obtain an entire population of proteins that is clean, high quality, and free of interfering substances, artifacts, non-proteinaceous contaminants, and proteases. This chapter shows that this is a challenging issue for highly lignified poplar tissues that are also rich in phenolic compounds and other secondary metabolites. Protein sample complexity is another important issue that influences project outcomes. For example, whole cell protein extracts will only reveal highly abundant proteins, whereas fractionation of a protein sample prior to MS diminishes its complexity, therefore allowing detection of low copy proteins and proteins with extreme physicochemical characteristics (Washburn et al. 2001; McCarthy et al. 2004; Essader et al. 2005). Complementary approaches such as organelle subproteomes that are highly enriched in organelle-specific proteins, the majority of which would not be detectable in whole cell protein extracts, could be developed for poplar.

As in other species, a challenging task in poplar proteomics is the detection of highly hydrophobic integral membrane proteins, which play essential roles in signal transduction, cell-to-cell recognition, transport, etc. Due to their poor solubility, low abundance, and usually alkaline pI these proteins remain a great challenge for both gel-based and gel-free techniques as they require specific and often tedious extraction approaches/strategies (Hurkman and Tanaka 1986; Wilkins et al. 1998; Seigneurin-Berny et al. 1999; Santoni et al. 1999; Ferro et al. 2000; Wu et al. 2003; Szponarski et al. 2004). Detection and analyses of post-translational modifications that control crucial regulatory processes (phosphorylation) or are associated with specific biochemical activities (acetylation, ubiquitination) remains a critical issue. Although specific dyes have been recently developed for detecting phosphorylated and glycosylated proteins on 2-D gels (Pro-Q Diamond and Pro-Q Emerald, respectively), they need to be used cautiously since unmodified proteins may also be stained. There also exists no single technology to detect, identify, catalogue, and classify as many proteins as possible, including proteins with extreme pIs and molecular weights, as well as hydrophobic characteristics. We have to instead use a combination of both gel-based and gel-free MS-based methods to maximize protein coverage, which increases cost. Additionally, although a draft of the poplar genome sequence and a resulting protein database have been developed, there is always more room to improve these invaluable resources by closing the existing gaps in the genome sequence, completing the gene models, providing expression evidence for the models, and improving structural and functional annotation.

Despite these challenges, the ultimate goal remains to develop a systems level understanding of poplar life history. The accomplishments and experience that have been reported in this chapter suggest that the time is ripe to undertake the following full-scale poplar proteome projects. *First*, with the help of mature proteomics technology, we can identify and quantify proteins present in poplar cells/tissues at various developmental stages and in response to agriculturally important biotic and abiotic signals. This may include a catalogue of all protein variants that results from splice variation and post-translational modification. *Second*, we can delineate protein-protein, protein-DNA, protein-RNA, or protein-metabolite interactions to identify a protein's partners to construct protein-interaction networks in signalling pathways. *Third*, we can develop one antibody for each poplar protein to use in identifying protein interaction networks or to track their locations in cells or tissues. *Fourth*, we can determine atomic resolution 3-dimensional structures of important poplar proteins, protein-protein complexes, and protein-metabolite complexes. Such a project would be helpful to identify core domains, predict protein folding, and understand the relationship between protein sequence, structure, and function, aiding us in assigning functions to unannotated proteins. *Fifth*, we can integrate large-scale proteome data with various types of genome-wide data (e.g., DNA sequence, transcript, transcription-factor binding site, QTL, metabolite, etc.) to build a computational model that allows us to simulate, predict, and control poplar growth and development. *Finally*, there is a profound lack in our community of appropriately trained scientists in proteomics. We can develop proteomics curricula for undergraduate and graduate students. This can be enhanced through collaboration among individual laboratories by exchanging researchers to shape the future of poplar proteomics. These projects could reveal which/how much protein is actually expressed in different cell/tissue types and the molecular networks of proteins, offering clues to the functions of thousands of poplar proteins. Such knowledge could ultimately provide insights into poplar life history with potential economic benefits.

## Acknowledgments

This work was funded by (A) ANR (Interpopger, Genoplante Popsec-GPLA06028G and Genoplante PROTICws -GPLA009001 projects) awarded to CP, DV, JJ, SD and DM (B) BioPollAtm (Primequal2/MEDD, ADEME, ACI INSU CNRS) and IFLOZ (AT INSU CNRS) awarded to YJ, and (**C**) the US National Science Foundation (CY).

## References

Adkins JN, Varnum SM, Auberry KJ, Moore RJ, Angells NH, Smith RD, Springer RD, Pounds JG (2002) Towards a human blood serum proteome. Mol Cell Proteom 1: 947–955.
Agrawal GK, Rakwal R (2006) Rice proteomics: a cornerstone for cereal food crop proteomes. Mass Spectrom Rev 25: 1–53.

Agrawal GK, Rakwal R, Yonekura M, Kubo A, Saji H (2002) Proteome analysis of differentially displayed proteins as a tool for investigating ozone stress in rice (*Oryza sativa* L.) seedlings. Proteomics 2: 947–959.

Aina R, Labra M, Fumagalli P, Vannini C, Marsoni M, Cucchi U, Bracale M, Sgorbati S, Citterio S (2007) Thiol-peptide level and proteomic changes in response to cadmium toxicity in *Oryza sativa* L. roots. Environ Exp Bot 59: 381–392.

Alexandersson E, Saalbach G, Larsson C, Kjellbom P (2004) *Arabidopsis* plasma membrane proteomics identifies components of transport, signal transduction and membrane trafficking. Plant Cell Physiol 45: 1543–1556.

Arisi ACM, Mocquot B, Lagriffoul A, Mench M, Foyer CH, Jouanin L (2000) Responses to cadmium in leaves of transformed poplars overexpressing γ-glutamylcysteine synthetase. Physiol Plant 109: 143–149.

Ashmore MR (2005) Assessing the future global impacts of ozone on vegetation. Plant Cell Env 28: 949–964.

Beck EH, Fettig S, Knake C, Hartig K, Bhattarai T (2007) Specific and unspecific responses of plants to cold and drought stress. J Biosci 32: 501–510.

Bittsanszky A, Komives T, Gullner G, Gyulai G, Kiss J, Heszky L, Radimszky L, Rennenberg H (2005) Ability of transgenic poplars with elevated glutathione content to tolerate zinc (2+) stress. Environ Int 31: 251–254.

Bogeat-Triboulot, MB, Brosché M, Renaut J, Jouve L, LeThiec D, Fayyaz P, Vinocur B, Witters E, Laukens K, Teichmann T, Altman A, Hausman JF, Polle A, Kangasjärvi J, Dreyer E (2007) Gradual soil water depletion results in reversible changes of gene expression, protein profiles, ecophysiology, and growth performance in *Populus euphratica*, a poplar growing in arid regions. Plant Physiol 143: 876–892.

Bohler S, Bagard M, Oufir M, Planchon S, Hoffmann L, Jolivet Y, Hausman JF, Dizengremel P, Renaut J (2007) A DIGE analysis of developing poplar leaves subjected to ozone reveals major changes in carbon metabolism. Proteomics 7: 1584–1599.

Bonhomme L, Barbaroux C, Monclus R, Morabito D, Berthelot A, Villar M, Dreyer E, Brignolas F (2008) Genetic variation in productivity, leaf traits and carbon isotope discrimination in hybrid poplars cultivated on contrasting sites. Ann For Sci 65: 503p1–503p9.

Bonhomme L, Monclus R, Vincent D, Carpin S, Claverol S, Lomenech AM, Labas V, Plomion C, Brignolas F, Morabito D (2009a) Genetic variation and drought response in two *Populus* x *euramericana* genotypes through 2-DE proteomic analysis of leaves from field and glasshouse cultivated plants. Phytochemistry 70: 988–1002.

Bonhomme L, Monclus R, Vincent D, Carpin S, Lomenech AM, Plomion C, Brignolas F, Morabito D (2009b) Leaf proteome analysis of eight *Populus* x *euramericana* genotypes: genetic variation in drought response and in water-use efficiency involves photosynthesis-related proteins. Proteomics 9: 4121–4142.

Bray EA (1997) Plant responses to water deficit. Trends Plant Sci 2: 48–54.

Bréda N, Granier A, Huc R, Dreyer E (2006) Temperate forest trees and stands under severe drought: a review of ecophysiological responses, adaptation processes and long-term consequences. Ann For Sci 63: 625–644.

Burnette NW (1981) 'Western blotting': electrophoretic transfer of proteins from sodium dodecyl sulfate—polyacrylamide gels to unmodified nitrocellulose and radiographic detection with antibody and radioiodinated protein A. Anal Biochem 112: 195–203.

Bylesjö M, Nilsson R, Srivastava V, Grönlund A, Johansson AI, Jansson S, Karlsson J, Moritz T, Wingsle G, Trygg J (2009) Integrated Analysis of Transcript, Protein and Metabolite Data to Study Lignin Biosynthesis in Hybrid Aspen. J Prot Res 8: 199–210.

Cabane M, Pireaux JC, Leger E, Weber E, Dizengremel P, Pollet B, Lapierre C (2004) Condensed lignins are synthesized in poplar leaves exposed to ozone. Plant Physiol 134: 586–594.

Carpentier SC, Panis B, Vertommen A, Swennen R, Sergeant K, Renaut J, Laukens K, Witters E, Samyn B, Devreese B (2008) Proteome analysis of non-model plants: a challenging but powerful approach. Mass Spectrom Rev 27: 354–77.

Ceulemans R, Impens I, Lemeur R, Moermans R, Samsuddin Z (1978) Water movement in the soil-poplar-atmosphere system I Comparative study of stomatal morphology and anatomy, and the influence of stomatal density and dimensions on the leaf diffusion characteristics in different poplar clones. Oecol Plant 13: 1–12.

Chefdor F, Bénédetti H, Depierreux C, Delmotte F, Morabito D, Carpin S (2006) Osmotic stress sensing in *Populus*: Components identification of a phosphorelay system. FEBS Lett 580: 77–81.

Chen S, Wang S, Altman A, Hüttermann A (1997) Genotypic variation in drought tolerance of poplar in relation to abscisic acid. Tree Physiol 17: 797–803.

Close TJ (1997) Dehydrins: a commonalty in the response of plants to dehydration and low temperature. Physiol Plant 100: 291–296.

Cobbet C, Goldsbrough P (2002) Phytochelatins and metallothioneins: roles in heavy metal detoxification and homeostasis. Annu Rev Plant Biol 53: 159–182.

Corthals GL, Wasinger VC, Hochstrasser DF, Sanchez JC (2000) The dynamic range of protein expression: a challenge for proteomic research. Electrophoresis 21: 1104–1115.

Cox SE, Stushnoff C (2001) Temperature- related shifts in soluble carbohydrate content during dormancy and cold acclimation in *Populus tremuloides*. Can J For Res 31: 730–737.

Damerval C, de Vienne D, Zivy M, Thiellement H (1986) Technical improvements in two-dimensional electrophoresis increase the level of genetic variation detected in wheat-seedling proteins. Electrophoresis 7: 52–54.

Das P, Samantaray S, Rout GR (1997) Studies on cadmium toxicity in plants: a review. Environ Pollut 98: 29–36.

Desprez-Loustau ML, Marçais B, Nageleisen LM, Piou D, Vannini A (2006) Interactive effects of drought and pathogens in forest trees. Ann For Sci 63: 597–612.

Dizengremel P (2001) Effects of ozone on the carbon metabolism of forest trees. Plant Physiol Biochnol 39: 729–742.

Dizengremel P, Le Thiec D, Hasenfratz-Sauder MP, Vaultier MN, Bagard M, Jolivet Y (2009) Metabolic-dependent changes in plant cell redox power after ozone exposure. Plant Biol 11: 35–42.

Dowkiw A, Bastien C (2004) Characterization of two major genetic factors controlling quantitative resistance to *Melampsora larici-populina* leaf rust in hybrid poplars: strain-specificity, field expression, combined effects, and relationship with a defeated qualitative resistance gene. Phytopathology 94: 1358–1367.

Dreyer E, Bogeat-Triboulot MB, Le Thiec D, Guehl JM, Brignolas F, Villar M, Bastien C, Martin F, Kohler A (2004) Drought tolerance of poplars: can we expect to improve it? Biofutur 247: 54–58.

Du J, Xie HL, Zhang DQ, He XQ, Wang MJ, Li YZ, Cui KM, Lu MZ (2006) Regeneration of the secondary vascular system in poplar as a novel system to investigate gene expression by a proteomic approach. Proteomics 6: 881–895.

Early P, Rogers J, Davis M, Calame K, Bond M, Wall R, Hood L (1980) Two mRNAs can be produced from a single immunoglobulin mu gene by alternative RNA processing pathways. Cell 20: 313–319.

Eckermann C, Eichel J, Schröder J (2000) Plant methionine synthase: new insights into properties and expression. Biol Chem 381: 695–703.

Edman P (1949) A method for the determination of amino acid sequence in peptides. Arch Biochem 22: 475.

El-Khatib RT, Hamerlynck EP, Gallardo F, Kirby EG (2004) Transgenic poplar characterized by ectopic expression of a pine cytosolic glutamine synthetase gene exhibits enhanced tolerance to water stress. Tree Physiol 24: 729–736.

Essader AS, Cargile BJ, Bundy JL, Stephenson Jr JL (2005) A comparison of immobilized pH gradient isoelectric focusing and strong-cation-exchange chromatography as a first dimension in shotgun proteomics. Proteomics 5: 24–34.

Ferro M, Seigneurin-Berny D, Rolland N, Chapel A, Salvi D, Garin J, Joyard J (2000) Organic solvent extraction as a versatile procedure to identify hydrophobic chloroplast membrane protein. Electrophoresis 21: 3517–3526.

Fiorani Celedon PA, de Andrade A, Meireles KG, Gallo de Carvalho MC, Caldas DG, Moon DH, Carneiro RT, Franceschini LM, Oda S, Labate CA (2007) Proteomic analysis of the cambial region in juvenile Eucalyptus grandis at three ages. Proteomics 7: 2258–2274.

Gebre GM, Kuhns MR, Brandle JR (1994) Organic solute accumulation and dehydration tolerance in tree water-stressed *Populus deltoides* clones. Tree Physiol 14: 575–587.

Gielen, B, Löw M, Deckmyn G, Metzger U, Franck F, Heerdt C, Matyssek R, Valcke R, Ceulemans R (2007) Chronic ozone exposure affects leaf senescence of adult beech trees: a chlorophyll fluorescence approach. J Exp Bot 58: 785–795.

Gion J-M, Lalanne C, Le Provost G, Ferry-Dumazet H, Paiva J, Chaumeil P, Frigerio JM, Brach J, Barré A, de Daruvar A, Claverol S, Bonneu M, Sommerer N, Negroni L, Plomion C (2005) The proteome of maritime pine wood forming tissue. *Proteomics* 5: 3731–3751.

Gion J-M, Lalanne C, Le Provost G, Ferry-Dumazet H, Paiva J, Frigerio JM, Chaumeil P, Barré A, de Daruvar A, Brach J, Claverol S, Bonneu M, Plomion C (2005) The proteome of maritime pine wood forming tissue. 31-*Proteomics* 5: 3731–3751.

Griffith M, Yaish MW (2004) Antifreeze proteins in overwintering plants: a tale of two activities. Trends Plant Sci 9: 399–405.

Griffith M, Lumb C, Wiseman SB, Wisniewski M, Johnson RW, Marangoni AG (2005) Antifreeze proteins modify the freezing process *in planta*. Plant Physiol 138: 330–340.

Gruis D, Schulze J, Jung R (2004) Storage protein accumulation in the absence of the vacuolar processing enzyme family of cysteine proteases. Plant Cell 16: 270–290.

Gupta P, Duplessis S, White H, Karnosky DF, Martin F, Podila GK (2005) Gene expressions patterns of trembling aspen trees following long term exposure to interacting elevated $CO_2$ and tropospheric $O_3$. New Phytol 167: 129–142.

Guy CS, Anderson JV, Haskell DW, Li QB (1994) CAPS, cors, dehydrins, and molecular chaperones: their relationship with low temperature responses in spinach In: JH Cherry (ed) Biochemical and Cellular Mechanisms of Stress Tolerance in Plants North Atlantic Treaty Organization, Advanced Research Workshop in Cell Biology, Springer, Berlin, Germany, pp 479–499.

Gygi SP, Rochon Y, Franza BR, R Aebersold (1999) Correlation between protein and mRNA abundance in yeast. Mol Cell Biol 19: 1720–1730.

Gygi SP, Corthals GL, Zhang Y, Rochon Y, Aebersold R (2000) Evaluation of two-dimensional gel electrophoresis-based proteome analysis technology. Proc Natl Acad Sci USA 97: 9390–9395.

Han ZM, Yin TM, Li CD, Huang MR, Wu RL (2000) Host effect on genetic variation of *Marssonina brunnea* pathogenic to poplars. Theor Appl Genet 100: 614–620.

Hanson AD, Gage DA, Shachar-Hill Y (2000) Plant one-carbon metabolism and its engineering. Trends Plant Sci 5: 206–213.

Hausman J-F, Evers D, Thiellement H, Jouve L (2000) Compared responses of poplar cuttings and *in vitro* raised shoots to short-term chilling treatments. Plant Cell Rep 19: 954–960.

Haynes PA, Roberts TH (2007) Subcellular shotgun proteomics in plants: Looking beyond the usual suspects Proteomics 7: 2963–2975.

He C, Zhang J, Duan A, S Zheng, Sun H, Fu L (2008) Proteins responding to drought and high-temperature stress in *Populus* x *euramericana* cv '74/76'. Trees 22: 803–813.

Heath RL (2008) Modification of the biochemical pathways of plants induced by ozone: what are the varied routes to change? Environ Pollut 155: 453–463.

Heath R, Taylor G (1997) Physiological processes and plant responses to ozone exposure In: H Sandermann, AR Wellburn, RL Heath (eds) Forest Decline and Ozone: A Comparison of Controlled Chamber and Field Experiments Ecological Studies, vol 127 Springer, Berlin, Germany, pp 317–368.

Hoogland C, Mostaguir K, Appel RD, Lisacek F (2008) The World-2DPAGE Constellation to promote and publish gel-based proteomics data through the ExPASy server. J Proteom 71: 245–248.

Hurkman, WJ, Tanaka CK (1986) Solubilization of plant membrane proteins for analysis by two-dimensional gel electrophoresis. Plant Physiol 81: 802–806.

Ibrahim L, Proe MF, Cameron AD (1997) Main effects of nitrogen supply and drought stress upon whole-plant carbon allocation in poplar. Can J For Res 27: 1412–1419.

Ingram J, Bartels D (1996) The molecular basis of dehydration tolerance in plants. Annu Rev Plant Physiol Plant Mol Biol 47: 377–403.

Inouye D (2001) The ecological and evolutionary significance of frost in the context of climate change. Ecol Lett 3: 457–463.

IPCC Summary for Policymakers (2007) In: S Solomon, D Qin, M Manning, Z Chen, M Marquis, KB Averyt, M Tignor, HL Miller (eds) Climate Change 2007: The Physical Science Basis. Contribution of working group I to the fourth assessment report of the intergovernmental panel on climate change. Cambridge Univ Press, New York, USA.

Jorrin JV, Maldonado AM, Castillejo MA (2007) Plant proteome analysis: a 2006 update. Proteomics 7: 2947–2962.

Käll L, Krogh A, Sonnhammer ELL (2004) A combined transmembrane topology and signal peptide prediction method. J Mol Biol 338: 1027–1036.

Kangasjärvi J, Talvinen J, Utriainen M, Karjalainen R (1994) Plant defence systems induced by ozone. Plant Cell Env 17: 783–794.

Kieffer P, Dommes J, Hoffmann L, Hausman JF, Renaut J (2008) Quantitative changes in protein expression of cadmium-exposed poplar plants. Proteomics 8: 2514–2530.

Kohler A, Rinaldi C, Duplessis S, Baucher M, Geelen D, Duchaussoy F, Meyers BC, Boerjan W, Martin F (2008) Genome-wide identification of *NBS* resistance genes in *Populus trichocarpa*. Plant Mol Biol 66: 619–636.

Koprivova A, Koprivova S, Jager D, Will B, Jouanin L, Rennenberg H (2002) Evaluation of transgenic poplars over-expressing enzymes of glutathione synthesis for phytoremediation of cadmium. Plant Biol 4: 664–670.

Laemmli UK (1970) Cleavage of structural proteins during the assembly of the head of bacteriophage T4. Nature 227: 680–685.

Lalanne C, Bastien C, Frey P, Dowkiw A, Faivre-Rampant P, Duplessis S, Plomion C (2007) Proteomic study of rust resistance in poplar. Poster abstract at the IUFRO Tree Biotechnology Meeting, Azores, 3–8 June 2007.

Langella O, Zivy M, Joets J (2007) The PROTICdb database for 2-DE proteomics. Meth Mol Biol 355: 279–303.

Li P, Mane SP, Sioson AA, Vasquez Robinet C, Heath LS, Bohnert HJ, Grene R (2006) Effects of chronic ozone exposure on gene expression in Arabidopsis thaliana ecotypes and in *Thellungiella halophila*. Plant Cell Env 29: 854–868.

Lippert D, Chowrira S, Ralph SG, Zhuang J, Aeschliman D, Ritland C, Ritland K, Bohlmann J (2007) Conifer defense against insects: Proteome analysis of Sitka spruce (*Picea sitchensis*) bark induced by mechanical wounding or feeding by white pine weevils (*Pissodes strobi*). Proteomics 7: 248–270.

Liu Z, Dickmann DI (1992) Abscisic acid accumulation in leaves of two contrasting hybrid poplar clones affected by nitrogen fertilization plus cyclic flooding and soil drying. Tree Physiol 11: 109–122.

Liu Z, Dickmann DI (1996) Effects of water and nitrogen interaction on net photosynthesis, stomatal conductance, and water-use efficiency in two hybrid poplar clones. Physiol Plant 97: 507–512.

Lopez-Matas MA, Nuñez P, Soto A, Allona I, Casado R, Collada C, Guevara MA, Aragoncillo C, Gomez L (2004) Protein cryoprotective activity of a cytosolic small heat shock protein that accumulates constitutively in chestnut stems and is up-regulated by low and high temperatures. Plant Physiol 134: 1708–1717.

Loustau A, Bosc A, Colin A, Ogée J, Davi H, François C, Dulrene E, Déqué M, Cloppet E, Arronays D, Le Bas C, Saby N, Pignard G, Hamza N, Granier A, Bréda N, Ciais P, Viovy N, Delage F (2005) Modeling climate change effects on the potential production of French plains forests at the sub-regional level. Tree Physiol 25: 813–823.

Lukatkin AS (2002) Contribution of oxidative stress to the development of cold-induced damage to leaves of chilling-sensitive plants: 1 Reactive oxygen species formation during plant chilling. Russ J Plant Physiol 49: 622–627.

Major IT, Constabel PC (2006) Molecular analysis of poplar defense against herbivory: comparison of wound- and insect elicitor-induced gene expression. New Phytol 172: 617–635.

Marron N, Delay D, Petit JM, Dreyer E, Kahlem G, Delmotte FM, Brignolas F (2002) Physiological traits of two *Populus × euramericana* clones, Luisa Avanzo and Dorskamp, during a water stress and re-watering cycle. Tree Physiol 22: 849–858.

Marron N, Dreyer E, Boudouresque E, Delay D, Petit JM, Delmotte FM, Brignolas F (2003) Impact of successive drought and re-watering cycles on growth and specific leaf area of two *Populus × canadensis* (Moench) clones, 'Dorskamp' and 'Luisa_Avanzo'. Tree Physiol 23: 1225–1235.

Martens L, Palazzi LM, Hermjakob H (2008) Data standards and controlled vocabularies for proteomics. Meth Mol Biol 484: 279–286.

McCarthy FM, Burgess SC, van den Berg BHJ, Koter MD, Pharr GT (2004) Differential detergent fractionation for non-electrophoretic eukaryote cell proteomics. J Proteom Res 4: 316–324.

Miller JD, Arteca RN, Pell EJ (1999) Senescence-associated gene expression during ozone-induced leaf senescence in *Arabidopsis*. Plant Physiol 120: 1015–1023.

Miranda M, Ralph SG, Mellway R, White R, Heath MC, Bohlmann J, Constabel CP (2007) The transcriptional response of hybrid poplar (*Populus trichocarpa* x *P deltoides*) to infection by *Melampsora medusae* leaf rust involves induction of flavonoid pathway genes leading to the accumulation of proanthocyanidins. Mol Plant Microb Interact 20: 816–831.

Mittler R (2006) Abiotic stress, the field environment and stress combination. Trends Plant Sci 11:15–19.

Monclus R, Dreyer E, Villar M, Delmotte FM, Delay D, Petit JM, Barbaroux C, Thiec D, Brechet C, Brignolas F (2006) Impact of drought on productivity and water use efficiency in 29 genotypes of *Populus deltoides* x *Populus nigra*. New Phytol 169: 765–777.

Narita Y, Taguchi H, Nakamura T, Ueda A, Shi W, Takabe T (2004) Characterization of the salt-inducible methionine synthase from barley leaves. Plant Sci 167: 1009–1016.

Nimchuk Z, Eulgem T, Holt BF III, Dangl JL (2003) Recognition and response in the plant immune system. Annu Rev Genet 37: 579–609.

O'Farrell PH (1975) High resolution two-dimensional electrophoresis of proteins. J Biol Chem 250: 4007–4021.

Oliver AE, Crowe LM, Crowe JH (1998) Methods for dehydration-tolerance: Depression of the phase transition temperature in dry membranes and carbohydrate vitrification. Seed Sci Res 8: 211–222.

Paiva JA, Garcés M, Alves A, Garnier-Géré P, Rodrigues JC, Lalanne C, Porcon S, Le Provost G, da silva Perez D, Brach J, Frigerio JM, Claverol S, Barré A, Fevereiro P, Plomion C (2008) Molecular and phenotypic profiling from the base to the crown in maritime pine wood-forming tissue. New Phytol 178: 283–301.

Pallardy SG, Kozlowski TT (1981) Water relations in *Populus* clones. Ecology 62: 159–169.

Pelah D, Shoseyov O, Altman A (1995) Characterization of BspA, a major boiling-stable, water-stress-responsive protein in aspen (*Populus tremula*). Tree Physiol 15: 673–678.

Pelah D, Wang W, Altman A, Shoseyov O, Bartels D (1997) Differential accumulation of water stress-related proteins, sucrose synthase and soluble sugars in *Populus* species that differ in their water stress response. Physiol Plant 99: 153–159.

Plomion C, Pionneau C, Brach J, Costa P, Baillères H (2000) Compression wood-responsive proteins in developing xylem of maritime pine (*Pinus pinaster* ait). Plant Physiol 123: 959–969.

Plomion C, Pionneau C, Bailleres H (2003) Analysis of protein expression along the normal to tension wood gradient in *Eucalyptus gunnii*. Holzforschung 57: 353–358.

Plomion C, Lalanne C, Claverol S, Meddour H, Kohler A, Bogeat-Triboulot MB, Barre A, Le Provost G, Dumazet H, Jacob D, Bastien C, Dreyer E, de Daruvar A, Guehl JM, Schmitter JM, Martin F, Bonneu M (2006) Mapping the proteome of poplar and application to the discovery of drought-stress responsive proteins. Proteomics 6: 6509–6527.

Prasad MN (1995) Cadmium toxicity and tolerance in vascular plants. Environ Exp Bot 35: 525–545.

Rabilloud T (1998) Use of thiourea to increase the solubility of membrane proteins in two-dimensional electrophoresis. Electrophoresis 19: 758–760.

Ralph S, Oddy C, Cooper D, Yueh H, Jancsik S, Kolosova N, Philippe RN, Aeschliman D, White R, Huber D, Ritland CE, Benoit F, Rigby T, Nantel A, Butterfield YSN, Kirkpatrick R, Chun E, Liu J, Palmquist D, Wynhoven B, Stott J, Yang G, Barber S, Holt RA, Siddiqui A, Jones SJM, Marra MA, Ellis B, Douglas CJ, Ritland K, Bohlmann J (2006) Genomics of hybrid poplar (*Populus trichocarpa* × *deltoides*) interacting with forest tent caterpillars (*Malacosoma disstria*): normalized and full-length cDNA libraries, expressed sequence tags, and a cDNA microarray for the study of insect-induced defences in poplar. Mol Ecol 15: 1275–1297.

Rao MV, Davies KR (2001) The physiology of ozone induced cell death. Planta 213: 682–690.

Renaut J, Lutts S, Hoffmann L, Hausman JF (2004) Responses of poplar to chilling temperatures: proteomic and physiological aspects. Plant Biol 6: 81–90.

Renaut J, Hoffmann L, Hausman JF (2005) Biochemical and physiological mechanisms related to cold acclimation and enhanced freezing tolerance in poplar plantlets. Physiol Plant 125: 82–94.

Renaut J, Hausman JF, Wisniewski ME (2006) Proteomics and low-temperature studies: bridging the gap between gene expression and metabolism. Physiol Plant 126: 97–109.

Renaut J, Bohler S, Hausman JF, Hoffmann L, Sergeant K, Jolivet Y, Dizengremel P (2008) The impact of atmospheric composition on plants A case study of ozone and poplar. Mass Spectrom Rev 28: 495–516.

Rinaldi C, Kohler A, Frey P, Duchaussoy F, Ningre N, Couloux A, Wincker P, Le Thiec D, Fluch S, Martin F, Duplessis S (2007) Transcript profiling of poplar leaves upon infection with compatible and incompatible strains of the foliar rust *Melampsora larici-populina*. Plant Physiol 144: 347–366.

Rinne PL, Kaikuranta PL, van der Plas LH, van der Schoot C (1999) Dehydrins in cold-acclimated apices of birch (*Betula pubescens* ehrh ): production, localization and potential role in rescuing enzyme function during dehydration. Planta 209: 377–388.

Rosenfeld MG, Lin C, Amara SG, Stolarsky L, Roos BA, Ong ES, Evans RM (1982) Calcitonin mRNA polymorphism: Peptide switching associated with alternative RNA splicing events. Proc Natl Acad Sci USA 79: 1717–1721.

Roth U, von Roepenack-Lahaye E, Clemens S (2006) Proteome changes in *Arabidopsis thaliana* roots upon exposure to Cd $^{2+}$. J Exp Bot 57: 4003–4013.

Rouhier N, Gelhaye E, Gualberto JM, Jordy MN, De Fay E, Hirasawa M, Duplessis S, Lemaire SD, Frey P, Martin F, Manieri W, Knaff DB, Jacquot JP (2004) Poplar peroxiredoxin Q A thioredoxin-linked chloroplast antioxidant functional in pathogen defense. Plant Physiol 134: 1027–1038.

Rozenberg P, Van Loo J, Hannrup B, Grabner M (2002) Clonal variation of wood density record of cambium reaction to water deficit in *Picea abies* (L.) Karst. Ann For Sci 59: 533–540.

Sandermann H, Wellburn AR, Heath RL (1997) Forest decline and ozone: a comparison of controlled chamber and field experiments. Ecological Studies, vol 127 Springer, Berlin, Germany.

Sanità di Toppi L, Gabbrielli R (1999) Response to cadmium in higher plants. Environ Exp Bot 41: 105–130.

Santoni V, Rabilloud T, Doumas P, Rouquie D, Mansion M, Kieffer S, Garin J, Rossignol M (1999) Towards the recovery of hydrophobic proteins on two-dimensional electrophoresis gels. Electrophoresis 20: 705–711.

Sarry JE, Kuhn L, Ducruix C, Lafaye A, Junot C, Hugouvieux V, Jourdain A, Bastien O, Fievet JB, Vailhen D, Amekraz B, Moulin C, Ezan E, Garin J, Bourguignon J (2006) The early responses of *Arabidopsis thaliana* cells to cadmium exposure explored by protein and metabolite profiling analyses. Proteomics 6: 2180–2198.

Sauter JJ, Wisniewski ME, Witt W (1996) Interrelationships between ultrastructure, sugar levels, and frost hardiness of ray parenchyma cells during frost acclimation and deacclimation in poplar (*Populus* x *canadensis* Moench 'robusta') wood. J Plant Physiol 149: 451–461.

Sauter JJ, Elle D, Witt W (1998) A starch granule bound endoamylase and its possible role during cold acclimation of parenchyma cells in Poplar wood (*Populus* x *canadensis* Moench 'robusta'). J Plant Physiol 153: 739–744 .

Sauter JJ, Westphal S, Wisniewski M (1999) Immunological identification of dehydrin-related proteins in the wood of five species of *Populus* and in *Salix caprea* L. J Plant Physiol 154: 781–788.

Saxe H, Cannell MGR, Johnsen Ø, Ryan MG, Vourlitis G (2001) Tree and forest functioning in response to global warming. New Phytol 149:369–400.

Schneider M, Bairoch A, Wu CH, Apweiler R (2005) Plant protein annotation in the UniProt knowledgebase. Plant Physiol 138: 59–66.

Seigneurin-Berny D, Rolland N, Garin J, Joyard J (1999) Differential extraction of hydrophobic proteins from chloroplast envelope membranes: a subcellular-specific proteomic approach to identify rare intrinsic membrane proteins. Plant J 19: 217–228.

Smalle J, Vierstra RD (2004) The ubiquitin 26S proteasome proteolytic pathway. Annu Rev Plant Biol 55: 555–590.

Smirnoff N (1998) Plant resistance to environmental stress. Curr Opin Biotechnol 9: 214–219.

Smith B, Ashburner M, Rosse C, Bard J, Bug W, Ceusters W, Goldberg LJ, Eilbeck K, Ireland A, Mungall CJ, OBI Consortium, Leontis N, Rocca-Serra P, Ruttenberg A, Sansone SA, Scheuermann RH, Shah N, Whetzel PL, Lewis S (2007) The OBO Foundry: coordinated evolution of ontologies to support biomedical data integration. Nat Biotechnol 25: 1251–1255.

Smith CM, Rodriguez-Buey M, Karlsson J, Campbell MM (2004) The response of the poplar transcriptome to wounding and subsequent infection by a viral pathogen. New Phytol 164: 123–136.

Sterky F, Bhalerao RR, Unneberg P, Segerman B, Nilsson P, Brunner AM, Campaa L, Jonsson J Lindvall, Tandre K, Strauss SH, Sundberg B, Gustafsson P, Uhlen M, Bhalerao RP, Nilsson O, Sandberg G, Karlsson J, Lundeberg J, Jansson S (2004) A Populus EST resource for plant functional genomics. Proc Natl Acad Sci USA 101: 13951–13956.

Szponarski W, Sommerer N, Boyer JC, Rossignol M, Gibrat R (2004) Large-scale characterization of integral proteins from *Arabidopsis* vacuolar membrane by two-dimensional liquid chromatography. Proteomics 4: 397–406.

Tamaoki M, Nakajima N, Kubo A, Aono M, Matsuyama T, Saji H (2003) Transcriptome analysis of $O_3$-exposed Arabidopsis reveals that multiple signal pathway act mutually antagonistically to induce gene expression. Plant Mol Biol 53: 443–456.

Tanner S, Shen Z, Ng J, Florea L, Guigó R, Briggs SP, Bafna V (2007) Improving gene annotation using peptide mass spectrometry. Genome Res 17: 231–239.

Thomashow MF (1999) Plant cold acclimation: freezing tolerance genes and regulatory mechanisms. Annu Rev Plant Physiol Plant Mol Biol 50: 571–599.

Thurston GT, Regan S, Rampitsch C, Xing T (2006) Proteomic and phosphoproteomic approaches to understand plant-pathogen interactions. Physiol Mol Plant Pathol 66: 3–11.

Timperio AM, Egidi MG, Zolla L (2008) Proteomics applied on plant abiotic stresses: role of heat shock proteins (HSP). J Proteom 71: 391–411.

Tschaplinski TJ, Blake TJ (1989) Water relations, photosynthetic capacity, and root/shoot partitioning of photosynthate as determinants of productivity in hybrid poplar. Can J Bot 67: 1689–1697.

Tschaplinski TJ, Tuskan GA, Gunderson CA (1994) Water-stress tolerance of black and eastern cottonwood clones and four hybrid progeny. I Growth, water relations and gas exchange. Can J For Res 24: 364–371.

Tschaplinski TJ, Tuskan GA, Gebre DE, Todd DE (1998) Drought resistance of two hybrid *Populus* clones grown in a large-scale plantation. Tree Physiol 18: 653–658.

Tuskan GA, DiFazio S, Jansson S, Bohlmann J, Grigoriev I, Hellsten U, Putnam N, Ralph S, Rombauts S, Salamov A, Schein J, Sterck L, Aerts A, Bhalerao RR, Bhalerao RP, Blaudez D, Boerjan W, Brun A, Brunner A, Busov V, Campbell M, Carlson J, Chalot M, Chapman J, Chen GL, Cooper D, Coutinho PM, Couturier J, Covert S, Cronk Q, Cunningham R, Davis J, Degroeve S, Déjardin A, dePamphilis C, Detter J, Dirks B, Dubchak I, Duplessis S, Ehlting J, Ellis B, Gendler K, Goodstein D, Gribskov M, Grimwood J, Groover A, Gunter L, Hamberger B, Heinze B, Helariutta Y, Henrissat B, Holligan D, Holt R, Huang W, Islam-Faridi N, Jones S, Jones-Rhoades M, Jorgensen R, Joshi C, Kangasjärvi J, Karlsson J, C Kelleher, Kirkpatrick R, Kirst M, Kohler A, Kalluri U, Larimer F, Leebens-Mack J, Leplé JC, Locascio P, Lou Y, Lucas S, Martin F, Montanini B, Napoli C, Nelson DR, Nelson C, Nieminen K, Nilsson O, Pereda V, Peter G, Philippe R, Pilate G, Poliakov A, Razumovskaya J, Richardson P, Rinaldi C, Ritland K, Rouzé P, Ryaboy D, Schmutz J, Schrader J, Segerman B, Shin H, Siddiqui A, Sterky F, Terry A, Tsai CJ, Uberbacher E, Unneberg P, Vahala J, Wall K, Wessler S, Yang G, Yin T, Douglas C, Marra M, Sandberg G, Van de Peer Y, Rokhsar D (2006) The Genome of black cottonwood, *Populus trichocarpa* (Torr & Gray). Science 313: 1596–1604.

Ukaji N, Kuwabara C, Takezawa D, Arakawa K, S Fujikawa (2004) Accumulation of pathogenesis-related (PR) 10/Bet v 1 protein homologues in mulberry (*Morus bombycis* Koidz) tree during winter. Plant Cell Environ 27: 1112–1121.

Unlü M, Morgan ME, Minden JS (1997) Difference gel electrophoresis: a single gel method for detecting changes in protein extracts. Electrophoresis 18: 2071–2077.

Vâlcu CM, Schlink K (2006a) Reduction of proteins during sample preparation and two-dimensional gel electrophoresis of woody plant samples. Proteomics 6: 1599–1605.

Vâlcu CM, Schlink K (2006b) Efficient extraction of proteins from woody plant samples for two-dimensional electrophoresis. Proteomics 6: 4166–4175.

van der Hoorn RA, Jones JD (2004) The plant proteolytic machinery and its role in defence. Curr Opin Plant Biol 7: 400–407.

van der Mijnsbrugge K, Meyermans H, van Montagu M, Bauw G, Boerjan W (2000) Wood formation in poplar: identification, characterization, and seasonal variation of xylem proteins. Planta 210: 589–598.

Vido K, Spector D, Lagniel G, Lopez S, Toledano MB, Labarre J (2001) A proteome analysis of the cadmium response in *Saccharomyces cerevisiae*. J Biol Chem 276: 8469–8474.

Washburn MP, Wolters D, Yates JR III (2001) Large-scale analysis of the yeast proteome by multidimensional protein identification technology. Nat Biotechnol 19: 242–247.

Welling A, Moritz T, Palva ET, Junttila O (2002) Independent activation of cold acclimation by low temperature and short photoperiod in hybrid aspen. Plant Physiol 129: 1633–1641.

Werneke JM, Chatfield JM, Ogren WL (1989) Alternative mRNA splicing generates the two ribulosebisphosphate carboxylase/oxygenase activase polypeptides in spinach and *Arabidopsis*. Plant Cell 1: 815–825.

Wilkins MR, Sanchez JC, Gooley AA, Apel RD, Humphery-Smith I, Hochstrasser DF, Williams KL (1996a) Progress with proteome projects: why all proteins expressed by a genome should be identified and how to do it. Biotechnol Genet Eng Rev 13: 19–50.

Wilkins MR, Pasquali C, Appel RD, Ou K, Golaz O, Sanchez JC, Yan JX, Gooley AA, Hughes G, Humphery-Smith I, Williams KL, Hochstrasser DF (1996b) From proteins to proteomes: large scale protein identification by two-dimensional electrophoresis and amino acid analysis. Biotechnology 14: 61–65.

Wilkins MR, E Gasteiger, JC Sanchez, A Bairoch, and DF Hochstrasser 1998 Two-dimensional gel electrophoresis for proteome projects: the effects of protein hydrophobicity and copy number Electrophoresis 19:1501–150.

Wisniewski M, Webb R, Balsamo R, Close TJ, Yu XM, Griffith M (1999) Purification, immunolocalization, cryoprotective, and antifreeze activity of PCA60: A dehydrin from peach (*Prunus persica*). Physiol Plant 105: 600–608.

Wu CC, MacCoss MJ, Howell KE, Yates JR III (2003) A method for the comprehensive proteomic analysis of membrane proteins. Nat Biotechnol 21: 532–538.

Yuan K, Zhang B, Zhang Y, Cheng Q, Wang W, Huang M (2008) Identification of differentially expressed proteins in poplar leaves induced by *Marssonina brunnea* f sp *Multigermtubi*. J Genet Genomics 35: 49–60.

Zahner R (1968) Water deficits and growth of trees. In: TT Kozlowski (ed). Water Deficits and Plant Growth. Academic Press, New York, USA, pp 191–254.

Zsuffa L, Giordano E, Pryor LD, Stettler RF (1996) Trends in poplar culture: some global and regional perspectives. In: RF Stettler, HD Bradshaw Jr, PE Heilman, TM Hinckley (eds) Biology of *Populus* and its Implications for Management and Conservation. NRC Research Press, National Research Council of Canada, Ottawa, Canada, pp 515–539.

[1]INRA, UMR1202 BIOGECO, 69 Route d'Arcachon, F-33612 Cestas, France;
[a]e-mail: *plomion@pierroton.inra.fr*
[b]e-mail: *vincent@pierroton.inra.fr*
[c]e-mail: *bedon@pierroton.inra.fr*
[2]UMR de GENETIQUE VEGETALE, INRA/Univ. Paris XI/CNRS/INA PG, Ferme du Moulon, Gif-sur-Yvette F-91190, France; e-mail: *joets@moulon.inra.fr*
[3]UMR de GENETIQUE VEGETALE, INRA/Univ Paris-Sud/CNRS/AgroParisTech, Ferme du Moulon, Gif-sur-Yvette, F-91190, France; e-mail: *bonhomme@moulon.inra.fr*
[4]UFR-Faculté des Sciences, Laboratoire de Biologie des Ligneux et des Grandes Cultures, Université d'Orléans, UPRES EA 1207, rue de Chartres, BP 6759 Orléans Cedex 45067 02, France; e-mail: *domenico.morabito@univ-orleans.fr*
[5]Interactions Arbres/Micro-organismes, Centre INRA de Nancy, UMR1136 INRA/Université Nancy, Champenoux 54280, France; e-mail: *duplessi@nancy.inra.fr*
[6]Department of Forest Genetics and Plant Physiology, Swedish University of Agricultural Sciences, Umeå Plant Science Centre, Umeå, 90183, Sweden;
[d]e-mail: *Robert.Nilsson@genfys.slu.se*
[e]e-mail: *Gunnar.Wingsle@genfys.slu.se*
[7]Department of Biochemistry, Lund University, Box 124, Lund S-22100, Sweden; e-mail: *Christer.Larsson@plantbio.lu.se*
[8]UMR1137 Ecologie et Ecophysiologie Forestières, INRA/UHP, Nancy-Université BP239, Vandoeuvre-lès-Nancy Cedex F-54506, France; e-mail: *jolivet@scbiol.uhp-nancy.fr*
[9]Centre de Recherche Public-Gabriel Lippmann, Department of Environment and Agrobiotechnologies (EVA), Proteomics Platform, 41 rue du Brill, Belvaux L-4422, Luxembourg; e-mail: *renaut@lippmann.lu*
[10]Department of Biochemistry and Molecular Biology, Mississippi State University, Starkville, MS 39762, USA; e-mail: *op2@msstate.edu*
[11]Department of Forestry, Mississippi State University, Starkville, MS 39762, USA; e-mail: *mcy1@msstate.edu*
*Corresponding author

# 7

# Metabolomics in Poplar

*Andrew R. Robinson*[1,a] and *Shawn D. Mansfield*[1,b,]*

## ABSTRACT

The sequencing of the poplar genome has opened the door to "high-throughput" genome-wide research aimed at understanding the underpinnings of the complex biological processes inherent to tree growth and development. In parallel to the significant transcriptomics efforts, which have successfully been employed in several functional studies in a variety of poplar tissues, the more recent development of the emerging post-transcriptional tools such as proteomics and metabolomics, which characterize the abundance of protein and the metabolites in tissues, should substantially enable systems biology assessments of tree development and response(s) to biotic and abiotic stresses. This chapter discusses the recent applications of metabolomics in poplar.

**Key words:** developmental biology; metabolite profiling; metabolomics; metabolic markers; GC-MS; LC-MS; high throughput biology; poplar wood formation

## 7.1 Introduction

Metabolomics represents one of the more recent additions to the plant functional genomics toolbox—it combines the measurement, identification and statistical interrogation of the small, soluble molecules (metabolites) that act as intermediates and end products of metabolic pathways in living tissue. The primary outcome of metabolomics analyses are metabolite "fingerprints" that can, for example, act as surrogate indicators of the metabolic state in vivo during development or response to external stimuli.

[1]Department of Wood Science, Faculty of Forestry, University of British Columbia, 4030-2424 Main Mall, Vancouver, BC, V6T 1Z4, Canada.
[a]e-mail: *andrewrobinsonnz@gmail.com*
[b]e-mail: *shawn.mansfield@ubc.ca*
*Corresponding author

The analysis of metabolites involved in primary and secondary metabolism is not itself a recent development, however, modern advances in analytical technologies and the accessibility of computing power in the recent research environment have made the broad-scale, collective analysis of living tissues' metabolite composition a practical endeavor.

At its inception, the promise of metabolomics to help bridge the "regulatory space" existing between genotype and phenotype in plants quickly raised the profile of this field. It is apparent, however, that despite substantial technological, methodological and conceptual advances, the field of metabolomics continues to be confronted with challenges, including comprehensive detection and identification of metabolites in crude sample extracts, high sample throughput (both in terms of sample processing and data analysis), and the effective interpretation of results. These technical limitations have meant that the emergence of increased biological understanding and diagnostic capabilities via metabolomics has not been as rapid and widespread as initially hoped. Consequently, metabolomics platforms in integrated plant functional genomics programs has been slower to be implemented in routine functional genomic evaluations of plants/trees. Indeed, in its first decade the field has undergone considerable technical advancement, yet has only just begun to realize its potential, and has yet to achieve a truly mainstream presence in tree biology.

To date, a relatively small number of metabolomics experiments have been reported for poplar species compared to the model plant system *Arabidopsis*, in which many metabolomics techniques were, and continue to be founded. Although metabolomics has not been extensively applied in poplar research to date, poplar's position as the model system angiosperm makes it an ideal target for metabolomics analysis of, for example, wood properties, development, injury response, and dormancy cycles in trees. This is especially true where integrated analyses involving data from several -omics approaches are concerned, given the availability of the genome sequence, genetic maps, expressed sequence tag (EST) libraries, proteomic tools and several microarray chips for poplar.

## 7.2 The Foundations of Plant Metabolomics

Metabolomics is founded on the premise that investigating the interrelationships between metabolites, as well as between metabolites and genetics, gene expression, environmental stimuli and phenotype, can aid in defining the biological underpinnings that control the establishment of phenotypic traits in biological systems. Theoretically, relationships between phenotypic traits and metabolite accumulation features have the potential for greater stability than relationships with other causal agents of phenotype that exist upstream of cellular metabolism. This is because metabolic features reflect the cellular activity immediately preceding the emergence of a physical phenotype and, presumably, integrate inputs from the upstream genotype, gene and protein expression patterns, temporal and spatial subcellular localization of gene products, as well as the influences of environmental and developmental factors.

## 7.2.1 Metabolomics vs. Metabolite Profiling

Metabolomics analyses have been broadly classified as either "targeted" or "non-targeted". Targeted analysis, otherwise known as "metabolite profiling", typically focuses on quantifying a defined group of metabolites that are related by either a metabolic pathway or a molecule class. These studies tend towards a higher degree of a priori knowledge as far as compound identity and interrelationship are concerned, and in their most refined form become "target analyses"—involving the measurement of one or very few metabolites to serve as, for example, phenotypic biomarkers. Conversely, non-targeted analysis aims to measure as broad of a range of metabolites as possible, with the intention of creating a global metabolic fingerprint. In the first instance, global fingerprinting is not as concerned with the metabolites' identity and absolute abundance as it is with their relative abundance and interrelationships, and aims primarily to classify samples based on metabolic "features". Ultimately, the reductive data mining approaches commonly employed in these analyses frequently lead to the identification of subsets of discriminating metabolites whose abundances correlate with specific treatments or genetic or phenotypic traits of interest. Subsequently, attempts can be made to identify those compounds so that their biological significance may be rationalized. Whereas broad-scale metabolomics is a recent development of the last decade, targeted analysis of metabolism has a much longer history. Although it is arguable that targeted analyses are not metabolomics in the strict sense because of their narrow focus, they do comprise the origin from which non-targeted, global metabolomics approaches have been derived with the assistance of advancing technology. In fact, analyses conducted in the metabolomics era have frequently occupied a middle ground in terms of the degree of prior knowledge of the identity and role of the metabolites being analyzed, the breadth of metabolites being analyzed and the basis for their inclusion. It is apparent that the scale and rationality of analyses are not criteria that allow a practical distinction between modern metabolomics and historical metabolic analyses to be made. In reality, contemporary metabolomics is defined by a new working environment—one in which powerful new analytical tools, abundant computing power and effective data-handling software have made it conceivable to tackle metabolic issues at the whole organism or tissue level, with an emphasis on deconvoluting biological complexity.

## 7.2.2 The Effective Incorporation of Metabolomics into Systems Biology

From the turn of the century, at the time when the first reports of broad-scale metabolite profiling were made, it has been suggested that metabolomics would evolve into a powerful, integrated branch of plant systems biology (Fiehn et al. 2001; Weckwerth 2003; Fernie et al. 2004). In the intervening time, it has become clear that the utility of metabolomic data can only be fully realized when it is interpreted in conjunction with its genomic, transcriptomic and proteomic counterparts. Fulfillment of this realization demands the concurrent analysis of sample sets by multiple "-omics" platforms, which brings with it increasingly complex experimental scenarios and the accompanying, substantive logistical demands. It is reassuring to see that despite the inherent difficulties, multi-omic analyses, and the data processing tools required to conduct them (Daub et al. 2003; Wurtele et al. 2003; Klukas et al. 2006; van Riel 2006; Bylesjo et al. 2007), are becoming more commonplace, and that data from one "-omics" platform can assist in the interpretation of that from others. To date, the most common combination has been metabolomics and transcriptomics analyses (Colebatch et al. 2004; Hirai et al. 2004; Urbanczyk-Wochniak et al. 2005; Nakamura et al. 2007; Osuna et al. 2007), with some excellent examples of combined data being presented in either correlation networks (Nikiforova et al. 2005a) or pathway scaffolds (Tohge et al. 2005). Combined metabolomics/genetics studies are also emerging (Morreel et al. 2006; Lisec et al. 2008). Possibly the most effective holistic approach to multi-omics presented to date is the orthogonal 2 partial least squares (O2PLS)-based integrated analysis of transcript, protein and metabolite data, for which lignin biosynthesis in hybrid aspen was selected as the trial system (Bylesjo et al. 2009). In any case, it is clear that the future advancement of metabolomics will be closely tied to these kinds of combined multi-platform analyses.

## 7.2.3 Tools for Measuring Metabolites

A variety of analytical tools are available for the generation of metabolite fingerprints or profiles, with specific tools being more appropriate for the determination of metabolites having particular physico/chemical properties. In this regard, the analysis of the plant metabolome requires no special consideration over that of other organisms, with chromatography, mass spectrometry and NMR spectroscopy being the analytical mainstays across the field.

Gas chromatography (GC) is the chromatographic technique of choice for the analysis of smaller (MW < ~ 1,000 Da) molecules, owing to its applicability to a broad range of molecular classes and high resolution. Furthermore, the recent emergence of ultra-fast gas chromatography offers a significant increase in sample processing efficiency that promises to assist the development of very high-throughput metabolomics. The usual approach to sample introduction for GC is the evaporation of liquid phase extracts in the injector, although other techniques such as headspace extraction can be effective in specific scenarios and may avoid the need for lengthy sample preparation (Kjalstrand et al. 1998, Wang et al. 2006). Alternatively, high pressure liquid chromatography (HPLC) is useful for the separation of molecules too large or too labile for GC. Furthermore, the advent of ultra-high pressure liquid chromatography (U-HPLC) has facilitated the much-needed increases in the resolution of liquid chromatography for metabolomics (Grata et al. 2008). Although the range of metabolites that may be analyzed by liquid chromatography is frequently limited to a specific polarity range in the "middle ground" (Roepenack-Lahaye et al. 2004), variant approaches including capillary-based and hydrophilic interaction chromatography (HILC) can be used to broaden this specificity (Tolstikov and Fiehn 2002, Tolstikov et al. 2003, Roepenack-Lahaye et al. 2004). Capillary electrophoresis is another emerging liquid-based separation technology with potential applications in plant metabolomics (Soga et al. 2003).

Chromatographic separation systems require an attached quantitative detector, and mass spectrometers have achieved widespread popularity in metabolomics research. Quadrupoles, ion traps, Fourier transform (FT) and time-of-flight (TOF) mass analyzers, combined with various sample introduction methods appropriate for the respective spectrometer and the preceding gas or liquid chromatography technique, have all been applied in various settings. The mass spectral data generated provides extensive molecular structural information, and can be used to deconvolute signals from co-eluting metabolites, effectively increasing the resolution of the chromatographic analysis (discussed below). Also popular are photodiode array (PDA) systems, which may be implemented as detectors for liquid chromatography—either alone or in combination with a mass spectrometer. PDAs measure light absorption across the ultraviolet and visible wavelengths, generating characteristic spectra for responsive analytes such as aromatics.

NMR spectroscopy is an alternative to chromatography/mass spectrometry for resolving compounds in metabolomics analyses (Ratcliffe and Shachar-Hill 2001, Ott et al. 2003, Charlton et al. 2004, Terskikh et al. 2005, Sanchez et al. 2008). A major benefit of NMR spectroscopy is that it is non-destructive, permitting living samples to be repeatedly analyzed over the course of an experiment, or studied by alternative approaches once NMR analysis is complete. Biological NMR spectroscopy usually exploits the magnetic properties of $^1$H or $^{13}$C nuclei, but may also target $^{31}$P or $^{15}$N (Bligny and Douce 2001). Thus, NMR spectra provide information on the number and type of atomic nuclei, for example, $^1$H nuclei in a mixture of metabolites, and it is possible to resolve the contribution of individual molecules to the spectra generated by complex metabolite mixtures. While this technique has been applied directly to intact plant tissues, or crude metabolite preparations, it has also seen some application as a detection tool in HPLC-based analyses (Wolfender et al. 2003). The application of NMR-based metabolomics has yet to be demonstrated in poplar, however.

The provision of structural information by the detection system is crucial to the success of metabolomics analyses. Without compound identification, all that can be provided by the analysis of metabolite extracts is a metabolic fingerprint, and while potentially useful for distinguishing between distinct metabolic systems, a fingerprint alone is not informative about underlying biological relationships. This fact has fuelled the popularity of photodiode array-, NMR-, and mass spectrometry-based detection in metabolomics, as the molecule- or molecular class-specific spectral patterns generated by these techniques can facilitate the identification of metabolites, via matches with the signatures of standard compounds. Spectral matches, combined with matches for retention times or indices (in analyses that include chromatographic separation), can provide a high level of certainty for positive identifications. Furthermore, with appropriate spectrometers, elemental composition calculations, soft chemical ionization techniques or MS$_n$ analysis can also contribute to the identification of compounds in the absence of verified standards (Fiehn et al. 2000b; Tolstikov and Fiehn 2002).

Obviously, libraries of the spectral and retention index data for biological molecules are of enormous utility when attempting to identify metabolite compounds, and extensive libraries are available for NMR and GC/(EI)MS. Although it is possible to assemble such libraries for LC/MS, the high degree of instrument- and eluent-dependent variation in analyte fragmentation patterns has meant that these libraries are not universally compatible. For GC/MS, however, inert carrier gases and the standard use of a 70eV potential for electron ionization (EI) and molecule fragmentation have meant that not only can libraries be constructed, but they can also be shared between instruments and research groups. This has led to the publishing of extensive commercial and freely distributed libraries of EI mass spectra. While commercial libraries represent an extremely broad range of molecules (e.g., the 2008 NIST library contains more than 190,000 compounds including various states of derivatization), smaller, free libraries such as those provided by the Gölm Metabolite Database (GMD) (Kopka et al. 2005) are tailored specifically to the needs of plant metabolomics. These libraries are less redundant and frequently have more utility. Despite the growing number of available libraries, many compounds resolved from metabolite continue to elude identification. The ongoing expansion of mass spectral library resources is of paramount importance, because the process of compound identification constitutes a major limiting factor in the plant metabolomics field.

### 7.2.4 The Analytical Process

Practical metabolomics is concerned with measuring and analyzing metabolite pools in an attempt to understand metabolic networks and develop biological markers. An expanding range of analytical and software tools are available to assist in this regard.

### 7.2.4.1 Sample Preparation

The source material for samples are selected for their relevance to the research objectives (e.g., developing xylem tissue is a good substrate for analysis of xylem biosynthetic metabolism), and may be comprised of whole plants (practical only for small species such as *Arabidopsis*), plant fluids (e.g., xylem or phloem sap), compounds released in gas exchange (e.g., volatile terpenoids), individual plant organs (e.g., root, leaf, stem or inflorescence), and now even laser capture microdissected cell groups (Schad et al. 2005). While some of the analytical tools employed in metabolomics permit the determination of metabolite composition with minimal sample preparation (e.g., nuclear magnetic resonance spectroscopy; NMR), others require active extraction of metabolites from the specific tissue, prior to analysis. This process typically involves –80°C flash freezing or freeze-drying of samples and tissue disruption (Fiehn et al. 2000a, Shepherd et al. 2007), followed by some form of liquid solvent extraction and, when required, further solvent partitioning of the crude extracts.

   Although metabolite extractions based on single solvents (e.g., methanol or chloroform) are applicable, the composition of the extract obtained will exhibit bias toward metabolites that are highly soluble in the chosen solvent, which may be either desirable or undesirable in particular analyses. Because of this, sample preparation for completely non-targeted chromatography-based metabolomics has frequently employed multi-solvent extractions, typically including water (very polar) and at least one other less-polar solvent. Variations of the extraction and derivatization protocols for *Arabidopsis* published by Fiehn et al. (2000a, b) are often employed. The extraction is based on a dual-phase water/methanol/chloroform extraction that yields polar metabolites in the water/methanol phase and less-polar metabolites in the methanol/chloroform phase. A method for single-phase extraction with these same solvents, combined in ratios that do not lead to phase separation, has also been established (Gullberg et al. 2004). In situations where specific metabolite classes are being targeted (e.g., phenolics), selective extraction and subsequent metabolite partitioning and enrichment can be used to refine samples to achieve better resolution and signal-to-noise ratios for the target metabolites. Such a method involving methanol extraction, followed by lyophilization and subsequent partitioning of the metabolites between water and cyclohexane, was employed for the concentration of phenolic metabolites in an aqueous phase (Damiani et al. 2005). Another example of an extraction protocol specifically tailored to a subject metabolite class is that employed for the specific extraction of membrane phospholipids developed for *Arabidopsis* (Welti et al. 2002, Yang et al. 2007). Isopropanol with butylated hydroxytoluene (BHT) is used as the primary solvent, with various mixtures of chloroform, water and methanol with BHT used for subsequent, exhaustive tissue extraction.

When samples are to be analyzed by gas chromatography (GC), it is common to derivatize the metabolites post-extraction as a way of increasing volatility and therefore the high mass cut-off of the analysis. In the classic approach to metabolite sample preparation (Fiehn et al. 2000a, b), this involves the protection of carbonyl moieties by reaction with an alkoxy-oxyamine hydrochloride, followed by the elimination of acidic protons by reaction with a trimethylsilylating agent (e.g., N-methyl-N-trimethylsilyltrifluoroacetamide (MSTFA)). Where appropriate, it is also recommended that a methanol/chloroform-based trans-methylation of hydrocarbon chains be carried out prior to the other derivatization reactions.

The documented optimization of conditions for metabolite extraction and derivatization in *Arabidopsis* leaves, stems and cell cultures (Gullberg et al. 2004; t'Kindt et al. 2008), developing xylem of loblolly pine (Morris et al. 2004) and potato tubers (Shepherd et al. 2007) clearly illustrate the importance of ensuring that the process has enough stringency to achieve good metabolite extraction, but is not harsh enough to cause degradation of labile compounds. The susceptibility of the metabolite profile to variations in sample handling and analytical conditions is a known limitation of metabolomics, which demands consistent processing in order for comparable datasets to be generated from individual samples or sample batches.

## 7.2.4.2 Data Processing and Analysis

In chromatograms of complex biological samples the partial or complete co-elution of metabolites is a frequent occurrence that, if not addressed, can limit biological resolution and introduce error into downstream data analyses. Additionally, the collation of metabolite profile data from multiple samples is required prior to statistical analysis, but manual collation becomes impractical in chromatography-based analyses involving large sets of samples and/or metabolites. This is because unavoidable fluctuations in temperature ramps, eluent gradients, column pressure or flow rates lead to inter-sample variation in the metabolite separation domain (which is time-based in most high-resolution chromatography), ensuring that the retention time of any given metabolite is seldom, if ever, a single exact value across all sample runs. Fortunately, the pressing need to resolve these issues has led to the development of algorithms that are capable of deconvoluting the signals from co-eluting metabolites. Both commercial and free software tools that semi-automate these tasks have emerged, with notable non-commercial offerings including NIST AMDIS (for deconvolution only), MSFACTs (Duran et al. 2003), metAlign (Tolstikov et al. 2003; Tikunov et al. 2005), correlation optimized warping (COW) (Christin et al. 2008; Nielsen et al. 1998), and the highly capable XCMS (Smith et al. 2006; Benton et al. 2008).

Data analysis in metabolomics has advanced at a considerable rate with the ongoing introduction of statistical analyses and other calculative tools to the field. Most statistical tools have been applied from exploratory or reductive perspectives. Classic univariate tests between means, such as Student's t-test, the F-test and more robust incarnations like Tukey's "Honestly Significant Difference" (HSD) test have been used to individually identify metabolites exhibiting genotype- or treatment-related differences in abundance (Fiehn et al. 2000a; Yeh et al. 2006). Although useful, these tests deal with each metabolite as an isolated entity, and are unable to take the interdependence of the components of metabolite profiles into account (i.e., the "network" paradigm). Multivariate analyses, however, are better suited to this task. The multivariate tools initially adopted in metabolomics were principal components analysis (PCA) (Fiehn et al. 2000a, Chen et al. 2003) and hierarchical cluster analysis (HCA) (Roessner et al. 2001a, b), and the scope of many classic and contemporary analyses is limited to these two techniques. Both are useful for comparing complete profiles from multiple samples, and generate diagrammatic outputs that are visually appealing and easily interpreted. Although PCA does provide some information regarding the particular metabolites responsible for any distinction between sample classes, neither PCA nor HCA are very diagnostic, as they are unable to provide calculated measures of the relationships between metabolite profiles and, for example, phenotypic traits. Canonical correlation analysis (CCA) is one method that can better assist in defining the relationships between two sets of variables, such as metabolites and quantitative phenotypic traits (Meyer et al. 2007). Essentially, CCA identifies groups of variables in one set that are correlated to groups of variables in the other, and indicates the relative contributions of individual variables to the relationship. However, in cases where diagnostics are an objective, techniques that generate models for the prediction of specific traits on the basis of metabolite profiles typically have more utility. To this end, multiple discriminant analysis (MDA) is useful for distinguishing samples by class (e.g., genotype, species), while partial least squares regression (PLSR) (Dijksterhuis et al. 2005; Meyer et al. 2007) and less conventional "stepwise" variable selection procedures (Li and Nyholt 2001; Klukas et al. 2006; Yamashita et al. 2007) are powerful techniques for modeling quantitative phenotypic traits (e.g., the total lignin content of wood, or any other quantifiable property).

The graphical presentation of biochemical pathways and molecular interactions, as supported by metabolomic data, is an important component of metabolomics, and can contribute considerably to data interpretation and the derived understanding of biological relationships at the molecular level. Neural networking, as conducted by the "Pajek" software (Batagelj and Mrvar 2003), is a graphical correlative statistical approach capable of effectively summarizing interactive networks, which uses marker size and proximity to visualize the interactions within sets of variables. In metabolomics, the networks generated by this process provide valuable insight into the interdependency between specific metabolites, which can reveal hubs or metabolic control points within the systems being analyzed (Batagelj and Mrvar 2003; Fiehn 2003; Steuer et al. 2003; Giuliani et al. 2004). Another option is to record the behavior of various metabolites on conceptual metabolic pathway scaffolds established using previous research. Obviously, this could be done manually, but scaffolding and annotation software such as MapMan (Thimm et al. 2004) can expedite the process. These scaffolds are annotated with the contributions of interesting metabolites to a given relationship with, for example, a phenotypic trait. These may be represented by numerical scores (Hirai et al. 2004) or color-coding and heatmap output markers (Nikiforova et al. 2005b).

## 7.3 The Application of Metabolomics Technology in the Study of Poplar Species

Concurrent with the rise of metabolomics technology over the last decade, metabolomics analyses have increasingly been carried out on a wide range of poplar species using a number of analytical and data processing techniques. Metabolomics is becoming a recognized research approach in poplar because of its usefulness in investigating the structure and dynamics of biomolecular systems and addressing specific biological questions.

### 7.3.1 Molecular Signaling Mechanisms

The analysis of *Arabidopsis*-derived gibberellic acid (GA) insensitive (*gai*) and repressor of *GA1-like1* (*rgl1*) modified *Populus tremula* × *P. alba* (Busov et al. 2006) is an excellent example of metabolomics techniques being applied to the study of molecular signaling mechanisms. Both GAI and RGL-1 are believed to be repressor transcription factors that regulate gibberellic acid metabolism, and the transgenes employed in this study contained a mutated DELLA domain, which would ordinarily be involved in mediating GA responses.

Relative to wild type controls, both *gai*- and *rgl1*-expressing poplar exhibited severe dwarfing of shoots and increased root generation. Gibberellic acid metabolism was clearly perturbed in the modified lines, with the leaf tissue exhibiting multi-fold increases in the abundance of bioactive GAs such as $GA_1$, $GA_2$ and $GA_4$, and decreases in $C_{20}$ GAs such as $GA_{53}$, $GA_{19}$ and $GA_{44}$. Also, an apparent lowering in activity of GA 2-oxidase(s) lead to further accumulation of $GA_1$, $GA_4$ and $GA_{20}$ at the expense of $C_2$-hydroxylated $GA_8$, $GA_{34}$ and $GA_{29}$, respectively. The general metabolic effects of this perturbation were extensive; in roots, both mutant types exhibited reduced monosaccharide (glucose, fructose and galactose) concentrations, increased concentrations of citric acid and amino acids (Glu, Arg, GABA and Asp), and the emergence of two unidentified glucosides not detected in the wild-type. Additionally, Asn and carbamoyl aspartate contents increased in *rgl1* transformants, while salicin and threonic acid contents increased in *gai* transformants. Metabolite profiles of leaf tissue from modified lines were characterized largely by changes in metabolites related to phenylpropanoid metabolism. This included increased accumulation of defense, storage, and detoxification compounds (e.g., phenolic glucosides and catechol) coupled with decreases in phenylalanine, quinic acid, salicylic acid, and phenolic conjugates including quinic acid conjugates. Furthermore, the abundances of many nitrogen-containing metabolites, including amino acids, were lower in leaves of modified plants, with some differences between *rgl1* and *gai* lines.

The changes observed in metabolite profiles suggested a mechanism by which the dwarf phenotype might possibly arise. The decrease in monosaccharides, and accumulation of organic and amino acids in roots could be ascribed to increased respiratory activity that is promoted by the modified gibberellic acid environment, which leads to increased root growth. The concurrent decline in nitrogen-containing metabolites (including phenylalanine) in leaves is consistent with increased partitioning of nitrogen to the roots in order to fuel higher root biomass allocation. Furthermore, the apparent diversion of carbon flux into phenolic storage and defense compounds, at the expense of cell wall biosynthesis via the phenylpropanoid pathway, is consistent with the reduction in shoot growth.

Whatever the mechanisms are, the metabolomics analysis of the *gai* and *rgl1* expressing poplar offered support to a role for GA-based regulation of defense response, wood biosynthesis, partitioning of resources between roots and shoots, morphological determination, and other ecophysiological adaptations in poplar species.

### 7.3.2 Pooling Phenomena in Secondary Metabolites

Poplar species are known for their constitutively high salicylic acid content, although the mechanisms by which this secondary metabolite is biosynthesized in these species are still not well understood. Morse et al. (2007) investigated the effects of over-expressing the bacterial *nahG* transgene, which catalyzes the conversion of salicylic acid to catechol (Gaffney et al. 1993), on the metabolite profiles of *Populus tremula* × *P. alba* leaves. Interestingly, this study found that the metabolic effects of *nahG* in poplar were quite different to those in tobacco, suggesting that differences exist between salicylic acid-associated metabolic networks in the *Populus* and *Nicotiana* genera. In contrast to tobacco, in which *nahG* expression leads to a significant reduction in the salicylic acid pool and increase in the catechol pool in leaves (Gaffney et al. 1993), *nahG* expression in poplar appears to be relatively silent in this regard—having only marginal effects on these metabolite pools—but has much more substantial effects on a series of related glycosides and shikimate and phenylpropanoid intermediates. Most notably, quinic acid and some of its conjugates were substantially decreased, while catechol glucoside increased. Based on these observations, it was suggested that there are mechanisms actively working to maintain the constitutively high levels of salicylic acid in poplar leaves, and that this high level is of some adaptive significance. In these *nahG* transgenics, flux from salicylic acid into catechol is apparently met with increased partitioning of substrate into salicylic acid production, and conversion of excess catechol into glycosidic storage/detoxification forms. As an extension of this work, it would certainly be interesting to see what effects the *nahG* construct would have on metabolite profiles in developing xylem and phloem, given the high level of shikimate and phenylpropanoid pathway activity in wood formation.

In another study that remains unique in poplar, Morreel et al. (2006) performed quantitative trait loci (QTL) analysis on the relative concentrations of the major flavonoid metabolites found in the apical tissues of two full-sib poplar families having a common maternal parent (*Populus deltoides* cv. S9–2 × *Populus nigra* cv. Ghoy and *Populus deltoides* cv. S9–2 × *Populus trichocarpa* cv. V24). In a multivariate analysis of 15 common flavonoids against a microsatellite-aligned amplified fragment length polymorphism (AFLP) linkage map, four robust metabolite-QTLs (mQTLs) associated with rate-limiting steps in flavonoid biosynthesis were identified; of these, three involved identifiable metabolites. Quercetin and quercetin 3-methyl ether were associated with linkage group III of *Populus nigra*, which is itself proximal to a series of chalcone synthase homologues. This supported other evidence (Burbulis and Winkel-Shirley 1999; Saslowsky and Winkel-Shirley 2001; Lukacin et al. 2003; Winkel 2004) that quercetin production may arise from flux through a chalcone synthase-controlled metabolic channel involving chalcone synthase, chalcone isomerase, flavonone 3-hydroxylase, and flavonol synthase activities. Pinostrobin levels were associated with linkage group XIII of *Populus deltoides*, indicating that this QTL was related to an unidentified 7-*O*-methyl transferase activity targeting pinocembrin as substrate. Interestingly, pinobanksin 3-acetate levels were associated with the same linkage group of *Populus nigra*. This indicated a relationship with 3-*O*-acetyltransferase activity targeting a pinobanksin substrate, possibly derived from QTL-linked BADH gene superfamily homologues. The approach followed in this work (termed "genetical metabolomics") demonstrated that metabolite profiles can be used for QTL analysis to reveal loci that control flux through complex, multi-branched pathways of related metabolites. There is clear potential for the use of similar analyses to identify mQTLs and the associated genes that are responsible for control points in other metabolic pathways.

### 7.3.3 Response and Adaptation to Environmental Conditions

The ability to respond to environmental factors that challenge homeostatic equilibrium has far reaching consequences for plant health and survival, and, in the case of cultivated crops, productivity. Such abiotic pressure demands coordinated, system-wide adjustment in order for equilibrium to be maintained, and metabolomics technologies have become popular tools for investigating the biochemical mechanisms of this process—be it in the short or long term, or on evolutionary timescales. Environmentally pressured systems continue to be key subjects in the development and application of tools for multi-omics analysis in plants, and relevant studies on poplar species have been published.

Under hypoxic conditions, such as those caused in the rhizosphere by brief or prolonged flooding, normally aerobic, respiration-dependent terrestrial plants must somehow maintain energy generation in order to survive. A combined metabolomics and transcript profiling study of Gray poplar (*Populus canescens*), which is a flood-tolerant species, investigated the broadscale molecular responses of hypoxic roots and normoxic "feeder" leaves, and the interrelationships between these organs in such scenarios (Kreuzwieser et al. 2009). On account of this study, the metabolic signatures of this transition are now known in detail and the core results of the analysis will be outlined here. Under hypoxic conditions metabolic changes are observed in both roots and leaves, but significant changes in gene expression are seen only in the roots. Hypoxia prompts a transient increase in lactate in the roots that is followed by a long-term increase in the ethanol pool, and this indicates that the initial response to anaerobic conditions is a switch to lactic acid fermentation that is replaced soon after by alcohol fermentation. Sucrose pools decrease in the leaves, but increase in the roots, and increase considerably in phloem sap. Concurrently, glucose, fructose and fructose 6-phosphate pools increase in the roots, while the starch pool decreases slightly. This suggests an increased transport of sucrose from leaves to roots, with subsequent degradation into monomeric sugars in order to satisfy an increased demand for carbohydrates in hypoxic roots maintaining fermentation processes. Increased succinate pools in hypoxic roots support transcript-based evidence for the induction of an enhanced glyoxylate cycle, which likely fulfills a gluconeogenic function by linking lipid degradation to carbohydrate production in these conditions.

Amino acid concentration changes in both leaves and roots indicate that hypoxia has widespread affects on nitrogen metabolism. In roots, a reduction in glutamine suggests decreased nitrogen assimilation via ammonium. Also, a reduction in glutamic acid combined with increased succinate, alanine, and GABA suggests the onset of the GABA shunt, which is a proton consuming process that acts to stabilize pH while fermentation is occurring. This is an interesting adaptation, as the GABA shunt is either only a transient process or non-existent in flood-intolerant species such as *Arabidopsis thaliana* (Klok et al. 2002; Liu et al. 2005). Many pyruvate-derived (alanine, valine, leucine) and glycolysis intermediate-derived amino acids are more strongly accumulated, whereas many derived from TCA cycle intermediates (glutamic acid, glutamine, aspartate, and asparagine) are downregulated. This is further support that during hypoxia, glycolytic processes are favored over respiration via the TCA cycle, with a resultant increase in the redirection of carbon into glycolytic intermediate-derived amino acids and a decrease in flux into TCA intermediate-derived amino acids.

Overall, the data suggest that Gray poplar derives its high flood tolerance from an ability to maintain energy metabolism during oxygen deficiency. Hypoxia necessitates a metabolic transition from aerobic respiration to anaerobic acid or alcoholic fermentation, and this species is able to maintain these anaerobic processes long term via the coordinated modulation of a variety of cellular processes related primarily to carbon supply and nitrogen metabolism.

A targeted metabolite profiling analysis coupled to transcript profiling investigated the foliar metabolic character of *Populus euphratica* (Brosche et al. 2005), which is a heat- and salt-tolerant species that survives in saline semi-arid conditions. Leaf samples were taken from plants grown under salinity and drought conditions of different severity, and solvent extracts analyzed by GC-MS. Although the number of metabolites for which data was analyzed was limited to only 22, the list did include common examples of organic acids, sugars, sugar alcohols, amino acids, and glycerol, with some shifts being observed in metabolites related to osmoprotection. The sugars fructose, glucose, raffinose, sucrose, trehalose and xylulose tended towards insignificant decreases under conditions of salinity and drought, with the exception that under the most severe conditions raffinose concentrations approximately doubled. By contrast, other metabolites including glyceric acid, glycerol, β-alanine, proline, and valine were more responsive and upregulated to a greater extent under the less severe conditions. Interestingly, the overall finding was that in both transcript and metabolite profiles, the response of *Populus euphratica* to a combination of high temperatures, salinity and drought was much less dramatic that has previously been observed in intolerant *Arabidopsis* (Kaplan et al. 2004; Rizhsky et al. 2004). This suggests that this species is constitutively prepared for stressful conditions and/or that other responsive mechanisms exist that were not uncovered by this study.

Poplar hybrids have been used to demonstrate the potential utility of metabolomic analyses in elucidating the control mechanisms of highly plastic traits that exhibit strong genotype × environment effects. Harding et al. (2005) analyzed the relationship between foliar flavonoid-derived condensed tannin (CT) and salicylate-derived phenolic glucoside (PG) contents, and primary metabolite profiles in several *Populus fremontii* L. × *Populus angustifolia* James hybrids, under conditions of either nitrogen surplus or deprivation. The lines studied were either low growth or high CT-PG; or high growth or low CT-PG. Principal components analysis of metabolites, resolved from solvent extracts by GC-MS, suggested a complex relationship between primary metabolism (amino acids, sugars, glycolytic and respiratory intermediates) and CT/PG abundance. In general, CT-PG concentration appeared to correlate negatively with amino acid concentration, while there was little correlation with Krebs cycle intermediates aside from fumarate. When plants were subjected to nitrogen deprivation, the adaptive CT-PG response was varied, with the up- or downregulation of CT, PG, and amino acid metabolite pools being hybrid line-specific. This was coupled to between-line differences in gene expression associated with branches of amino acid and secondary metabolism. The conclusion of this analysis was that the mechanisms by which CT-PG contents are regulated in poplar hybrids are highly variable, but that combined metabolomic/transcriptomic analysis showed promise as a way of elucidating such complex control mechanisms of plasticity in growth trait phenotypes.

### 7.3.4 Plant Development and Seasonal Cycles

One approach to understanding plant developmental is to track the behavior of metabolites through the natural developmental process. Metabolomics techniques allow broad observation of metabolism and any unexpected relationships therein, as associated with a developmental process. For example, the metabolic sink-to-source transition of developing quaking aspen (*Populus tremuloides*) leaves was followed in the work of Jeong et al. (2004), who observed clear distinctions between young, expanded and mature leaves. Through ontogeny, multi-fold changes in two-thirds of the identified metabolites were observed, with major trends seen in carbohydrate and amino acid metabolism. This conformed to the photosynthetic and respiratory shifts associated with a transition from carbon heterotrophy to carbon autotrophy, and from rapid synthesis to maturation of cell structure.

Another aspect of developmental biology that has received attention in poplar metabolomics is the process of tissues' transition through activity-dormancy cycles. Metabolomic analyses are being included as part of multi-"omic" time course studies in which the aim is to associate comprehensive molecular modulation data with developmental transitions in cellular anatomy and physiology which occur in response to environmental (seasonal) and hormonal cues. These high-resolution molecular "timetables" promise to increase our understanding of the multi-leveled regulation and mechanisms of activity-dormancy transitions in poplar, and provide reference frameworks on which future studies may be based.

In a study of aspen (*Populus tremula*) grown under natural conditions, Druart et al. (2007) observed dynamic modulation in the cambial transcriptome and metabolome during the course of this lateral meristem's activity-dormancy cycle. This work brought considerable insight into the molecular mechanisms involved in both the cessation of cambial cell division and induction of cold hardiness in autumn, and the subsequent reactivation of the meristem and resumption of cell division in spring. The analysis indicated that distinct temporal patterns in metabolite pools were associated with the discrete stages of the activity/dormancy cycle, and that approximately 20% of GC-MS-resolved cambial metabolites modulated significantly over the course of the cycle. Specifically, the peaking of abscisic acid (ABA) and gamma amino butyric acid (GABA) pools during the latter phase (i.e., combined short-day/low temperature phase) of cold hardiness development suggested that these metabolites play important regulatory roles in this process—possibly functioning as links between environmental stimuli and downstream transcription factors. Also, declining starch reserves and increased soluble carbohydrate (sucrose, raffinose, and galactinol) and fatty acid related (glyceric acid and free fatty acids) metabolite pools coincided with vacuolar phospholipid membrane reorganization and the development of short day-induced cold hardiness and dormancy induction. These metabolic shifts, combined with supporting transcript-based evidence, suggested that starch breakdown plays a key role in the autumn transition by supporting protein, cryoprotectant, and fatty acid biosynthesis, as well as energy generation for the development of cold hardiness even while photosynthesis declines under shortening days. Concurrently, amino acid pools rise as various proteins and enzymes are recycled into bark storage protein, a major nitrogen storage facility during winter dormancy in poplar (Clausen and Apel 1991).

During the initial phase of cambial reactivation in early spring, there is a need for energy and carbon skeletons prior to the restoration of photosynthetic activity. Evidence from transcript profiling supports the induction of sucrose catabolism at this time, which can generate the hexoses that fuel glycolysis. Metabolite profile data indicate a reduction in trisaccharides, sterol and fatty acids, and an increase in *cis*-aconitic acid and itaconitic acid at this stage, which suggests the mobilization of stored fats to fuel metabolism via induction of the glyoxylate cycle. Cambial reactivation also overlaps with the degradation of bark storage protein, and increases in amino acid pools are observed during this period. Liberated amino acids may be used for the synthesis of new proteins as well as other roles. Glutamate may be converted into alpha-ketoglutarate and aspartate to feed the TCA cycle during early cambial reactivation, and to provide carbon skeletons that enable further nitrogen assimilation. It also appears that increased levels of glutamate-derived GABA may act as a signal promoting the activation of stress response genes during this time.

In another study, similarly involving a time course analysis based on complementary metabolic and transcriptomic data, Ruttink et al. (2007) traced the onset of apical bud dormancy in poplar (*Populus tremula* × *P. alba*), scrutinizing the molecular program of bud development, dormancy induction and fixation that results from exposure to short days. This work provided a detailed picture of the multi-phase reconfiguration of gene expression and metabolism through the course of bud development. Seventeen percent of the ~ 1,000 apical metabolites resolved by GC/MS underwent significant changes in abundance over the course of development. In the initial phase spanning the initiation of bud formation (the first two weeks), decreases in major and minor carbohydrate pools were observed, apparently resulting from the reduction in photosynthetic activity under short days and the consequent shift from a starch surplus to a daily deficit. The depletion of sugar pools appears to prompt an induction of the glyoxylate cycle, possibly as a mechanism for maintaining the supply of carbohydrate intermediates for gluconeogenesis or energy generation under conditions where glycolysis is reduced (as suggested by changes in gene expression). The second response phase involves the morphological transition to a closed bud, cessation of cell proliferation, and meristem inactivation, beginning after 3 to 4 weeks of short days. At this time a general downregulation of primary metabolism was observed, apparently under the control of abscisic acid. Interestingly, exposure to short day length appears to be all that is required for the elicitation of not only bud formation and dormancy, but also other winter-hardening responses including acclimation to dehydration and cold, via gene expression and metabolic shifts. However, shifts in gene expression and metabolic activity do not themselves seem to be the factors that trigger the final fixation of dormancy.

### 7.3.5 Xylem Biosynthesis

The nature of the metabolism that gives rise to reaction wood was investigated in an analysis of the metabolic and gene expression profiles in developing tension wood of poplar (Andersson-Gunneras et al. 2006). Although this work was dominated by transcript analysis, the multivariate analyses in the metabolomic component did reveal 26 metabolites that differed significantly between normal secondary cell wall and reactionary G-layer biosynthesis. Linoleic and oleic fatty acids were increased. Xylose and xylitol increased, whereas other sugars and sugar alcohols such as sucrose, arabinose and inositol decreased. Notably, the monolignol precursor shikimate was also decreased, as were other organic and amino acids including phosphate, citric acid, pentonic acid, aspartic acid, and galactaric acid. When viewed in conjunction with the extensive gene expression data, these metabolic shifts suggested the reprogramming of mechanisms for cellulose, lignin and cell wall matrix carbohydrate biosynthesis, amongst others. In particular, the apparent decrease in the activity of the pentose phosphate and shikimate pathways, and the concurrent increase in UDP-D-glucose biosynthesis, were certainly in keeping with the decreased lignification and cellulose enrichment typically observed in the G-layer.

Ferulate 5-hydroxylase (F5H) catalyzes the conversion of coniferaldehyde to 5-hydroxyconiferaldehyde, thereby playing an important role in the partitioning of flux between coniferyl alcohol (the precursor of the guaiacyl lignin monomer) and sinapyl alcohol (the precursor of the syringyl lignin monomer) in the monolignol-specific branch of the phenylpropanoid pathway. Robinson et al. (2005) conducted a GC/MS-based metabolomics analysis of hybrid poplar (*Populus tremula* × *P. alba*) transformed with the *Arabidopsis* F5H gene under the control of the cinnamate 4-hydroxylase (C4H) promoter, which had previously been shown to exhibit increased syringyl: guaiacyl lignin monomer content (Huntley et al. 2003). It was found that modified lines could be differentiated from wild type on the basis of metabolite profiles, whether the profiles were derived from sterile tissue suspension cultures or developing xylem taken from greenhouse-grown trees. The only shift in modified trees directly related to phenylpropanoid metabolism was an increase in sinapyl alcohol, which was in keeping with a transgene driven increase in flux through the branch of the pathway leading to sinapyl alcohol, and the observed increase in syringyl lignin. Interestingly, most differential metabolites were related to carbohydrate metabolism, which suggested that even though the transgene targeted lignin biosynthesis, the shift in lignin composition necessitated adjustments to the metabolic processes responsible for synthesizing the carbohydrate component of xylem. This analysis demonstrated the viability of using metabolomics to differentiate between tree lines exhibiting differences in physico-chemical wood traits.

Cinnamoyl-CoA reductase (CCR) catalyzes the conversion of feruloyl-CoA to coniferaldehyde, in what is considered to be the first committed reaction step in the monolignol-specific branch of the phenylpropanoid pathway. An analysis of CCR-downregulated poplar noted a dramatic decrease in lignin content and an increase in the incorporation of ferulic acid into lignin, with an approximate doubling of the ratio between ferulic acid or sinapic acid, and coniferaldehyde or sinapaldehyde. This suggests that the downregulation caused a shift in flux from monolignol biosynthesis toward ferulic acid (Leple et al. 2007). LC/MS analysis revealed an increase in the production of the glucosylated phenolics, glucopyranosyl sinapic acid and glucopyranosyl vanillic acid, while GC/MS analysis identified 20 known metabolites that accumulated differentially due to CCR downregulation, with strong representation from participants in respiration, ascorbic acid, sugar (e.g., glucose, mannose and *myo*-inositol) and hemicellulose and pectin metabolism. Thus, it was confirmed that the misregulation had affected not only phenylpropanoid metabolism, but also various other pathways associated with primary metabolism and secondary cell wall biosynthesis. The transcriptomic and metabolomic data from this study (in concert with another involving CCR downregulated tobacco (Dauwe et al. 2007)), indicated that a downregulation of general carbohydrate metabolism and reduction and remodeling of hemicellulose and pectin glycans that cross-link lignin monomers took place in response to signals arising from the lignin-related changes in chemical and structural properties of the developing secondary wall. Some of these changes in carbohydrate metabolism could have been part of a stress response in the modified lines, as also suggested by metabolite and transcript shifts that suggested the emergence of a stressed state involving increases in photo-oxidative stress and photorespiration in CCR-tobacco (Dauwe et al. 2007). Furthermore, the accumulation of glycosylated and quinylated derivatives of feruloyl-CoA, the usual substrate of CCR, suggests the existence of detoxification mechanisms that work to limit the accumulation of this metabolite, and may be the sink for carbon made available from the degradation of starch in a situation of reduced cell wall biosynthesis.

### 7.3.6 Advanced -omics Data Integration in the Study of Xylem Development

In the present era of metabolomics and integrated genomics, novel approaches to the concurrent analysis of data from multiple -omics platforms are being developed. The current forefront of data integration technology was redefined by the recent work of Bylesjo et al. (2009), in which an orthogonal 2 partial least squares (O2PLS)-based analysis was applied in the concurrent processing of transcript, protein and metabolite data from wild type and two antisense *PttMYB21a* transformed *Populus tremula* × *P. tremuloides* hybrid genotypes. The antisense MYB lines exhibited slower growth and reduced lignin biosynthesis. Developing xylem tissue from all three genotypes, and from three internode positions corresponding to an approximate growth gradient, was collected from greenhouse grown plants. Transcript, protein, and metabolite data were generated by cDNA microarray, UPLC-MS peptide profiling, and GC/TOFMS metabolite profiling, respectively. The ensuing O2PLS analysis was able to partition the joint covariance in transcript, protein and metabolite data (including all effects related to experimental design), as well as the variation specific to each of these profiling platforms, and the residual variance. Approximately 40% of the total variance was covariant across platforms, indicating that a considerable part of the variation in the steady-state system can be linked through transcript, protein and metabolite levels.

Within the platform-independent joint covariance, two main effects were captured that were visible both in the two-dimensional graphical projections of joint score vectors for each sample, and in the variables from each dataset. The first was a developmental gradient independent of genotype (known as the *internode effect*), and the second was a distinction between genotypes independent of developmental stage (known as the *genotype effect*). Consequently, the integrated analysis not only confirms the link between transcripts, proteins and metabolites, but also suggests which elements from each molecular class are connected. As far as the *internode effect* was concerned, the general trend was for younger, more actively growing tissue to exhibit higher levels of transcripts related to protein translation and photosynthesis, and proteins related to translation elongation and glucose metabolism. Differential metabolite pools included, among other things, amino acids and myo-inositol, which are generally implicated in plant growth processes. In the case of the *genotype effect*, the retarded growth rate of the more severe antisense MYB genotype was accompanied by decreases in growth-essential transcripts including translation elongation factors, tubulin, and actin-depolymerizing factor, as well as numerous transcripts centrally or peripherally involved in lignin biosynthesis, and some related metabolites. Interestingly, a counterintuitive, inverse relationship existed between associated transcript and protein abundances, suggesting the involvement of post-translational regulatory mechanisms in this scenario. While the underlying biology relating to the results of this analysis were only discussed in general terms, this study has made it clear that the growing ability to routinely perform combined profiling analyses of this kind has profound implications for the field of tree biotechnology and, indeed, plant biology as a whole.

## 7.4 Concluding Remarks

The metabolomics analyses conducted in poplar species to date demonstrate the effective use of this approach to rapidly identify the distinguishing metabolic components in differential biological systems. Poplar species have and continue to provide well-characterized woody plant systems with which the technology of metabolomics may be refined. In fact, with the depth of poplar-based resources currently available, this genus now appears to be taking a primary position as a system on which the development of metabolite-inclusive, integrated genomic analyses for woody plants will be based. As a final remark, it is apparent that despite the progress made to date, there is an aspect of metabolomics that has yet to be exploited extensively in any plant system—let alone in poplar. Metabolomic analyses can not only help to improve our understanding of the molecular mechanisms underlying plant growth, development and adaptation, but there is potential for the generation of accurate metabolic markers for physico-chemical traits based on metabolomics analyses, and the application of those markers in trait monitoring and prediction scenarios. The current lack of progress in poplar with regard to this approach should be viewed with optimism, however. As with the current rapid advancement of the metabolomics and integrated genomics fields, it is likely that extensive application of such technology, as well as other related technologies, will be seen in the near future.

## References

Andersson-Gunneras S, Mellerowicz EJ, Love J, Segerman B, Ohmiya Y, Coutinho PM, Nilsson P, Henrissat B, Moritz T, Sundberg B (2006) Biosynthesis of cellulose-enriched tension wood in Populus: global analysis of transcripts and metabolites identifies biochemical and developmental regulators in secondary wall biosynthesis. Plant J 45: 144–165.
Batagelj V, Mrvar A (2003) Pajek—Analysis and visualization of large networks. In: M Juenger, P Mutzel (eds) Graph Drawing Software. Springer, Berlin, Germany, pp 77–103.

Benton HP, Wong DM, Trauger SA, Siuzdak G (2008) XCMS2: Processing tandem mass spectrometry data for metabolite identification and structural characterization. Anal Chem 80: 6382–6389.

Bligny R, Douce R (2001) NMR and plant metabolism. Curr Opin Plant Biol 4: 191–196.

Brosche M, Vinocur B, Alatalo ER, Lamminmaki A, Teichmann T, Ottow EA, Djilianov D, Afif D, Bogeat-Triboulot MB, Altman A, Polle A, Dreyer E, Rudd S, Lars P, Auvinen P, Kangasjarvi J (2005) Gene expression and metabolite profiling of *Populus euphratica* growing in the Negev desert. Genome Biol 6: R101.

Burbulis IE, Winkel-Shirley B (1999) Interactions among enzymes of the Arabidopsis flavonoid biosynthetic pathway. Proc Natl Acad Sci USA 96: 12929–12934.

Busov V, Meilan R, Pearce DW, Rood SB, Ma CP, Tschaplinski TJ, Strauss SH (2006) Transgenic modification of gai or rgl1 causes dwarfing and alters gibberellins, root growth, and metabolite profiles in *Populus*. Planta 224: 288–299.

Bylesjo M, Eriksson D, Kusano M, Moritz T, Trygg J (2007) Data integration in plant biology: the O2PLS method for combined modeling of transcript and metabolite data. Plant J 52: 1181–1191.

Bylesjo M, Nilsson R, Srivastava V, Gronlund A, Johansson AI, Jansson S, Karlsson J, Moritz T, Wingsle G, Trygg J (2009) Integrated analysis of transcript, protein and metabolite data to study lignin biosynthesis in hybrid aspen. J Proteome Res 8: 199–210.

Charlton A, Allnutt T, Holmes S, Chisholm J, Bean S, Ellis N, Mullineaux P, Oehlschlager S (2004) NMR profiling of transgenic peas. Plant Biotechnol J 2: 27–35.

Chen F, Duran AL, Blount JW, Sumner LW, Dixon RA (2003) Profiling phenolic metabolites in transgenic alfalfa modified in lignin biosynthesis. Phytochemistry 64: 1013–1021.

Christin C, Smilde AK, Hoefsloot HCJ, Suits F, Bischoff R, Horvatovich PL (2008) Optimized time alignment algorithm for LC-MS data: Correlation optimized warping using component detection algorithm-selected mass chromatograms. Ann Chem 80: 7012–7021.

Clausen S, Apel K (1991) Seasonal changes in the concentration of the major storage protein and its messenger-RNA in xylem ray cells of poplar trees. Plant Mol Biol 17: 669–678.

Colebatch G, Desbrosses G, Ott T, Krusell L, Montanari O, Kloska S, Kopka J, Udvardi MK (2004) Global changes in transcription orchestrate metabolic differentiation during symbiotic nitrogen fixation in *Lotus japonicus*. Plant J 39: 487–512.

Damiani I, Morreel K, Danoun S, Goeminne G, Yahiaoui N, Marque C, Kopka J, Messens E, Goffner D, Boerjan W, Boudet AM, Rochange S (2005) Metabolite profiling reveals a role for atypical cinnamyl alcohol dehydrogenase CAD1 in the synthesis of coniferyl alcohol in tobacco xylem. Plant Mol Biol 59: 753–769.

Daub CO, Kloska S, Selbig J (2003) MetaGeneAlyse: Analysis of integrated transcriptional and metabolite data. Bioinformatics 19: 2332–2333.

Dauwe R, Morreel K, Goeminne G, Gielen B, Rohde A, van Beeumen J, Ralph J, Boudet AM, Kopka J, Rochange SF, Halpin C, Messens E, Boerjan W (2007) Molecular phenotyping of lignin-modified tobacco reveals associated changes in cell-wall metabolism, primary metabolism, stress metabolism and photorespiration. Plant J 52: 263–285.

Dijksterhuis G, Martens H, Martens M (2005) Combined Procrustes analysis and PLSR for internal and external mapping of data from multiple sources. Comput Stat Data Ann 48: 47–62.

Druart N, Johansson A, Baba K, Schrader J, Sjodin A, Bhalerao RR, Resman L, Trygg J, Moritz T, Bhalerao RP (2007) Environmental and hormonal regulation of the activity-dormancy cycle in the cambial meristem involves stage-specific modulation of transcriptional and metabolic networks. Plant J 50: 557–573.

Duran AL, Yang J, Wang LJ, Sumner LW (2003) Metabolomics spectral formatting, alignment and conversion tools (MSFACTs). Bioinformatics 19: 2283–2293.

Fernie AR, Trethewey RN, Krotzky AJ, Willmitzer L (2004) Metabolite profiling: from diagnostics to systems biology. Nat Rev Mol Cell Biol 5: 763–769.

Fiehn O (2003) Metabolic networks of *Cucurbita maxima* phloem. Phytochemistry 62: 875–886.

Fiehn O, Kopka J, Doermann P, Altmann T, Trethewey RN, Willmitzer L (2000a) Metabolite profiling for plant functional genomics. Nat Biotechnol 18: 1157–1161.

Fiehn O, Kopka J, Trethewey RN, Willmitzer L (2000b) Identification of uncommon plant metabolites based on calculation of elemental compositions using gas chromatography and quadrupole mass spectrometry. Ann Chem 72: 3573–3580.

Fiehn O, Kloska S, Altmann T (2001) Integrated studies on plant biology using multiparallel techniques. Curr Opin Biotechnol 12: 82–86.

Gaffney T, Friedrich L, Vernooij B, Negrotto D, Nye G, Uknes S, Ward E, Kessmann H, Ryals J (1993) Requirement of salicylic acid for the induction of systemic acquired resistance. Science 261: 754–756.

Giuliani A, Zbilut JP, Conti F, Manetti C, Miccheli A (2004) Invariant features of metabolic networks: a data analysis application on scaling properties of biochemical pathways. Physica A 337: 157–170.

Grata E, Boccard J, Guillarme D, Glauser G, Carrupt PA, Farmer EE, Wolfender JL, Rudaz S (2008) UPLC-TOF-MS for plant metabolomics: A sequential approach for wound marker analysis in *Arabidopsis thaliana*. J Chromatogr B 871: 261–270.

Gullberg J, Jonsson P, Nordstrom A, Sjostrom M, Moritz T (2004) Design of experiments: an efficient strategy to identify factors influencing extraction and derivatization of *Arabidopsis thaliana* samples in metabolomic studies with gas chromatography/mass spectrometry. Ann Biochem 331: 283–295.

Harding SA, Jiang HY, Jeong ML, Casado FL, Lin HW, Tsai CJ (2005) Functional genomics analysis of foliar condensed tannin and phenolic glycoside regulation in natural cottonwood hybrids. Tree Physiol 25: 1475–1486.

Hirai MY, Yano M, Goodenowe DB, Kanaya S, Kimura T, Awazuhara M, Arita M, Fujiwara T, Saito K (2004) Integration of transcriptomics and metabolomics for understanding of global responses to nutritional stresses in *Arabidopsis thaliana*. Proc Natl Acad Sci USA 101: 10205–10210.

Huntley SK, Ellis D, Gilbert M, Chapple C, Mansfield SD (2003) Significant increases in pulping efficiency in C4H-F5H-transformed poplars: Improved chemical savings and reduced environmental toxins. J Agri Food Chem 51: 6178–6183.

Jeong ML, Jiang HY, Chen HS, Tsai CJ, Harding SA (2004) Metabolic profiling of the sink-to-source transition in developing leaves of quaking aspen. Plant Physiol 136: 3364–3375.

Kaplan F, Kopka J, Haskell DW, Zhao W, Schiller KC, Gatzke N, Sung DY, Guy CL (2004) Exploring the temperature-stress metabolome of Arabidopsis. Plant Physiol 136: 4159–4168.

Kjalstrand J, Ramnas O, Petersson G (1998) Gas chromatographic and mass spectrometric analysis of 36 lignin-related methoxyphenols from uncontrolled combustion of wood. J Chromatogr A 824: 205–210.

Klok EJ, Wilson IW, Wilson D, Chapman SC, Ewing RM, Somerville SC, Peacock WJ, Dolferus R, Dennis ES (2002) Expression profile analysis of the low-oxygen response in Arabidopsis root cultures. Plant Cell 14: 2481–2494.

Klukas C, Junker BH, Schreiber F (2006) The VANTED software system for transcriptomics, proteomics and metabolomics analysis. J Pestic Sci 31: 289–292.

Kopka J, Schauer N, Krueger S, Birkemeyer C, Usadel B, Bergmuller E, Dormann P, Weckwerth W, Gibon, Y, Stitt M, Willmitzer L, Fernie AR, Steinhauser D (2005) GMD@CSB.DB: The Golm Metabolome Database. Bioinformatics 21: 1635–1638.

Kreuzwieser J, Hauberg J, Howell K, Carroll A, Rennenberg H, Millar A, Whelan J (2009) Differential response of gray poplar leaves and roots underpins stress adaptation during hypoxia. Plant Physiol 149: 461–473.

Leple JC, Dauwe R, Morreel K, Storme V, Lapierre C, Pollet B, Naumann A, Kang KY, Kim H, Ruel K, Lefebvre A, Joseleau JP, Grima-Pettenati J, de Rycke R, Andersson-Gunneras S, Erban A, Fehrle I, Petit-Conil M, Kopka J, Polle A, Messens E, Sundberg B, Mansfield SD, Ralph J, Pilate G, Boerjan W (2007) Downregulation of cinnamoyl-coenzyme a reductase in poplar: Multiple-level phenotyping reveals effects on cell wall polymer metabolism and structure. Plant Cell 19: 3669–3691.

Li WT, Nyholt DR (2001) Marker selection by Akaike information criterion and Bayesian information criterion. Genet Epidemiol 21: S272–S277.

Lisec J, Meyer RC, Steinfath M, Redestig H, Becher M, Witucka-Wall H, Fiehn O, Torjek O, Selbig J, Altmann T, Willmitzer L (2008) Identification of metabolic and biomass QTL in *Arabidopsis thaliana* in a parallel analysis of RIL and IL populations. Plant J 53: 960–972.

Liu FL, VanToai T, Moy LP, Bock G, Linford LD, Quackenbush J (2005) Global transcription profiling reveals comprehensive insights into hypoxic response in Arabidopsis. Plant Physiol 137: 1115–1129.

Lukacin R, Wellmann F, Britsch L, Martens S, Matern U (2003) Flavonol synthase from *Citrus unshiu* is a bifunctional dioxygenase. Phytochemistry 62: 287–292.

Meyer RC, Steinfath M, Lisec J, Becher M, Witucka-Wall H, Torjek O, Fiehn O, Eckardt A, Willmitzer L, Selbig J, Altmann T (2007) The metabolic signature related to high plant growth rate in *Arabidopsis thaliana*. Proc Natl Acad Sci USA 104: 4759–4764.

Morreel K, Goeminne G, Storme V, Sterck L, Ralph J, Coppieters W, Breyne P, Steenackers M, Georges M, Messens E, Boerjan W (2006) Genetical metabolomics of flavonoid biosynthesis in Populus: a case study. Plant J 47: 224–237.

Morris CR, Scott JT, Chang H-M, Sederoff RR, O'Malley D, Kadla JF (2004) Metabolic profiling: A new tool in the study of wood formation. J Agri Food Chem 52: 1427–1434.

Morse AM, Tschaplinski TJ, Dervinis C, Pijut PM, Schmelz EA, Day W, Davis JM (2007) Salicylate and catechol levels are maintained in nahG transgenic poplar. Phytochemistry 68: 2043–2052.

Nakamura Y, Kimura A, Saga H, Oikawa A, Shinbo Y, Kai K, Sakurai N, Suzuki H, Kitayama M, Shibata D, Kanaya S, Ohta D (2007) Differential metabolomics unraveling light/dark regulation of metabolic activities in Arabidopsis cell culture. Planta 227: 57–66.

Nielsen NPV, Carstensen JM, Smedsgaard J (1998) Aligning of single and multiple wavelength chromatographic profiles for chemometric data analysis using correlation optimised warping. J Chromatogr A. 805: 17–35.

Nikiforova VJ, Daub CO, Hesse H, Willmitzer L, Hoefgen R (2005a) Integrative gene-metabolite network with implemented causality deciphers informational fluxes of sulphur stress response. J Exp Bot 56: 1887–1896.

Nikiforova VJ, Kopka J, Tolstikov V, Fiehn O, Hopkins L, Hawkesford MJ, Hesse H, Hoefgen R (2005b) Systems rebalancing of metabolism in response to sulfur deprivation, as revealed by metabolome analysis of Arabidopsis plants. Plant Physiol 138: 304–318.

Osuna D, Usadel B, Morcuende R, Gibon Y, Blasing OE, Hohne M, Gunter M, Kamlage B, Trethewey R, Scheible WR, Stitt M (2007) Temporal responses of transcripts, enzyme activities and metabolites after adding sucrose to carbon-deprived Arabidopsis seedlings. Plant J 49: 463–491.

Ott K-H, Aranibar N, Singh B, Stockton GW (2003) Metabonomics classifies pathways affected by bioactive compounds. Artificial neural network classification of NMR spectra of plant extracts. Phytochemistry 62: 971–985.

Ratcliffe RG, Shachar-Hill Y (2001) Probing plant metabolism with NMR. In: Rl Jones, HJ Bohnert, DP Delmar (eds) Annu Rev Plant Phys, Palo Alto. pp 499–526.

Rizhsky L, Liang HJ, Shuman J, Shulaev V, Davletova S, Mittler R (2004) When defense pathways collide. The response of Arabidopsis to a combination of drought and heat stress. Plant Physiol 134: 1683–1696.

Robinson AR, Gheneim R, Kozak RA, Ellis DD, Mansfield SD (2005) The potential of metabolite profiling as a selection tool for genotype discrimination in Populus. J Exp Bot 56: 2807–2819.

Roepenack-Lahaye EV, Degenkolb T, Zerjeski M, Franz M, Roth U, Wessjohann L, Schmidt J, Scheel D, Clemens S (2004) Profiling of Arabidopsis secondary metabolites by capillary liquid chromatography coupled to electrospray ionization quadrupole time-of-flight mass spectrometry. Plant Physiol 134: 548–559.

Roessner U, Luedemann A, Brust D, Fiehn O, Linke T, Willmitzer L, Fernie AR (2001a) Metabolic profiling allows comprehensive phenotyping of genetically or environmentally modified plant systems. Plant Cell 13: 11–29.

Roessner U, Willmitzer L, Fernie AR (2001b) High-resolution metabolic phenotyping of genetically and environmentally diverse potato tuber systems. Identification of phenocopies. Plant Physiol 127: 749–764.

Ruttink T, Arend M, Morreel K, Storme V, Rombauts S, Fromm J, Bhalerao RP, Boerjan W, Rohde A (2007) A molecular timetable for apical bud formation and dormancy induction in poplar. Plant Cell 19: 2370–2390.

Sanchez DH, Siahpoosh MR, Roessner U, Udvardi M, Kopka J (2008) Plant metabolomics reveals conserved and divergent metabolic responses to salinity. Physiol Plant 132: 209–219.

Saslowsky D, Winkel-Shirley B (2001) Localization of flavonoid enzymes in Arabidopsis roots. Plant J 27: 37–48.

Schad M, Mungur R, Fiehn O, Kehr J (2005) Metabolic profiling of laser microdissected vascular bundles of *Arabidopsis thaliana*. Plant Meth 1:2.

Shepherd T, Dobson G, Verrall SR, Conner S, Griffiths DW, McNicol JW, Davies HV, Stewart D (2007) Potato metabolomics by GC-MS: what are the limiting factors? Metabolomics 3: 475–488.

Smith CA, Want EJ, O'Maille G, Abagyan R, Siuzdak G (2006) XCMS: Processing mass spectrometry data for metabolite profiling using nonlinear peak alignment, matching, and identification. Ann Chem 78: 779–787.

Soga T, Ohashi Y, Ueno Y, Naraoka H, Tomita M, Nishioka T (2003) Quantitative metabolome analysis using capillary electrophoresis mass spectrometry. J Proteome Res 2: 488–494.

Steuer R, Kurths J, Fiehn O, Weckwerth W (2003) Observing and interpreting correlations in metabolomic networks. Bioinformatics 19: 1019–1026.

t'Kindt R, de Veylder L, Storme M, Deforce D, van Bocxlaer J (2008) LC-MS metabolic profiling of Arabidopsis thaliana plant leaves and cell cultures: Optimization of pre-LC-MS procedure parameters. J Chromatogr B 871: 37–43.

Terskikh VV, Feurtado JA, Borchardt S, Giblin M, Abrams SR, Kermode AR (2005) In vivo C-13 NMR metabolite profiling: Potential for understanding and assessing conifer seed quality. J Exp Bot 56: 2253–2265.

Thimm O, Blasing O, Gibon Y, Nagel A, Meyer S, Kruger P, Selbig J, Muller LA, Rhee SY, Stitt M (2004) MAPMAN: A user-driven tool to display genomics data sets onto diagrams of metabolic pathways and other biological processes. Plant J 37: 914–939.

Tikunov Y, Lommen A, de Vos CHR, Verhoeven HA, Bino RJ, Hall RD, Bovy AG (2005) A novel approach for nontargeted data analysis for metabolomics. Large-scale profiling of tomato fruit volatiles. Plant Physiol 139: 1125–1137.

Tohge T, Nishiyama Y, Hirai MY, Yano M, Nakajima J, Awazuhara M, Inoue E, Takahashi H, Goodenowe DB, Kitayama M, Noji M, Yamazaki M, Saito K (2005) Functional genomics by integrated analysis of metabolome and transcriptome of Arabidopsis plants over-expressing an MYB transcription factor. Plant J 42: 218–235.

Tolstikov VV, Fiehn O (2002) Analysis of highly polar compounds of plant origin: combination of hydrophilic interaction chromatography and electrospray ion trap mass spectrometry. Anal Biochem 301: 298–307.

Tolstikov VV, Lommen A, Nakanishi K, Tanaka N, Fiehn O (2003) Monolithic silica-based capillary reversed-phase liquid chromatography/electrospray mass spectrometry for plant metabolomics. Ann Chem 75: 6737–6740.

Urbanczyk-Wochniak E, Baxter C, Kolbe A, Kopka J, Sweetlove LJ, Fernie AR (2005) Profiling of diurnal patterns of metabolite and transcript abundance in potato (*Solanum tuberosum*) leaves. Planta 221: 891–903.

van Riel NAW (2006) Dynamic modelling and analysis of biochemical networks: Mechanism-based models and model-based experiments. Brief Bioinformat 7: 364–374.

Wang SY, Wang YS, Tseng YH, Lin CT, Liu CP (2006) Analysis of fragrance compositions of precious coniferous woods grown in Taiwan. Holzforschung 60: 528–532.

Weckwerth W (2003) Metabolomics in systems biology. Annu Rev Plant Biol 54: 669–689.

Welti R, Li WQ, Li MY, Sang YM, Biesiada H, Zhou HE, Rajashekar CB, Williams TD, Wang XM (2002) Profiling membrane lipids in plant stress responses—Role of phospholipase D alpha in freezing-induced lipid changes in Arabidopsis. J Biol Chem 277: 31994–32002.

Winkel BSJ (2004) Metabolic channeling in plants. Annu Rev Plant Biol 55: 85–107.

Wolfender J-L, Ndjoko K, Hostettmann K (2003) Liquid chromatography with ultraviolet absorbance-mass spectrometric detection and with nuclear magnetic resonance spectroscopy: A powerful combination for the on-line structural investigation of plant metabolites. J Chromatogr A 1000: 437–455.

Wurtele ES, Li J, Diao L, Zhang H, Foster CM, Fatland B, Dickerson J, Brown A, Cox Z, Cook D, Lee E-K, Hofmann H (2003) MetNet: software to build and model the biogenetic lattice of Arabidopsis. Comp Funct Genom 4: 239–245.

Yamashita T, Yamashita K, Kamimura R (2007) A stepwise AIC method for variable selection in linear regression. Commun Stat-Theory Meth 36: 2395–2403.

Yang S, Qiao B, Lu SH, Yuan YJ (2007) Comparative lipidomics analysis of cellular development and apoptosis in two Taxus cell lines. Biochim Biophys Acta Mol Cell Biol. Lipids 1771: 600–612.

Yeh TF, Morris CR, Goldfarb B, Chang H-M, Kadla JF (2006) Utilization of polar metabolite profiling in the comparison of juvenile wood and compression wood in loblolly pine (*Pinus taeda*). Tree Physiol 26: 1497–1503.

# 8

# Transcription Factors in Poplar Growth and Development

*Amy M. Brunner[1],* and *Eric P. Beers[2]*

## ABSTRACT

Because of their central role in regulating gene expression, transcription factors are key components of most signal transduction pathways and regulate innumerable subcellular to organism-level biological processes. Transcription factors bind directly to the promoters of target genes in a sequence-specific manner to either activate or repress transcription. Accumulating evidence indicates that genetic variation in the genes encoding transcription factors and in transcription factor binding sites plays a major role in phenotypic variation within species and divergence between species. Hence, genes encoding transcription factors are likely to be of major importance for both tree domestication and adaptation. The *Populus* molecular biology and genomics toolkit is enabling the identification of transcription factors important for tree growth and development as well as the development of strategies to use transcription factors for tree improvement. Here we discuss transcription factors in the context of the growth and development of the shoot system and the secondary plant body—processes that are directly relevant to economically and ecologically important traits in trees. In addition, we use comparative genomics of the relevant transcription factors to illustrate both conservation with and divergence from annual plants and differential patterns of gene duplication and subsequent retention or loss.

**Keywords:** Transcription factor, gene regulation, shoot system, secondary growth

[1]Department of Forest Resources and Environmental Conservation, Virginia Polytechnic Institute and State University, Blacksburg, VA 24061-0324; e-mail: *abrunner@vt.edu*
[2]Department of Horticulture, Virginia Polytechnic Institute and State University, Blacksburg, VA 24061-0327; e-mail: *ebeers@vt.edu*
*Corresponding author

## 8.1 Introduction

Transcription factors (TFs) bind directly to the promoters of target genes in a sequence-specific manner to either activate or repress transcription. TFs carry out their function as part of multi-protein complexes; thus, proteins which do not by themselves bind DNA, but are part of multi-protein complexes that bind specific DNA sequences, are often considered to be TFs. Because of their central role in regulating gene expression, they are key components of most signal transduction pathways and regulate innumerable subcellular to organism-level biological processes. Moreover, changes in gene expression play a major role in phenotypic plasticity, variation within species, and divergence between species (Babu et al. 2004; Wittkopp 2007). Doebley and Lukens (1998) proposed that changes in the *cis*-regulatory elements of TFs represent a predominant mechanism for the evolution of plant form. Although the genes and mechanisms underlying plant phenotypic diversity are still largely unknown, considerable evidence supports that TFs are likely to have a major role. Most plant TFs are part of large, multi-gene families defined by the type of DNA-binding domain that they encode, and TF family or subfamily size can vary significantly among plant taxa (Riechmann et al. 2000; Qu and Zhu 2006; Tuskan et al. 2006; Ming et al. 2008). Whole genome duplications have been prevalent in angiosperm evolution, and studies indicate that TFs have been preferentially retained following genome duplication (De Bodt et al. 2005; Shiu et al. 2005; Semon and Wolfe 2007).

TFs have clearly been important for crop domestication (Doebley et al. 2006). Of six genes known to control classic domestication traits, five encode TFs. Interestingly, 10 of 17 genes controlling complex phenotypes of agronomic importance are TFs and the remainders are various types of regulatory genes. In contrast, TFs account for only three of nine genes controlling simple phenotypes and the rest are biosynthetic enzymes. While these selections were based solely on phenotype without prior knowledge of the underlying genes, a number of crop biotechnology companies are focusing on the reverse approach of selecting and evaluating TFs for advanced crop improvement (e.g., *http://www.mendelbio.com/*).

In contrast to the major crops, forest trees have undergone minimal domestication (Bradshaw and Strauss 2001). However, there is an increasing need to sustainably provide more wood—for traditional wood products as well as for newer end uses such as bioenergy and carbon sequestration—from trees grown on a smaller proportion of land (Boerjan 2005; Groover 2007). Thus, we are challenged to accomplish forest tree domestication and more advanced breeding tailored to specific environments and end uses in a vastly shorter time frame than was achieved for food crops. At the same time, natural forest populations are increasingly threatened by a combination of factors, including climate change, urbanization and exotic pests. Genomics and biotechnology offer the potential to make the necessary advances in tree improvement and to discover the genes underlying adaptive traits that could aid forest conservation (Bradshaw and Strauss 2001; Boerjan 2005; Busov et al. 2005; Groover 2007; Neale 2007; Aitken et al. 2008). Genes encoding TFs are likely to be of major importance for both tree domestication and adaptation, but identifying these genes and discovering how to manipulate them for a predictive outcome are challenging tasks. Among tree taxa, the *Populus* system has the most complete molecular biology and genomics toolkit (reviewed in Brunner et al. 2004; Jansson and Douglas 2007); and thus, the most opportunity to advance our knowledge of the TFs important for tree growth and development as well as to develop strategies to use TFs for tree improvement.

There are a number of TF databases, and at least two of these (*http://plntfdb.bio.uni-potsdam.de/v2.0/*; *http://dptf.cbi.pku.edu.cn/*) include TFs identified in the poplar genome sequence. We will not present a genome-wide review of poplar TFs, but rather discuss TFs in the context of the growth and development of the shoot system and the secondary plant body—processes that are directly relevant to economically and ecologically important traits in trees. In addition, the comparative genomics of the relevant TFs illustrate both conservation with and divergence from annual plants and differential patterns of gene duplication and subsequent retention/loss.

## 8.2 Shoot System Growth and Development

The spatiotemporal pattern of shoot meristem activity and phytomer growth determines plant shoot architecture (reviewed in Sussex and Kerk 2001; McSteen and Leyser 2005). Tree crown architecture develops over more complex and much larger spatial and temporal scales than the shoot system of herbaceous plants. In poplar and other woody plants of the temperate and boreal zones, shoot development exhibits a recurrent seasonal pattern of growth and winter dormancy (reviewed in Howe et al. 2003; Welling and Palva 2006; Rohde and Bhalerao 2007). The timings of these seasonal transitions show adaptive variation correlated with climatic and geographic gradients. In addition, trees exhibit developmental phase change or maturation that occurs over years in a number of shoot architecture traits such as branch angle and frequency, and some of these traits also show within-tree gradients (Greenwood 1995; Brunner et al. 2003). Field performance of poplar clones is closely tied to the time of fall bud set, and spring bud break and shoot phenology interacts with crown structure to determine leaf production, display and duration, and thus, photosynthetic capacity, over the growing season throughout the rotation (reviewed in Heilman et al. 1996). The physiology and morphology of poplars in relation to wood yield has been extensively studied, leading to the identification of a crown ideotype associated with high productivity (reviewed in Dickmann et al. 2001). Ideotypic features include abundant sylleptic branching, a high ratio of long to short shoots in the upper crown and upturned branches forming a long, narrow crown.

*Populus* exhibits abundant natural genetic variation in virtually all aspects of shoot phenology and architecture (reviewed in Dickmann et al. 2001) that can be exploited in breeding programs. Although phenotypic selection has been effective, knowledge of the genes and mechanisms underlying these processes could enable major advances using association genetics and genetic engineering. The genetic pathways are just beginning to be revealed in poplar, but studies have already shown key roles for transcription factors. It is also clear that because of the conservation of genes and pathways during plant evolution, much can be learned from work in annual plants and we discuss shoot system development in poplar and annuals in a comparative context.

## 8.2.1 Seasonal Shoot Development

Bud endodormancy or winter dormancy enables temperate and boreal woody plant shoot meristems to survive the severe freezing and dehydration stress of winter conditions (reviewed in Howe et al. 2003; Welling and Palva 2006; Rohde and Bhalerao 2007). Decreasing daylength and temperature in fall induce growth cessation, cold acclimation, dormancy, and finally, maximum cold hardiness. Following an extended chilling period to release dormancy, increasing temperature and photoperiods in spring stimulate the resumption of growth and bud flush. Although bud endodormancy is unique to woody plants, photoperiod and temperature control various processes in all plants, and recent studies indicate that genes and pathways present in annual plants have been redeployed to regulate endodormancy in poplar (Bohlenius et al. 2006; Ruttink et al. 2007; Rohde et al. 2007).

Long daylengths (LDs) promote flowering in *Arabidopsis*, and the B-box-type zinc-finger transcriptional activator CONSTANS (CO) is a key component of the photoperiodic flowering pathway (reviewed in Imaizumi and Kay 2006; Kobayashi and Weigel 2007). The circadian clock acts to establish the rhythm of *CO* gene expression and light regulates the stability and activity of CO protein. Under LDs, *CO* mRNA peaks in the light and because light stabilizes the CO protein, CO accumulates to a level sufficient to activate *FLOWERING LOCUS T (FT)*. The FT protein moves via the phloem to the shoot meristem where FT interacts with the bZIP transcription factor FD, and the FT-FD complex is able to activate floral identity genes such as those encoding the MADS-domain transcription factors APETALA1 (AP1) and FRUITFULL (FUL).

Recent work has shown that components of the photoperiodic flowering pathway have been co-opted to regulate short-day (SD)-induced bud set in poplar. The critical daylength for growth cessation and bud set in poplar varies with latitude; the more northern populations set bud at a longer daylength than southern populations. Bohlenius et al. (2006) showed that the *PtCO2/PtFT1* regulon mediates SD-induced bud set in poplar and differences in the rhythm of *PtCO2* expression could explain the natural variation in the timing of bud set. Under 19-hour days, trees from the most northern latitudes ceased growth, but southern trees continued to grow; the peak expression of *PtCO2* occurs during the dark and *PtFT1* is not induced in northern trees, whereas the peak of *PtCO2* expression is still in the light for southern trees and *PtFT1* is expressed. Overexpression of *PtFT1* prevented SD-induced growth cessation and bud set (Bohlenius et al. 2006). Work by Gary Coleman and colleagues (pers. comm.) has shown that expression of *PtFD1*, one of two poplar *FD* homologs, increases in shoot apices and developing apical buds within 3 weeks of SD treatment, reaches a maximum after 4 to 6 weeks of SD treatment and then declines with continued SD exposure beyond 6 weeks. Moreover, vegetative apical bud development does not occur in SD-grown transgenic poplars overexpressing *PtFD1*. Conversely, downregulation of *PtFD1* by RNAi appears to accelerate apical bud formation during SD. Genome-wide gene expression studies suggest that additional TFs acting in the photoperiodic flowering pathway, such as *AP1*, are also involved in regulating SD-induced bud formation and dormancy induction in poplar (Ruttink et al. 2007).

Temporal gene expression patterns indicate that different signaling pathways are sequentially activated to achieve the overlapping stages of growth cessation, bud formation, adaptation to cold and dehydration and endodormancy (Ruttink et al. 2007). Subsequent to the photoperiodic pathway, multiple members of the ETHYLENE RESPONSE FACTOR (ERF) transcription factor subfamily as well as other components of ethylene biosynthesis and signal transduction pathways are transiently activated in apices. To form a bud, primordia differentiate into bud scales and embryonic leaves and internodes cease to elongate. Accordingly, poplar homologs of transcription factors regulating cell proliferation, polarity and elongation are differentially expressed, including AP2/ERF family member *AINTEGUMENTA* (*ANT*) and the DELLA-domain protein GA INSENSITIVE (GAI) that negatively regulates GA signaling. Twelve TFs were differentially expressed in both bud and cambium growth-to-dormancy transitions, including a homolog of the NAC family member *ATAF1* that is induced by dehydration and abscisic acid (ABA) treatment (Lu et al. 2007).

The prolonged chilling to release endodormancy is reminiscent of vernalization, the acquisition of flowering competence by prolonged exposure of seed or plant to cold (Horvath et al. 2003). In *Arabidopsis*, the MADS-domain transcription factor FLOWERING LOCUS C (FLC), a strong repressor of flowering, is central to the vernalization response (reviewed in Schmitz and Amasino 2007; Dennis and Peacock 2007). Vernalization results in the mitotically stable epigenetic silencing of *FLC* via histone modification, and *FLC* expression is reset in the next generation. Six poplar *FLC*-like genes have been identified in the poplar genome, and at least one of these has a role in regulating bud endodormancy (G. Coleman, pers. comm.). *PtFLC2* is expressed in shoot apices of LD-grown poplars and during the early stages of SD-initiated bud development. During continued SD exposure in combination with low temperature, transcript levels of *PtFLC2* decline. The decline of *PtFLC2* expression appears to be associated with cold-mediated dormancy release similar to the downregulation of *FLC* during vernalization. A number of alternatively spliced *PtFLC2* transcripts were detected and the expressions of some of the splice variants were associated with the later stages of bud dormancy. Constitutive overexpression of one *PtFLC2* splicing variant in poplar altered the response to photoperiod and cold, resulting in delayed growth cessation and bud formation, reduced depth of bud dormancy, reduced amount of chilling required to overcome dormancy, and delayed leaf senescence and abscission. The altered dormancy responses could be overcome by prolonged exposure to SD and/or low temperatures. Although the involvement of *PtFLC2* in bud endodormancy is an initial indication of shared regulatory elements with the vernalization pathway, there are key differences between vernalization and endodormancy release that suggest commonalities might be limited (Rohde and Bhalerao 2007). For example, if epigenetic regulation was involved in bud endodormancy, it would need to be reset seasonally rather than be mitotically stable throughout the remainder of the plant's life cycle.

## *8.2.2 Branching*

The first step in branching is the initiation of new meristems in the axils of leaves, and plant architecture is largely determined by the fate and activity of these axillary meristems (reviewed in Schmitz and Theres 2005; McSteen and Leyser 2005; Dun et al. 2006). A central component of axillary meristem control is bud dormancy that allows plants to adapt their branching to a variety of conditions. In addition to endodormancy that is unique to temperate/boreal woody perennial plants, other types of dormancy that occur in most higher plants are: 1) paradormancy, the inhibition of growth by distal organs (e.g., apical dominance); and 2) ecodormancy, induced by limitations in environmental factors (e.g., nutrient stress) (Lang

1987). Regardless of the type of dormancy, models posit multiple stages, including stages of bud dormancy establishment, maintenance, transition and sustained growth (Dun et al. 2006; Rohde and Bhalerao 2007). These stages may enable multiple lateral buds to simultaneously respond to endogenous or environmental cues for outgrowth, yet remain responsive to a homeostatic control system that limits the number of buds that continue to grow into shoots.

Studies of branching in annual plants have focused on apical dominance, and in general, annual plants exhibit sylleptic branching (i.e., shoots develop from lateral meristems during the same season that they are formed). In contrast, temperate zone trees typically produce proleptic branches that develop from lateral buds that have undergone endodormancy. Poplars are one of the relatively few temperate zone woody taxa that exhibit both proleptic and sylleptic branching, and thus, provide an excellent model system for comparing the genetic regulation of branching in herbaceous annuals and trees and ultimately how these two types of branching are regulated. Sylleptic branching is the major factor giving rise to the high leaf area index of poplars, and sylleptic branches translocate a larger proportion of carbon to the stem than proleptics (Wu and Stettler 1998; Scarascia-Mugnozza et al. 1999; Rae et al. 2004). Poplars also produce a range of proleptic branches that set bud at different times, and thus, vary in length. These include long, indeterminate shoots that contain both preformed and neoformed leaves that can grow to several meters in length as well as short, determinate shoots that set terminal buds early in the growing season (reviewed in Dickmann et al., 2001). There is both natural variation in the proportion of long shoots and a strong age-related trend—older trees have fewer long shoots. Compared to SD-induced bud set, very little is known about what regulates the proportion of shoots that set terminal buds in the LD of early summer.

The tomato genes *BLIND* (*Bl*), which encodes a MYB domain transcription factor, and *LATERAL SUPPRESSOR* (*Ls*), which encodes a GRAS domain transcription factor, control the initiation of lateral meristems (Schumacher et al. 1999; Schmitz et al. 2002). Moreover, *Ls* orthologs from rice (*MOC1*; Li et al. 2003) and *Arabidopsis* (*LAS*; Greb et al. 2003) and the *Arabidopsis Bl* co-orthologs *REGULATOR OF AXILLARY MERISTEMS* (*RAX*) *1/2/3* (Muller et al. 2006; Keller et al. 2006) have similar functions. However, the orthologs also exhibit some functional diversification. For example, *ls* mutants lack branches during the vegetative phase, but produce branches after the plant has transitioned to flowering. In contrast, *moc1* mutants show large reductions in branching in both vegetative and inflorescence shoots. *Bl* regulates axillary meristem initiation during both vegetative and reproductive development, whereas *RAX1/2/3* regulate branching along overlapping zones of the shoot with *RAX1* acting early in vegetative development and *RAX2/3* primarily acting later during inflorescence development. These results demonstrate both the effects that gene duplication/loss and subsequent functional divergence can have in altering regulatory networks, and that the regulation and activity of axillary meristems can change after the plant undergoes developmental phase transitions.

The production of fewer tillers to concentrate resources to the main shoot was central to the domestication of maize from teosinte, and studies support that this resulted from changes in expression of *TEOSINTE BRANCHED1 (TB1)*, a TCP family transcription factor (Doebley et al. 1997). Rice *TB1* and the *Arabidopsis* homolog *BRANCHED1 (BRC1)* also suppress outgrowth of lateral buds (Takeda et al. 2003; Aguilar-Martinez et al. 2007). *BRC1* acts locally within the bud and both apical dominance and planting density regulate *BRC1* expression, suggesting that it may integrate endogenous and environmental signaling pathways (Aguilar-Martinez et al. 2007). A related gene, *BRC2*, seemed to have a minor role in regulating branch outgrowth, and in contrast to *BRC1*, *BRC2* expression was not altered by planting density.

Relative to *Arabidopsis* and rice, the *Ls*, *Bl* and *TB1* subfamilies are expanded in poplar (Fig. 8-1). *LAS* transcripts accumulate in the axils of developing leaves and decline as the lateral meristem develops (Greb et al. 2003). The three poplar *LAS* co-orthologs differ in expression pattern, suggesting that they may have diverged in function (Fig. 8-1d). Only *PoptrGRAS45* showed an expression pattern similar to *LAS* with highest expression near the shoot tip where axillary meristems are beginning to establish, decreasing expression in more developed lateral meristems at nodes farther down from the shoot apex, and low expression in other tissues. Compared to annual plants, poplar has much longer juvenile and adult vegetative phases, and in adult trees, floral meristems develop in the axils of bracts on lateral inflorescence shoots at the same time that vegetative meristems develop in leaf axils. Moreover, axillary vegetative meristems can initiate within a poplar bud before endodormancy and on actively growing shoots. Thus, one possibility is that the multiple poplar *Ls*, *Bl* and *TB1* homologs have diversified to regulate development and outgrowth of specific types of lateral meristems and/or at different seasonal or age-related developmental phases that reflect the more complex spatial and temporal scales of lateral meristem development in trees. Moreover, genes such as *BRC1* that control lateral bud outgrowth in annual plants could potentially have a role in regulating spring outgrowth (bud flush) of terminal as well as lateral buds in poplar and/or prevent outgrowth of terminal buds set in summer on short shoots.

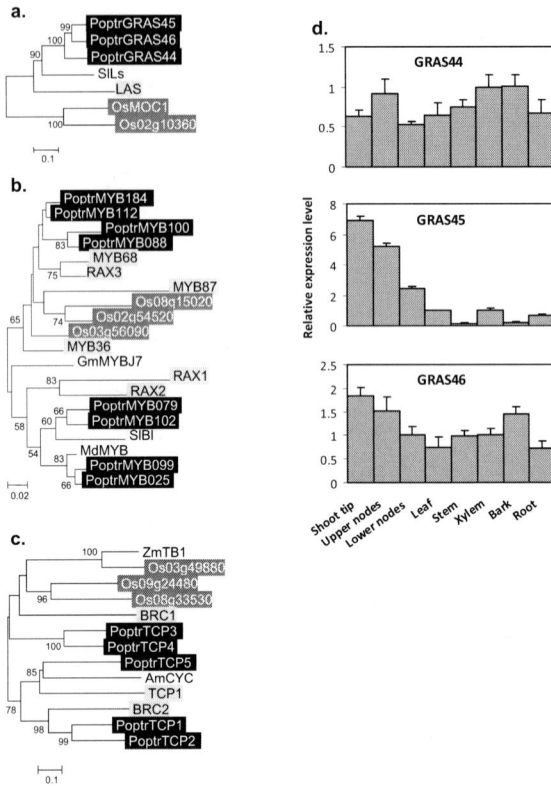

**Figure 8-1** Neighbor-joining phylogenetic analyses and gene expression of transcription factor subfamilies involved in lateral shoot development (a) Ls subgroup of the GRAS family (b) Bl subgroup of the MYB family (c) TB1 subgroup of the TCP family. Poplar proteins are in black boxes, *Arabidopsis* proteins are in light gray boxes, and rice proteins are in dark gray boxes. Names for poplar GRAS and MYB proteins follow the manual annotation of A. Brunner and M. Campbell, respectively, that are entered in the poplar genome database (*http://genome.jgi-psf. org/Poptr1_1/Poptr1_1.home.html*). Species for the additional sequences are *Solanum lycopersicum* (SlLs, SlBl), *Antirrhinum majus* (AmCYC), *Zea mays* (ZmTB1), *Glycine max* (GmMYBJ7), *Malus x domestica* (MdMYB). Bootstrap support of 50% or higher is shown at nodes. The bar at the base of each tree represents a tree branch length equivalent to amino acid changes per residue as indicated. Protein ID (for poplar proteins), AGI locus (*http://www.arabidopsis.org/index.jsp*), rice locus (*http://rice.plantbiology.msu.edu/index.shtml*), or GenBank accession (*http://www.ncbi.nlm. nih.gov/*) numbers for amino acid sequences are as follows: SlLs, AAD05242; LAS, At1g55580; PoptrGRAS44, 550683; PoptrGRAS45, 806287; PoptrGRAS46, 574701; OsMOC1, Os06g40780; SlBl, AAL69334; RAX1, At5G23000; RAX2, At2G36890; RAX3, At3G49690; MYB36, At5G57620; MYB68, At5G65790; MYB87, At4G37780; PoptrMYB025, 558303; PoptrMYB099, 584440; PoptrMYB179, 674550; PoptrMYB102, 676067; PoptrMYB088, 578627; PoptrMYB100, 585164; PoptrMYB112, 408761; PoptrMYB184, 563328; GmMYBJ7, ABI73973; MdMYB, AAZ20440; ZmTB1, AAL17055; BRC2, At1G68800; TCP1, At1G67260; BRC1, At3G18550; AmCYC, O49250; PoptrTCP1, 658950; PoptrTCP2, 656764; PoptrTCP3, 772741; PoptrTCP4, 574834; PoptrTCP5, 580771 (d) Relative transcript levels of poplar Ls homologs determined by quantitative PCR. Values were normalized to ubiquitin levels and calibrated to the xylem sample.

### 8.2.3 Shoot Growth

One of the major functions of GA is to promote stem elongation, and the dramatic yield increases in rice and wheat accomplished by the "Green Revolution" were due to mutations that interfered with GA production or GA signaling, resulting in semi-dwarf phenotypes (reviewed in Hedden 2003). The wheat *Reduced height1* allele as well as various *Arabidopsis* dwarf phenotypes are due to mutations in genes encoding DELLA-domain proteins that belong to the GRAS transcription factor family. DELLA proteins are key repressors of GA signaling and semi-dominant gain-of-function alleles are produced by deleting or altering the sequence of the 27-amino acid DELLA domain to produce proteins that function as constitutive (GA-insensitive) repressors of growth (reviewed in Fleet and Sun 2005; Schwechheimer 2008). Whereas rice contains only one member of this subfamily, there are five *Arabidopsis* genes [*GAI, REPRESSOR OF ga1-3 (RGA), RGA-LIKE1 (RGL1), RGL2* and *RGL3*] and four poplar genes (Busov et al. 2005). The *Arabidopsis* genes exhibit a degree of functional diversification with *RGA* and *GAI* primarily acting to repress GA signaling during vegetative growth and floral induction. DELLA proteins are highly conserved and heterologous expression of *gai* or *rgl1* in poplar reduces internode elongation and overall plant height (Busov et al. 2005, 2006). Under field conditions, the multiple independent poplar transgenic events varied widely in the degree of height reduction and also exhibited alterations in crown form. These results indicate that manipulation of GA signaling can be used to alter tree growth and form in useful ways. A yield and carbon sequestration field study of selected transgenic events showing a modest reduction in height and in some cases, narrower, denser crown structure is underway at Oregon State University (Fig. 8-2).

Other TFs appear to have roles in maintaining GA homeostasis via feedback regulation of genes encoding the GA biosynthesis genes, *GA20-oxidase (GA20ox)* and *GA3ox*. The tobacco bZIP transcriptional activator REPRESSION OF SHOOT GROWTH (RSG) binds to the promoter of the GA biosynthesis gene *GA3* (Fukazawa et al. 2000). Constitutive expression of a dominant negative form of RSG blocks activation of the *GA3* promoter, reduces GA levels and causes a dwarf phenotype. Moreover, RSG is sequestered in the cytoplasm via binding to 14-3-3 proteins and translocated to the nucleus in response to a reduction in GA levels (Ishida et al. 2004). The RSG dominant negative form does not bind 14-3-3 proteins, remains in the nucleus and represses the upregulation of *GA20ox* observed in GA-deficient control plants. The *Arabidopsis* AT-HOOK PROTEIN OF GA FEEDBACK (AGF1) binds to a 43-bp sequence in the promoter of the *GA3ox1* gene and mutation of this *cis*-element eliminates AGF1 binding and GA negative feedback regulation (Matsushita et al. 2007).

**Figure 8-2** Field trial of poplar transgenics containing wild type (*GAI*) and mutant (*rgl1*) *Arabidopsis* DELLA-domain encoding transgenes. Trees are in their second growing season in Corvallis, Oregon (USA). Taller trees on the left are ramets of a transgenic event carrying the *35S::GAI* transgene; on the right are *35S::rgl1* trees showing substantially reduced stature and narrower, denser crowns (photo courtesy of S. Strauss and V. Busov). The man is approximately 1.75 m in height.

## 8.3 Secondary Growth and Development

Trees produce an extensive secondary plant body through the addition of cells dividing at the vascular cambium and the cork cambium. The vascular cambium is a sheath of pluripotent stem cells located between the secondary xylem and secondary phloem, dividing on its inner face to produce secondary xylem (wood) and on its outer face to produce secondary phloem. Secondary xylem transports water, dissolved minerals, amino acids and signaling molecules and contains storage parenchyma important to perennial growth. Secondary phloem is important to the storage and transport of sugars in support of heterotrophic tissues. Also transported by phloem conducting cells are polypeptides, mRNA and small signaling molecules involved in the regulation of growth, development and defense (Lough and Lucas 2006). In addition to their conduction and storage functions, both secondary phloem and secondary xylem contain fibers that provide physical support.

An additional lateral meristem, the cork cambium or phellogen, forms outside of the vascular cambium and produces phelloderm toward the inner stem or root and cork (phellum) toward the surface. Together, the phelloderm, cork cambium and cork constitute the periderm, which, along with the secondary phloem, comprises the bark. Cork cambium-derived phellum is suberized and therefore limits water loss as well as damage to the stem and root due to injuries and pathogens. In many species the outer bark is also a rich source of secondary metabolites derived from phenylpropanoid and terpenoid pathway activity. The antimicrobial and insecticidal activities of these metabolites enhance the efficacy of the physical barrier conferred by suberized cells, e.g., against both primary and secondary damage due to bark beetles (reviewed by Franceschi et al. 2005).

It is clear from the preceding overview that the tissues produced by the vascular and cork cambia are of great fundamental importance to plant growth and survival. Humans as well depend on wood and bark as the raw material for fuel and shelter (e.g., Krasutsky 2006). This traditional dependence is poised to increase significantly as a result of intensive efforts to improve the efficiency of cellulosic ethanol production towards the development of sustainable, renewable energy (DOE US 2006). Beyond these roles, wood and bark are processed to yield a vast array of products from paper and textiles to food additives, bioactive compounds and cosmetics (e.g., Krasutsky 2006; Whistler and Bemiller 1997; Mahdi et al. 2006; Ogawa et al. 2006). Despite our long history of dependence on products derived from wood and bark, we have only recently begun to identify genes that control the development of these tissues. Not surprisingly, TFs have figured prominently among these newly discovered regulators of wood formation. These TFs are considered here in the context of additional, uncharacterized TFs that are expressed in the periderm or vascular tissues and may therefore be good candidates for reverse genetics experiments aimed at modifying secondary growth.

### 8.3.1 Periderm

Two genomic studies have specifically targeted the periderm in *Arabidopsis* (Zhao et al. 2005) and cork oak (*Quercus suber*) (Soler et al. 2007), and they provide some interesting clues regarding the involvement of TFs in the periderm. Zhao et al. (2005) isolated periderm (designated non-vascular tissue) from the root-hypocotyl of eight-week-old *Arabidopsis* plants and profiled gene expression in that tissue compared to isolated secondary xylem and phloem-cambium. The 12 TFs reported to be upregulated in *Arabidopsis* periderm included members of AP2, bHLH, bZIP, MYB, NAC and WRKY families. Among the periderm-associated bHLH genes is *LONG HYPOCOTYL IN FAR-RED1* (*HFR1*), which promotes phytochrome-mediated signaling via interaction with another bHLH protein, PHYTOCHROME INTERACTING FACTOR3 (PIF3) (Fairchild et al. 2000). This finding points to a role for phytochrome-mediated development of the secondary plant body. The superfamily of R2R3-MYB TFs consists of 198 genes in *Arabidopsis*, 183 in rice (Yanhui et al. 2006), and 224 in poplar (http://genome.jgi-psf. org/Poptr1_1/Poptr1_1.home.html). Both of the *Arabidopsis* periderm-associated MYBs (At3g49690, MYB84 and At5g65790, MYB68) share high levels of identity with the R2R3-MYB protein Bl from tomato (Schmitz et al. 2002; Muller et al. 2006). As discussed previously (section 8.2.2), *MYB84*, also known as *RAX3*, is a redundant promoter of axillary bud formation along with *RAX1* (At5g23000, *MYB37*) and *RAX2* (At2g36890, *MYB38*) (Muller et al. 2006). Promoter-reporter experiments for both *MYB68* and *MYB84/RAX3* reflected coexpression in the root pericycle, especially in association with emerging lateral roots (Feng et al. 2004). A putative ortholog of *MYB84* was upregulated in cork of *Quercus suber* (Soler et al. 2007). Thus, in addition to its role in shoot axillary meristem production, *MYB84/RAX3* may have roles in promoting meristematic activity in the cork cambium of stems and roots and pericycle cells specialized for lateral root production. *WRKY38* (At5g22570), *WRKY65* (At1g29280) and *WRKY71* (At1g29860) were associated with *Arabidopsis* periderm (Zhao et al. 2005). *WRKY38* (At5g22570) expression was also upregulated in response to treatment with benzothiadiazole (BTH), a functional analog of salicylic acid (SA) and inducer of systemic acquired resistance (SAR), and was also dependent on *NONEXPRESSOR OF PR GENES1* (*NPR1*) expression (Wang et al. 2006). These features of *WRKY38* gene expression suggest that *WRKY38* was upregulated in *Arabidopsis* root periderm in response to microbes in the rhizosphere. These findings from recent work with *Arabidopsis* and cork oak indicate that genome-wide profiling of periderm isolated from poplar would lead to novel discoveries regarding this important interface between secondary vascular tissues and the environment.

### 8.3.2 Phloem and Cambium

Phloem sap has been the subject of proteomic analyses in a range of species, including, for example, cucumber (*Cucumis sativus*) and pumpkin (*Cucurbita maxima*; Walz et al. 2004), castor bean (*Ricinus communis*; Barnes et al. 2004), rape (*Brassica napus*; Giavalisco et al. 2006) and rice (Aki et al. 2008). Phloem sap samples are collected as exudates from incisions, punctures, or aphid or leafhopper stylets following ablation of the insect. Positive and negative marker gene/protein and metabolite analyses have shown that it is possible to obtain high quality, uncontaminated samples using these methods (e.g., Giavalisco et al. 2006). The major advantage of proteomic analysis of tissue- or cell-specific extracts is that phloem conducting-cell proteins are identified directly, not inferred from transcript profiles. However, only a relatively small number of proteins, e.g., 18 from cucurbits (Walz et al. 2004), 33 from castor bean (Barnes et al. 2004), 107 from rice (Aki et al. 2008) and 140 from rape (Giavalisco et al. 2006), were present at sufficient levels for positive identification. Among these, only one putative TF, a zinc finger protein with homology to RAN binding protein (Os02g0203700), was identified in the rice phloem extract (Aki et al. 2008).

A depth of genome coverage similar to that obtained from the aforementioned phloem proteomics studies has been reported for two EST surveys (< 1,000 sequenced clones) based on high quality phloem sap mRNA from castor bean (Doering-Saad et al. 2006) and laser capture microdissection (LCM) of phloem sieve element-companion cell complexes from rice (Asano et al. 2002) and *Arabidopsis* (Ivashikina et al. 2003). A WRKY TF most similar to *WRKY15* (At2g23320) and a bHLH TF most similar to At1g59640/*BIGPETAL* from *Arabidopsis* were identified among the cDNAs from castor bean phloem sap, while laser-captured rice phloem cells yielded cDNA for a NAC domain gene homologous to At4g10350. From these studies it is not possible to conclude whether these three TFs are phloem-specific, as limited comparisons with gene expression in adjacent tissues were presented. The discovery of so few proteins or transcripts for TFs associated with collected phloem sap or following LCM of phloem conducting cells may be due to the unique enucleate nature of mature sieve tube elements. However, even when celery (*Apium graviolens*) procambium, phloem parenchyma and phloem conducting cells were manually dissected as a unit and profiled using a cDNA macroarray (1,326 probes) and EST sequencing (989 ESTs), in a study that included careful comparisons with gene expression in other tissues, phloem-specific TF discovery was limited to one MYB-related and one SCARECROW-like TF (Vilaine et al. 2003).

Efforts to profile phloem- and cambium-biased gene expression in two other models, *Arabidopsis* and poplar, have benefited from the existence of cDNA and oligo arrays providing wider and in some cases genome-wide coverage. The genome-wide studies were conducted with samples obtained from poplar inner bark scrapings (A. Brunner and colleagues, unpublished) or manual dissection of *Arabidopsis* secondary tissues (Zhao et al. 2005), and yielded tissue-level resolution of transcripts in phloem-cambium as a unit. In contrast, higher resolution, yet lower-percentage genome coverage (13,526-gene cDNA microarray), was achieved by profiling a thin-sectioned series from poplar wood-forming tissue (Schrader et al. 2004b). The TFs exhibiting xylem-, phloem- or cambium-biased expression, as determined by these three investigations, are summarized in Table 8-1. Most of the TFs in Table 8-1 are discussed below; however, several vascular-associated TFs are noted only in Table 8-1 and readers should see references cited there for further information. For an example of genome-wide profiling of the

**Table 8-1** Transcription factors associated with poplar and *Arabidopsis* secondary tissues.

| [1]JGI model, poplar gene name | [2]AGI number | Transcription factor family/ *Arabidopsis* gene name(s) | [3]Tissue-biased expression (Poptr/At) | Loss-of-function phenotype | Gain-of-function phenotype | Comments/References |
|---|---|---|---|---|---|---|
| gw1.III.2100.1 (grail3.0018034201), *PoptrMYB128*; gw1.I.8836.1 (grail3.0008045101), *PoptrMYB010* | At1g63910 | R2R3-MYB/ *MYB103* | [4]X/X | | | |
| gw1.127.114.1 (eugene3.01270035), *PoptrMYB092*; gw1.III.861.1, *PoptrMYB125*; fgenesh4_pg.C_LG_XII001269 (eugene3.00121239), *PoptrMYB199*; gw1.XV.2932.1 (estExt_fgenesh4_pg.C_LG_XV0997), *PoptrMYB075*; fgenesh4_pg.C_LG_XII001270, *PoptrMYB200* | At4g22680 | R2R3-MYB/ *MYB85* | X/X | | | |
| gw1.122.261.1, *PoptrMYB090*; eugene3.00070127, *PoptrMYB161* | At1g17950 | R2R3-MYB/ *MYB52* | X/X | *PttMYB21a* antisense plants contained higher concentrations of acid soluble lignin | | Karpinska et al. 2004 |
| gw1.IX.3536.1, *PoptrMYB021*; gw1.I.1547.1 (fgenesh4_pm.C_LG_I000734), *PoptrMYB002* | At5g12870 | R2R3-MYB/ *MYB46* | X/X | Dominant repression yields reduced thickness of fiber and vessel walls | Ectopic deposition of secondary walls | Expression regulated by NAC domain protein SND1; Zhong et al. 2007a |

| | | | | | | |
|---|---|---|---|---|---|---|
| eugene3.00190735, PopMYB168 | At1g22640 | R2R3-MYB/MYB3 | X/NB | | | |
| gw1.VII.1441.1 (grail3.0011003601), PoptrMYB031; gw1.V.2460.1 (grail3.0026002401), PopMYB026 | At4g33450 | R2R3 MYB/MYB69 | X/X | | | Stracke et al. 2007 |
| fgenesh4_pm.C_LG_I000710, PoptrMYB206 gw1.IX.4341.1, PoptrMYB023 | At3g46130 | R2R3-MYB/MYB111 | X/X | As a triple knockout with *myb11* and *myb12*, lacks flavonols but anthocyanin accumulation was normal | | |
| gw1.VII.3881.1 (fgenesh4_pm.C_LG_VII000469), PoptrNAC025; w1.57.51.1 (gw1.57.170.1), PoptrNAC039 | At2g18060 | NAC/ANAC037 VND1 | X/NB | | | Kubo et al. 2005 |
| fgenesh4_pg.C_LG_IV000476, PoptrNAC157; eugene3.00070120 (estExt_fgenesh4_pg.C_LG_VII0110), PoptrNAC156; eugene3.01240095, PoptrNAC105 | At4g28500 | NAC/ANAC073 | X/X | | | |
| gw1.145.134.1 (estExt_fgenesh4_pm.C_1450025), PoptrNAC150 | At4g29230 | NAC/ANAC075 | X/X | | | |
| gw1.III.864.1 (fgenesh4_pm.C_LG_III000335), PoptrNAC046 | At1g12260 | NAC/ANAC007 VND4 | X/X | | | Kubo et al. 2005 |
| gw1.I.5485.1 (estExt_fgenesh4_pg.C_LG_I3128), PoptrNAC063 | At1g32770 | NAC/ANAC012 SND1 NST3 | X/X | As a double knockout with *nst1* is deficient in secondary walls of secondary xylem in hypocotyls and interfascicular fibers in stems | Ectopic secondary wall thickening | Mitsuda et al. 2007; Zhong et al. 2007b; Ko et al. 2007 |

*Table 8-1 contd....*

*Table 8-1 contd....*

| ¹JGI model, poplar gene name | ²AGI number | Transcription factor family/ Arabidopsis gene name(s) | ³Tissue-biased expression (Poptr/At) | Loss-of-function phenotype | Gain-of-function phenotype | Comments/ References |
|---|---|---|---|---|---|---|
| estExt_fgenesh4_pg.C_LG_I2905 (*PtrHB4*, AY919619); estExt_fgenesh4_pg.C_2360002 (*PtrHB3*, AY919618) | At2g34710 | HD-Zip III/ *PHB* | X/X | Triple knockout *phb phv cna* has ectopic vascular bundles in stems | | Prigge et al. 2005; Genbank accession numbers follow PtrHB gene names, according to Ko et al. 2006 |
| fgenesh4_pm.C_LG_I000560 (*PtrHB5*, AY919620) | At1g52150 | HD-Zip III/ *CNA* | X/X | | | Prigge et al. 2005 |
| gw1.IX.4748.1 (*PtrHB1*, AY919616) | At5g06690 | HD-Zip III/ *REV*//*IFL1* | X/X | Disrupted fiber and vessel formation in stems | *rev10-9* mutant has radialized vascular bundles with xylem on both sides of phloem | Emery et al. 2003; Zhong et al. 1997; Zhong and Ye 1999 |
| estExt_Genewise1_v1.C_LG_II2783 | At2g44745 | WRKY/ WRKY12 | X/X | | | |
| gw1.127.187.1 | At2g30590 | WRKY/ WRKY21 | X/NB | | | |
| fgenesh4_pg.C_LG_III000900 grail3.0011025901 | At4g39410 | WRKY/ WRKY13 | X/X | | | |

| Poplar gene | Arabidopsis gene | Name | X designation | Description | Reference |
|---|---|---|---|---|---|
| gw1.14518.1.1<br>gw1.3296.2.1<br>gw1.16923.1.1<br>gw1.X.25.1 | At1g68200 | CCCH ZF | X/X | | |
| eugene3.00002371 | At1g66810 | CCCH ZF | X/X | | |
| gw1.I.9208.1 | At1g62990 | Knotted HD/<br>KNAT7<br>IXR11 | X/NB | Moderately irregular xylem | Brown et al. 2005 |
| gw1.XVII.1082.1 | At3g28920 | ZF HD/ | X/NB | | |
| estExt_fgenesh4_pg.C_LG_II0076 | At4g35550 | ZIP HD/ | X/NB | | |
| gw1.VII.3291.1<br>gw1.V.3738.1 | At2g23760 | BEL1-like HD/<br>BLH4 | X/X | | |
| fgenesh4_pg.C_LG_II000285 | At1g75410 | BEL1-like HD/<br>BLH3 | X/NB | | |
| grail3.0022019601 | At5g02030 | BEL1-like HD/<br>BLH9<br>RPL<br>PNY<br>LSN | X/NB | Ectopic lignification of the replum; reduced interfascicular fibers and increased vasculature in stems | Roeder et al. 2003; Smith and Hake 2003 |
| gw1.IV.685.1<br>estExt_Genewise1_v1.C_LG_IX1644 | At2g16400 | BEL1-like HD/<br>BLH7 | X/NB | | |
| eugene3.00070251 | At4g15248 | ZF (B-Box) | X/NR | | |
| grail3.0022029401 | At2g28710 | ZF (C2H2) | X/NB | | |
| fgenesh4_pg.C_LG_XIV000113 | At1g02040 | ZF (C2H2) | X/NR | | |
| eugene3.00091367 | At5g59820 | ZF (C2H2) | X/NB | | |
| estExt_fgenesh4_pm.C_LG_II0638 | At3g60520 | ZF (PHD) | X/NB | | |
| eugene3.00161312 | At4g32010 | B3 | X/NB | | |
| gw1.VI.985.1 | At4g32010 | B3 | X/NB | | |
| grail3.0003020501 | At1g19850 | Auxin Response Factor/<br>MP | X/XPC | Defective in embryo axis and vascular strand formation | Hardke and Berleth 1998 |

*Table 8-1 contd....*

*Table 8-1 contd....*

| [1]JGI model, poplar gene name | [2]AGI number | Transcription factor family/ *Arabidopsis* gene name(s) | [3]Tissue-biased expression (Poptr/At) | Loss-of-function phenotype | Gain-of-function phenotype | Comments/ References |
|---|---|---|---|---|---|---|
| eugene3.00181144 | At4g32280 | Aux/IAA/ IAA29 | X/XPC | | | |
| eugene3.00091186 | At5g60440 | MADS-box, AGL62 | X/NB | | | |
| estExt_fgenesh4_pg.C_LG_ XII1207 | At5g62165 | MADS-box, AGL42 | X/NB | | | A quiescent center marker; Nawy et al. 2005 |
| estExt_fgenesh4_pg.C_LG_ VIII0151 | At5g04840 | bZIP | X/NB | | | |
| estExt_Genewise1_v1.C_LG_ X5603 | At5g43700 | Aux/IAA/ ATAUX2-11 IAA4 | PC/NB | | | |
| gw1.XIX.1883.1 | At1g71692 | MADS-box/ AGL12 XAL1 | PC/PC | Short root, small root apical meristem, late flowering | | Tapia-Lopez et al. 2008 |
| gw1.XII.993.1 | At1g26310 | MADS-box/ AGL10 CAL | PC/NB | As double knockout with *ap1*, massive proliferation of inflorescence meristem resembling cauliflower head | | Kempin et al. 1995 |
| estExt_Genewise1_v1.C_LG_ II1820 | At4g08150 | Knotted HD/ KNAT1 BP | PC/NB | Premature lignin deposition in stems | Delayed and decreased lignin in stems | Mele et al. 2003 |

| Poplar gene model | Arabidopsis gene | TF family / name | Cluster | Phenotype | Additional phenotype | Reference |
|---|---|---|---|---|---|---|
| gw1.V.825.1 | At4g09820 | bHLH/ *TT8* | PC/NB | Reduced pigmentation correlated with reduced expression of two late flavonoid biosynthesis genes, *DHF* and *BAN* | | |
| gw1.XII.1246.1, *PoptrNAC020* | At3g17730 | NAC | PC/NB | | | |
| gw1.VII.1479.1 (fgenesh4_pm.C_LG_VII000145), *PoptrNAC129* | At5g64530 | NAC/ *XND1* | PC/X | Short plants, short tracheary elements | Parenchyma-like cells replace tracheary elements | Zhao et al. 2008 |
| fgenesh4_pg.C_LG_II001601 | At3g61850 | ZF (DOF)/ *DAG1* | PC/PC | Higher percentage germination than wt, reduced dependence on red light for germination | Lower germination rate and percentage | Papi et al. 2002; Gualberti et al. 2002 |
| aspen_CL12292Contig1 | At1g07640 | ZF (DOF)/ *OBP2* | PC/PC | | Increased *CYP83B1* expression and auxin concentration | Skirycz et al. 2006 |
| gw1.261.5.1 | At2g37590 | ZF (DOF) | PC/PC | | | |
| fgenesh4_pg.C_LG_VI000651 | | | | | | |
| estExt_fgenesh4_pg.C_LG_VIII0451 | At3g55370 | ZF (DOF)/ *OBP3* | PC/PC | Reduced response to red light | | Ward et al. 2005 |
| gw1.XII.1092.1 | At5g62940 | ZF (DOF) | PC/PC | | | |
| gw1.V.3854.1 | At5g60200 | ZF (DOF) | PC/XPC | | | |
| gw1.VII.3109.1 | | | | | | |
| fgenesh4_pg.C_LG_I001774 | At2g28810 | ZF (DOF) | PC/PC | | | |

*Table 8-1 contd.....*

*Table 8-1 contd....*

| [1]JGI model, poplar gene name | [2]AGI number | Transcription factor family / Arabidopsis gene name(s) | [3]Tissue-biased expression (Poptr/At) | Loss-of-function phenotype | Gain-of-function phenotype | Comments/References |
|---|---|---|---|---|---|---|
| gw1.II.4141.1 | At4g37750 | AP2/EREBP/ANT | PC/PC | Decreased cell number, and hence organ size, in shoot organs | Increased cell number in embryo and organs | Mizukami and Fischer 2000; Expression localized to cambium by Schrader et al. 2004b |
| gw1.57.294.1 gw1.VII.3405.1 | At4g36710 | SCF/SCF6 | PC/PC | | | |
| eugene3.00410032 | At1g68810 | bHLH | PC/X | | | |
| gw1.III.2581.1, PoptrMYB132 | At3g13540 | R2R3-MYB/MYB5 | PC/NB | | | |
| gw1.XIX.2908.1 (grail3.0094009001), PoptrMYB063; gw1.XIII.43.1 (estExt_fgenesh4_pm.C_LG_XIII0436), PoptrMYB049 | At2g31180 | R2R3-MYB/MYB14 | PC/NB | | | |
| estExt_fgenesh4_pm.C_LG_X0664, fgenesh4_pg.C_LG_VIII000710, | At1g79430 | GARP/APL WDY | PC/PC | Xylem-like cells replace phloem sieve elements | | Bonke et al. 2003 |
| eugene3.00130592 gw1.XIX.991.1 | At3g04030 | GARP/MYR2 | PC/PC | Double knockout with *myr1* is early flowering | | C. Zhao and E. Beers, unpublished |
| eugene3.00030689, PoptrGARP1 | At1g32240 | GARP/KAN2 | PC/PCNV | | | |
| estExt_fgenesh4_pg.C_1220055, PoptrGARP2 | At5g16560 | GARP/KAN1 | PC/NB | | | |
| grail3.0031002201, PoptrGARP3 | At5g16560 | GARP/KAN1 | PC/NB | | | |

| Model | AGI | Family | Expression | Comments | | References |
|---|---|---|---|---|---|---|
| gw1.VII.712.1 | At4g37650 | GRAS/ SHR | PC/NB | Radial patterning of the root is altered and stem cell activity is lost | | Levesque et al. 2006; Helariutta et al. 2000 |
| eugene3.00180812 | At5g57390 | AP2/EREBP/ AIL5 | Phloem mother cells | AIL5 is similar to PLT1. *plt1 plt2* double knockout has reduced root meristem cell number and meristem is eventually lost to differentiation | Ectopic PLT expression results in ectopic basal organs, roots and hypocotyls, and root stem cell niche formation | Schrader et al. 2004b; Aida et al. 2004 |
| eugene3.00640055, *PoptrMYB174* | At2g46410 | R3-MYB CPC | Phloem mother cells | Reduce root hair differentiation | Excess root hairs/reduced trichomes | Schrader et al. 2004b; Wada et al. 1997 |
| eugene3.00151209, *PoptrMYB077* | At3g61250 | R2R3-MYB/ MYB17 LMI2 | Phloem mother cells | Flower to inflorescence meristem convesion, similar to *ap1* mutant | | Schrader et al. 2004b; J. Pastore and D. Wagner, pers.comm. |

[1]Models in parentheses are alternatives to JGI models for R2R3-MYB (M. Campbell, pers. comm.) and NAC domain (S. Covert, pers. comm.) family members. [2]AGI, *Arabidopsis* Genome Initiative locus number for the homolog sharing the highest identity (amino acid level) with the poplar proteins in the same row. [3]Expression bias information for poplar genes (≥ 4-fold higher in xylem *vs.* phloem or phloem *vs.* xylem) is from A. Brunner and colleagues (unpublished) and for *Arabidopsis* genes as reported by Zhao et al. (2005), unless otherwise indicated in the *Comments/References* column. [4]X, xylem-biased expression; PC, phloem-cambium-biased expression; NV, nonvascular (periderm)-biased expression; XPC, vascular tissue-biased expression; PCNV, bark-biased expression; NB, not biased.

primary phloem transcriptome in *Arabidopsis*, readers are referred to an interesting comparison of transcripts found in phloem exudate (mobile signals from phloem conducting cells) versus those found in phloem parenchyma plus sieve element-companion cell complexes captured by LCM (Deeken et al. 2008).

The high-resolution map developed by Schrader et al. (2004b) included two poplar genes (Genbank accession numbers, BU820600 and BU823291) that are most similar to the AP2/EREBP family members *AINTEGUMENTA* (*ANT*) and *ANT-LIKE5* (*AIL5*), respectively, and exhibited peak expression in active cambium stem cells and phloem mother cells (Table 8-1). *ANT* participates in the control of cell proliferation in primary tissues (Mizukami and Fischer 2000). That the poplar ortholog of *ANT* is downregulated in dormant cambium further supports a role for *ANT* in regulating cambium cell proliferation (Schrader et al. 2004a). Analogous to the findings for *ANT* expression in active poplar cambium, *ANT* mRNA localized to the phloem-cambium in secondary tissues of *Arabidopsis* (Zhao et al. 2005). Also upregulated in poplar phloem mother cells were two MYBs homologous to *CAPRICE* (*CPC*) and *MYB17* (Table 8-1). *CPC* is a regulator of epidermal cell differentiation (Wada et al. 1997). *MYB17* was recently identified as a target of LFY- a floral homeotic gene activator (Weigel and Meyerowitz 1993)—and named *LATE MERISTEM IDENTITY2* (*LMI2*) for its loss-of-function phenotype, which included a partial conversion of floral meristems to inflorescence meristems (J. Pastore and D. Wagner, pers. comm.) similar to the *AP1* loss-of-function phenotype (Irish and Sussex, 1990; Mandel et al. 1992). Two GARP family (Riechmann et al. 2000) genes homologous to *ALTERED PHLOEM* (*APL*) (Bonke et al. 2003) also exhibit phloem-biased expression in poplar (Table 8-1). In *Arabidopsis* roots, loss of *APL* resulted in replacement of phloem sieve elements with tracheary element-like cells, indicating that *APL* is required to promote phloem cell identity and block formation of xylem cells in phloem. These results from phloem-cambium profiling suggest that some genes regulating meristem activity and differentiation of cells in the primary plant body also control the activity and fate of lateral meristem cells in the secondary plant body.

The phloem-mobile product of *FT* expression that regulates flowering time in *Arabidopsis* and bud set in poplar was discussed above (Section 8.2.1). Three additional phloem-cambium-associated genes deserve consideration as TFs that may regulate poplar shoot development. They are the poplar orthologs of the *Arabidopsis* GARP family TF *MYR2* and the MADS-box gene *AGL12/XAL1* (Table 8-1). In *Arabidopsis*, *MYR1* and its paralog *MYR2* are redundant negative regulators of flowering (C. Zhao and E. Beers, unpublished) and *AGL12/XAL1* is a photoperiod responsive gene that promotes flowering in *Arabidopsis* (Tapia-Lopez et al. 2008).

In addition to identifying TFs potentially involved in cambium cell proliferation, differentiation of phloem cells and shoot development, poplar and *Arabidopsis* expression profiling experiments have identified several TFs likely to be regulators of phloem-localized specialized metabolism. For example, a bHLH TF homologous to *TRANSPARENT TESTA8* (*TT8*) from *Arabidopsis* was upregulated in poplar phloem (Table 8-1). In *Arabidopsis*, *TT8* is required for expression of the late flavonoid biosynthesis genes dihydroflavonol 4-reductase (*DFR*) and *BANYULS* (*BAN*) (Nesi et al. 2000). The phloem-associated expression of a putative poplar *TT8* ortholog is consistent with the observation that eight flavonoid biosynthesis genes, including *DFR*, were upregulated in poplar phloem mother cells (Schrader et al. 2004b). Suppression of the lignin branch of phenylpropanoid metabolism mediated by a poplar homolog of the *Arabidopsis* knotted-HD TF, *BREVIPEDICELLUS* (*BP/KNAT1*) may also be important to overall regulation of flavonoid metabolism in poplar phloem. It has been suggested that *BP* plays a role in regulating vascular cell fate, in part through its apparent ability to negatively regulate lignification (Mele et al. 2003). A total of nine DOF genes exhibited phloem-biased expression in poplar (Table 8-1). The *Arabidopsis* DOF known as *OBF BINDING PROTEIN2* (*OBP2*) is a regulator of indole glucosinolate and auxin production via its promotion of expression of the cytochrome P450 monooxygenase *CYP83B1* (Skirycz et al. 2006). Thus, the poplar DOFs may be important links to a more detailed understanding of secondary metabolism in poplar phloem. Results from these genome-wide profiling projects set the stage for large-scale phloem-cambium functional genomics experiments. Investigations of promoter activity should be included in such studies to identify those transcripts resulting from phloem cell-specific gene expression versus those present as a result of mRNA trafficking through conducting cells as non-cell-autonomous signals (Lough and Lucas 2006).

### 8.3.3 Xylem

Compared to phloem- and periderm-associated TFs, much more emphasis has been placed on discovery and characterization of regulatory genes that impact xylem cell differentiation and lignocellulose synthesis. Studies of eucalyptus, pine, spruce, poplar and *Arabidopsis* have demonstrated that four families of TFs include members with important roles in xylem development: R2R3-MYB, NAC, HB-Zip III and GARP (Riechmann et al. 2000). Sixteen poplar R2R3-MYB genes exhibited xylem-biased expression (Table 8-1), yet information on how these genes affect xylem differentiation or function is available for only a few xylem-associated MYBs. One report implicates *PttMYB21a*, a putative ortholog of *Arabidopsis MYB52*, as a repressor of lignification (Karpinska et al. 2004), and additional work indicates that poplar MYBs most similar to the *MYB46/PtMYB4/EgMYB2/PgMYB4* subgroup of MYBs from *Arabidopsis* (Zhong et al. 2007a), *Pinus taeda* (loblolly pine; Patzlaff et al. 2003), *Eucalyptus grandis* (Goicoechea et al. 2005) and *Picea glauca*, (white spruce; Bedon et al. 2007), respectively, are promoters of lignification. Each of these genes exhibits preferential or exclusive expression in lignifying xylem cells, and two co-orthologs are expressed in poplar wood-forming tissue (Table 8-1). From ectopic expression studies, *EgMYB2*, *PtMYB4* and *MYB46* are known to promote expression of genes involved in both lignin and cellulose synthesis. In eucalyptus, *EgMYB2* colocalizes with a QTL that accounts for 4.5% of the variation in lignin content (Goicoechea et al. 2005). Overexpression in tobacco of *EgMYB2* did not have a strong impact on expression of genes acting early in phenylpropanoid metabolism (*PAL, C4H*). However, *4CL* and other genes (*HCT, CCR, CAD, CH3, CCoAOMT, F5H, COMT*) required for the production of coumaryl, coniferyl and sinapyl monolignols were all upregulated to varying degrees ranging from 3- to 40-fold (Goicoechea et al. 2005). Although the analysis of lignin pathway gene expression in tobacco plants overexpressing *PtMYB4* was not as comprehensive as for *EgMYB2*, a similar trend in increased expression of lignin pathway genes was observed. Zhong et al. (2007a) tested the role of *MYB46* and found that overexpression increased expression of lignin biosynthesis genes, *4CL* and *CCoAOMT* as well as genes involved in cellulose and xylan synthesis (*CesA7, CesA8* and *FRA8*). These latter findings are consistent with the observation that overexpression of *MYB46* led to a > 40% increase in vessel wall thickness. Moreover, *PtMYB4* and *MYB46* expression promoted secondary thickening in cells that normally produce only primary walls (Patzlaff et al. 2003; Zhong et al. 2007a). While it is not yet known how many of the lignin, cellulose and xylan synthesis genes are direct targets of the aforementioned MYB genes, it has been shown that PtMYB4 and EgMYB2 can directly transactivate lignin biosynthesis genes by binding to AC elements in their promoters (Patzlaff et al. 2003; Goicoechea et al. 2005). R2R3-MYB genes are not the only TFs known to interact with regulatory regions of lignin pathway genes, as the previously mentioned knotted-HD protein BP/KNAT1 can bind promoters of *CCoAOMT* and *COMT* (Mele et al. 2003). That homologous MYB genes from pine and eucalyptus exhibit both xylem-localized expression and the ability to promote lignin formation in tobacco points to a conservation of function for some MYB family members despite approximately 300 million years of separation of gymnosperms and angiosperms.

The NAC domain family of TFs is a plant-specific family of at least 105 genes in *Arabidopsis* (Ooka et al. 2003) and 160 genes in poplar (http://genome.jgi-psf.org/Poptr1_1/Poptr1_1.home.html). Recent investigations using *Arabidopsis* have established that at least half of the members of a ten-gene subgroup (subgroup IIb, Mitsuda et al. 2005; Fig. 8-3a) of NACs

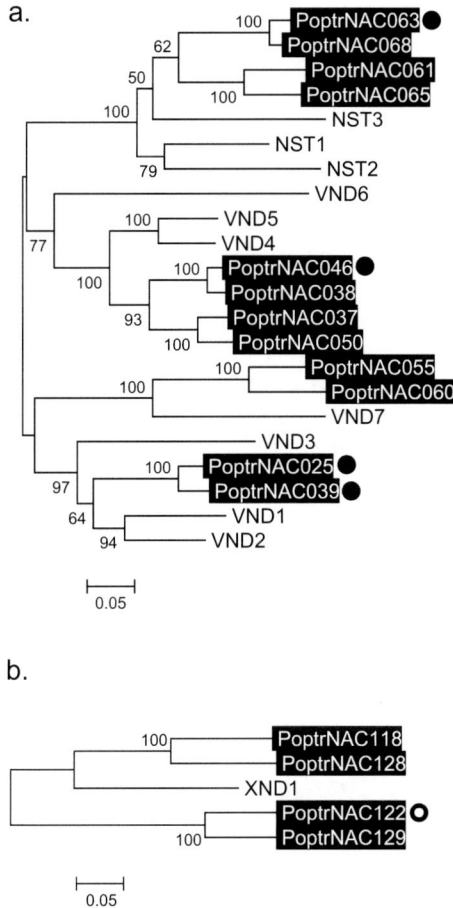

**Figure 8-3** Neighbor-joining phylogenetic analysis of selected vascular-associated members of the NAC domain family of transcription factors from *Arabidopsis* and poplar (a) VND1 through VND7 and the closely-related NST1 through NST3 proteins from *Arabidopsis* are grouped with the 12 proteins from poplar (black boxes). Four of the poplar VND or NST homologs exhibited xylem-biased expression (filled circles). (b) XND1 from *Arabidopsis* is shown with the four proteins from poplar (black boxes), one of which exhibited phloem-cambium-biased expression (empty circle). Names for NAC genes are according to the manual annotation of S. Covert (*http://genome.jgi-psf.org/Poptr1_1/Poptr1_1.home.html*). Bootstrap and scale bar details are as indicated for Figure 8-1. Protein ID (for poplar proteins) or AGI locus (*http://www.arabidopsis.org/index.jsp*) numbers for amino acid sequences used for trees are as follows: NST1, At2g46770; NST2, At3g61910; NST3, At1g32770; VND1, At2g18060; VND2, At4g36160; VND3, At5g66300; VND4, At1g12260; VND5, At1g62700; VND6, At5g62380; VND7, At1g71930; PoptrNAC025, 802983; PoptrNAC037, 773505; PoptrNAC038, 784383; PoptrNAC039, 424203; PoptrNAC046, 799870; PoptrNAC050, 776558; PoptrNAC055, 592235; PoptrNAC060, 249692; PoptrNAC061, 755466; PoptrNAC063, 815836; PoptrNAC065, 245162; PoptrNAC068, 569285; XND1, At5g64530; PoptrNAC118, 756574; PoptrNAC122, 800928; PoptrNAC128, 797511; PoptrNAC129, 802659.

comprising *VASCULAR NAC DOMAIN1* (*VND1*) through *VND7* and *NAC SECONDARY WALL THICKENING PROMOTING FACTOR1* (*NST1*) through *NST3* have distinct roles as promoters of lignocellulose production in multiple cell types. The lignocellulose-promoting roles for *NST1/NST2* and *VND6/VND7* were discovered as a result of reverse genetics experiments looking for anther dehiscence and xylem cell differentiation phenotypes, respectively. While the connection between these two processes may not be readily apparent, both feature thickened secondary cell walls, with those of the anther endothecium necessary for dehiscence (Keijzer et al. 1987) and those of xylem vessels and fibers contributing to physical support of stems. Four poplar NACs putatively orthologous to members of the *VND/NST* subgroup are upregulated in xylem (Table 8-1). Based on work with *Arabidopsis* (discussed below), it is reasonable to hypothesize that these genes promote ligocellulose synthesis.

The functions of the three *Arabidopsis NST* genes have been chronicled in several recent reports (Mitsuda et al. 2005, 2007; Zhong et al. 2006; Ko et al. 2007; Zhong et al. 2007a, 2007b). *NST1* and *NST2* perform redundant roles as positive regulators of secondary thickening of endothecium cell walls (Mitsuda et al. 2005), while the combination of *NST1* and *NST3* promote secondary wall formation in fibers (Zhong et al. 2006; Zhong et al. 2007b; Mitsuda et al. 2007). In contrast, Ko et al. (2007) concluded that *NST3* acts a repressor of secondary wall synthesis in fibers. Their view was based on the observation that *35S:NST3* plants exhibited reduced secondary wall deposition in fibers, the same phenotype observed by Zhong et al. (2006) in dominant repression experiments using an NST3-SRDX fusion (Hiratsu et al. 2003) and for *nst1 nst3* double loss-of-function plants (Mitsuda et al. 2007; Zhong et al. 2007b). Perhaps overexpression of *NST3* can, in some cases, lead to dominant repression of lignocellulose formation in fibers, thereby phenocopying *NST1/NST3* loss-of-function results. In summary, most studies reported to date link overexpression of *NST* genes with ectopic formation of secondary cell walls and support a model where *NST* genes are promoters of lignocellulose synthesis.

The expression of all seven *VND* genes localizes to procambium or xylem cells (Kubo et al. 2005). This co-expression of *VND* genes combined with their high degree of amino acid identity suggests that they act redundantly, as observed for certain *NST* gene combinations. Indeed, loss-of-function lines for *VND6* or *VND7* did not differ from wild type plants (Kubo et al. 2005). However, dominant repression of *VND6* or *VND7*, via expression of VND6- or VND7-SRDX fusions blocked the formation of secondary cell walls in the vessels of roots (Kubo et al. 2005) and shoots (Yamaguchi et al. 2008). Similar to results from overexpression experiments with *NST* genes, both *VND6* and *VND7* promote ectopic secondary cell wall formation when overexpressed in *Arabidopsis* and poplar (Kubo et al. 2005). Thus, while cell-type specialization is suggested by these studies—with *NST1/NST2* acting on the endothecium (Mitsuda et al. 2005), *NST1/NST3* functioning in fiber cells (Zhong et al. 2006; Mitsuda et al. 2007; Zhong et al. 2007a) and *VND6/VND7* affecting vessel member cell walls - these five closely related NAC genes all have the ability to promote secondary cell wall synthesis. Consequently, they and their orthologs in poplar and other biomass crops may become very important elements of strategies for increasing biomass quantity and quality.

While some progress has been reported regarding the identification of primary targets of TFs that promote lignocellulose synthesis (Patzlaff et al. 2003; Goicoechea et al. 2005), a full understanding of the gene networks that regulate wood formation will require the identification of additional direct targets of the MYB and NAC genes that promote or repress lignocellulose synthesis. Secondary cell wall thickening induced by *MYB46, NST1* or *NST3* was linked with increased expression of several marker genes representing lignocellulose synthesis (Mitsuda et al. 2005; Zhong et al. 2006; Zhong et al. 2007a) and promoters of these genes could be tested for interaction with the relevant TFs. Conspicuously absent from the list of genes upregulated in concert with ectopic secondary wall formation were genes encoding hydrolases associated with the autolytic component of programmed cell death in xylem (e.g., Ito and Fukuda 2002; Avci et al. 2008). Apparently, ectopic lignocellulose synthesis driven by the MYBs and NACs discussed here occurs independent of the cell suicide component of tracheary element differentiation, making these TFs especially attractive as tools for increasing biomass of living cells and illustrating that developmentally programmed cell death can be uncoupled from the differentiation of ectopic tracheary element-like cells.

In addition to identifying downstream targets of lignocellulose promoting TFs, it is also important to identify the upstream regulators of the lignocellulose-promoting NACs and MYBs. Zhong et al. (2007a) recently reported that the NAC domain protein NST3/SND1 is a transactivator of *MYB46* and also identified the corresponding *cis*-element in the promoter of *MYB46*. The identification of *MYB46* as a direct target of NST3/SND1 provides an important link between the MYB TFs that are able to activate expression of lignin structural genes and a subgroup of NAC domain TFs that have been described as master regulators of xylem cell differentiation (Kubo et al. 2005). *MYB26/MALESTERILE35* (*MS35*) is another positive regulator of secondary cell wall synthesis that acts specifically to promote thickening of walls of the anther endothecium rather than xylem cells (Yang et al. 2007). Like *MYB46* (Zhong et al. 2007a), *PtMYB4* (Patzlaff et al. 2003) and the *VND/NST* genes (Kubo et al. 2005; Mitsuda et al. 2005, 2007) described above, *MYB26* was able to promote ectopic formation of tracheary element-like cells when overexpressed. Notably, in addition to increasing expression of structural genes for lignocellulose synthesis, overexpression of *MYB26* also increased expression of *NST1* and *NST2*. The results obtained by Zhong et al. (2007a) and Yang et al. (2007) provide just a glimpse of the potential for MYB-NAC interactions to form key components of networks required for regulating cell type-specific lignocellulose synthesis.

The *NST/VND* NACs are not the only members of the NAC domain family that are capable of regulating xylem cell differentiation. In contrast to the phenotypes resulting from overexpression experiments with *NST/VND* genes, overexpression of *XYLEM NAC DOMAIN1* (*XND1*) (Fig. 8-3b) from *Arabidopsis* was sufficient to completely block lignocellulose deposition and programmed cell death in xylem, apparently without preventing phloem differentiation (Zhao et al. 2008). *XND1* loss-of-function plants were smaller than wild type and produced shorter tracheary elements. Together these loss- and gain-of-function phenotypes suggest that *XND1* regulates xylem cell size through negative regulation of the terminal steps in tracheary element differentiation. *XND1* is single-copy gene in *Arabidopsis*, but is represented by four putative co-orthologs in poplar: *NAC118, NAC122, NAC128* and *NAC129*. Each of the poplar *XND1* co-orthologs exhibits a distinct expression pattern (Grant et al. 2010). The apparent ability of *XND1* to negatively regulate both lignocellulose synthesis and programmed death of tracheary elements may reflect a more general role in regulation of proliferation versus differentiation. It will be interesting to see if the *XND1* co-orthologs in poplar, with their varied expression patterns, have diversified to perform similar roles within multiple contexts or have acquired entirely new functions.

There are five class-III homeobox-leucine zipper domain (HB-Zip III) TFs in *Arabidopsis* (Sessa et al. 1998), and they function in embryo patterning and development and post-embryonically in meristem function, organ polarity and vascular development (Prigge et al. 2005). Reflecting their roles in the latter, all five *Arabidopsis* HB-Zip III genes (Fig. 8-4a) exhibited xylem-biased expression in secondary vascular tissues (Zhao et al. 2005). Ko et al. (2006) reported that eight HB-Zip III genes were predicted by the *P. trichocarpa* genome sequence. Four of these exhibited xylem-biased expression in poplar (Table 8-1; Fig. 8-4a), and an HB-Zip III from *P. tremula x P. alba* (*PtaHB1*) has been linked with wood formation (Ko et al. 2006). Despite their vascular co-expression patterns, results obtained with plants harboring loss-of-function mutations in single or multiple HB-Zip III genes revealed that the HB-Zip III genes differentially impact vascular tissue development (Prigge et al. 2005). For example, loss of *REVOLUTA/ INTERFASCICULAR FIBER1* (*IFL1/REV*) expression led to disruptions in fiber and vessel element formation in stems (Zhong et al. 1997; Zhong and Ye 1999), while *PHABULOSA* (*PHB*) *PHAVOLUTA* (*PHV*) *CORONA* (*CNA*) triple knockouts (*phb phv cna*) produced ectopic vascular bundles in stems

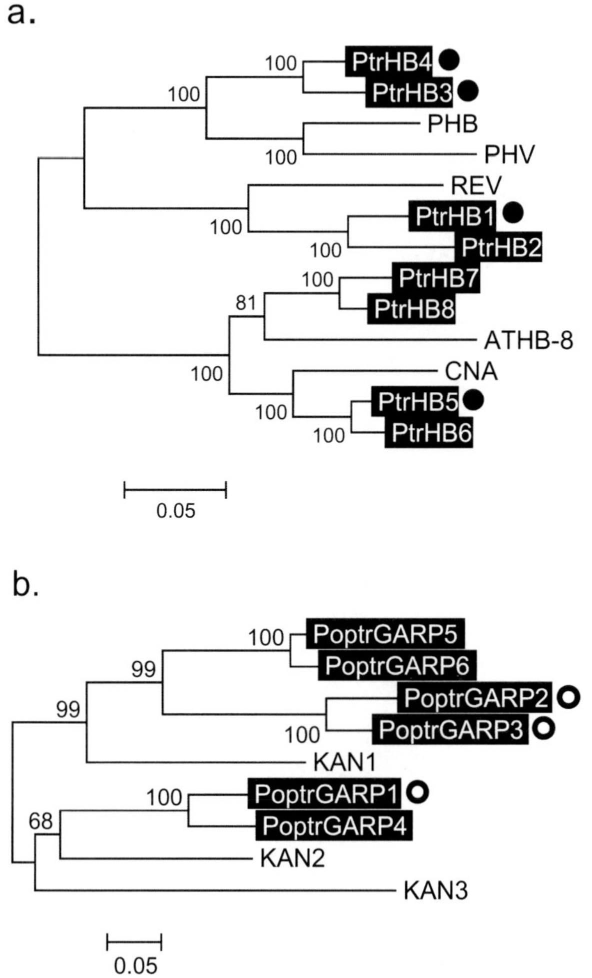

**Figure 8-4** Neighbor-joining phylogenetic analysis of vascular-associated HB-Zip III and GARP family transcription factors from *Arabidopsis* and poplar (a) Proteins encoded by the five HB-Zip III genes from *Arabidopsis* are grouped with the eight proteins from poplar (black boxes). Four of the poplar HB-Zip III genes exhibited xylem-biased expression (filled circles) (b) The GARP family members KAN1 through KAN3 are grouped with the six proteins from poplar (black boxes). Three of the poplar GARP genes exhibited phloem-cambium-biased expression (empty circles). Names for the poplar HB-Zip III genes are according to Ko et al. (2006). Bootstrap and scale bar details are as indicated for Figure 8-1. Protein ID (PoptrGARP genes) (*http://genome.jgi-psf.org/Poptr1_1/Poptr1_1.home.html*), AGI locus (*http://www.arabidopsis.org/index.jsp*) or GenBank accession (PtrHB genes) numbers for amino acid sequences used for trees are as follows: ATHB-8, At4g32880; CNA, At1g52150; PHB, At2g34710; PHV, At1g30490; REV, At5g60690; PtrHB1–PtrHB8, AY919616–AY919623; KAN1, AT5G16560; KAN2, AT1G32240; KAN3, AT4G17695; PoptrGARP1, 554056; PoptrGARP2, 827763; PoptrGARP3, 661354; PoptrGARP4, 590054; PoptrGARP5, 755990; PoptrGARP6, 556019.

(Prigge et al. 2005). Although there is evidence that at least one HB-Zip III gene (*AtHB8*) acts in the procambium and cambium to promote xylem cell fate (Baima et al. 2001), it is difficult to identify specific roles played by the HB-Zip III genes in vascular tissue differentiation in the context of their broader roles in meristem cell fate and organ polarity. For example, the ectopic vascular bundles in *phb phv cna* plants may have formed due to the accompanying enlarged meristem rather than as a result of altered cell fate in the pith (Prigge et al. 2005) and the radialized vascular bundle—with xylem surrounding phloem—observed in the *rev-10d* gain-of-function mutant (Emery et al. 2003) may be a consequence of the reported ability to HB-Zip III genes to promote adaxial/central organ polarity rather than a specific effect on vascular cell fate.

The ability of HB-Zip III genes to promote adaxial identity is counteracted by three members of the GARP family of MYB-related genes, *KANADI1* (*KAN1*) through *KAN3* (Fig. 8-4b), that promote abaxial/peripheral organ polarity (Eshed et al. 2001; Emery et al. 2003). Poplar stems also express putative KAN orthologs in a phloem-biased pattern (Table 8-1; Fig. 8-4b), as reported for *Arabidopsis* (Emery et al. 2003; Zhao et al. 2005). Conservation in poplar of the central vs. peripheral expression pattern for the HB-Zip III genes vs. KAN-like GARPs suggests that these genes are regulators of organ and tissue polarity in poplar. Compared to *Arabidopsis*, however, both gene families have expanded and may therefore regulate additional aspects of polarity, meristem activity and vascular cell fate specific to woody perennials.

## 8.4 Perspectives

The poplar genome sequence, genome-wide gene expression studies, and comparative analyses with *Arabidopsis* and other plants are allowing inferences about TF evolution and function in poplar. However, very few TFs have been functionally characterized in poplar at the whole plant level. For obvious reasons, most poplar transgenic studies have been limited to growth chambers and greenhouses and thus capture only a small fraction of the tree growth habit and life cycle. Because comprehensive studies are feasible for only a subset of the poplar TFs, we are challenged to judiciously select poplar TFs for functional analyses and association genetic approaches. Currently, choices are made mostly based on comparative genomics and gene expression. But to understand the role of TFs in adaptation and to realize the potential of TFs for poplar domestication, we need to understand the place of TFs in regulatory networks. Fortunately, we can build on methods developed for other organisms to discover the TF-DNA and protein-protein interactions that can ultimately reveal the TF interactions and regulatory networks important for tree growth and development.

# Acknowledgments

We thank the Office of Science (BER and BES), US Department of Energy, Grant Nos. DE-FG02-04ER63788, DE-FG02-06ER64185, DE-FG02-04ER15627 and DE-FG02-07ER64449, and the USDA National Institute of Food and Agriculture Grant no. 2008-35301-19167 for supporting our research on shoot and secondary vascular tissue development and Dr. James Tokuhisa for critical reading of the manuscript.

# References

Aguilar-Martinez JA, Poza-Carrion C, Cubas P (2007) Arabidopsis *BRANCHED1* acts as an integrator of branching signals within axillary buds. Plant Cell 19: 458–472.

Aida M, Beis D, Heidstra R, Willemsen V, Blilou I, Galinha C, Nussaume L, Noh YS, Amasino R, Scheres B (2004) The *PLETHORA* genes mediate patterning of the Arabidopsis root stem cell niche. Cell 119: 109–120.

Aitken SN, Yeaman S, Holliday J, Wang T, Sierra C-M (2008) Adaptation, migration or extirpation: climate change outcomes for tree populations. Evol Appl 1: 95–111.

Aki T, Shigyo M, Nakano R, Yoneyama T, Yanagisawa S (2008) Nano scale proteomics revealed the presence of regulatory proteins including three FT-Like proteins in phloem and xylem saps from rice. Plant Cell Physiol 49: 767–790.

Asano T, Masumura T, Kusano H, Kikuchi S, Kurita A, Shimada H, Kadowaki K (2002) Construction of a specialized cDNA library from plant cells isolated by laser capture microdissection: toward comprehensive analysis of the genes expressed in the rice phloem. Plant J 32: 401–408.

Avci U, Petzold HE, Ismail IO, Beers EP, Haigler CH (2008) Cysteine proteases XCP1 and XCP2 aid micro-autolysis within the intact central vacuole during xylogenesis in Arabidopsis roots. Plant J 56: 303–315.

Babu MM, Luscombe NM, Aravind L, Gerstein M, Teichmann SA (2004) Structure and evolution of transcriptional regulatory networks. Curr Opin Struct Biol 14: 283–291.

Baima S, Possenti M, Matteucci A, Wisman E, Altamura MM, Ruberti I, Morelli G (2001) The Arabidopsis ATHB-8 HD-zip protein acts as a differentiation-promoting transcription factor of the vascular meristems. Plant Physiol 126: 643–655.

Barnes A, Bale J, Constantinidou C, Ashton P, Jones A, Pritchard J (2004) Determining protein identity from sieve element sap in *Ricinus communis* L. by quadrupole time of flight (Q-TOF) mass spectrometry. J Exp Bot 55: 1473–1481.

Bedon F, Grima-Pettenati J, Mackay J (2007) Conifer R2R3-MYB transcription factors: sequence analyses and gene expression in wood-forming tissues of white spruce (*Picea glauca*). BMC Plant Biol 7: 17.

Boerjan W (2005) Biotechnology and the domestication of forest trees. Curr Opin Biotechnol 16: 159–166.

Bohlenius H, Huang T, Charbonnel-Campaa L, Brunner AM, Jansson S, Strauss SH, Nilsson O (2006) CO/FT regulatory module controls timing of flowering and seasonal growth cessation in trees. Science 312: 1040–1043.

Bonke M, Thitamadee S, Mahonen AP, Hauser M, Helariutta Y (2003) APL regulates vascular tissue identity in Arabidopsis. Nature 426: 181–186.

Bradshaw HD Jr, Strauss SH (2001) Breeding strategies for the 21st century: Domestication of poplar. In: DI Dickmann, JG Isebrands, JE Eckenwalder, J Richardson (eds) Poplar Culture in North America, part 2. NRC Research Press, National Research Council of Canada, Ottawa, ON, Canada, pp 383–394.

Brown DM, Zeef LAH, Ellis J, Goodacre R, Turner SR (2005) Identification of novel genes in Arabidopsis involved in secondary cell wall formation using expression profiling and reverse genetics. Plant Cell 17: 2281–2295.

Brunner AM, Goldfarb B, Strauss SH (2003) Controlling maturation and flowering for forest tree domestication. In: CN Stewart (ed) Transgenic Plants: Current Innovations and Future Trends. Horizon Press, Norwich, UK, pp 9–44.

Brunner AM, Busov VB, Strauss SH (2004) Poplar genome sequence: functional genomics in an ecologically dominant plant species. Trends Plant Sci 9: 49–56.

Busov VB, Brunner AM, Meilan R, Filichkin S, Ganio L, Gandhi S, Strauss SH (2005) Genetic transformation: a powerful tool for dissection of adaptive traits in trees. New Phytol. 167: 9–18.

Busov V, Meilan R, Pearce DW, Rood SB, Ma C, Tschaplinski TJ, Strauss SH (2006) Transgenic modification of *gai* or *rgl1* causes dwarfing and alters gibberellins, root growth, and metabolite profiles in *Populus*. Planta 224: 288–299.

De Bodt S, Maere S, Van de Peer Y (2005) Genome duplication and the origin of angiosperms. Trends Ecol Evol 20: 591–597.

Deeken R, Ache P, Kajahn I, Klinkenberg J, Bringmann G, Hedrich R (2008) Identification of *Arabidopsis thaliana* phloem RNAs provides a search criterion for phloem-based transcripts hidden in complex datasets of microarray experiments. Plant J: 55: 746–759.

Dennis ES, Peacock WJ (2007) Epigenetic regulation of flowering. Curr Opin Plant Biol 10: 520–527.

Dickmann DI, Isebrands JG, Blake TJ, Kosola K, Kort J (2001) Physiology ecology of poplars. In: DI Dickmann, JG Isebrands,JE Eckenwalder, J Richardson (eds) Poplar Culture in North America, Part 2. NRC Research Press, National Research Council of Canada, Ottowa, ON, Canada, pp 77–118.

DOE US (2006) Breaking the biological barriers to cellulosic ethanol: a joint research agenda, DOE/SC-0095, US Department of Energy Office of Science and Office of Energy Efficiency and Renewable Energy.

Doebley J, Lukens L (1998) Transcriptional regulators and the evolution of plant form. Plant Cell 10: 1075–1082.

Doebley J, Stec A, Hubbard L (1997) The evolution of apical dominance in maize. Nature 386: 485–488.

Doebley JF, Gaut BS, Smith BD (2006) The molecular genetics of crop domestication. Cell 127: 1309–1321.

Doering-Saad C, Newbury HJ, Couldridge CE, Bale JS, Pritchard J (2006) A phloem-enriched cDNA library from Ricinus: insights into phloem function. J Exp Bot 57: 3183–3193.

Dun EA, Ferguson BJ, Beveridge CA (2006) Apical dominance and shoot branching: Divergent opinions or divergent mechanisms? Plant Physiol 142: 812–819.

Emery JF, Floyd SK, Alvarez J, Eshed Y, Hawker NP, Izhaki A, Baum SF, Bowman JL (2003) Radial patterning of Arabidopsis shoots by class III HD-ZIP and KANADI genes. Curr Biol 13: 1768–1774.

Eshed Y, Baum SF, Perea JV, Bowman JL (2001) Establishment of polarity in lateral organs of plants. Curr Biol 11: 1251–1260.

Fairchild CD, Schumaker MA, Quail PH (2000) *HFR1* encodes an atypical bHLH protein that acts in phytochrome A signal transduction. Genes Dev 14: 2377–2391.

Feng C, Andreasson E, Maslak A, Mock HP, Mattsson O, Mundy J (2004) Arabidopsis *MYB68* in development and responses to environmental cues. Plant Sci 167: 1099–1107.

Fleet CM, Sun TP (2005) A DELLAcate balance: the role of gibberellin in plant morphogenesis. Curr Opin Plant Biol 8: 77–85.

Franceschi VR, Krokene P, Christiansen E, Krekling T (2005) Anatomical and chemical defenses of conifer bark against bark beetles and other pests. New Phytol 167: 353–375.

Fukazawa J, Sakai T, Ishida S, Yamaguchi I, Kamiya Y, Takahashi Y (2000) Repression of shoot growth, a bZIP transcriptional activator, regulates cell elongation by controlling the level of gibberellins. Plant Cell 12: 901–915.

Giavalisco P, Kapitza K, Kolasa A, Buhtz A, Kehr J (2006) Towards the proteome of *Brassica napus* phloem sap. Proteomics 6: 896–909.

Goicoechea M, Lacombe E, Legay S, Mihaljevic S, Rech P, Jauneau A, Lapierre C, Pollet B, Verhaegen D, Chaubet-Gigot N, Grima-Pettenati J (2005) EgMYB2, a new transcriptional activator from Eucalyptus xylem, regulates secondary cell wall formation and lignin biosynthesis. Plant J 43: 553–567.

Grant EH, Fujino T, Beers EP, Brunner AM (2010) Characterization of NAC domain transcription factors implicated in control of vascular cell differentiation in Arabidopsis and Poplus. Planta 232: 337–352.

Greb T, Clarenz O, Schafer E, Muller D, Herrero R, Schmitz G, Theres K (2003) Molecular analysis of the *LATERAL SUPPRESSOR* gene in Arabidopsis reveals a conserved control mechanism for axillary meristem formation. Genes Dev 17: 1175–1187.

Greenwood MS (1995) Juvenility and maturation in conifers: current concepts. Tree Physiol 15: 433–438.

Groover AT (2007) Will genomics guide a greener forest biotech? Trends Plant Sci 12: 234–238.

Gualberti G, Papi M, Bellucci L, Ricci I, Bouchez D, Camilleri C, Costantino P, Vittorioso P (2002) Mutations in the Dof zinc finger genes *DAG2* and *DAG1* influence with opposite effects the germination of Arabidopsis seeds. Plant Cell 14: 1253–1263.

Hardtke CS, and Berleth T (1998) The Arabidopsis gene *MONOPTEROS* encodes a transcription factor mediating embryo axis formation and vascular development. EMBO J 17: 1405–1411.

Hedden P (2003) The genes of the Green Revolution. Trends Genet 19: 5–9.

Heilman PE, . Hinckley TM, Roberts DA, Ceulemans R (1996) Production physiology. In: RF Stettler, HD Bradshaw Jr, PE Heilman, TM Hinckley (eds) Biology of *Populus* and its Implications for Management and Conservation. NRC Research Press, National Research Council of Canada, Ottawa, ON, Canada, pp 459–489.

Helariutta Y, Fukaki H, Wysocka-Diller J, Nakajima K, Jung J, Sena G, Hauser MT, Benfey PN (2000) The *SHORT-ROOT* gene controls radial patterning of the Arabidopsis root through radial signaling. Cell 101: 555–567.

Hiratsu K, Matsui K, Koyama T, Ohme-Takagi M (2003) Dominant repression of target genes by chimeric repressors that include the EAR motif, a repression domain, in Arabidopsis. Plant J 34: 733–739.

Horvath DP, Anderson JV, Chao WS, Foley ME (2003) Knowing when to grow: signals regulating bud dormancy. Trends Plant Sci 8: 534–40.

Howe GT, Aitken SN, Neale DB, Jermstad KD, Wheeler NC, Chen THH (2003) From genotype to phenotype: unraveling the complexities of cold adaptation in forest trees. Can J Bot 81: 1247–1266.

Imaizumi T, Kay SA (2006) Photoperiodic control of flowering: not only by coincidence. Trends Plant Sci 11: 550–558.

Irish VF, Sussex IM (1990) Function of the *apetala-1* gene during Arabidopsis floral development. Plant Cell 2: 741–753.

Ishida S, Fukazawa J, Yuasa T, Takahashi Y (2004) Involvement of 14-3-3 signaling protein binding in the functional regulation of the transcriptional activator REPRESSION OF SHOOT GROWTH by gibberellins. Plant Cell 16: 2641–2651.

Ito J, Fukuda H (2002) ZEN1 is a key enzyme in the degradation of nuclear DNA during programmed cell death of tracheary elements. Plant Cell 14: 3201–3211.

Ivashikina N, Deeken R, Ache P, Kranz E, Pommerrenig B, Sauer N, Hedrich R (2003) Isolation of AtSUC2 promoter–GFP-marked companion cells for patch–clamp studies and expression profiling. Plant J 36: 931–945.

Jansson S, Douglas CJ (2007) *Populus*: a model system for plant biology. Annu Rev Plant Biol 58: 435–458.

Karpinska B, Karlsson M, Srivastava M, Stenberg A, Schrader J, Sterky F, Bhalerao R, Wingsle G (2004) MYB transcription factors are differentially expressed and regulated during secondary vascular tissue development in hybrid aspen. Plant Mol Biol 56: 255–270.

Keijzer CJ (1987) The processes of anther dehiscence and pollen dispersonal—the opening mechanism of longitudinally dehiscing anthers. New Phytol 105: 487–498.

Keller T, Abbott J, Moritz T, Doerner P (2006) Arabidopsis *REGULATOR OF AXILLARY MERISTEMS1* controls a leaf axil stem cell niche and modulates vegetative development. Plant Cell 18: 598–611.

Kempin SA, Savidge B, Yanofsky MF (1995) Molecular basis of the cauliflower phenotype in Arabidopsis. Science 267: 522–525.

Ko JH, Prassinos C, Han KH (2006) Developmental and seasonal expression of *PtaHB1*, a Populus gene encoding a class III HD-Zip protein, is closely associated with secondary growth and inversely correlated with the level of microRNA (miR166). New Phytol 169: 469–478.

Ko JH, Yang SH, Park AH, Lerouxel O, Han KH (2007) *ANAC012*, a member of the plant-specific NAC transcription factor family, negatively regulates xylary fiber development in *Arabidopsis thaliana*. Plant J 50: 1035–1048.

Kobayashi Y, Weigel D (2007) Move on up, it's time for change—mobile signals controlling photoperiod-dependent flowering. Genes Dev 21: 2371–2384.

Krasutsky PA (2006) Birch bark research and development. Nat Prod Rep 23: 919–942.

Kubo M, Udagawa M, Nishikubo N, Horiguchi G, Yamaguchi M, Ito J, Mimura T, Fukuda H, Demura T (2005) Transcription switches for protoxylem and metaxylem vessel formation. Genes Dev 19: 1855–1860.

Lang GA (1987) Dormancy: a new universal terminology. HortScience 22: 817–820.

Levesque MP, Vernoux T, Busch W, Cui H, Wang JY, Blilou I, Hassan H, Nakajima K, Matsumoto N, Lohmann JU, Scheres B, Benfey PN (2006) Whole-genome analysis of the SHORT-ROOT developmental pathway in Arabidopsis. PLoS Biol 4: e143.

Li X, Qian Q, Fu Z, Wang Y, Xiong G, Zeng D, Wang X, Liu X, Teng S, Hiroshi F, Yuan M, Luo D, Han B, Li J (2003) Control of tillering in rice. Nature 422: 618–621.

Lough TJ, Lucas WJ (2006) Integrative plant biology: role of phloem long-distance macromolecular trafficking. Annu Rev Plant Biol 57: 203–232.

Lu PL, Chen NZ, An R, . Su Z, Qi BS, Ren F, Chen J, Wang XC (2007) A novel drought-inducible gene, *ATAF1*, encodes a NAC family protein that negatively regulates the expression of stress-responsive genes in Arabidopsis. Plant Mol Biol 63: 289–305.

Mahdi JG, Mahdi AJ, Bowen ID (2006) The historical analysis of aspirin discovery, its relation to the willow tree and antiproliferative and anticancer potential. Cell Prolif 39: 147–155.

Mandel MA, Gustafson-Brown C, Savidge B, Yanofsky MF (1992) Molecular characterization of the Arabidopsis floral homeotic gene *APETALA1*. Nature 360: 273–277.

Matsushita A, Furumoto T, Ishida S, Takahashi Y (2007) AGF1, an AT-hook protein, is necessary for the negative feedback of *AtGA3ox1* encoding GA 3-oxidase. Plant Physiol 143: 1152–1162.

McSteen P, Leyser O (2005) Shoot branching. Annu. Rev Plant Biol 56: 353–374.

Mele G, Ori N, Sato Y, Hake S (2003) The knotted1-like homeobox gene *BREVIPEDICELLUS* regulates cell differentiation by modulating metabolic pathways. Genes Dev 17: 2088–2093.

Ming R, Hou S, Feng Y, Yu Q, Dionne-Laporte A, Saw JH, Senin P, Wang W, Ly BV, Lewis KL, Salzberg SL, Feng L, Jones MR, Skelton RL, Murray JE, Chen C, Qian W, Shen J, Du P, Eustice M, Tong E, Tang H, Lyons E, Paull RE, Michael TP, Wall K, Rice DW, Albert H, Wang ML, Zhu YJ, Schatz M, Nagarajan N, Acob RA, Guan P, Blas A, Wai CM, Ackerman CM, Ren Y, Liu C, Wang J, Na JK, Shakirov EV, Haas B, Thimmapuram J, Nelson D, Wang X, Bowers JE, Gschwend AR, Delcher AL, Singh R, Suzuki JY, Tripathi S, Neupane K, Wei H, Irikura B, Paidi M, Jiang N, Zhang W, Presting G, Windsor A, Navajas-Perez R, Torres MJ, Feltus FA, Porter B, Li Y, Burroughs AM, Luo MC, Liu L, Christopher DA,

Mount SM, Moore PH, Sugimura T, Jiang J, Schuler MA, Friedman V, Mitchell-Olds T, Shippen DE, dePamphilis CW, Palmer JD, Freeling M, Paterson AH, Gonsalves D, Wang L, Alam M (2008) The draft genome of the transgenic tropical fruit tree papaya (*Carica papaya Linnaeus*). Nature 452: 991–996.

Mitsuda N, Seki M, Shinozaki K, Ohme-Takagi M (2005) The NAC transcription factors NST1 and NST2 of Arabidopsis regulate secondary wall thickenings and are required for anther dehiscence. Plant Cell 17: 2993–3006.

Mitsuda N, Iwase A, Yamamoto H, Yoshida M, Seki M, Shinozaki K, Ohme-Takagi M (2007) NAC transcription factors, NST1 and NST3, are key regulators of the formation of secondary walls in woody tissues of Arabidopsis. Plant Cell 19: 270–280.

Mizukami Y, Fischer RL (2000) Plant organ size control: *AINTEGUMENTA* regulates growth and cell numbers during organogenesis. Proc Natl Acad Sci USA 97: 942–947.

Muller D, Schmitz G, Theres K (2006) *Blind* homologous R2R3 Myb genes control the pattern of lateral meristem initiation in Arabidopsis. Plant Cell 18: 586–597.

Nawy T, Lee JY, Colinas J, Wang JY, Thongrod SC, Malamy JE, Birnbaum K, Benfey PN (2005) Transcriptional profile of the Arabidopsis root quiescent center. Plant Cell 17: 1908–1925.

Neale DB (2007) Genomics to tree breeding and forest health. Curr Opin Genet Dev 17: 539–544.

Nesi N, Debeaujon I, Jond C, Pelletier G, Caboche M, Lepiniec L (2000) The *TT8* gene encodes a basic helix-loop-helix domain protein required for expression of *DFR* and *BAN* genes in Arabidopsis siliques. Plant Cell 12: 1863–1878.

Ogawa Y, Oku H, Iwaoka E, Iinuma M, Ishiguro K (2006) Allergy-preventive phenolic glycosides from *Populus sieboldii*. J Nat Prod 69: 1215–1217.

Ooka H, Satoh K, Doi K, Nagata T, Otomo Y, Murakami K, Matsubara K, Osato N, Kawai J, Carninci P, Hayashizaki Y, Suzuki K, Kojima K, Takahara Y, Yamamoto K, Kikuchi S (2003) Comprehensive analysis of NAC family genes in *Oryza sativa* and *Arabidopsis thaliana*. DNA Res 10: 239–247.

Papi M, Sabatini S, Altamura MM, Hennig L, Schafer E, Costantino P, Vittorioso P (2002) Inactivation of the phloem-specific Dof zinc finger gene *DAG1* affects response to light and integrity of the testa of Arabidopsis seeds. Plant Physiol 128: 411–417.

Patzlaff A, McInnis S, Courtenay A, Surman C, Newman LJ, Smith C, Bevan MW, Mansfield S, Whetten RW, Sederoff RR, Campbell MM (2003) Characterisation of a pine MYB that regulates lignification. Plant J 36: 743–754.

Prigge MJ, Otsuga D, Alonso JM, Ecker JR, Drews GN, Clark SE (2005) Class III homeodomain-leucine zipper gene family members have overlapping, antagonistic, and distinct roles in Arabidopsis development. Plant Cell 17: 61–76.

Qu LJ, Zhu YX (2006) Transcription factor families in Arabidopsis: major progress and outstanding issues for future research. Curr. Opin. Plant Biol 9: 544–549.

Rae AM, Robinson KM, Street NR, Taylor G (2004) Morphological and physiological traits influencing biomass productivity in short-rotation coppice poplar. Can J For Res 34: 1488–1498.

Riechmann JL, Heard J, Martin G, Reuber L, Jiang C, Keddie J, Adam L, Pineda O, Ratcliffe OJ, Samaha RR, Creelman R, Pilgrim M, Broun P, Zhang JZ, Ghandehari D, Sherman BK, Yu G (2000) Arabidopsis transcription factors: genome-wide comparative analysis among eukaryotes. Science 290: 2105–2110.

Roeder AH, Ferrandiz C, Yanofsky MF (2003) The role of the REPLUMLESS homeodomain protein in patterning the Arabidopsis fruit. Curr Biol 13: 1630–1635.

Rohde A, Bhalerao RP (2007) Plant dormancy in the perennial context. Trends Plant Sci 12: 217–223.

Rohde A, Ruttink T, Hostyn V, Sterck L, Van Driessche K, Boerjan W (2007) Gene expression during the induction, maintenance, and release of dormancy in apical buds of poplar. J Exp Bot 58: 4047–4060.

Ruttink T, Arend M, Morreel K, Storme V, Rombauts S, Fromm J, Bhalerao RP, Boerjan W, Rohde A (2007) A molecular timetable for apical bud formation and dormancy induction in poplar. Plant Cell 19: 2370–2390.

Scarascia-Mugnozza GE, Hinckley TM, Stettler RF, Heilman PE, Isebands JB (1999) Production physiology and morphology of Populus species and their hybrids grown under short rotation III Seasonal carbon allocation pattern from branches. Can J For Res 29: 1419–1432.

Schmitz G, Theres K (2005) Shoot and inflorescence branching. Curr Opin Plant Biol 8: 506–511.

Schmitz G, Tillmann E, Carriero F, Fiore C, Cellini F, Theres K (2002) The tomato *Blind* gene encodes a MYB transcription factor that controls the formation of lateral meristems. Proc Natl Acad Sci USA 99: 1064–1069.

Schmitz RJ, Amasino RM (2007) Vernalization: a model for investigating epigenetics and eukaryotic gene regulation in plants. Biochim Biophys Acta 1769: 269–275.

Schrader J, Moyle R, Bhalerao R, Hertzberg M, Lundeberg J, Nilsson P, Bhalerao RP (2004a) Cambial meristem dormancy in trees involves extensive remodelling of the transcriptome. Plant J 40: 173–187.

Schrader J, Nilsson J, Mellerowicz E, Berglund A, Nilsson P, Hertzberg M, Sandberg G (2004b) A high-resolution transcript profile across the wood-forming meristem of poplar identifies potential regulators of cambial stem cell identity. Plant Cell 16: 2278–2292.

Schumacher K, Schmitt T, Rossberg M, Schmitz G, Theres K (1999) The *Lateral suppressor* (*Ls*) gene of tomato encodes a new member of the VHIID protein family. Proc Natl Acad Sci USA 96: 290–295.

Schwechheimer C (2008) Understanding gibberellic acid signaling—are we there yet? Curr Opin Plant Biol 11: 9–15.

Semon M, Wolfe KH (2007) Consequences of genome duplication. Curr Opin Gene Dev 17: 505–512.

Sessa G, Steindler C, Morelli G, Ruberti I (1998) The Arabidopsis *Athb-8, -9* and *-14* genes are members of a small gene family coding for highly related HD-ZIP proteins. Plant Mol Biol 38: 609–622.

Shiu SH, Shih MC, Li WH (2005) Transcription factor families have much higher expansion rates in plants than in animals. Plant Physiol 139: 18–26.

Skirycz A, Reichelt M, Burow M, Birkemeyer C, Rolcik J, Kopka J, Zanor MI, Gershenzon J, Strnad M, Szopa J, Mueller-Roeber B, Witt I (2006) DOF transcription factor *AtDof1.1* (*OBP2*) is part of a regulatory network controlling glucosinolate biosynthesis in Arabidopsis. Plant J 47: 10–24.

Smith HM, Hake S (2003) The interaction of two homeobox genes, *BREVIPEDICELLUS* and *PENNYWISE*, regulates internode patterning in the Arabidopsis inflorescence. Plant Cell 15: 1717–1727.

Soler M, Serra O, Molinas M, Huguet G, Fluch S, Figueras M (2007) A genomic approach to suberin biosynthesis and cork differentiation. Plant Physiol. 144: 419–431.

Stracke R, Ishihara H, Huep G, Barsch A, Mehrtens F, Niehaus K, Weisshaar B (2007) Differential regulation of closely related R2R3-MYB transcription factors controls flavonol accumulation in different parts of the *Arabidopsis thaliana* seedling. Plant J 50: 660–677.

Sussex IM, Kerk NM (2001) The evolution of plant architecture. Curr Opin Plant Biol 4: 33–37.

Takeda T, Suwa Y, Suzuki M, Kitano H, Ueguchi-Tanaka M, Ashikari M, Matsuoka M, Ueguchi C (2003) The *OsTB1* gene negatively regulates lateral branching in rice. Plant J 33: 513–520.

Tapia-Lopez RB, Garcia-Ponce B, Dubrovsky JB, Garay-Arroyo A, Perez-Ruiz RV, Kim SH, Acevedo F, Pelaz S, Alvarez-Buylla ER (2008) An *AGAMOUS*-related MADS-box gene, *XAL1* (*AGL12*), regulates root meristem cell proliferation and flowering transition in Arabidopsis. Plant Physiol 146: 1182–1192.

Tuskan GA, DiFazio S, Jansson S, Bohlmann J, Grigoriev I, Hellsten U, Putnam N, Ralph S, Rombauts S, Salamov A, Schein J, Sterck L, Aerts A, Bhalerao RR, Bhalerao RP, Blaudez D, Boerjan W, Brun A, Brunner A, Busov V, Campbell M, Carlson J, Chalot M, Chapman J, Chen GL, Cooper D, Coutinho PM, Couturier J, Covert S, Cronk Q, Cunningham R, Davis J, Degroeve S, Dejardin A, Depamphilis C, Detter J, Dirks B, Dubchak I, Duplessis S, Ehlting J, Ellis B, Gendler K, Goodstein D, Gribskov M, Grimwood J, Groover A, Gunter L, Hamberger B, Heinze B, Helariutta Y, Henrissat B, Holligan D, Holt R, Huang W, Islam-Faridi N, Jones S, Jones-Rhoades M, Jorgensen R, Joshi C, Kangasjarvi J, Karlsson J, Kelleher C, Kirkpatrick R, Kirst M, Kohler A, Kalluri U, Larimer F, Leebens-Mack J, Leple JC, Locascio P, Lou Y, Lucas S, Martin F, Montanini B, Napoli C, Nelson DR, Nelson C, Nieminen K, Nilsson O, Pereda V, Peter G, Philippe R, Pilate G, Poliakov A, Razumovskaya J, Richardson P, Rinaldi C, Ritland K, Rouze P, Ryaboy D, Schmutz J, Schrader J, Segerman B, Shin H, Siddiqui A, Sterky F, Terry A, Tsai CJ, Uberbacher E, Unneberg P, Vahala J, Wall K, Wessler S, Yang G, Yin T, Douglas C, Marra M, Sandberg G, Van de Peer Y, Rokhsar D (2006) The genome of black cottonwood, *Populus trichocarpa* (Torr. & Gray). Science 313: 1596–1604.

Vilaine F, Palauqui JC, Amselem J, Kusiak C, Lemoine R, Dinant S (2003) Towards deciphering phloem: a transcriptome analysis of the phloem of Apium graveolens. Plant J 36: 67–81.

Wada T, Tachibana T, Shimura Y, Okada K (1997) Epidermal cell differentiation in Arabidopsis determined by a Myb homolog, CPC. Science 277: 1113–1116.

Walz C, Giavalisco P, Schad M, Juenger M, Klose J, Kehr J (2004) Proteomics of curcurbit phloem exudate reveals a network of defence proteins. Phytochemistry 65: 1795–1804.

Wang D, Amornsiripanitch N, Dong X (2006) A genomic approach to identify regulatory nodes in the transcriptional network of systemic acquired resistance in plants. PLoS Pathol 2: e123.

Ward JM, Cufr CA, Denzel MA, Neff MM (2005) The Dof transcription factor OBP3 modulates phytochrome and cryptochrome signaling in Arabidopis. 17: 475–485.

Weigel D, Meyerowitz EM (1993) Activation of floral homeotic genes in Arabidopsis. Science 261: 1723–1726.

Welling A, Palva ET (2006) Molecular control of cold acclimation in trees. Physiol Plant 127: 167–181.

Whistler R, BeMiller JN (1997) Carbohydrate chemistry for food scientists. Cellulosics Eagen Press, St Paul, MN, USA.

Wittkopp PJ (2007) Variable gene expression in eukaryotes: a network perspective. J Exp Biol 210: 1567–1575.

Wu R, Stettler RF (1998) Quantitative genetics of growth and development in Populus III Phenotypical plasticity of crown structure and function. Heredity 81: 299–310.

Yamaguchi M, Kubo M, Fukuda H, Demura T (2008) VASCULAR-RELATED NAC-DOMAIN7 is involved in differentiation of all types of xylem vessels in Arabidopsis roots and shoots. Plant J 55: 652–664.

Yang C, Xu Z, Song J, Conner K, Vizcay Barrena G, Wilson ZA (2007) Arabidopsis MYB26/ MALE STERILE35 regulates secondary thickening in the endothecium and is essential for anther dehiscence. Plant Cell 19: 534–548.

Yanhui C, Xiaoyuan Y, Kun H, Meihua L, Jigang L, Zhaofeng G, Zhiqiang L, Yunfei Z, Xiaoxiao W, Xiaoming Q, Yunping S, Li Z, Xiaohui D, Jingchu L, Xing-Wang D, Zhangliang C, Hongya G, Li-Jia Q (2006) The MYB transcription factor superfamily of Arabidopsis: expression analysis and phylogenetic comparison with the rice MYB family. Plant Mol Biol 60: 107–124.

Zhao CS, Craig JC, Petzold HE, Dickerman AW, Beers EP (2005) The xylem and phloem transcriptomes from secondary tissues of the Arabidopsis root-hypocotyl. Plant Physiol 138: 803–818.

Zhao C, Avci U, Grant EH, Haigler CH, Beers EP (2008) XND1, a member of the NAC domain family in *Arabidopsis thaliana*, negatively regulates lignocellulose synthesis and programmed cell death in xylem. Plant J 53: 425–436.

Zhong R, Ye ZH (1999) *IFL1*, a gene regulating interfascicular fiber differentiation in Arabidopsis, encodes a homeodomain-leucine zipper protein. Plant Cell 11: 2139–2152.

Zhong R, Taylor JJ, Ye ZH (1997) Disruption of interfascicular fiber differentiation in an Arabidopsis mutant. Plant Cell 9: 2159–2170.

Zhong R, Demura T, Ye ZH (2006) SND1, a NAC domain transcription factor, is a key regulator of secondary wall synthesis in fibers of Arabidopsis. Plant Cell 18: 3158–3170.

Zhong R, Richardson EA, Ye ZH (2007a) The *MYB46* transcription factor is a direct target of SND1 and regulates secondary wall biosynthesis in Arabidopsis. Plant Cell 19: 2776–2792.

Zhong R, Richardson EA, Ye ZH (2007b) Two NAC domain transcription factors, SND1 and NST1, function redundantly in regulation of secondary wall synthesis in fibers of Arabidopsis. Planta 225: 1603–1611.

# 9

# Auxin Signaling and Response Mechanisms and Roles in Plant Growth and Development

*Udaya C. Kalluri,[1,a,]\* Manojit M. Basu,[1,b] Sara S. Jawdy[1,c]
and Gerald A. Tuskan[1,d]*

## ABSTRACT

Auxin, Indole-3-Acetic Acid (IAA), is known to play major roles in various plant developmental events and physiological responses. The key checkpoints in activity of this phytohormone have been reported at various steps during auxin biosynthesis, conjugation, transport, perception, signaling and response regulation. The present chapter summarizes the general aspects of auxin action in plants and paints an emerging picture from *Populus* genomics studies. An overview of the genomics of auxin action highlighting the insights from expression and mutant studies involving *PIN, AUX, GH3, SAUR, Aux/IAA, ARF* gene family members and auxin response expression profiles in *Populus* is provided here. The availability of genome sequences from several plant species and higher-precision technologies to undertake functional genomics investigations is rapidly expanding our fundamental understanding of auxin action and in turn, is expected to open new doors for applications in the agricultural, forestry, environmental and energy sectors.

**Keywords:** auxin, *Populus*, development, genomics, phytohormone action

[1]Environmental Sciences Division, PO BOX 2008, Oak Ridge National Laboratory, TN 37831, USA;
[a]e-mail: *kalluriudayc@ornl.gov*
[b]e-mail: *basumm@ornl.gov*
[c]e-mail: *jawdys@ornl.gov*
[d]e-mail: *tuskanga@ornl.gov*
\*Corresponding author

## 9.1 Introduction

The phytohormone auxin plays a major role in plant growth, development and responses to various external and internal signals. Indole-3-acetic acid (IAA), the most common natural auxin, has been linked to plant characteristics such as apical dominance; tropic responses; lateral and adventitious root growth; tissue and organ patterning; vascular, flower and fruit development; and abscission. At the cellular level, auxin functions as a key signal for division, expansion and differentiation across the entire life cycle of a plant (Guilfoyle et al. 1998; Fig. 9-1) Auxin levels and gradients are important in establishing and achieving several above- and belowground plant growth features, leading to its label as a "morphogen" (Bhalerao and Bennett 2003). IAA activity as well as the plant's response to IAA may be subject to control at various checkpoints including biosynthesis, conjugation, long-distance transport, cellular and subcellular transport, perception and signal transduction. These checkpoints are coordinated

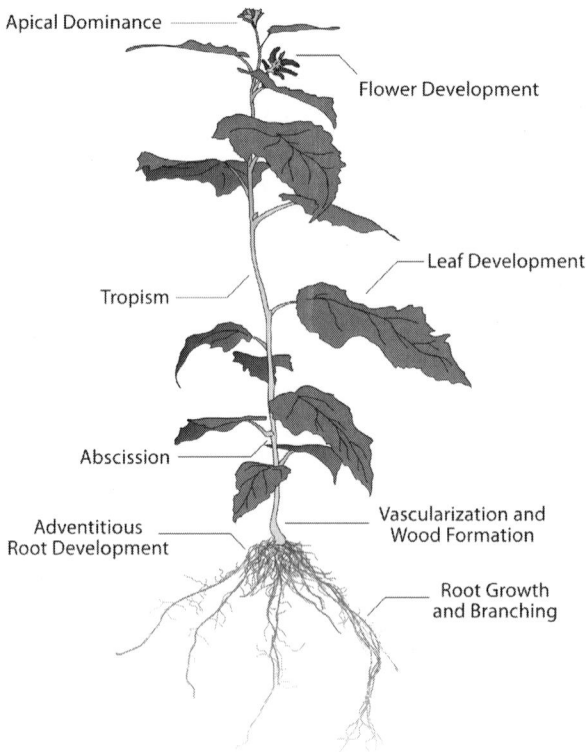

**Figure 9-1** Cartoon of a *Populus* plant depicting roles of auxin in various plant developmental and physiological stages. Note the floral organs depicted in the cartoon do not typically develop in Populus until a tree reaches reproductive maturity [5-15 years of vegetative growth].

by complex feedback and feedforward mechanisms that determine the regulation of auxin response genes and proteins and the resultant auxin response phenotype (Woodward and Bartel 2005). The mystery about how such a simple molecule regulates a variety of responses within an assortment of cells, tissues and organs of plants is being uncovered through forward and reverse genetic studies. The following sections provide a condensed summary of general aspects of auxin action in plants and portray an emerging picture from *Populus* genomics studies.

## 9.2 Auxin Biosynthesis

Auxin is typically synthesized in young, actively growing and highly dividing cells such as those found in young leaves, shoot and root apices and cambial regions (Woodward and Bartel 2005). There are multiple auxin or IAA biosynthesis pathways that plants use, including the indole-3-pyruvic acid (IPA) pathway, the tryptamine (TAM) pathway, the indole-3-acetaldoxime (IAOx) pathway and the indole-3-acetamide (IAM) pathway, in each of which tryptophan (Trp), an amino acid with an indole ring (Srivastava 2002), is the primary precursor of IAA. The most common pathway, the IPA pathway, involves the deamination of Trp to indole-3-pyryvic acid (Normanly et al. 1995; Taiz and Zeiger 1998). Decarboxylation of IPA produces indole-3-acetaldehyde, which then is oxidized to IAA (Srivastava 2002). The second most common TAM pathway involves the decarboxylation of Trp to tryptamine, which ultimately is used to produce IAA (Normanly et al. 1995).

In addition to tryptophan-dependent IAA biosynthesis, tryptophan-independent IAA biosynthesis pathways have been discovered in Trp mutant plants (Woodward and Bartel 2005). Recent studies have shown that for some plants the importance of Trp as an IAA precursor is minor and they are still capable of accumulating adequate amounts of IAA de novo (Normanly et al. 1993). For example, the double recessive Trp mutant, *orange pericarp* (*orp*) of maize, retains the ability to produce IAA de novo and accumulates up to fifty-times more IAA in its seedlings than its wild type counterpart (Wright et al. 1991). *Arabidopsis* Trp mutants, *trp1*, *trp2*, and *trp3*, have helped in elucidation of the Trp-independent pathway. Analyses of these mutants implies that the Trp-independent IAA biosynthesis pathway branches from indole-3-glycerol phosphate or indole during Trp biosynthesis. Both *trp2* and *trp3* mutants cannot synthesize Trp, but still accumulate IAA. The *trp1* mutant accumulates anthranilate (a relatively early precursor in the Trp biosynthesis pathway) but displays a phenotype similar to IAA deficient plants. This indicates that the mutation in *trp1* occurs upstream to the Trp-independent IAA biosynthesis branch point (Srivastava 2002).

Despite extensive efforts to obtain mutants that are unable to synthesize IAA, no group has thus far been successful. An explanation as to why this is the case remains uncertain. However, there are at least two reasonable explanations: first, because there are so many possible pathways to synthesize IAA, mutation in one gene does not halt IAA biosynthesis and second, because IAA regulates many vital processes within plants throughout development, mutations that cause an inability to synthesize IAA may be lethal or unrecoverable (Srivastava 2002).

## 9.3 Auxin Conjugation

Higher plants keep "free" (active) IAA levels low and store most IAA either in a conjugated form known as "bound" (inactive) IAA or in the form of IBA (Tam et al. 2000). IAA can be conjugated with sugars and sugar alcohols via ester linkages or with amino acids, peptides, and possibly proteins via an amide linkage. "Free" IAA and IBA are made available from conjugates via, respectively, hydrolysis (possibly in the endoplasmic reticulum) and peroxisomal B-oxidation (Cohen and Bandurski 1982). Auxin bioassays have been used to determine that certain IAA conjugates are active, whereas others are not (Woodward and Bartel 2005). However, the general belief is that conjugates are inactive until they are hydrolyzed by the plant tissue with the resulting free IAA as the dominant active form. IAA conjugates that are inactive in auxin bioassays are hypothesized to be intermediates targeted for IAA degradation. Plants are also known to regulate levels of free auxin within a cell via compartmentalization and active transport (Woodward and Bartel 2005).

## 9.4 Auxin Transport

Auxin moves into the cell through the process of diffusion, which is enhanced by a proton co-transporter (Swarup et al. 2001) and is secreted at the lower end of the cell by a carrier-dependent, ATP-utilizing efflux mechanism. The proton co-transporters aid in the transport of hydrogen ions and/or change in pH in response to auxin. As such, this auxin action has been referred to as the acid growth hypothesis, where changes in acidification of the cell wall leads to wall loosening and cell expansion due to turgor growth (Rayle and Cleland 1992).

In the traditional notion, polar auxin transport (PAT) occurs in a basipetal (apex to base) direction (Goldsmith 1977). However, it has also been observed that auxin moves in acropetal (base to apex) and basipetal directions as well as in a non-PAT mode of auxin redistribution in tissues such as roots in *Arabidopsis thaliana* (Ljung 2001; Swarup et al. 2005; Fig. 9-2). The speed and distance along with directionality often required for moving auxin from the shoot apex to root tip, particularly in large woody plants like *Populus,* can be achieved via the vascular transport system where auxin carrier proteins are responsible for loading and unloading IAA in and out of the phloem, therefore providing a constant stream of information from cell to cell (Baker 2000).

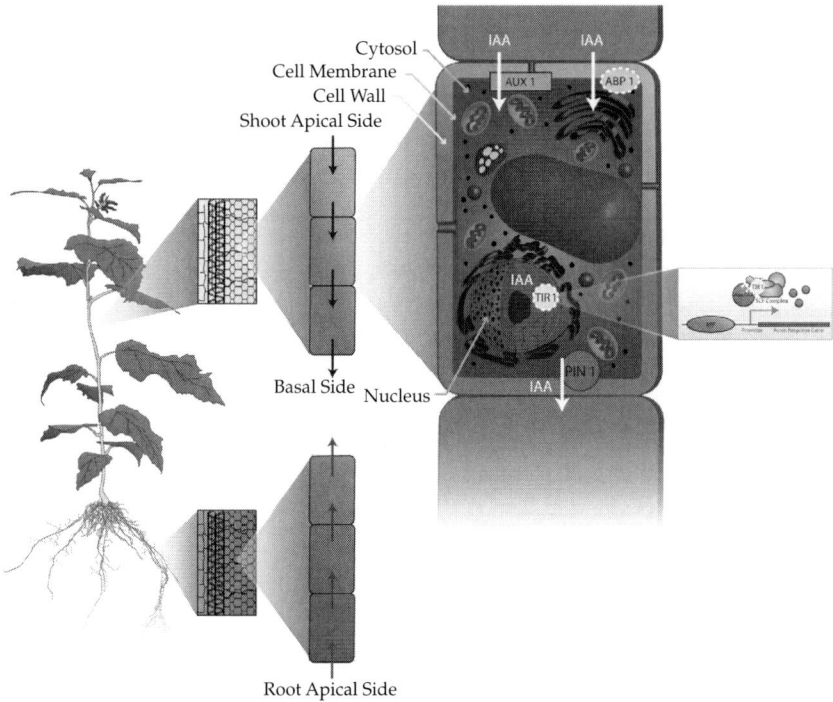

**Figure 9-2** Depiction of auxin transport mechanism at whole plant, tissue and cellular levels. The polar transport of auxin facilitated by AUX1-like influx and PIN1-like efflux proteins can occur in acropetal [from shoot apex to roots] and basipetal manner [mainly roots, upwards from root apex]. The cell structure conceptually illustrates membrane localized auxin pathway proteins, AUX1, ABP1 and PIN1, and the zoomed view of the nucleus conceptually illustrates the action of auxin-receptor, TIR-mediated control of auxin response transcription. Note: The various components in the cartoon are not drawn to scale. Auxin transport occurs in Basipetal (from lateral root cap through root epidermis, depicted here) as well as acropetal (through central tissues of the root toward the root tip) directions in roots. ABP1 is also known to be localized in endoplasmic reticulum).

*Color image of this figure appears in the color plate section at the end of the book.*

## 9.4.1 Auxin Influx and Efflux Carriers

Over the past few years several classes of membrane proteins have been identified as putative auxin transport carriers, however only recently with the advances in the field of functional genomics have these carriers been confirmed to have a major role in cell-to-cell auxin transport (Bennett et al. 1996; Swarup et al. 2004; Petrasek et al. 2006; Bainbridge et al. 2008; Fig. 9-2).

Evidence for the existence of an auxin influx carrier was first reported by Rubery and Sheldrake (1974) and was first demonstrated by Lomax (1986) in zucchini hypocotyl protoplasts, where it was shown that carrier binding is specific to active auxin and that it is a saturable process. Substantial research on auxin influx carriers has been carried out in the model plant *Arabidopsis* since the identification of the *aux1* mutation (in the gene locus coding for the AUX1 influx protein) by Maher and Martindale (1980). Root elongation in *aux1* mutants was reported to be less sensitive to exogenous auxin (2,4-D) application compared to wild type plants. The delay in response to exogenous auxin application was comparable to other auxin signaling mutants, *axr1* and *axr2* (Evans et al. 1994). The putative auxin influx carrier AUX1 belongs to a family of four highly homologous proteins in *Arabidopsis* that are members of the auxin amino acid permease family of proton-driven transporters (Parry et al. 2001). Three other members of this family have been named LAX1, LAX2 and LAX3 (Like-AUX1) and share between 73–82% similarity at the protein level in *Arabidopsis*. Several studies carried out on AUX1 and LAX proteins have provided insights into the molecular basis of auxin transport and the developmental processes regulated by these influx carriers (Blakeslee et al. 2005).

The mode of action of auxin efflux is also an active subject of research as blockage of auxin efflux leads to pronounced effects on plant growth and development. Research efforts during the last decade have helped identify the genes encoding auxin efflux carriers, PINs and p-glycoproteins (PGP), enabling a better understanding of the molecular mechanism of auxin efflux. Asymmetric localization of PIN proteins in different cell types influences the directionality of auxin flow via a tissue or plant part (Friml 2003). In the case of *Arabidopsis*, each PIN protein shows tissue-specific expression and mutations in any of the *PIN* genes generally results in the growth of phenotypes that are consistent with the loss of auxin efflux in the corresponding tissue (Blakeslee et al. 2005). *PIN1* is expressed primarily in the xylem parenchyma and is essential for basipetal transport in shoot tissues and for acropetal transport in root tissues. *PIN2* is required for gravitropic response in roots and basipetal auxin transport. *PIN3* appears to function in the lateral redistribution of auxin and is involved in tropic responses in both hypocotyl and root. Both *PIN2* and *PIN3* are expressed in root columella and pericycle cells (Blakeslee et al. 2005).

In addition to PIN proteins, biochemical evidence from *Arabidopsis* supports a role for PGPs in auxin transport. The PGP family of membrane-localized efflux carriers was first identified in mammalian cancer cells, where they conferred multi-drug resistance (MDR) to chemotherapeutic cancer treatments (Geisler and Murphy 2006). PGPs/MDRs/ABCBs (adenosine triphosphate (ATP)–binding cassette (ABC) transporters have been characterized in various plant species such as *Arabidopsis*, potato, wheat, barley, maize and *Coptis japonica* (Ambudkar et al. 1999; Geisler and Murphy 2006). *AtPGP1* was the first plant PGP identified and overexpression lines were found to elongate under dim light similar to that of wild type plants treated with a low concentration of auxin. *AtPGP1* mutants mirror the phenotype demonstrated by auxin transport inhibitors like NPA (1-naphthylphthalamic acid) (Sidler et al. 1998). Further studies showed partial dwarfing and reduced PAT in hypocotyls, roots and inflorescences in *AtPGP19* mutants, auxin inducible homologs of *AtPGP1* (Noh et al. 2001). Characterization of *AtPGP1* and *AtPGP19* suggests that these genes act as ATP-dependent hydrophobic anion carriers capable of exporting auxin *in planta*. Mutations in genes encoding PGPs have also been studied in rice and maize (Geisler and Murphy 2006). The role of PGPs as potential auxin importers has also been demonstrated with the use of the gene *AtPGP4/ABCB4* ( Geisler and Murphy 2006).

### 9.4.2 *Auxin Influx and Efflux Genes in Populus*

Expression studies in *Populus* involving expressed sequence tag (EST) and cDNA sequences homologous to *Arabidopsis* auxin transport carriers have suggested that *PIN*-like genes have differential functional contexts (Schrader et al. 2003). Experiments conducted on excised stem segments grown in vitro and subjected to endogenous IAA depletion showed a continuous decline in expression of PAT genes. When exposed to exogenous IAA, these stem segments showed a near normal expression pattern of *PttLAX1* (*Populus tremula* × *P. tremuloides*) and *PttLAX3* genes. However, the efflux carrier genes showed twice the level of expression when compared to wild types. It has been observed that *PttLAX2* and *PttPIN1* show weak expression throughout leaf growth unlike *PttLAX3* and *PttPIN2* which are expressed mostly in young leaves. *PttLAX1* and *PttPIN3* are found to be highly expressed in mature leaves (Schrader et al. 2003). Based on these expression patterns it has been suggested that *PttLAX3* and *PttPIN1* are involved in downward auxin transport as the expression of these genes can be detected in the cambial region throughout the mature stem (Schrader et al. 2003).

More recently, altered *PttPIN1* gene expression was reportedly observed in *Populus IAA3* mutant lines, which overexpress an auxin responsive gene, *Aux/IAA* (refer to the section entitled auxin response regulators in *Populus*) (Nilsson et al. 2008). The mutant plants did not show inhibition of root elongation when grown in Murashige and Skoog (MS) medium with IBA supplement at 0.5 µM unlike the wild type control plants. These results suggest that members of the auxin response and auxin transport family may work together in response to exogenous application or depletion of auxin (Nilsson et al. 2008).

Another interesting aspect of auxin gradient and PAT is observed during dormancy, a physiological process that allows perennials to survive harsh winter conditions. During dormancy the leaves senesce followed by abscission and suspension of activities including response to the growth promoting activities of auxin. During this stage exogenously applied auxin does not penetrate into the underlying tissues of the cambium and analysis of *Populus* PAT gene expression shows a reduction in expression level. However, by the end of dormancy and return of favorable temperature and day length conditions, the auxin transport system is reactivated and genes partaking in PAT are reactivated in the cambium region (Schrader et al. 2003).

## 9.5 Auxin Perception

Several molecular components of the auxin signal transduction mechanism are known to be regulated via protein turnover carried out by the auxin-triggered ubiquitination-proteasome pathway. Central to the cellular events of auxin signal perception and transduction are TIR1 and TIR1-like F-box proteins. TIR1, the substrate specificity component of the SCF[TIR1] E3 ubiquitin ligase complex, has recently been shown to function as an auxin-receptor and auxin signal transducer to Auxin/Indole-3-Acetic Acid (Aux/IAA) proteins (Dharmasiri et al. 2005; Kepinski and Leyser 2005; Tan et al. 2007; Fig. 9-2). Role of the TIR group of proteins in the auxin response pathway was discovered through mutant analysis and later confirmed through structural biology studies. In the auxin activated status, Aux/IAA repressor proteins are ubiquitylated via interaction with the auxin-modified SCF[TIR1] complex and are then degraded by 26S proteasome action (Kepinski and Leyser 2004), thus mediating the auxin signal. Interestingly, the F-box gene family at large is nearly half the size reported from *Arabidopsis* and rice but the *TIR1*-like gene set is expanded in *Populus* (Yang et al. 2008).

Auxin binding protein (ABP1) belongs to another class of proteins that have been experimentally verified to possess auxin-binding affinity (Lobler and Klambt 1985; David et al. 2007). Other than evidence for membrane localization and auxin-binding, no conclusive structural evidence is currently available to unequivocally designate ABP1 as yet another auxin receptor.

## 9.6 Auxin Response Regulation

At the cellular level, auxin response is primarily governed by two gene families, viz. Aux/IAA and auxin response factors (ARFs). The genes encoding for Aux/IAAs are known as early responsive genes and are short-lived nuclear proteins which mediate auxin response by interacting with ARF transcription factors (Abel and Theologies 1995). Auxin-mediated degradation of the Aux/IAA proteins via the ubiquitin-26S proteasome proteolytic pathway relieves ARF proteins for interaction and dimerization with other partner ARF proteins to carry out regulation of the downstream auxin-responsive genes including *Aux/IAA* genes.

Aux/IAA proteins have four characteristic domains or motifs; an N-terminal repression domain, an adjacent domain involved in protein stability, and two C-terminal domains III and IV (CTD) through which homo- and heterodimers of Aux/IAAs and/or ARFs are formed (Hagen and Guilfoyle 2002). The ARFs on the other hand are transcription factors containing four conserved domains: an N-terminal B3-like DNA binding domain that includes an additional ARF family specific domain, a variable middle region that confers activator or repressor activity and domains III and IV which are similar to those found in Aux/IAAs (Ulmasov et al. 1999a). ARFs, however, can bind to auxin responsive cis-elements (AuxRE; TGTCTC) present upstream to the coding sequence of auxin responsive genes irrespective of auxin status in the cell, which has been demonstrated using gel-mobility shift and yeast one-hybrid studies (Ulmasov et al. 1999b).

In addition to the existence of large gene families and differing heterodimerization affinities that regulate ARFs, the presence of a characteristic Q-(glutamine) rich middle region induces an activator activity, compared to a P/S/T-rich middle region which incorporates repressor activity (Ulmasov et al. 1999a; Tiwari et al. 2003). Protoplast transfection studies have been used to demonstrate that genes like ARF1 (Ulmasov et al. 1999b; Tiwari et al. 2003), ARF2 (Tiwari et al. 2003) and possibly ARF 3, 4 and 9 are transcriptional repressors as they are comprised of a P/S/T-rich middle region, whereas ARF 5, 6, 7 and 8 are activators of downstream genes required for auxin response (Tiwari et al. 2003). C-terminal domains in ARFs are required for auxin response whereas middle regions function in an auxin-independent manner (Tiwari et al. 2003). Other proteins or cofactors also play a role in determining the efficiency and specificity with which ARFs bind to their target genes (Ulmasov et al. 1999a).

Most of the knowledge on auxin signal transduction events to date has been obtained through functional genomic studies of *Arabidopsis*, but inferences are now being made in other model species such as *Populus* (Tuskan et al. 2006) for which whole-genome sequences are now available. The complexity of auxin regulatory activity is due to the large sizes of the *ARF* and *Aux/IAA* gene families and the inherent variation in activation or repression activity among ARFs, heterodimerization affinities, expression patterns, and auxin-mediated transcriptional and post-transcriptional regulation.

### 9.6.1 *Auxin Response Regulators in Populus*

Comparative genomics studies predicted a total of 35 *Aux/IAA* and 39 *ARF* genes in the *Populus* genome (Tuskan et al. 2006; Kalluri et al. 2007). This represents an expanded repertoire of auxin response regulatory genes in comparison with *Arabidopsis* where 29 *Aux/IAA* and 23 *ARF* genes are reported (Hagen and Guilfoyle 2002; Liscum and Reed 2002) as well as with rice that has 31 *Aux/IAA* and 25 *ARF* genes (Jain et al. 2006a; Wang et al. 2007). Closer analysis showed that the activator set of *ARF* genes are two-fold overrepresented in the *Populus* genome.

It has also been found that certain predicted *Populus* Aux/IAA proteins carry modifications in the conserved domain architecture. For example, domain I is missing in PoptrIAA29 and its *Arabidopsis* ortholog. Similar modifications have also been noticed in PoptrIAA3, PoptrIAA33 and PoptrIAA34 where domains I and II appear to be missing. Among the modifications observed in *Populus* Aux/IAAs, one is particularly interesting. PoptrIAA7.1 includes a unique tandem duplication of the domain II region, which is required for the SCFTIR1-dependent proteosome mediated degradation of Aux/IAA (Ramos et al. 2001; Kalluri et al. 2007). This duplication of domain II has not been observed in any other plant groups other than those belonging to the genus *Populus* (Kalluri et al. 2007). Down-regulation of this gene leads to a severe dwarf phenotype in *Populus* (U. Kalluri et al. unpublished).

Expression studies based on EST analysis, RT-PCR and microarray analysis show that there is a considerable difference in expression patterns of various co-orthologs or subgroup members. A detailed study of the *PoptrIAA3* subgroup suggests the occurrence of subfunctionalization or neofunctionalization in duplicates (Kalluri et al. 2007). Among all *PoptrIAA3* subgroup genes, *PoptrIAA3.2* was found to have greater expression in stem tissue than in roots or leaves and *PoptrIAA3.4* was found to have higher expression in male and female catkins, and in floral buds.

Unlike *Arabidopsis*, which contains a group of seven tandemly duplicated *ARF* genes, *Populus* contains just one tandem duplication consisting of two genes (Kalluri et al. 2007). Phylogenetic analysis between *Populus*, *Arabidopsis* and rice ARF predicted protein sequences suggests differing gene family histories. Overall, *Aux/IAA* and *ARF* gene families in *Populus*, *Arabidopsis* and rice have a common origin, including the conservation of activator groups in ARF, however, the gene family structures of *Populus* and *Arabidopsis* display a greater degree of conservation with each other compared to monocot rice. Interestingly, some *Populus ARF* genes (*PoptrARF7.3* and *PoptrARF7.4*) have been found to group closely with rice and apparently lack a sequence ortholog in *Arabidopsis* (Kalluri et al. 2007).

### 9.6.2 Role of Small Noncoding RNA Species in Auxin Signaling

In the past few years yet another layer of complexity in the auxin response mechanism has begun to surface. At least two different classes of small noncoding RNA species, microRNA (miRNA) and trans-acting short-interfering RNA (ta-siRNA), have been implicated in the auxin response pathway. Such regulation has been reported for *ARF* and *TIR1*-like F-box genes. miR160, a highly conserved plant miRNA group, is known to regulate *AtARF10, 16* and *17* genes (Axtell and Bartel 2005; Mallory et al. 2005) and miR167 has been shown to regulate *AtARF6* and *AtARF8* genes (Wu et al. 2006). Interestingly, these two miRNA groups are predicted to be two-fold overrepresented in *Populus* (Tuskan et al. 2006). miR160 target sequences are found in *PoptrARF10.1–10.2*, *PoptrARF16.1–16.5* and *PoptrARF17.1–17.2* and target sequences for miR167 are found in *PoptrARF6.1–6.3* and *PoptrARF8.1–8.2*.

During juvenile to adult phase transition in *Arabidopsis*, ARF3 and ARF4 genes have been shown to be post-transcriptionally regulated by TAS3 ta-siRNA (Allen et al. 2005; Williams et al. 2005; Hunter et al. 2006; Fahlgren et al. 2006). As reported for *Arabidopsis*, ARF2, ARF3 and ARF4 genes as well as related rice and wheat sequences, and ta-siRNA target site sequences were also found once in the *Populus* genes, *PoptrARF2.2* and *PoptrARF2.6*, and twice in *PoptrARF3.1*, *PoptrARF3.2* and *PoptrARF4*. Further analysis showed the presence of such sites in ARF3-like genes from grape, tobacco, *Medicago* and tomato. Finally, data suggests that miR393 negatively regulates TIR1 and TIR1-like AFB genes (Jones-Rhoades and Bartel 2004; Sunkar and Zhu 2004). The miR393 group, which potentially may regulate several members of the large F-box gene family, consists of two members in *Arabidopsis* and four members in *Populus* (Tuskan et al. 2006). Certainly, the roles of small noncoding RNA species in auxin signaling and response and the expanded repertoire of the *ARF* and *Aux/IAA* gene families provide myriad opportunities for complex regulatory interactions of auxin-related transcription in *Populus*.

## 9.7 Primary Auxin Response Genes

Along with *Aux/IAA* genes, several other classes of genes, such as *SAUR* and *GH3* are known to be transcriptionally up-regulated as part of primary auxin response.

*Small auxin-induced RNA (SAUR)* genes were first identified from auxin-treated soybean hypocotyl sections (Hagen and Guilfoyle 2002). It was observed that these genes are activated within two to five minutes of auxin application. Further investigation revealed SAURs are expressed in the cell elongation zone in soybean hypocotyls and are strongly expressed in the epidermal and cortical cells, leading to elongation of the hypocotyl in response to exogenous auxin treatment (Hagen and Guilfoyle 2002). Sequence analysis of the three soybean *SAUR* transcripts and genomic DNA revealed that these genes lack introns and the predicted protein structure is not homologous to any other known amino acid sequence (Hagen and Guilfoyle 2002). It is believed that the *SAUR* genes are involved in the auxin signal transduction pathway involving calcium and calmodulin (Hagen and Guilfoyle 2002). SAURs have been reported from mung bean (Yamamoto et al. 1992), pea (Guilfoyle 1995), *Arabidopsis* (Gil et al. 1994), rice (Jain et al. 2006b), radish (Anai et al. 1998) and *Zea mays* (Yang and Poovaiah 2000).

*GH3* genes, encoding auxin conjugating enzymes, have been identified from auxin treated, etiolated soybean seedlings (Hagen et al. 1984). Induction of *GH3* mRNA was observed as early as five minutes post-auxin treatment. However, unlike *Aux/IAA* and *SAUR* genes, *GH3* gene expression is not affected by protein synthesis inhibitors (Hagen and Guilfoyle 2002). Under normal conditions *GH3* genes show very low expression and are mostly found in the vascular system, but application of exogenous auxin triggers high expression of the transcripts. The *GH3* gene family has been identified as consisting of 19 members with expression support from all major tissues of *Arabidopsis* (Guilfoyle 1999). Detailed studies of two *Arabidopsis GH3* genes suggest a functional role in photomorphogenesis particularly as a link between phytochrome signaling and auxin responses (Hagen and Guilfoyle 2002). An auxin responsive *GH3* gene has also been characterized in tobacco as well as in several other dicot and monocot plants (Hagen and Guilfoyle 2002; Jain et al. 2006c)

## 9.8 Genomics of Auxin Responses in *Populus*

In order to learn more about downstream signaling components of auxin response on a whole-plant level in *Populus*, a study was conducted in which the *Populus* hybrid H11-11 clone (*P. trichocarpa* x *P. deltoids)* was treated with foliar applications of auxin (Jawdy 2006). ESTs were obtained from subtracted cDNA libraries derived from three major plant organs of treated and control plants: the leaves, stems and roots. Five plants were foliar sprayed with 100μM IAA every 24 hours and the same number of control plants were sprayed with the base solution devoid of auxin. At the end of the nine-day treatment period, RNA was extracted from each tissue type and equal amounts of total RNA were bulked across tissue types within treatments. Poly(A)+RNA (mRNA) was isolated from the two bulked RNA samples and reverse transcription was used to create cDNA populations. Suppression subtractive hybridization (SSH) was performed on the cDNA populations with two subtractions resulting in two libraries, one with genes up-regulated in response to auxin and one with genes down-regulated in response to auxin. The subtracted products were shotgun cloned, transformed and selected to create ten 96-well plates for each library resulting in 960 up-regulated ESTs and 960 down-regulated ESTs. Rolling circle amplification was used to amplify the ESTs, which were then sequenced. Custom Perl scripts were used to generate fasta files which were BLASTed against the Joint Genome Institute (JGI) *Populus trichocarpa* genome database v1.1. The top hit to each *Populus* gene model was then BLASTed against an *Arabidopsis* peptide database to try to gain functional insight into the gene being expressed. An electronic hybridization was performed to further purge the EST libraries of redundant sequence. This was done by removing an EST in one library if it had the same gene model hit in the other library, resulting in 731 ESTs up-regulated in response to auxin and 756 ESTs down-regulated in response to auxin. Two functional characterization tools, MIPS (*http://mips.gsf.de/proj/funcatDB/search_main_frame.html*) and TAIR (*www.Arabidopsis.org*), were used to categorize ESTs from each library into general functional groups. The functional categories differ significantly between the two libraries (Table 9-1). The table shows functional categories that differ significantly between the two libraries. Percentages within a row followed by a letter in common are not significantly different [P ≤ 0.05]. Calculations were based on the significance tests of Audic and Claverie (1997).

Of the 26 functional categories shown, eight appear to have a significantly different number of ESTs represented in them. Two of the functional categories, metabolism and energy, were significantly higher in the down-regulated library as compared to the up-regulated library. The remaining six categories, protein synthesis, interaction with the cell environment, interaction with the environment, cell fate, cell type localization and cell defense were significantly higher in the up-regulated library as compared to the down-regulated library. According to MIPS and TAIR gene function classification databases, 15.9% of the ESTs in the up-regulated library play a predicted role in protein synthesis, destination, regulation and binding. Of the ESTs included in this category, 25.7% were ribosomal genes. Putative heat shock genes (HSPs) were also highly

**Table 9-1** MIPS-based functional categorization of auxin-response ESTs.

| MIPS No. | Functional category | Number and % ESTs in library | | | |
|---|---|---|---|---|---|
| | | Up-regulated | | Down-regulated | |
| 1 | Metabolism | 60 | 5.1a | 114 | 11.1b |
| 2 | Energy | 40 | 3.4a | 64 | 6.2b |
| 4 | Storage proteins | 0 | 0.0a | 1 | 0.1a |
| 10 | Cell growth and division | 19 | 1.6a | 19 | 1.8a |
| 11 | Transcription | 17 | 1.4a | 12 | 1.2a |
| 12 | Protein synthesis | 65 | 5.5a | 36 | 3.5b |
| 14 | Protein destination | 54 | 4.6a | 33 | 3.2a |
| 16 | Protein with binding function | 55 | 4.6a | 34 | 3.3a |
| 18 | Protein activity regulation | 12 | 1.0a | 4 | 0.4a |
| 20 | Cell transport | 33 | 2.8a | 35 | 3.4a |
| 30 | Cell communication/Signal transduction | 22 | 1.9a | 14 | 1.4a |
| 32 | Cell defense, death, aging | 69 | 5.8a | 28 | 2.7b |
| 34 | Interaction with cell environment | 30 | 2.5a | 11 | 1.1b |
| 36 | Interaction with environment | 27 | 2.3a | 6 | 0.6b |
| 40 | Cell fate | 23 | 1.9a | 8 | 0.8b |
| 41 | Development | 24 | 2.0a | 13 | 1.3a |
| 42 | Biogenesis of cell components | 55 | 4.6a | 47 | 4.6a |
| 43 | Cell type differentiation | 1 | 0.1a | 1 | 0.1a |
| 45 | Tissue differentiation | 2 | 0.2a | 0 | 0.0a |
| 70 | Subcellular localization | 85 | 7.2a | 67 | 6.5a |
| 73 | Cell type localization | 18 | 1.5a | 4 | 0.4b |
| 75 | Tissue localization | 1 | 0.1a | 2 | 0.2a |
| 77 | Organ localization | 12 | 1.0a | 4 | 0.4a |
| 98 | Class not clear-cut | 55 | 4.6a | 163 | 15.8a |
| 99 | Unclassified proteins | 337 | 28.4a | 288 | 27.9a |
| | No hits to *Arabidopsis* | 70 | 5.9a | 23 | 2.2a |
| | Total | 1186 | 100.0 | 1031 | 100.0 |

abundant (23.0%) as were cyclophilin genes (10.8%). Approximately 10% of the ESTs in the down-regulated library play a predicted role in protein synthesis, destination, regulation and binding. As in the up-regulated library, the majority of the ESTs in this functional category (31.9%) were ribosomal genes. However, in contrast to the up-regulated library, there were relatively few down-regulated putative HSPS genes (5.9%) and no cyclophilin genes. Therefore, it is possible that the up-regulated putative HSPS genes are involved in auxin response rather than temperature shock response as presumed.

In addition to the most abundant ESTs up- and down-regulated in response to IAA treatment, there were also several other genes of interest in each library (Table 9-2). For example, an auxin response regulator gene, *Aux/IAA16*, was found to be up-regulated. WRKY transcription factors were the most abundant type of transcription factor in the up-regulated library. Members of this family are exclusively found in plants. Most WRKY proteins induce gene expression by binding to upstream W-box

**Table 9-2** List of highly-represented ESTs observed in either down-regulated or up-regulated auxin- response libraries.

| *Populus* gene model | Response | # of ESTs | Definition |
|---|---|---|---|
| estExt_fgenesh4_pm.C_LG_VIII0195 | up-regulated | 9 | heat shock protein, putative |
| eugene3.00091335 | up-regulated | 8 | metallothionein-like protein |
| eugene3.01070053 | up-regulated | 6 | Expressed protein similar to cell wall-plasma |
| gw1.VI.1805.1 | up-regulated | 5 | membrane linker protein |
| eugene3.00013054 | up-regulated | 5 | heat-shock protein |
| grail3.0009049102 | up-regulated | 5 | aluminum-induced protein-like |
| grail3.0032000401 | up-regulated | 4 | putative protein |
| estExt_fgenesh4_pg.C_1290056 | up-regulated | 4 | phytochelatin synthetase-like protein |
| gw1.XVIII.2006.1 | up-regulated | 4 | nucleoid DNA-binding protein cnd41-like protein |
| gw1.VIII.677.1 | up-regulated | 4 | hydroxyproline-rich glycoprotein family |
| gw1.X.3496.1 | up-regulated | 4 | copper homeostasis factor, putative |
| estExt_fgenesh4_pm.C_LG_V0171 | up-regulated | 4 | catalase 2/identical to catalase 2 |
| estExt_Genewise1_v1.C_LG_IX3694 | up-regulated | 4 | alanine-glyoxylate aminotransferase |
| gw1.XVI.771.1 | up-regulated | 3 | unknown protein |
| grail3.0111002302 | Down-regulated | 114 | peroxidase, putative |
| estExt_Genewise1_v1.C_LG_X3955 | Down-regulated | 26 | translationally controlled tumor protein-like protein |
| estExt_fgenesh4_pg.C_LG_II1020 | Down-regulated | 10 | tubulin alpha-2/alpha-4 chain, putative |
| gw1.VIII.2613.1 | Down-regulated | 10 | auxin-regulated protein |
| estExt_Genewise1_v1.C_LG_VIII0155 | Down-regulated | 7 | major latex protein (MLP)-related |
| grail3.0028002001 | Down-regulated | 6 | cysteine proteinase AALP, putative |
| fgenesh4_pg.C_scaffold_40000333 | Down-regulated | 5 | cysteine proteinase RD21A |
| grail3.0045008602 | Down-regulated | 5 | cysteine proteinase RD19A |
| gw1.X.2081.1 | Down-regulated | 5 | auxin-regulated protein |
| eugene3.00880022 | Down-regulated | 5 | adenosylhomocysteinase |
| estExt_Genewise1_v1.C_LG_I4174 | Down-regulated | 4 | tubulin alpha-6 chain (TUA6) |
| estExt_fgenesh4_pg.C_280066 | Down-regulated | 4 | sucrose synthase (sucrose-UDP glucosyltransferase), putative |
| grail3.0001046601 | Down-regulated | 4 | specific tissue protein 2 |
| estExt_fgenesh4_pm.C_LG_XIV0257 | Down-regulated | 4 | S-adenosylmethionine synthase 2 |

motifs typically found in defense related genes (Eulgem et al. 2000). Three of the four WRKY genes identified in the up-regulated library, WRKY 70 (Li et al. 2004), WRKY 69 and WRKY 39 (Dong et al. 2003), are thought to induce expression of defense-related genes because of their ability to bind to W-box domains. However, they have not yet been fully characterized. A fourth WRKY family member, WRKY 51, identified in the up-regulated library is thought to play a possible role in leaf senescence (Guo et al. 2004). According to the *Arabidopsis* e-FP browser, *Arabidopsis* homologs of WRKY 70 and WRKY 39 are shown to be up-regulated in response to IAA treatment in *Arabidopsis (http://www.bar.utoronto.ca/efp/cgi-bin/efpWeb.cgi)*.

Three down-regulated transcription factor genes of interest include *ARF9*, *IAA7*-like and *SCARECROW-LIKE 1 (SCR1)*, which were all previously known to be regulated by auxin (Ulmasov et al. 1999b; Nagpal et al. 2000; Gao et al. 2004). ARF9 does not appear to have either of the characteristic Q or P/S/T-rich middle regions typical of ARF activators and repressors, respectively. Additionally, over-expression of ARF9 has not been shown to activate or repress transcription (Ulmasov et al. 1999b). Therefore, it is thought that apparently inert ARF9 proteins may function by simply binding to target DNA elements and serving as scaffolds for other ARFs or Aux/IAAs. IAA7 has been shown to be a crucial element in several distinct cellular processes at all stages of development of *Arabidopsis*. For example, it is thought to be involved in tissue patterning in roots, cell enlargement, gravitropic response and seedling shoot development in light (Nagpal et al. 2000). Studies have shown that a mutation in IAA7 leads to auxin insensitivity (Liscum and Reed 2002). Down-regulation of an *IAA7*-like gene in *Populus* leads to a severe dwarf phenotype (U. Kalluri et al. unpublished).

In *Brassica*, SCR1 is a transcriptional activator thought to be regulated by auxin that interacts with histone deacetylase (HDA). It contains domains that are conserved in the GRAS (GAI, RGA, SCR) family of proteins and interacts with HDA19 through a VHIID domain. *SCR1* is expressed predominantly in the roots, but also in shoots and mature leaves where it might play a role in radial patterning (Gao et al. 2004). Since *ARF9*, *IAA7*-like and *SCR1*-like genes were found to be down-regulated in response to long-term auxin treatment of *Populus* plants, it is possible that they are involved in repression of the auxin response pathway when auxin levels are low and are therefore down-regulated when auxin levels are elevated.

## 9.9 Auxin and Wood Formation in *Populus*

There is evidence in support of changes in auxin levels during xylogenesis and existence of an auxin gradient across the various wood formation zones (Sundberg et al. 2000). During secondary xylem development, IAA concentrations are believed to decline in a spatial gradient from cell division zones, the source of auxin, to cell maturation and secondary wall formation zones as cambial cells divide and produce xylem initials. This gradient potentially contributes to transcriptome remodeling that leads to development of the secondary xylem (Nilsson et al. 2008). An auxin concentration gradient has been observed in the cambial region of Scots pine implying the existence of a mechanism that creates a gradation in the amount of available auxin in these tissues (Uggla et al. 1998). Considering the homeostatic tendency of plants to maintain auxin levels, polar auxin transport is believed to play a key role in controlling auxin gradient in cambial and secondary xylem tissues (Sundberg et al. 1994). Wood-forming trees provide an ideal platform for auxin transport-related studies of differentiated cell types. They also allow for study of various wood developmental stages for cell-specific measurements of hormone concentration and gene expression (Hertzberg et al. 2001). Additionally, wood forming tissues are one of the few tissues where transport mechanisms are believed to play a major role in maintaining auxin gradient—a particular point of interest in auxin transport studies (Schrader et al. 2003). Expression studies conducted during basipetal auxin transport between internodes 1–10 showed that the strongest expression among the efflux transporter genes, *PttLAX1-3* and *PttPIN1-3*, was displayed by *PttPIN2*, while *PttLAX1*, *PttLAX3* and *PttPIN1* were mainly expressed between internode 7–11, which coincides with the initiation of secondary meristem (Schrader et al. 2003). Expression of the influx carrier, *PttLAX2*, appeared to correspond with the onset of large-scale secondary wall formation in xylem cells between internodes 11-16 (Schrader et al. 2003).

Expression studies based on EST, RT-PCR and microarray techniques reveal that certain members of the auxin response regulator gene families, *Aux/IAA* and *ARF*, are preferentially expressed in stem tissues or in the developmental context of wood formation (Schrader et al. 2003; Kalluri et al. 2007). Comparative genomics analyses show that a pair of *Populus* ARF7-like genes, *PoptrARF7.3* and *PoptrARF 7.4*, group closely with rice ARF genes but seemingly lack a sequence ortholog in *Arabidopsis*. *PoptrARF7.3* ESTs were found in wild-type *Populus* stems as well as stems under tension stress (Schrader et al. 2003; Kalluri et al. 2007). Among the *Aux/IAA* genes, *PoptrIAA3.1* and *PoptrIAA3.2* appear to be expressed in the context of wood formation (Schrader et al. 2003; Kalluri et al. 2007). The precise functional roles of these and other auxin pathway genes during wood formation in *Populus* remains to be determined.

## 9.10 Auxin and Root Development in *Populus*

Auxin is an important signaling cue during root development, influencing, for example, initiation of primary and adventitious roots, lateral or secondary root branching and root growth. Although this fact may be common knowledge, there is little published literature for *Populus* relating auxin dynamics to root initiation, growth rate and tendency to branch. Knowledge of the underlying molecular mechanisms of auxin-mediated root development in *Populus* is even more limited.

Because of the clonal nature of *Populus*, it may regenerate through vegetative suckers from the existing root system or by adventitious rooting of stem cuttings. Growth of the vegetative suckers is very rapid given that they are able to absorb more nutrition than similar, unsuckered growth by using the parent root system (Zahner and DeByle 1965). New root suckers are dependent on the parent root system for as long as 25 years, although some new roots develop within the first few years at the base of each sucker. Eventually a separate, complete root system is formed to support the new growth. The parent root system may remain alive and active for about 40–50 years in total (DeByle 1964).

Since most *Populus* research has been carried out using plants grown in soil, the roots have been inaccessible for observation. This poses a challenge for evaluating the extent of growth and monitoring, especially in a dynamic manner, various root parameters such as number and order of roots, gravitropic set point angles, average root length, and average diameter. Unlike *Arabidopsis*, *Populus* roots are slow growing and the architecture is complex. Even five days post-seed germination, Mahoney and Rood (1991) found root length to be only 1.5 mm, compared to four to five cm in *Arabidopsis*. However, after 46 days and once the seedlings had a leaf area of one $cm^2$, root length averaged 17 cm. Little is known, however, about the molecular events guiding the architecture of the *Populus* root system.

Basu et al. (unpublished) have found that *Populus* cuttings, grown in sterile, solid MS media with supplements of sucrose and auxin, root well when grown with an exogenous auxin (NAA/IBA) concentration between $10^{-5}$ to $10^{-3}$ mM. Higher concentrations of both NAA and IBA were deleterious for rooting whereas concentrations of $10^{-7}$ mM and lower did not significantly affect root growth. Greenwood cuttings required about a week for meristem activation and another four weeks for initiation of rooting. However, rooting initiated within two weeks when dormant cuttings were placed in growth media containing auxin. Top shoot cuttings (cuttings with intact shoot apex) have also been successfully grown and rooted in sterile half-strength MS media with 0.5 µM IBA supplement (Nilsson et al. 2008).

Complementation of physiologic and morphometric research approaches with molecular analysis will shed light on underlying mechanisms of *Populus* root growth and development. Using *Arabidopsis* mutants, several research groups have demonstrated that certain Aux/IAAs and ARFs are important to root development (Hamann et al. 2002; Knox et al. 2003; Okushima et al. 2005). The observed redundancy and similarity in these mutant phenotypes has given rise to the proposal that IAA12 negatively regulates and ARF5 positively regulates control of development of root meristem in an auxin-dependent manner (Hamann et al. 2002). Expression studies in *Populus* show that PoptrARF5 is highly expressed in roots while PoptrIAA12 is lowly expressed in roots, suggesting that they may co-regulate root development in *Populus* (Kalluri et al. 2007). Moreover, studies using the *Arabidopsis* IAA3 gain-of function mutant (*shy2*) showed that this gene is important in lateral root development and gravitropism (Tian et al. 2002). *Populus* has an expanded set of *IAA3*-like genes among which *PoptrIAA3.1* has the highest expression in roots based on RT-PCR studies. Additional functional understanding of auxin control of root development will propel applied research towards improving carbon biosequestration capability (via enhancement of belowground carbon allocation and recalcitrance) and lignocellulosic feedstock traits of *Populus* plants (via enhancement of aboveground carbon allocation in stems and decreased recalcitrance).

## 9.11 Concluding Remarks

The availability of genome sequences of several different plant species is creating a new wave of science based on comparative genomics. Such a cross-species investigative approach will be very useful in identifying the genetic features and molecular processes underlying conserved, as well as divergent plant properties. Such an understanding will provide new opportunities to facilitate postulation and validation of hypotheses on mechanisms linking auxin signaling to auxin response phenotypes. The influence of plant endophytic and rhizosphere microbiota on plant-level properties and the underlying role of auxin in establishing these relationships is another emerging area of research expected to garner greater prominence in the next few years. Further revelations on fundamental aspects of auxin-mediated plant responses will enable new applications in agricultural, environmental and energy sectors.

## Acknowledgements

The authors would like to thank Dr. Stan Wullschleger and Dr. Poornima Sukumar for their helpful comments on the manuscript. The unpublished data included in this chapter were obtained through a project funded by the US Department of Energy, Office of Science, Biological and Environmental Research. Oak Ridge National Laboratory is managed by UT-Battelle, LLC, for the US Department of Energy under contract DE-AC05-00OR22725.

# References

Abel S, Theologis A (1995) A polymorphic bipartite motif signals nuclear targeting of early auxin-inducible proteins related to PS-IAA4 from pea (*Pisum sativum*). Plant J 8: 87–96.

Allen E, Xie Z, Gustafson AM, Carrington JC (2005) MicroRNA-directed phasing during trans-acting siRNA biogenesis in plants. Cell 121: 207–221.

Ambudkar SV, Dey S, Hrycyna CA, Ramachandra M, Pastan I, Gottesman MM (1999) Biochemical, cellular, and pharmacological aspects of the multidrug transporter. Annu Rev Pharmacol Toxicol 39: 361–398.

Anai T, Kono N, Kosemura S, Yamamura S, Hasegawa K (1998) Isolation and characterization of an auxin-inducible SAUR gene from radish seedlings. DNA Seq 9: 329–333.

Audic S, Claverie JM (1997) The significance of digital gene expression profiles. Genome Res 7: 986–995.

Axtell MJ, Bartel DP (2005) Antiquity of microRNAs and their targets in land plants. Plant Cell 17: 1658–1673.

Bainbridge K, Guyomarc'h S, Bayer E, Swarup R, Bennett M, Mandel T, Kuhlemeier C (2008) Auxin influx carriers stabilize phyllotactic patterning. Genes Dev 22: 810–823.

Baker DA (2000) Long-distance vascular transport of endogenous hormones in plants and their role in source: Sink regulation. Isr J Plant Sci 48: 199–203.

Bennett MJ, Marchant A, Green HG, May ST, Ward SP, Millner PA, Walker AR, Schulz B, Feldmann KA (1996) *Arabidopsis AUX1* gene: a permease-like regulator of root gravitropism. Science 273: 948–950.

Bhalerao RP, Bennett MJ (2003) The case for morphogens in plants. Nat Cell Biol 5: 939–943.

Blakeslee JJ, Peer WA, Murphy AS (2005) Auxin transport. Curr Opin Plant Biol 8: 494–500.

Cohen JD, Bandurski RS (1982) Chemistry and physiology of the bound auxins. Annu Rev Plant Physiol Plant Mol Biol 33: 403–430.

David KM, Couch D, Braun N, Brown S, Grosclaude J, Perrot-Rechenmann C (2007) The auxin-binding protein 1 is essential for the control of cell cycle. Plant J 50: 197–206.

DeByle NV (1964) Detection of functional intraclonal aspen root connections by tracers and excavation. For Sci 10: 386–396.

Dharmasiri N, Dharmasiri S, Estelle M (2005) The F-box protein TIR1 is an auxin receptor. Nature 435: 441–445.

Dong J, Chen C, Chen Z (2003) Expression profiles of the *Arabidopsis* WRKY gene superfamily during plant defense response. Plant Mol Biol 51: 21–37.

Eulgem T, Rushton PJ, Robatzek S, Somssich IE (2000) The WRKY superfamily of plant transcription factors. Trends Plant Sci 5: 199–206.

Evans ML, Ishikawa H, Estelle MA (1994) Responses of *Arabidopsis* roots to auxin studied with high temporal resolution: Comparison of wild type and auxin response mutants. Planta 194: 215–222.

Fahlgren N, Montgomery TA, Howell MD, Allen E, Dvorak SK, Alexander AL, Carrington JC (2006) Regulation of auxin response factor3 by TAS3 ta-siRNA affects developmental timing and patterning in *Arabidopsis*. Curr Biol 16: 939–944.

Friml J (2003) Auxin transport—shaping the plant. Curr Opin Plant Biol 6: 7–12.

Gao MJ, Parkin I, Lydiate D, and Hannoufa A (2004. An auxin-responsive SCARECROW-like transcriptional activator interacts with histone deacetylase. Plant Mol Biol 55: 417–431.

Geisler M, Murphy A (2006) The ABC of auxin transport: The role of P-glycoproteins in plant development. FEBS Lett 580: 1094–1102.

Gil P, Liu Y, Orbovic V, Verkamp E, Poff KL, Green P (1994) Characterization of the auxin-inducible SAUR-AC1 gene for use as a genetic tool in *Arabidopsis*. Plant Physiol 104: 777–784.

Goldsmith MHM (1977) The polar transport of auxin. Annual Review of Plant Physiol 28: 439–478.

Guilfoyle TJ (1995) Auxin regulated gene expression and gravitropism in plants. ASGSB Bull 8: 39–45.

Guilfoyle TJ (1999) Auxin-regulated genes and promoters. In: PJJ Hooykaas, M Hall, KL Libbenga (eds) Biochemistry and Molecular Biology of Plant Hormones. Elsevier, Leiden, The Netherlands, pp 423–459.

Guilfoyle TJ, Hagen G, Ulmasov T, Murfett J (1998) How does auxin turn on genes? Plant Physiol 118: 341–347.

Guo Y, Cai Z, Gan S (2004) Transcriptome of *Arabidopsis* leaf senescence. Plant Cell Environ 27: 521–549.

Hagen G, Kleinschmidt A, Guilfoyle T (1984) Auxin-regulated gene expression in intact soybean hypocotyls and excised hypocotyl sections. Planta 162: 147–153.

Hagen G, Guilfoyle T (2002) Auxin-responsive gene expression: Genes, promoters and regulatory factors. Plant Mol Biol 49: 373–385.

Hamann T, Benkova E, Baurle Kientz IM, Jurgens G (2002) The *Arabidopsis BODENLOS* gene encodes an auxin response protein inhibiting *MONOPTEROS*-mediated embryo patterning. Genes Dev 16: 1610–1615.

Hardtke CS, Berleth T (1998) The *Arabidopsis* gene *MONOPTEROS* encodes a transcription factor mediating embryo axis formation and vascular development. Embo J 17: 1405–1411.

Hertzberg M, Aspeborg H, Schraderm J, Andersson A, Erlandsson R, Blomqvist K, Bhalerao R, Uhlen M, Teeri TT, Lundeberg J, et al. (2001) A transcriptional roadmap to wood formation. Proc Natl Acad Sci US 98: 14732–14737.

Hunter C, Willmann MR, Wu G, Yoshikawa M, de la Luz Gutierrez-Nava M, Poethig SR (2006) Trans-acting siRNA-mediated repression of *ETTIN* and *ARF4* regulates heteroblasty in *Arabidopsis*. Development 133: 2973–2981.

Jain M, Kaur N, Garg R, Thakur JK, Tyagi AK, Khurana JP (2006a) Structure and expression analysis of early auxin-responsive Aux/IAA gene family in rice (*Oryza sativa*). Funct Integr Genom 6: 47–59.

Jain M, Tyagi AK, Khurana JP (2006b) Genome-wide analysis, evolutionary expansion, and expression of early auxin-responsive *SAUR* gene family in rice (*Oryza sativa*). Genomics 88: 360–371.

Jain M, Kaur N, Tyagi AK, Khurana JP (2006c) The auxin-responsive *GH3* gene family in rice (*Oryza sativa*). Funct Integr Genom 6: 36–46.

Jawdy S (2006) Expression analysis of genes in *Populus* induced by exogenous auxin treatment. MS Thesis, Univ of Tennessee–Knoxville, TN, USA.

Jones-Rhoades MW, Bartel DP (2004) Computational identification of plant microRNAs and their targets, including a stress-induced miRNA. Mol Cell 14: 787–799.

Kalluri UC, DiFazio SP, Brunner AM, and Tuskan GA (2007) Genome-wide analysis of Aux/IAA and ARF gene families in *Populus trichocarpa*. BMC Plant Biol 7: 59.

Kepinski S, Leyser O (2004) Auxin-induced SCFTIR1-Aux/IAA interaction involves stable modification of the SCFTIR1 complex. Proc Natl Acad Sci USA 101: 12381–12386.

Kepinski S, Leyser O (2005) The *Arabidopsis* F-box protein TIR1 is an auxin receptor. Nature 435: 446–451.

Knox K, Grierson CS, Leyser O (2003) AXR3 and SHY2 interact to regulate root hair development. Development 130: 5769–5777.

Li J, Brader G, Palva ET (2004) The WRKY70 transcription factor: A node of convergence for jasmonate-mediated and salicylate-mediated signals in plant defense. Plant Cell 16(2):319–331.

Liscum E, Reed JW (2002) Genetics of Aux/IAA and ARF action in plant growth and development. Plant Mol Biol 49: 387–400.

Ljung K, Bhalerao RP, Sandberg G (2001) Sites and homeostatic control of auxin biosynthesis in *Arabidopsis* during vegetative growth. Plant J 28: 465–474.

Lobler M, Klambt D (1985) Auxin-binding protein from coleoptile membranes of corn (*Zea mays* L.) I. Purification by immunological methods and characterization. J Biol Chem 260: 9848–9853.

Lomax TL (1986) Active auxin uptake by specific plasma membrane carriers. In: M Bopp (ed) Plant Growth Substances. Springer, Berlin, Germany, pp 209–213.

Maher EP, Martindale SJ (1980) Mutants of *Arabidopsis thaliana* with altered responses to auxins and gravity. Biochem Genet 18: 1041–1053.

Mahoney JM, Rood SB (1991) A device for studying the influence of declining water table on poplar growth and survival. Tree Physiol 8: 305–314.

Mallory AC, Bartel DP, Bartel B (2005) MicroRNA-directed regulation of *Arabidopsis auxin response factor17* is essential for proper development and modulates expression of early auxin response genes. Plant Cell 17: 1360–1375.

Nagpal P, Walker LM, Young JC, Sonawala A, Timpte C, Estelle M, Reed JW (2000) AXR2 encodes a member of the Aux/IAA protein family. Plant Physiol 123: 563–573.

Nilsson J, Karlberg A, Antti H, Lopez-Vernaza M, Mellerowicz E, Perrot-Rechenmann C, Sandberg G, Bhalerao RP (2008) Dissecting the molecular basis of the regulation of wood formation by auxin in hybrid aspen. Plant Cell 20: 843–855.

Noh B, Murphy AS, Spalding EP (2001) Multidrug resistance-like genes of *Arabidopsis* required for auxin transport and auxin-mediated development. Plant Cell 16: 1898–1911.

Normanly J, Cohen JD, Fink GR (1993) *Arabidopsis thaliana* auxotrophs reveal a tryptophan-independent biosynthetic-pathway for indole-3-acetic-acid. Proc Natl Acad Sci USA 90: 10355–10359.

Normanly J, Slovin JP, Cohen JD (1995) Rethinking auxin biosynthesis and metabolism. Plant Physiol 107: 323–329.

Okushima Y, Overvoorde PJ, Arima K, Alonso JM, Chan A, Chang C, Ecker JR, Hughes B, Lui A, Nguyen D, Onodera C, Quach H, Smith A, Yu G, Theologis A (2005) Functional genomic analysis of the auxin response factor gene family members in *Arabidopsis thaliana*: Unique and overlapping functions of ARF7 and ARF19. Plant Cell 17: 444–463.

Parry G, Delbarre A, Marchant A, Swarup R, Napier R, Perrot-Rechenmann C, Bennett MJ (2001) Novel auxin transport inhibitors phenocopy the auxin influx carrier mutation aux1. Plant J 25: 399–406.

Petrasek J, Mravec J, Bouchard R, Blakeslee JJ, Abas M, Seifertova D, Wisniewska J, Tadele Z, Kubes M, Covanova M, Dhonukshe P, Skupa P, Benkova E, Perry L, Krecek P, Lee OR, Fink GR, Geisler M, Murphy AS, Luschnig C, Zazimalova E, Friml J (2006) PIN proteins perform a rate-limiting function in cellular auxin efflux. Science 312: 914–918.

Ramos JA, Zenser N, Leyser O, Callis J (2001) Rapid degradation of auxin/indoleacetic acid proteins requires conserved amino acids of domain II and is proteasome dependent. Plant Cell 13: 2349–2360.

Rayle DL, Cleland RE (1992) The Acid Growth Theory of auxin-induced cell elongation is alive and well. Plant Physiol 99: 1271–1274.

Rubery PH, Sheldrake AR (1974) Carrier mediated auxin transport. Planta 88: 101–121.

Schrader J, Baba K, May ST, Palme K, Bennett M, Bhalerao RP, Sandberg G (2003) Polar auxin transport in the wood-forming tissues of hybrid aspen is under simultaneous control of developmental and environmental signals. Proc Natl Acad Sci USA 100: 10096–10101.

Sidler M, Hassa P, Hasan S, Ringli C, Dudler R (1998) Involvement of an ABC transporter in a developmental pathway regulating hypocotyl cell elongation in the light. Plant Cell 10: 1623–1636.

Srivastava LM (2002) Auxins In: LM Srivastava (ed) Plant Growth and Development: Hormones and Environment. Academic Press, San Diego, USA, pp 155–169.

Sundberg B, Tuominen H, Little C (1994) Effects of the indole-3-acetic acid (IAA) transport inhibitors N-1-naphthylphthalamic acid and morphactin on endogenous IAA dynamics in relation to compression wood formation in 1-year-old Pinus sylvestris (L.) shoots. Plant Physiol 106: 469–476.

Sundberg B, Uggla C, Tuominen H (2000) Cambial growth and auxin gradients. In: R Savidge, J Barnett, R Napier R [eds.] Cell and Molecular Biology of Wood Formation. BIOS Scientific Publishers, Oxford, UK, pp 169–188.

Sunkar R, Zhu JK (2004) Novel and stress-regulated microRNAs and other small RNAs from *Arabidopsis*. Plant Cell 16: 2001–2019.

Swarup R, Friml J, Marchant A, Ljung K, Sandberg G, Palme K, Bennett M (2001) Localization of the auxin permease AUX1 suggests two functionally distinct hormone transport pathways operate in the *Arabidopsis* root apex. Genes Dev 15: 2648–2653.

Swarup R, Kargul J, Marchant A, Zadik D, Rahman A, Mills R, Yemm A, May S, Williams L, Millner P, Tsurumi S, Moore I, Napier R, Kerr ID, Bennett MJ (2004) Structure-function analysis of the presumptive *Arabidopsis* auxin permease AUX1. Plant Cell 16: 3069–3083.

Swarup R, Kramer EM, Perry P, Knox K, Leyser HM, Haseloff J, Beemster GT, Bhalerao R, Bennett MJ (2005) Root gravitropism requires lateral root cap and epidermal cells for transport and response to a mobile auxin signal. Nat Cell Biol 7: 1057–1065.

Taiz L, Zeiger E (1998) Auxins In: L Taiz, E Zeiger (eds) Plant Physiology.2nd edn. Sinauer Assoc, Sunderland, USA, pp 543–589.

Tam YY, Epstein E, Normanly J (2000) Characterization of auxin conjugates in *Arabidopsis*. Low steady-state levels of indole-3-acetyl-aspartate, indole-3-acetyl-glutamate, and indole-3-acetyl-glucose. Plant Physiol 123: 589–596.

Tan X, Calderon-Villalobos LI, Sharon M, Zheng C, Robinson CV, Estelle M, Zheng N (2007) Mechanism of auxin perception by the TIR1 ubiquitin ligase. Nature 446: 640–645.

Tian Q, Uhlir NJ, Reed JW (2002) *Arabidopsis* SHY2/IAA3 inhibits auxin-regulated gene expression. Plant Cell 14: 301–319.

Tiwari SB, Hagen G, Guilfoyle T (2003) The roles of auxin response factor domains in auxin-responsive transcription. Plant Cell 15: 533–543.

Tuskan GA, DiFazio S, Jansson S, Bohlmann J, Grigoriev I, Hellsten U, Putnam N, Ralph S, Rombauts S, Salamov A, Schein J, Sterck L, Aerts A, Bhalerao RR, Bhalerao RP, Blaudez D, Boerjan W, Brun A, Brunner A, Busov V, Campbell M, Carlson J, Chalot M, Chapman J, Chen GL, Cooper D, Coutinho PM, Couturier J, Covert S, Cronk Q, Cunningham R, Davis J, Degroeve S, Dejardin A, Depamphilis C, Detter J, Dirks B, Dubchak I, Duplessis S, Ehlting J, Ellis B, Gendler K, Goodstein D, Gribskov M, Grimwood J, Groover A, Gunter L, Hamberger B, Heinze B, Helariutta Y, Henrissat B, Holligan D, Holt R, Huang W, Islam-Faridi N, Jones S, Jones-Rhoades M, Jorgensen R, Joshi C, Kangasjarvi J, Karlsson J, Kelleher C, Kirkpatrick R, Kirst M, Kohler A, Kalluri U, Larimer F, Leebens-Mack J, Leple JC, Locascio P, Lou Y, Lucas S, Martin F, Montanini B, Napoli C, Nelson DR, Nelson C, Nieminen K, Nilsson O, Pereda V, Peter G, Philippe R, Pilate G, Poliakov A, Razumovskaya J, Richardson P, Rinaldi C, Ritland K, Rouze P, Ryaboy D, Schmutz J, Schrader J, Segerman B, Shin H, Siddiqui A, Sterky F, Terry A, Tsai CJ, Uberbacher E, Unneberg P, Vahala J, Wall K, Wessler S, Yang G, Yin T, Douglas C, Marra M, Sandberg G, Van de Peerand Y, Rokhsar D (2006) The genome of black cottonwood, *Populus trichocarpa* (Torr. and Gray). Science 313: 1596–1604.

Uggla C, Mellerowicz EJ, Sundberg B (1998) Indole-3-acetic acid controls cambial growth in scots pine by positional signaling. Plant Physiol 117: 113–121.

Ulmasov T, Hagen G, Guilfoyle TJ (1999a) Dimerization and DNA binding of auxin response factors. Plant J 19: 309–319.

Ulmasov T, Hagen G, Guilfoyle TJ (1999b) Activation and repression of transcription by auxin-response factors. Proc Natl Acad Sci USA, 96: 5844–5849.

Wang D, Pei K, Fu Y, Sun Z, Li S, Liu H, Tang K, Han B, Tao Y (2007) Genome-wide analysis of the auxin response factors (ARF) gene family in rice (*Oryza sativa*). Gene 394: 13–24.

Williams L, Carles CC, Osmont KS, Fletcher JC (2005) A database analysis method identifies an endogenous trans-acting short-interfering RNA that targets the *Arabidopsis* ARF2, ARF3, and ARF4 genes. Proc Natl Acad Sci USA, 102: 9703–9708.

Woodward AW, Bartel B (2005) Auxin: regulation, action, and interaction. Ann Bot (Lond) 95: 707–735.

Wright AD, Sampson MB, Neuffer MG, Michalczuk L, Slovin JP, Cohen JD (1991) Indole-3-acetic-acid biosynthesis in the mutant maize orange pericarp, a tryptophan auxotroph. Science 254(5034): 998–1000.

Wu MF, Tian Q, Reed JW (2006) *Arabidopsis* microRNA167 controls patterns of ARF6 and ARF8 expression, and regulates both female and male reproduction. Development 133: 4211–4218.

Yamamoto KT, Mori H, Imaseki H (1992) cDNA cloning of indole-3-acetic acid-regulated genes:Aux22 and SAUR from mung bean (*Vigna radiata*) hypocotyl tissue. Plant Cell Physiol 33: 93–97.

Yang T, Poovaiah BW (2000) Molecular and biochemical evidence for the involvement of calcium/calmodulin in auxin action. J Biol Chem 275: 3137–3143.

Yang X, Kalluri UC, Jawdy S, Gunter LE, Yin T, Tschaplinski TJ, Weston DJ, Ranjan P, Tuskan GA (2008) The F-box gene family is expanded in herbaceous annual plants relative to woody perennial plants. Plant Physiol 148:1189–1200.

Zahner R, DeByle NV (1965) Effect of pruning the parent root on growth of aspen suckers. Ecology 46: 373–375.

# 10

# Genetic Control of the Annual Growth Cycle in Woody Plants

*Jae-Heung Ko,[1] Sunchung Park,[2] Daniel E. Keathley[2] and Kyung-Hwan Han[3,]\**

## ABSTRACT

Proper regulation of the annual growth cycle between vegetative growth and dormancy is an important adaptive mechanism in temperate woody plants for winter survival. It appears that the control of meristematic activity constitutes an integral component of the annual growth cycle. It is important to understand how seasonal changes in the local climate affect the growth and differentiation of meristem cells. Changes in daylength serve as an environmental cue for seasonal rhythms. In most woody plants, the cessation of shoot growth and the induction of dormancy are either induced or accelerated by short days, and prevented or delayed by long days. In addition, cold plays a crucial role in both the entering and breaking of dormancy. In response to the fluctuations in daylength and temperature, plants are thought to entrain an internal biochemical oscillator, the circadian clock, to rhythmic changes in the external environment to make necessary physiological adjustments for winter survival. Regulation of these complex physiological processes appears to be achieved through the myriad genes that are up and down regulated in concert due to the interplay between various environmental signals. Poplar is especially well suited for the study of the annual growth cycle at both physiological and molecular levels.

**Keywords:** dormancy, growth cycle, microarrays; *Populus*, transcriptome

[1]Department of Plant & Environmental New Resources, Kyung Hee University, Yongin, 446-701, Republic of Korea; e-mail: *ko@msu.edu*
[2]Department of Forestry, Michigan State University, East Lansing, MI 48824-1222, USA.
[3]Department of Horticulture and Department of Forestry, Michigan State University, East Lansing, MI 48824-1222, USA; and, Department of Bioenergy Science and Technology, Chonnam National University, 300 Yongbong-dong, Bukgu, Gwangju 500-757, Republic of Korea.
*Corresponding author

## 10.1 Introduction

Temperate perennials, such as trees, must have a physiological mechanism for anticipating seasonal changes in order to enter dormancy prior to winter and resume growth in the spring. Proper timing of the onset and release of dormancy impacts the productivity, survival, and spatial distribution of temperate perennials. Models of continued global warming (Serreze et al. 2000) and climate change predict that temperate regions will experience the greatest changes in winter temperature (IPGC 1997). This increase in winter temperature will cause delayed and erratic flower development in tree species with a large chilling requirement, as well as early spring bud break and an increased risk of frost damage in species with low chilling requirements (Heide 2003). In addition, anticipated north/south plant migration in response to changing temperatures will challenge ecotypes because this movement also imposes different seasonal daylengths. Dormancy also plays a critical role in determining the degree to which invasive perennials survive the winter and thus these changes may also affect agricultural productivity and ecosystem biodiversity.

In addition to their ecological significance, trees also serve as a primary feedstock for biofuel, fiber, solid wood products, and various natural compounds. The need for construction lumber, fuelwood, and fiber by an expanding global population has grown 36% in the past 25 years, while available forestland has been rapidly diminishing. Meeting this growing demand in a sustainable fashion dictates achieving higher biological productivity in trees. The length of the dormancy period determines the growing season and thus affects forest productivity and wood quality. Primary productivity may be increased by a relatively short dormancy period, although prolonging the growth period may represent a trade-off between winter survival and biomass production. A better understanding of the mechanisms underlying dormancy can help land managers to mitigate future climate change impact by providing the necessary information and tools needed to monitor population trends, redefine seed zones, and select appropriate genotypes for reforestation based on molecular markers. In agriculture, inappropriate dormancy responses in fruit tree crops results in reduced productivity due to winter damage in flower buds. The bottleneck in genetic manipulation of dormancy responses is the lack of discrete traits or genes controlling the induction and release of dormancy. Identifying key regulatory genes in dormancy regulation can lead to a technology that can resolve the dilemma of achieving greater environmental protection of forest ecosystems while meeting the increasing demand for forest utilization.

Our understanding of how temperate perennials utilize environmental cues to develop adaptive mechanisms is limited. Measurement of daylength and perception of temperature are critical features of dormancy regulation.

However, the underlying mechanisms remain to be elucidated. In this chapter, we briefly review the literature in the field and discuss recent insights into the genetic mechanisms that control the annual alternation between secondary growth and dormancy in woody plants.

## 10.2 Vascular Cambium Activity

In woody plant growth, the vascular cambium increases the stem diameter by periclinal divisions and the circumference by anticlinal divisions, resulting in the developmental continuum of secondary phloem and xylem (Chaffey 1999). The control of meristematic activity constitutes an integral component of the annual growth cycle (Shimizu-Sato and Mori 2001; Rohde et al. 2002). Schrader et al. (2004a) obtained a high-resolution transcript profile across the wood-forming meristem of poplar, which identified potential regulators of cambial cell identity. Recently, poplar cambial cell-specific cDNA libraries and transcriptome profiles were produced from active and dormant cambium cells (Schrader et al. 2004b). This report described a significant reduction in the complexity of transcriptome during dormancy. We have also generated monthly expression profiles of about 2,000 poplar expressed sequence tags (ESTs) (Park et al. 2008). However, the information gained from these published studies is limited in the number of genes interrogated or dormancy status examined (only winter/summer comparison in Schrader et al. 2004b).

## 10.3 Vegetative Dormancy

Lang et al. (1987) described three types of vegetative dormancy: 1) paradormancy, also known as apical dominance, is the suppression of lateral bud growth by the actively growing portion such as the apical meristem; 2) in ecodormancy, growth is arrested by the environmental conditions that are not conducive to growth but resumes when conditions become favorable; 3) endodormancy is caused by plant endogenous factors and requires a sustained exposure to low temperatures for spring regrowth. The alternations between active growth and dormancy are closely timed with seasonal changes in the local climate. In most woody plants, the cessation of shoot growth and the induction of endodormancy are either induced or accelerated by short-days (SDs), and prevented or delayed by long-days (LDs) (Nitsch 1957; Vince-Prue 1975; Olsen et al. 1997). In *Populus*, growth cessation and endodormancy initiation are induced by SDs (Bañados 1992; Howe et al. 1995; Jeknic and Chen 1999) and controlled by photoreceptors (e.g., phytochromes) (Howe et al. 1996). Although SDs and low temperature regulate endodormancy-related traits, SDs seem to be the primary regulatory signal that prepares plants for the coming cold (Weiser 1970; Fuchigami et al. 1971). For example, SDs typically induce bud set under warm (non-inductive) temperatures, but low night temperatures usually fail to cause bud set under LDs (non-inductive) photoperiods (Håbjørg 1972). This relationship has an adaptive significance. First, by responding to photoperiodic cues, plants are able to begin acclimating to winter conditions prior to the onset of cold. Second, photoperiodic cues are much more reliable signals of the approaching winter compared to temperatures, which can vary from year to year. Cold acclimation by low temperature seems to be uncoupled from dormancy, suggesting independent signaling pathways for SD and cold (Welling et al. 2002). Because SDs and cold probably regulate endodormancy-related responses through non-identical sets of genes, it is beneficial to study these processes independently. Once the "chilling requirement" is met, the cambium enters the quiescence stage of dormancy, which is imposed solely by adverse external factors, typically low temperatures. Upon the return of warmer temperatures in spring, the plants are gradually dehardened and the cambial cells regain their ability to produce xylem in response to IAA (Mellerowicz et al. 1992). Many studies on phenological traits in *Populus* suggest that the timing of bud flush may be controlled by a modest number of major genes with low environmental variation (Bradshaw and Stettler 1995; Chen et al. 2002). However, no such genes have been identified.

## 10.4 Regulation of Annual Growth Cycle

### 10.4.1 Daylength

Light resets the circadian clock to the solar cycle (Harmer et al. 2000) and subsequently controls clock-regulated physiological rhythms. The short-day (SD) signal triggers the initiation of dormancy and cold hardiness development. This daylength-dependent behavior is an important adaptive mechanism because the ultimate survival of woody plants is dependent on not only the maximal capacity of cold hardening, but also on the timing and rate of cold acclimation and the stability of cold hardiness against unseasonably warm periods during winter (Weiser 1970). By controlling cold hardiness via developmental stage, which is regulated by predictable photoperiod, plants are able to distinguish between unseasonable and seasonable cold. The first step in photoperiodic time measurement is carried out at the level of photoreceptors, the pigments that can discriminate days from nights (Izawa et al. 2002). The phytochrome photoreceptors (Phy A-E) play important roles in the photoperiodic control of the annual growth cycle and cold-hardiness in trees (Howe et al. 1996, 1998; Olsen and Junttila 2002). SD-induced growth cessation of poplar trees is mediated by PhyA (Welling et al. 2002). Transgenic poplars overexpressing the oat PhyA were not able to detect SDs and therefore failed to stop growing in response to them (Olsen et al. 1997; Welling et al. 2002), suggesting that PhyA may play an important role in regulation of the dormancy cycle.

## 10.4.2 Biological Clock

Plants recognize fluctuations in daylength and temperature (Eriksson and Millar 2003) and accordingly adjust their physiological state in anticipation of seasonal changes (Yanovsky and Kay 2002). An internal biochemical oscillator, the circadian clock, makes this possible by entraining the circadian clock to rhythmic changes in the external environment (Somers et al. 1998; Yanovsky and Kay 2003). Such resetting of the internal clock to local time provides the plants with an adaptive advantage. Although oversimplified, the circadian system is often described as consisting of three general parts: a central oscillator that generates rhythmicity, input pathways that receive and relay the environmental cues that synchronize the oscillator, and output pathways that are controlled by the oscillator to create a range of biochemical and developmental pathways (Harmer et al. 2001). Observations in chronobiology strongly support the hypothesis that daylength measurement in seasonal responses relies on a circadian oscillator (Eriksson and Millar 2003; Hayama and Coupland 2003), known as "the external coincidence model". This model proposes that photoperiodic responses are triggered when the illuminated part of the day coincides with a photoinducible phase of the circadian cycle. The integration of temporal information from the circadian clock and light reception at specific photoreceptors allows the plant to correctly measure daylength during the seasonal changes in the environment. Do the same diurnal circadian clocks regulate the annual growth cycle? The only currently available observation relevant to this question is that the performance of the circadian clock in chestnut is disrupted during cold-induced ecodormancy, suggesting that circadian clock alteration may be necessary to reach advanced dormancy stages (Ramos et al. 2005).

The *Arabidopsis CONSTANS* (*CO*) encodes a putative transcriptional regulator that is directly regulated by the circadian clock (Putterill et al. 1995; Suarez-Lopez et al. 2001). *CO* acts on clock outputs to induce flowering (Suarez-Lopez et al. 2001) by regulating the two floral integrators *FT* and *SOC1* (Yoo et al. 2005). Interestingly, grafting experiments have shown that *CO* acts non-cell autonomously (An et al. 2004) through the movement of *FT* mRNA from leaves to the shoot apex to induce flowering (Huang et al. 2005). Thus, *FT* mRNA may represent the long sought after "florigen" signal. Currently, it is not known if the same genes or mechanisms might regulate both flowering time and dormancy, but it is interesting to note that *CO* regulates both flowering time and photoperiod-regulated tuberization in potato. The presence of *CO*-like genes in *Arabidopsis* raises the possibility that some of these paralogs could act as clock output genes to regulate downstream processes other than flowering. A possible mechanism integrating the environmental inputs of temperature and daylength in regulating dormancy is illustrated by the strong cross-talk between vernalization and photoperiod in vernalization pathways. This is illustrated by the vernalization-related MADS box transcription factor (*FLC*), which suppresses the ability of *CO* to activate target genes (Hepworth et al. 2002).

### 10.4.3 Low Temperature

Cold plays a crucial role in both the entering and breaking of dormancy. It is well known that both low temperature and SD signals are required to enter dormancy and achieve maximum cold hardiness. Cold-associated genes had much higher expression in SD conditions than LD conditions (Park et al. 2008). Thus, the responsiveness to a cold signal appeared to depend on how the dormancy development proceeded. In annual plants, the expression level of cold-associated genes is regulated primarily by response to temperature (Thomashow 1999) while in woody plants, they are induced differentially depending on different development stages. This suggests that woody plants may use different mechanisms than annual plants to attain the extreme freezing tolerance necessary for their survival. Extended exposure to cold temperatures is a prerequisite in many temperate perennials for breaking endodormancy ("chilling requirement") (Horvath et al. 2003; Anderson et al. 2005) and for promoting flowering ("vernalization") (Sung and Amasino 2004). The remarkable aspect in these processes is that plants are capable of measuring the duration of winter cold ("cold clock") and remembering this prior cold exposure in the spring. Recently, genetic screens for vernalization mutants led to the discovery of three genes that are required for the maintenance of the vernalized state (*VRN1* and *VRN2*) and the repression of *FLC* under cold conditions (*VIN3*) (Gendall et al. 2001; Levy et al. 2002; Sung and Amasino 2004). Vernalization appears to be different from cold acclimation in that the memory of the past winter remains stable for long periods through epigenetic changes (mitotically stable) whereas cold acclimation responses disappear upon return to warm conditions (Sung and Amasino 2004). A memory of prolonged cold exposure appears to be realized through chromatin modifications. The hypothesis that chromatin remodeling mechanisms may be involved in annual growth cycles remains to be tested.

## 10.5 Carbohydrate Physiology and Dormancy Cycle

Critical steps in the conversion of photosynthates to harvestable materials (woody biomass) can affect the photosynthetic process, as well as the amount of fixed carbohydrates allocated to yield. But these processes undergo profound changes during the dormancy cycle. Key control points in this regulation offer exciting avenues for manipulation; however, underlying mechanisms remain unclear, particularly in woody species like *Populus*. A more complete understanding of the control of processes such as sucrose synthesis and transport, starch biosynthesis, and the formation of cell wall constituents could provide invaluable tools for their adjustment. This is particularly important in the annual secondary growth cycle of a woody plant where wood formation requires sucrose metabolites and where sucrose and derivatives like raffinose may have other roles (e.g., as osmoprotectants in cold hardiness). Recent work indicates several especially promising avenues for regulation of photosynthate partitioning. A complicating factor (Koch et al. 2000; Koch 2004; Lunn and MacRae 2003; Ransom-Hodgkins et al. 2003), but also an important tool, is the critical role that sugars and sugar-related enzymes play in signaling (Cakir et al. 2003; Gibson 2005; Harrington and Bush 2003; Roitsch and Gonzalez 2004; Liu et al 2005). Sugar signaling in plants may also involve concurrent input along the transduction paths by effecters such as energy charge, P status, and phytohormones. The genes for most key carbohydrate-related enzymes and a number of transporters have been identified and characterized in herbaceous models; however, although there are exceptions such as recent studies of jasmonic acid (Babst et al. 2005) and elevated $CO_2$ effects in *Populus* (Taylor et al. 2005), very little work has been done in woody or perennial systems.

## 10.6 Poplar as a Model System for Studying the Annual Growth Cycle

The vascular cambium of poplar offers an excellent experimental system for the study of the annual growth cycle because: 1) its size and organization permits sampling of pure meristematic tissues using tangential cryosections (Uggla et al. 1996); 2) mitotically active regions, such as meristems, are the sites of cold perception and tissues that remember their prior cold exposure in the spring (Sung and Amasino 2004); 3) it provides year-round uniform samples that do not senesce (like leaves); and 4) its growth and differentiation result in secondary growth (e.g., wood formation), which has significant economic and environmental implications. In addition to its economic value as an important tree crop, poplar is the most well developed and widely accepted model system for tree biology. Along with the recent completion of genome sequencing and the availability of whole-transcriptome genechip arrays, it offers many attributes especially well suited for this type of research, including small easy clonal propagation which allows for replication of experiments, destructive sampling and available high-throughput transgenic technology.

## 10.7 Global Gene Expression Changes Associated with the Annual Growth Cycle in Poplar

### 10.7.1 Seasonal Growth Cycle

In a recent study conducted at Michigan (Park et al. 2008), *P. deltoides* trees were propagated by dormant cutting. Height growth was highly correlated with the diameter growth, indicating positive correlation of primary and cambial meristem activities. The first measurable increase in diameter growth in these trees was observed between April and May, when spring budbreak takes place, and peaked between July and August. Diameter growth stopped around mid-September, when the temperature tops 18°C on average and daylength is around 12-hours. This indicates that poplar plants perceive any daylength of less than 12-hours as SD and enter dormancy even if the temperature is still favorable for growth. On the other hand, the spring budbreak appeared to be correlated with the rising temperature rather than photoperiod because the plants did not resume their growth in April even when the daylength exceeded 13-hours.

### 10.7.2 Major Alterations in Gene Expression Reflect the Different Stages of Cambial Dormancy

In order to gain insights into the genetic regulation of the annual growth cycle, Park et al. (2008) carried out a series of transcriptome analyses using cDNA microarrays carrying 1,953 unigenes derived from the bark and xylem tissues of the field grown trees. For microarray hybridization, a woven loop design was used with two technical replications and a dye-swap. For statistical analysis, systemic errors were corrected using the LOWESS method after log2-transformation of the raw data. The normalized signal values were used in ANOVA analyses as implemented in the MAANOVA package in the statistical language R (Yang et al. 2002). Finally, the signal values were further normalized to remove the variances due to the array and dye (Cui and Churchill 2003). Hierarchical and k-means clustering were performed on the normalized data by using the CLUSTER program and were visualized with the TREEVIEW program (Eisen et al. 1998). A high proportion of the genes investigated were differentially expressed: 37% for the bark and 38% for the xylem tissues. The differentially expressed genes were further clustered hierarchically into three groups for the bark (Fig. 10.1) and two for the xylem (data not shown). In the bark, 138 genes (Cluster A) were up-regulated during the endodormancy, and 133 genes (Cluster B) for ecodormancy.

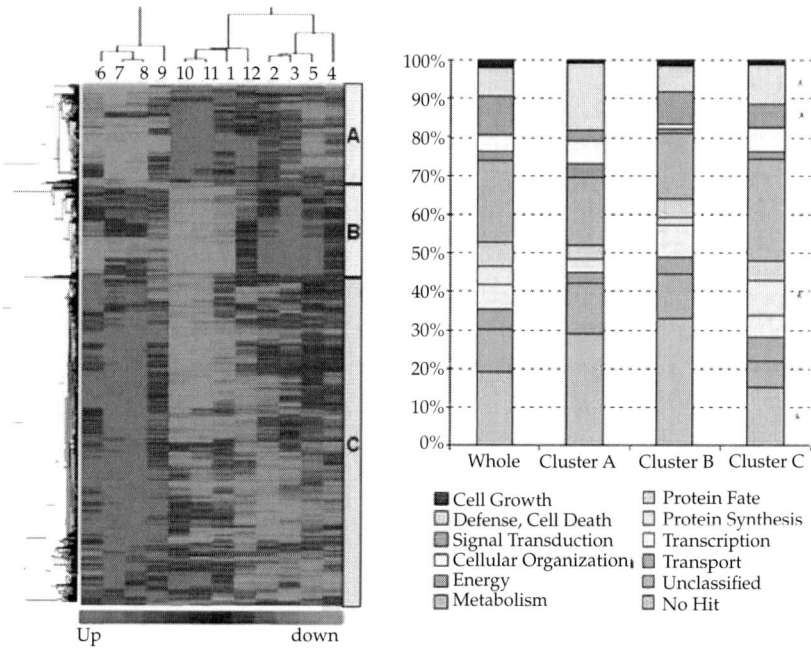

**Figure 10-1** Hierarchical clustering (left) and functional categorization (right) of 714 differentially expressed genes in bark tissue. In the hierarchical clustering, months are indicated numerically on each column. In functional characterization, the differentially expressed genes are presented as a percentage of the 1,953 unigenes on the array. The asterisk indicates statistically significant functional category (contingency test, p < 0.001) (Park et al. 2008).

*Color image of this figure appears in the color plate section at the end of the book.*

In Cluster A, defense/cell rescue-associated genes and no-hits were over-represented, while signal transduction and protein synthesis-associated genes were under-represented. Some genes in this cluster encode pathogenesis-associated proteins that are known to have antifreeze activity (Hon et al. 1995) and antioxidant proteins that protect cells during budbreak from free radicals generated by the winter chilling (Wang et al. 1991; Wang and Faust 1994). Functional classification of Cluster B genes indicates that ecodormancy is characterized by the activation of cellular signaling with reduced cellular activity. For example, proteins encoded by those genes include receptor kinases and proteins involved in cytokinin and auxin signaling. Both of the growth regulators are known to be involved in breaking bud dormancy (Cutting et al. 1991). Biological activities during winter have not received much research attention. There may be considerable metabolic activities during late winter, for instance, genes encoding fermentation-related proteins (e.g., pyruvate decarboxylases and alcohol dehydrogenase) were up-regulated, suggesting that the plants might be under oxygen deficiency stress, likely caused by ice encasement formed during freeze-thaw cycles in late winter (Bertland et al. 2003). It is notable that fatty acid oxidation-associated genes are up-regulated during winter, considering that lipid hydrolysis could allow the plants to preserve cryoprotective sugars during late winter (Erez et al. 1998).

### 10.7.3 Effects of Temperature and Photoperiod on Cold Acclimation and Dormancy Cycle

Bertland et al. (2003) and Erez et al. (1998) also studied transcriptional profiling with the same clonal materials subjected to five different controlled environmental treatments: two daylengths (8-hours for SD and 16-hours for LD), two temperatures (25°C and 4°C), and a drought (with LD and 25°C). The plants grown under long-day and cold conditions (LDC) did not show any visible change in the shoot but had reduced vessel size in the xylem, indicating reduced activity of cambial cells. Short-day and warm (SDW) grown plants induced terminal bud set and showed compressed layers of cambial zone, indicating that they were in dormancy. However, cambial cells in SDW-grown plants had thickened cell walls, which is characteristic of the rest (Chaffey et al. 1998). It is likely that LDC conditions are needed for inducing cold acclimation without dormancy, SDW conditions are needed for the first stage of cold acclimation and endodormancy, and short-day and cold conditions (SDC) are needed for the second stage of cold acclimation and ecodormancy. This hypothesis was further supported by a hierarchical clustering based on the gene expression profiles obtained from both of the controlled environment- and field-grown trees.

Park et al. (2008) reports on the use of k-means clustering to categorize the differentially expressed genes into eight groups of expression patterns (Fig. 10-2). There appear to be multiple signaling pathways involved in the regulation of the growth cycle. For example, the genes in Groups 1-4 responded negatively to SDC conditions (i.e., down-regulated compared to long-day warm grown plants (LDW)) while showing differential responses to long-day drought (LDD) or SDW conditions. G1 genes were up-regulated during active growth but down-regulated in all four non-optimum growing conditions, suggesting that their main functions may be in support of active growth. This hypothesis is supported by the fact that G1 includes active growth-related genes such as photosynthesis-, cell division-, and auxin signaling-associated genes. Genes in G3 and G4 responded positively to SDW (i.e., up-regulated compared to LDW) and seemed to have high levels of expression in September, suggesting that they might be involved in the dormancy onset process. Interestingly, G3 included starch and amino acid synthesis genes as well as dehydration-response genes. These genes are expressed prior to the onset of dormancy (Perry 1971). Moreover, the fact that the G3 genes respond positively to both SD and drought is notable in terms of understanding the molecular basis for drought stress-induced dormancy (Rinne et al. 1994). Likewise, the fact that G4 genes were up-regulated solely by SDW conditions and had higher expression in September than G3 genes suggested that they might be involved in dormancy onset rather than cold acclimation; indeed, G4 includes many cell wall-associated

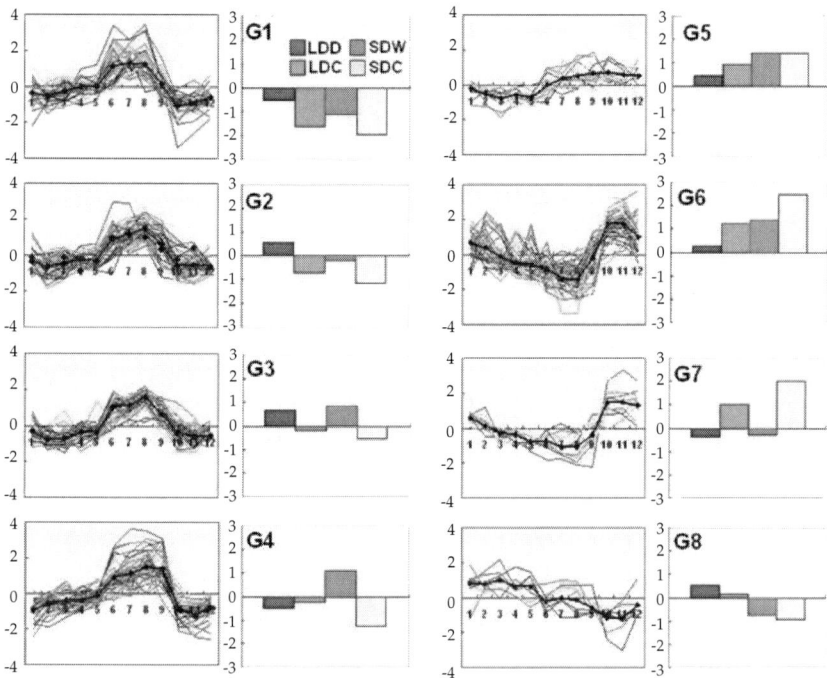

**Figure 10-2** K-means clustering of 274 differentially expressed genes. The line graphs on the left show seasonal expression changes, where the x-axis represents 12 months (1=Jan, 2=Feb, etc.) and the y-axis represents the fold change on a log2 scale (compared to the annual average). The bar graphs on the right show their environmental responsiveness, where mean fold changes (log2 scale) for each of four treatments (LDD, LDC, SDW, and SDC) over LDW are shown on the y-axis (Park et al. 2008).

*Color image of this figure appears in the color plate section at the end of the book.*

genes. Therefore, we hypothesize that the cell wall thickening during dormancy development may be regulated primarily by SD signal.

The genes up-regulated during late fall and winter in these studies (G5-7 in Fig. 10-2) appear to be of the most immediate interest in terms of providing an integrative view of the roles of environmental and developmental factors in dormancy/cold hardiness. Both SD and cold can induce cold acclimation independently (Welling et al. 2002). Consistent with this observation, many of the genes in the groups were up-regulated by either SD or cold. Trees often require both cold and SD to achieve high levels of cold hardiness. The transition from rest to quiescence in the cambium is gradually accomplished by chilling (Little and Bonga 1974). One of the activities associated with the satisfaction of chilling requirement is a restoration of cell-to-cell signaling networks in the meristem that had been disrupted during dormancy induction. This resumption of cell-to-cell communication via plasmodesmata may allow symplastic movement of small signaling molecules, hormones, or proteins responsible for dormancy release.

## 10.8 Concluding Remarks

Clearly the regulation of dormancy in woody plants is an extremely complex physiological process that allows for the necessary establishment and breaking of both ecodormancy and endodormancy, while buffering these long-lived species against the random, seasonal fluctuations in temperature that are common in the temperate region. As has long been expected, daylength plays a critical role in this process, but by itself cannot explain the patterns observed in these studies. Similarly, cold exposure is important, especially in fulfilling the chilling requirements for the breaking of dormancy. The true answer, however, to the regulation of these processes rests in the myriad genes that are up and down regulated in concert due to the interplay between these two environmental signals. The identification of groups of genes showing similar expression patterns during the seasonal shifts in dormancy offers new insights into understanding dormancy in woody plants. Continued study of these gene clusters may reasonably be expected to result in both understanding and control of these environmentally and economically important physiological processes that enable the survival of woody plants in temperate climates.

# References

An HL, Roussot C, Suarez-Lopez P, Corbesler L, Vincent C, Pineiro M, Hepworth S, Mouradov A, Justin S, Turnbull C, Coupland G (2004) CONSTANS acts in the phloem to regulate a systemic signal that induces photoperiodic flowering of Arabidopsis. Development 131: 3615–3626.

Anderson JV, Gesch RW, Jia Y, Chao WS, Horvath DP (2005) Seasonal shifts in dormancy status, carbohydrate metabolism, and related gene expression in crown buds of leafy spurge. Plant Cell Environ 28: 1567–1578.

Arora R, Rowland LJ, Tanino K (2003) Induction and release of bud dormancy in woody perennials: A science comes of age. HortScience 38: 911–921.

Bañados MP (1992) Nitrogen and environmental factors affect bark storage protein gene expression in poplar. MS Thesis, Oregon State Univ, Corvallis, OR, USA.

Bailey TL, Elkan C (1994) Fitting a mixture model by expectation maximization to discover motifs in biopolymers. Proc 2nd Int Conf on Intelligent Systems for Molecular Biology, AAAI Press, Menlo Park, California, USA, pp 28–36.

Babst BA, Ferrieri RA, Gray DW, Lerdau M, Schlyer DJ, Schueller M, Thorpe MR, Orians CM (2005) Jasmonic acid induces rapid changes in carbon transport and partitioning in Populus. New Phytol 167: 63–72.

Bertrand A, Castonguay Y, Nadeau P, Laberge S, Michaud R, Belanger G, Rochette P (2003) Oxygen deficiency affects carbohydrate reserves in overwintering forage crops. J Exp Bot 54: 1721–30.

Bradshaw HD, Stettler RF (1995) Molecular genetics of growth and development in *Populus*. IV. Mapping QTLs with large effects on growth, form, and phenology traits in a forest tree. Genetics 139: 963–973.

Cakir B, Agasse A, Gaillard C, Saumonneau A, Delrot S, Atanassova R (2003) A grape ASR protein involved in sugar and abscisic acid signaling. Plant Cell 15: 2165–2180.

Chaffey NJ, Barlow PW, Barnett JR (1998) A seasonal cycle of cell wall structure is accompanied by a cyclical rearrangement of cortical microtubules in fusiform cambial cells within taproots of Aesculus hippocastanum (Hippocastanaceae). New Phytol 139: 623–635.

Chaffey N (1999) Cambium: old challenges—new opportunities. Trees 13: 138–151.

Chen THH, Howe GT, Bradshaw HD (2002) Molecular genetic analysis of dormancy-related traits in poplars. Weed Sci 50: 232–240.

Clark SE (2001). Cell signalling at the shoot meristem. Nat Rev Mol Cell Biol 2: 276–284.

Contento AL, Kim SJ, Bassham DC (2004) Transcriptome profiling of the response of Arabidopsis suspension culture cells to Suc starvation. Plant Physiol 135: 2330–2347.

Cui X, Churchill GA (2003) Statistical tests for differential expression in cDNA microarray experiments. Genome Biol 4: 210.

Cutting JGM, Strydom DK, Jacobs G, Bellstedt DU, Vandermerwe KJ, Weiler EW (1991) Changes in Xylem Constituents in Response to Rest-Breaking Agents Applied to Apple before Budbreak. J Am Soc Hort Sci 116: 680–683.

Dickmann D, Keathley DE (1996) Linking physiology, molecular genetics, and the *Populus* ideotype In: RF Stettler, HD Bradshaw, PE Heilman Jr, TM Hinckley (eds) Biology of *Populus* and Its Implication for Management and Conservation. National Research Council, Ontario, Canada, pp 491–514.

Doerner P (2003) Plant meristems: A merry-go-round of signals. Curr Biol 13: R368–R374.

Eisen MB, Spellman PT, Brown PO, Botstein D (1998) Cluster analysis and display of genome-wide expression patterns. Proc Natl Acad Sci USA 95: 14863–14868.

Epperson BK (2003) Geographical Genetics. Monographs in Population Biology, Princeton Univ Press, Princeton, NJ, USA.

Erez A, Faust M, Line MJ (1998) Changes in water status in peach buds on induction, development and release from dormancy. Sci Hort 73: 111–123.

Eriksson ME, Millar AJ (2003) The circadian clock A plant's best friend in a spinning world. Plant Physiol 132: 732–738.

Farrar JJ, Evert RF (1997) Ultrastructure of cell division in the fusiform cells of the vascular cambium of *Robinia pseudoacacia*. Trees Struct Funct 11: 203–215.

Fuchigami LH, Weiser CJ, Evert DR (1971) Induction of cold acclimation in *Cornus stolonifera*. Michx. Plant Physiol 47: 98–103.

Gao Z, Loescher WH (2003) Expression of a celery mannose 6-phosphate reductase in *Arabidopsis thaliana* enhances salt tolerance and induces biosynthesis of both mannitol and a glucosyl-mannitol dimer. Plant Cell Environ 26: 275–283.

Gendall AR, Levy YY, Wilson A, Dean C (2001) The VERNALIZATION 2 gene mediates the epigenetic regulation of vernalization in Arabidopsis. Cell 107: 525–535.

Gibson SI (2005) Control of plant development and gene expression by sugar signaling. Curr Opin Plant Biol 8: 93–102.

Håbjørg A (1972) Effects of photoperiod and temperature on growth and development of three latitudinal and three altitudinal populations of *Betula pubescens* Ehrh. Meld Norg Landbr-Høgsk 51: 1–27.

Han K-H, Davis JM, Keathley DE (1990) Differential responses persist in shoot explants regenerated from callus of two mature black locust trees. Tree Physiol 6: 235–240.

Han K-H, Meilan R, Ma C, Strauss SH (2000) An *Agrobacterium* transformation protocol effective on a variety of cottonwood hybrids (genus *Populus*). Plant Cell Rep 19: 315–320.

Harmer SL, Hogenesch LB, Straume M, Chang HS, Han B, Zhu T, Wang X, Kreps JA, Kay SA (2000) Orchestrated transcription of key pathways in *Arabidopsis* by the circadian clock. Science 290: 2110–2113.

Harmer SL, Panda S, Kay SA (2001) Molecular bases of circadian rhythms. Annu Rev Cell Dev Biol 17: 215–253.

Harrington GN, Bush DR (2003) The bifunctional role of hexokinase in metabolism and glucose signaling. Plant Cell 15: 2493–2496.

Hayama R, Coupland G (2003) Shedding light on the circadian clock and the photoperiodic control of flowering. Curr Opin Plant Biol 6: 13–19.

Heide OM (2003) High autumn temperature delays spring bud burst in boreal trees, counterbalancing the effect of climatic warming. Tree Physiol 23: 931–936.

Hepworth SR, Valverde F, Ravenscroft D, Mouradov A, Coupland G (2002) Antagonistic regulation of flowering-time gene SOC1 by CONSTANS and FLC via separate promoter motifs. EMBO J 21: 4327–4337.

Hon WC, Griffith M, Mlynarz A, Kwok YC, Yang DS (1995) Antifreeze proteins in winter rye are similar to pathogenesis-related proteins. Plant Physiol 109: 879–889.

Horvath DP, Anderson JV, Chao WS, Foley ME (2003) Knowing when to grow: signals regulating bud dormancy. Trends Plant Sci 8: 534–540.

Howe GT, Hackett WP, Furnier GR, Klevorn RE (1995) Photoperiodic responses of a northern and southern ecotype of black cottonwood. Physiol Plant 93: 695–708.

Howe GT, Gardner G, Hackett WP, Furnier GR (1996) Phytochrome control of shortday-induced bud set in black cottonwood. Physiol. Planta 97: 95–103.

Howe GT, Bucciaglia PA, Hackett WP, Furnier GR, Cordonnier-Pratt MM, Gardner G (1998) Evidence that the phytochrome gene family in black cottonwood has one PHYA locus and two PHYB loci but lacks members of the PHYC/F and PHYE subfamilies. Mol Biol Evol 15: 160–175.

Huang T, Bohlenius H, Eriksson S, Parcy F, Nilsson O (2005) The mRNA of the Arabidopsis gene FT moves from leaf to shoot apex and induces flowering. Science 309: 1694–1696.

Hudson ME, Quail PH (2003) Identification of promoter motifs involved in the network of phytochrome A-regulated gene expression by combined analysis of genomic sequence and microarray data. Plant Physiol 133: 1605–1616.

Hughes JD, Estep PW, Tavazoie S, Church GM (2000) Computational identification of cis-regulatory elements associated with groups of functionally related genes in Saccharomyces cerevisiae. J Mol Biol 296: 1205–1214.

IPCC (1997) The Regional Impact of Climatic Change: An Assessment of Vulnerability. Cambridge Univ Press, Cambridge, UK.

Izawa T, Oikawa T, Sugiyama N, Tanisaka T, Yano M, Shimamoto K (2002) Phytochrome mediates the external light signal to repress FT orthologs in photoperiodic flowering of rice. Genes Dev 16: 2006–2020.

Jeknic Z, Chen THH (1999) Changes in protein profiles of poplar tissues during the induction of bud dormancy by short-day photoperiods. Plant Cell Physiol 40: 25–35.

Johanson U, West J, Lister C, Michaels S, Amasino R, Dean C (2000) Molecular analysis of FRIGIDA, a major determinant of natural variation in Arabidopsis flowering time. Science 290: 344–347.

Keller JD, Loescher W (1989) Nonstructural carbohydrate partitioning in perennial parts of sweet cherry (*Prunus avium* L.). J Am Soc Hort Sci 114: 969–975.

Ko JH, Han K-H (2004) *Arabidopsis* whole-transcriptome profiling defines the features of coordinated regulations that occur during wood formation. Plant Mol Biol (in press).

Ko JH, Han K-H, Park S, Yang J (2004) Plant body weight-induced secondary growth in Arabidopsis and its transcription phenotype revealed by whole-transcriptome profiling. Plant Physiol 135: 1069–1083.

Koch K (2004) Sucrose metabolism: regulatory mechanisms and pivotal roles in sugar sensing and plant development. Curr Opin Plant Biol 7: 235–246.

Koch KE, Ying Z, Wu Y, Avigne WT (2000) Multiple paths of sugar-sensing and a sugar/oxygen overlap for genes of sucrose and ethanol metabolism. J Expt Bot 51: 417–427.

Lang GA, Early JD, Martin GC, Darnell RL (1987) Endodormancy, Paradormancy, and Ecodormancy—Physiological Terminology and Classification for Dormancy Research. HortScience 22: 371–377.

Laux T (2003) The stem cell concept in plants: A matter of debate. Cell 113: 281–283.

Leple JC, Brasileiro ACM, Michel MF, Delmotte F, Jouanin L (1992) Transgenic poplars: expression of chimeric genes using four different constructs. Plant Cell Rep 11: 137–141.

Levy YY, Mesnage S, Mylne JS, Gendall AR, Dean C (2002) Multiple roles of Arabidopsis VRN1 in vernalization and flowering time control. Science 297: 243–246.

Little C H A, Bonga J M (1974) Rest in Cambium of *Abies-Balsamea*. Can J Bot 52: 1723–1730.

Liu J, Samac DA, Bucciarelli B, Allan DL, Vance CP (2005) Signaling of phosphorus deficiency-induced gene expression in white lupin requires sugar and phloem transport. Plant J 41: 257–268.

Lunn JE, Macrae E (2003) New complexities in the synthesis of sucrose. Curr Opin Plant Biol 6: 208–214.

Ma C, Strauss S, Meilan R (2004) Agrobacterium-mediated transformation of the genome-sequenced poplar clone, Nisqually-1 (*Populus trichocarpa*). Plant Mol Biol Rep 22: 1–9.

Martinez-Garcia J, Virgos-Soler A, Prat S (2002) Control of photoperiod-regulated tuberization in potato by the Arabidopsis flowering-time gene CONSTANS. Proc Natl Acad Sci USA 99: 15211–15216.

Mauseth J (1998) Botany: An Introduction to Plant Biology. Jones and Bartlett Publ, Sudbury, Massachusetts, USA.

Mellerowicz E J, Coleman WK, Riding RT, Little CHA (1992) Periodicity of cambial activity in *Abies balsamea* I Effects of temperature and photoperiod on cambial dormancy and frost hardiness. Physiol Plant 85: 515–525.

Meilan R, Han K-H, Ma C, DiFazio SP, Eaton JA, Hoien E, Stanton BJ, Crockett RP, Taylor ML, James RR, Skinner JS, Pilate G, Strauss SH (2002) The *CP4* transgene provides high levels of tolerance to Roundup® herbicide in field-grown hybrid poplars. Can J For Res 32: 967–976.

Nakajima K, Benfey PN (2002) Signaling in and out: Control of cell division and differentiation in the shoot and root. Plant Cell 14: S265–S276.

Nitsch JP (1957) Growth responses of woody plants to photoperiodic stimuli. Proc Am Soc Hort Sci 70: 512-525.

Oh S, Park S, Han K-H (2003) Transcriptional regulation of secondary growth in *Arabidopsis thaliana*. J Exp Bot 54: 2709–2922.

Olsen JE, Junttila O (2002) Far red end-of-day treatment restores wild type-like plant length in hybrid aspen overexpressing phytochrome A. Physiol Plant 115: 448–457.

Olsen JE, Junttila O, Nilsen J, Eriksson ME, Martinussen I, Olsson O, Sandberg G, Moritz T (1997) Ectopic expression of oat phytochrome A in hybrid aspen changes critical daylength for growth and prevents cold acclimatization. Plant J 12: 1339–1350.

Park SC, Oh S, Han K-H (2004) Large-scale computational analysis of poplar ESTs reveals the repertoire and unique features of expressed genes in poplar genome. Mol Breed 14: 429–440.

Park SC, Keathley DD, Han K-H (2008) Transcriptional profiles of the annual growth cycle in *Populus deltoides*. Tree Physiol 28: 321–329.

Perry TO (1971) Dormancy of Trees in Winter. Science 171: 29–36.

Price J, Laxmi A, St Martin SK, Jang JC (2004) Global transcription profiling reveals multiple sugar signal transduction mechanisms in Arabidopsis. Plant Cell 16: 2128–2150.

Putterill J, Robson F, Lee K, Simon R, Coupland G (1995) The CONSTANS gene of Arabidopsis promotes flowering and encodes a protein showing similarities to zinc finger transcription factors. Cell 80: 847–857.

Ramos A, Perez-Solis E, Ibanez C, Casado R, Collada C, Gomez L, Aragoncillo C, Allona I (2005) Winter disruption of the circadian clock in chestnut. Proc Natl Acad Sci USA, 102: 7037–7042.

Ransom-Hodgkins WD, Vaughn MW, Bush DR (2003) Protein phosphorylation plays a key role in sucrose-mediated transcriptional regulation of a phloem-specific proton–sucrose symporter. Planta 217: 483–489.

Rinne P, Saarelainen A, Junttila O (1994) Growth Cessation and Bud Dormancy in Relation to Aba Level in Seedlings and Coppice Shoots of Betula-Pubescens as Affected by a Short Photoperiod, Water-Stress and Chilling. Physiol Plant 90: 451–458.

Rohde A, Prinsen E, De Rycke R, Engler G, Van Montagu M, Boerjan W (2002) PtABI3 impinges on the growth and differentiation of embryonic leaves during bud set in poplar. Plant Cell 14: 2975–2975.

Roitsch T, Gonzalez MC (2004) Function and regulation of plant invertases: sweet sensations. Trends in Plant Sci 9: 606–613.

Rombauts S, Florquin K, Lescot M, Marchal K, Rouze P, Van de Peer Y (2003) Computational approaches to identify promoters and *cis*-regulatory elements in plant genomes. Plant Physiol 132: 1162–1176.

Roth FR, Hughes JD, Estep PE, Church GM (1998) Finding DNA regulatory motifs within unaligned non-coding sequences clustered by whole-genome mRNA quantitation. Nat Biotechnol 16: 939–945.

Rouillard J-M, Herbert CJ, Zuker M (2002) Genome-scale oligonucleotide design for microarrays. Bioinformatics 18: 486–487.

Rowland LJ, Arora R (1997) Proteins related to endodormancy (rest) in woody perennials. Plant Sci 126: 119–144.

Schrader J, Moyle R, Bhalerao R, Hertzberg M, Lundeberg J, Nilsson P, Bhalerao RP (2004a) Cambial meristem dormancy in trees involves extensive remodelling of the transcriptome. Plant J 40: 173–187.

Schrader J, Nilsson J, Mellerowicz E, Berglund A, Nilsson P, Hertzberg M, Sandberg G (2004b) A high-resolution transcript profile across the wood-forming meristem of poplar identifies potential regulators of cambial stem cell identity. Plant Cell 16: 2278–2292.

Serreze MC, Walsh JE, Chapin FS, Osterkamp T, Dyurgerov M, Romanovsky V, Oechel WC, Morison J, Zhang T, Barry RG (2000) Observational evidence of recent change in the northern high-latitude environment. Clim Chang 46: 159–207.

Shimizu-Sato S, Mori H (2001) Control of outgrowth and dormancy in axillary buds. Plant Physiol 127: 1405–1413.

Shindo C, Aranzana M, Lister C, Baxter C, Nicholls C, Nordborg M, Dean C (2005) Role of FRIGIDA and FLOWERING LOCUS C in determining variation in flowering time of Arabidopsis. Plant Physiol 138: 1163–1173.

Somers DE, Devlin PF, Kay SA (1998) Phytochromes and cryptochromes in the entrainment of the Arabidopsis circadian clock. Science 282: 1488–1490.

Suarez-Lopez P, Wheatley K, Robson F, Onouchi H, Valverde F, Coupland G (2001) CONSTANS mediates between the circadian clock and the control of flowering in Arabidopsis. Nature 410: 1116–1120.

Sung S, Amasino RM (2005). Remembering winter: Toward a molecular understanding of vernalization. Annu Rev Plant Biol 56: 491–508.

Taylor G Street NR, Tricker PJ, Sjodin A, Graham L, Skogstrom O, Calfapietra C, Scarascia-Mugnozza G, Jansson S (2005) The transcriptome of Populus in elevated CO2. New Phytol 167: 143–154

Thijs G, Moreau Y, De Smet F, Mathys J, Lescot M, Rombauts S, Rouze P, De Moor B, Marchal K (2002) INCLUSive: INtegrated clustering, upstream of sequence retrieval and motif sampling. Bioinformatics 18: 331–332.

Thomashow MF (1999) Plant cold acclimation: Freezing tolerance genes and regulatory mechanisms. Annu Rev Plant Physiol Plant Mol Biol 50: 571–599.

Uggla C, Moritz T, Sandberg G, Sundberg B (1996) Auxin as a positional signal in pattern formation in plants. Proc Natl Acad Sci USA 93: 9282–9286.

Vince-Prue D (1975) Photoperiodism in Plants. McGraw Hill, London, UK.

Wang E, Miller LD, Ohnmacht GA, Liu ET, Marincola FM (2000) High-fidelity mRNA amplification for gene profiling. Nat Biotechnol 18: 457–459.

Wang SY, Faust M (1994) Changes in the antioxidant system associated with budbreak in Anna apple (*Malus domestica* Borkh) Buds. J Am Soc Hort Sci 119: 735–741.

Wang SY, Jiao HJ, Faust M (1991) Changes in ascorbate, glutathione, and related enzyme-activities during thidiazuron-induced bud break of apple. Physiol Plant 82: 231–236.

Weiser CJ (1970) Cold resistance and injury in woody plants. Science 169: 1269–1278.

Welling A, Moritz T, Palva ET, Junttila O (2002) Independent activation of cold acclimation by low temperature and short photoperiod in hybrid aspen. Plant Physiol 129: 1633–1641.

Winter H, Huber SC (2000) Regulation of sucrose metabolism in higher plants: Localization and regulation of activity of key enzymes. Crit Rev Biochem Mol Biol 35: 253–289.

Yang J, Park S, Kamdem DP, Keathley DE, Retzel E, Paule C, Kapur V, Han K-H (2003) Novel gene expression profiles define the metabolic and physiological processes characteristic of wood and its extractive formation in a hardwood tree species, *Robinia pseudoacacia*. Plant Mol Biol 52: 935–956.

Yang J, Kamdem DP, Keathley DE, Han K-H (2004) Seasonal gene expression changes at the sapwood-heartwood transition zone of black locust (*Robinia pseudoacacia* L.) revealed by cDNA microarray analysis. Tree Physiol (in press).

Yang YH, Dudoit S, Luu P, Lin DM, Peng V, Ngai J, Speed TP (2002) Normalization for cDNA microarray data: a robust composite method addressing single and multiple slide systematic variation. Nucl Acids Res 30: e15.

Yanovsky MJ, Kay SA (2002) Molecular basis of seasonal time measurement in *Arabidopsis*. Nature 419: 308–312.

Yanovsky MJ, Kay SA (2003) Living by the calendar: How plants know when to flower. Nat Rev Mol Cell Biol 4: 265–275.

Yoo SK, Chung KS, Kim J, Lee J, Lee JH, Hong S, Yoo SJ, Yoo SY, Lee JS, Ahn JH (2005) CONSTANS activates SUPPRESSOR OF OVEREXPRESSION OF CONSTANS 1 through FLOWERING LOCUS T to promote flowering in *Arabidopsis*. Plant Physiol 139: 770–778.

# 11

# Regulation of Flowering Time in Poplar

*Cetin Yuceer,[1,a,]\* Chuan-Yu Hsu,[1,b] Amy M. Brunner[2] and Steven H. Strauss[3]*

## ABSTRACT

Trees have provided and will continue to provide shelter, energy, fiber, food, and numerous other benefits for society. However, the lengthy juvenile period is a major obstacle to early and frequent sexual reproduction for development of pedigreed offspring to accelerate tree domestication. Although much is known about the factors regulating the onset of sexual reproduction in the annual model plant *Arabidopsis*, far less is known about this transition in trees. Recent advances in poplar are beginning to provide a fundamental understanding of the signaling mechanism by which the onset of sexual reproduction is determined in trees. This chapter provides an overview of knowledge about the genetic, physiological, and environmental factors that regulate first time and seasonal reproduction poplar, making reference to insights from *Arabidopsis*. Furthermore, we discuss the potential for practical applications of knowledge in trees gained from fundamental flowering research.

**Keywords:** flowering, reproduction, development, juvenility, maturity, poplar, *Populus*

[1]Department of Forestry, Mississippi State University, PO Box 9681, Mississippi State, MS 39762.
[a]e-mail: *mcy1@msstate.edu*
[b]e-mail: *ch11@msstate.edu*
[2]Department of Forest Resources and Environmental Conservation, Virginia Polytechnic Institute and State University, Blacksburg, VA 24061-0324; e-mail: *abrunner@vt.edu*
[3]Department of Forest Ecosystems and Society, Oregon State University, Corvallis, OR 97331-5752; e-mail: *steve.strauss@oregonstate.edu*
\*Corresponding author

## 11.1 Introduction

Compared to food crops, forest tree improvement is in its infancy. Innate features of trees provide major barriers to breeding progress, most significant of which is the lengthy juvenile phase of 5 to 20 years before they are developmentally capable of flowering. The long delay in flowering and typically high genetic load of trees makes it infeasible to use advanced methods such as inbreeding and introgression of rare or exotic alleles. The net result is a very slow rate of domestication for all breeding goals. Transgenic approaches can potentially advance tree domestication, but concerns over the dispersal of transgenic pollen or seed, in addition to a number of other social and technical factors, have prevented most commercial uses of transgenic forest trees in the world (Brunner et al. 2007). Thus, understanding the factors that regulate tree flowering and discovering ways to manipulate it could enhance tree improvement by speeding breeding and research to develop effective means for genetic containment. Moreover, because flowering time is an adaptive trait that is affected by global warming (Fitter and Fitter 2002), discovery of the genes important for control of tree flowering might also aid in the development of strategies for maintaining healthy forest tree populations in the world with rapidly changing climates.

Poplar (*Populus* spp.) is economically and ecologically important, and is a model system for deciphering the molecular and physiological processes that regulate flowering time in trees. The main advantages of poplar compared to other trees include the rich variety of genomic resources available for it (e.g., whole-genome sequence; Tuskan et al. 2006), its amenability to *Agrobacterium*-mediated transformation (Han et al. 2000; Song et al. 2006; Cseke et al. 2007), and its well-studied developmental processes. Poplar and the annual herbaceous plant *Arabidopsis thaliana* are both angiosperms and eudicots, facilitating comparative genomics between these two taxa (Soltis et al. 1999; Wikstrom et al. 2001). Comparison of flowering genes and gene function and pathways between poplar and *Arabidopsis* will advance our understanding of how changes in gene number, expression, and interactions have resulted in drastically different floral morphologies and flowering habits.

This chapter will provide an overview of current knowledge about the genetic and physiological factors that control flowering time in poplar. We also discuss the potential for practical applications of knowledge gained from molecular flowering research.

## 11.2 Development and Architecture

Poplar has a life span of more than 100 years and a juvenile phase of approximately five years to more than a decade prior to the onset of flowering (Braatne et al. 1996), indicating slow maturation. Juvenile trees form vegetative buds, leaves, and internodes. A terminal bud is formed at the end of each shoot every season and is enclosed by several layers of bud scales that are formed by the enlargement of stipules to protect the foliage primordia of the following season's growth (Goffinet and Larson 1981). Juvenile trees exhibit rapid growth rates with long internodes, continuous shoot growth throughout the growing season and terminal bud formation at the end of the growing season when the critical daylength for bud set occurs. Following the first annual production of reproductive buds, seasonal production of both vegetative and reproductive buds occurs during the reproductive developmental phase. Thus, poplar has developed a shoot architecture that accommodates both vegetative and reproductive growth throughout its life cycle.

The developmental state of leaves, the positions of axillary buds, and seasonal timing of axillary meristem initiation on a shoot are important factors in flower initiation. Thus, a model for the development and architecture of axillary bud meristems and their temporal and spatial formation in shoots of mature *P. deltoides* was developed (Yuceer et al. 2003). Shoots with flower buds in mature trees tend to have short internodes and early cessation of primary vegetative growth. Consequently, shoots begin forming a terminal bud approximately two months following spring bud flush. It is currently unknown why, or how, shoots that produce flower buds cease growth prematurely.

Mature shoots possess a defined developmental pattern that includes specific locations for vegetative and reproductive buds and distinct leaf types (Critchfield 1960; Boes and Strauss 1994; Yuceer et al. 2003). Shoots on adult trees produce buds in a sequential manner, each with an associated leaf type. Early vegetative buds (Vegetative Zone I) are produced in axils of early preformed leaves, reproductive buds (Floral Zone) are produced in axils of late preformed leaves, and late vegetative buds (Vegetative Zone II) are produced in axils of neoformed leaves. During the first growing season (Year 1), the terminal bud forms and contains the early preformed leaves and the late preformed leaf primordia. Early preformed leaves are initiated early in the development of the terminal bud during Year 1 and have a long developmental period which is interrupted by a cold period (vernalization) prior to expansion in the second growing season (Year 2). The preformed buds that develop in the axils of the early preformed leaves (Vegetative Zone I) never develop into reproductive buds and form vegetative shoots with true leaf primordia. Late preformed leaf primordia develop during the advanced stage of terminal bud development and stay in a primordial stage during vernalization. The buds that develop in axils of these leaves are reproductive.

Spring flowering phenology varies among species, genotypes, and populations, but the sequence of events is the same in all cases. A typical phenology for *Populus deltoidies* in Mississippi, USA is described below. The terminal bud opens in late March of Year 2 following the formation in Year 1. Reproductive buds in the Floral Zone, numbering from two to 10 on shoots, subsequently become visible in late-leaf axils. Examination of the spring bud meristems in the Floral Zone indicates morphological changes that have led to inflorescence shoot formation, floral meristem development, and organ formation. On the developing inflorescence (catkin) beginning late spring (May), bracts and then axillary floral meristems develop acropetally. By the winter of Year 2, the floral meristems form a cup-like, reduced perianth with stigmas or tetrasporangiate anthers in the axils of fully-elongated bracts. As an adaptation to wind pollination, reproductive bud flush occurs before vegetative bud flush in March of Year 3; catkins rapidly elongate and floral anthesis occurs. Female trees continue to form seeds until May of Year 3.

After all preformed leaves have expanded in spring of Year 2, some shoots may produce neoformed leaves that initiate and expand entirely within the current growing season. Thus, the neoformed leaves have not undergone vernalization. These leaves comprise Vegetative Zone II and bear vegetative buds in their axils. Following the formation of reproductive buds, as many as 40 vegetative buds form in Vegetative Zone II.

Although it is unknown when exactly the floral induction occurs, the flowering process may begin as signal perception in early vernalized preformed leaves of the first growing season, prior to flower bud formation during the second growing season. Equally important is the question of whether the floral signal is translocated to the shoot apical meristem (SAM) where bud fate is determined or to developing axillary buds. It is possible that the floral signal translocates to the developing buds in the late-leaf axils through direct vascular connections, given that a specific repeating pattern of primary vascular tissues exists between leaves and the nodes where buds form (Larson and Pizzolato 1977; Pizzolato and Larson 1977; Dickson 1986). The primary vascular connections are formed in the primordial stem tissues of the overwintering terminal bud as a continuation of acropetal elongation of the shoot (Larson 1975).

## 11.3 Flowering-Time Genes

### 11.3.1 *Arabidopsis thaliana*

*Arabidopsis* is the best studied annual plant model, particularly for its reproductive biology. *Arabidopsis* completes its life cycle in two months, with a short juvenile period followed by the production of flowers (Somerville and Koornneef 2002). The SAM initially gives rise to vegetative organs such as leaves, but at some point the SAM is transformed into an indeterminate inflorescence meristem that produces floral buds on its flanks (Levy and Dean 1998). The main inflorescence shoot and axillary buds continuously produce reproductive organs, and then the plant dies. *Arabidopsis* does not undergo stages of seasonal vegetative and floral development at the reproductive phase, nor does it revert to the vegetative phase once it begins the reproductive phase (Boss et al. 2004).

A combination of environmental and developmental signals trigger flowering in *Arabidopsis*. The four major linked pathways that control flowering time are the photoperiodic, developmental, vernalization, and gibberellin pathways (Fig. 11-1). They transmit signals that regulate the expression of floral meristem identity genes *LEAFY* (*LFY*) and *APETALA1* (*AP1*), which control the formation of floral meristems (Weigel et al. 1992; Mandel and Yanofsky 1995).

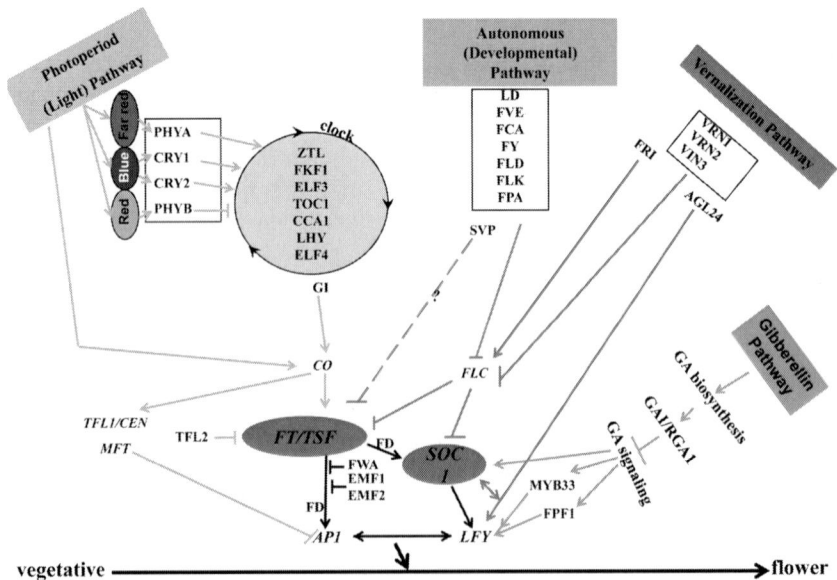

**Figure 11-1** A simplified genetic network that controls flowering time in the annual plant *Arabidopsis*. FT/TSF and SOC1 are floral integrators. Arrows indicate promotion and bars indicate repression.

*Color image of this figure appears in the color plate section at the end of the book.*

### 11.3.1.1 *Photoperiodic Pathway*

Duration of light period or photoperiod is one of the important environmental factors that control flowering in temperate plant species (Bernier et al. 1993;

Martinez-Garcia et al. 2002; Mouradov et al. 2002). Light signal is perceived by leaves and transported as a systemic signal or "florigen" to the shoot apex where floral development is induced (Knott 1934; Zeevaart 1976; Bernier and Perilleux 2005; Corbesier and Coupland 2005). *Arabidopsis* is a facultative long-day plant. Photoreceptors such as phytochromes and cryptochromes are involved in perception of light and mediate light input to the circadian clock (Goto et al. 1991; Johnson et al. 1994; Guo et al. 1998; Somers et al. 1998; Devlin and Kay 2000; Lin 2000). The far-red light sensor *PHYA* promotes flowering, but the red-light sensor *PHYB* inhibits flowering (Reed et al. 1993). PHYB is involved in degradation of CONSTANS (CO) protein early in the day (Valverde et al. 2004). Cryptochromes, CRY1 and CRY2, are blue light photoreceptors and encode flavoproteins (Lin et al. 1998; Cashmore et al. 1999). CRY2 is the main photoreceptor mediating day-length and flowering responses, perhaps by inhibiting *PHYB* signaling (Guo et al. 1998; Mockler et al. 1999; Mas et al. 2000).

The response of photoreceptors is integrated with clock entrainment factors such as *ZTL*, *FKF1*, and *ELF3* (Hicks et al. 1996; Zagotta et al. 1996; Somers et al. 2000). This results in the coordinated expression of the circadian-regulated genes such as *TOC1*, *CCA1*, *LHY*, and *ELF4*, which are the central components of the clock (Schaffer et al. 1998; Somers et al. 1998; Wang and Tobin 1998; Strayer et al. 2000). The clock then exerts its control of photoperiodic response by setting the rhythm of the flowering time genes *GIGANTEA* (*GI*) and *CO* (Putterill et al. 1995; Fowler et al. 1999; Park et al. 1999; Suarez-Lopez et al. 2001; Yanovsky and Kay 2002; Mizoguchi et al. 2005). Regulation of *CO* expression and activity is important for photoperiodic flowering. *Arabidopsis co* mutants are late flowering under long days, but flower at a similar time to wild-type under short days. Thus, *CO* promotes flowering under long days. High *CO mRNA* levels coincide with light in long days, but are largely confined to darkness in short days (Suarez-Lopez et al. 2001; Roden et al. 2002; Yanovsky and Kay 2002). Consequently, CO protein may not accumulate in darkness. Direct light activation of the encoded protein of *CO* also influences CO abundance or activity (Suarez-Lopez et al. 2001; Yanovsky and Kay 2002). CO protein is degraded in darkness, but light stabilizes it in the evening through cryptochromes and PHYA (Valverde et al. 2004). The promotion of flowering by *CO* requires *FLOWERING LOCUS T* (*FT*) and *SUPPRESSOR OF OVEREXPRESSION OF CONSTANS* (*SOC1*), previously described as AGL20) (Putterill et al. 1995; Borner et al. 2000; Lee et al. 2000; Onouchi et al. 2000; Samach et al. 2000; Wigge et al. 2005; Yoo et al. 2005).

The *FT* gene is activated by CO only under long days at the end of the day and promotes the transition from vegetative to reproductive phase (Kardailsky et al. 1999; Kobayashi et al. 1999; Samach et al. 2000; Suarez-Lopez et al. 2001). *FT* encodes a protein with similarity to mammalian phosphatidylethanolamine binding proteins, indicating that *FT* plays a role in signaling (Kardailsky et al. 1999; Kobayashi et al. 1999). CO protein activates *FT* in the leaf phloem companion cells (Takada and Goto 2003; An et al. 2004; Ayre and Turgeon 2004). The FT protein then moves out of the phloem to the SAM where floral development is induced (Corbesier et al. 2007; Jaeger and Wigge 2007; Mathieu et al. 2007). In the nucleus of the SAM, FT forms a complex with FD (bZIP transcription factor), which upregulates the MADS-box transcription factor *AP1* to induce floral development (Abe et al. 2005; Wigge et al. 2005).

*TWIN SISTER of FT* (*TSF*) is the closest homolog of *FT* in *Arabidopsis* (82% amino acid similarity), and perhaps being products of a duplication event. *TSF* acts redundantly with *FT* in the same molecular pathway (Yamaguchi et al. 2005) and both are involved in flower induction, show similar patterns of mRNA diurnal oscillation, and respond to long-day photoperiods (Kardailsky et al. 1999; Kobayashi et al. 1999; Suarez-Lopez et al. 2001; Yanovsky and Kay 2002; Yamaguchi et al. 2005). However, *TSF* and *FT* do not appear to affect each other's transcription.

Although *TERMINAL FLOWER1* (*TFL1*) is closely related to *FT*, it determines the potential for continuous growth of the shoot apex, prolonging the vegetative stage (Alvarez et al. 1992; Bradley et al. 1997). Loss-of-function mutation in *TFL1* promotes earlier flowering, whereas constitutive overexpression ($Pro_{35S}$:*TFL1*) delays flowering under long days with a prolonged vegetative stage (Ohshima et al. 1997; Ratcliffe et al. 1998). CO upregulates *TFL1* in the inflorescence meristem in the center of the shoot apex (Simon et al. 1996). However, the TFL protein also moves into other parts of the meristem (Conti and Bradley 2007). Given that CO activates both *FT* and *TFL1*, and that both genes are highly similar, the function of *TFL1* may be to compete with *FT* in the shoot apex to prevent the conversion of the apex into a source of floral meristems (Ahn et al. 2006). This might occur via competitive binding of FT and TFL1 to FD.

*TFL1* inhibits the activity of meristem identity genes *LFY* or *AP1* at the center of the shoot apex by delaying their upregulation and preventing the meristem from responding to *LFY* or *AP1* (Shannon and Meeks-Wagner 1991; Alvarez et al. 1992; Weigel et al. 1992; Bradley et al. 1997; Ratcliffe et al. 1998, 1999). In contrast, *LFY* and *AP1* prevent *TFL1* transcription in floral meristems on the apex periphery. *TFL2* represses *CO*-dependent activation of *FT* to restrict flowering in response to transient changes in *CO* activity if the long-day signal has not yet been perceived. The *SOC1* gene encodes a MADS-box protein and integrates the photoperiodic, autonomous, vernalization, and gibberellin pathways (Borner et al. 2000; Lee et al. 2000; Samach et al. 2000).

## 11.3.1.2 Autonomous (Developmental) Pathway

The autonomous pathway mediates flowering by monitoring the developmental stages of the plant. The homeodomain protein *LUMINIDEPENDENS* (LD) promotes flowering by reducing the levels of the floral repressor and MADS-box transcription factor *FLOWERING LOCUS C* (*FLC*) (Lee et al. 1994; Michaels and Amasino 1999). Other genes in the developmental pathway that primarily target *FLC* and that positively regulate flowering include *FVE, FCA, FY, FPA, FLOWERING LOCUS D* (*FLD*), and *FLOWERING LATE KH MOTIF* (*FLK*). *FCA, FPA,* and *FLK* are all RNA binding proteins (Macknight et al. 1997; Schomburg et al. 2001), whereas FY is a polyadenylation factor (Simpson et al. 2003; Lim et al. 2004; Mockler et al. 2004; Henderson et al. 2005; Metzger et al. 2005). *FCA* and *FY* regulate RNA processing of *FLC* (Simpson et al. 2003). FLD and FVE might play a role in histone deacetylation, because *FLD* is similar to the lysine-specific histone demethylase LSD1 (He et al. 2003; Ausin et al. 2004; Shi et al. 2004).

## 11.3.1.3 Vernalization Pathway

The vernalization pathway mediates low temperature signals that alter gene expression and induce flowering by reducing the levels of the floral repressor *FLC* (Michaels and Amasino 1999; Sheldon et al. 1999; Sheldon et al. 2000, 2002; Bastow et al. 2004; Searle et al. 2006). *FLC* appears to act at the shoot apex and in leaves to delay flowering, and is downregulated by *VIN3, VRN1,* and *VRN2* (Gendall et al. 2001; Levy et al. 2002; Sung and Amasino 2004). Conversely, *FRIGIDA (FRI)* upregulates *FLC*, which in turn delays flowering by reducing the expression of *FT* (Michaels and Amasino 2001). FLC protein directly binds to the regulatory regions of *FT* and *SOC1* prior to vernalization (Searle et al. 2006). This interaction appears to inhibit the formation of the systemic signal that is required to activate *SOC1*, which initiates the switch from vegetative to floral development. These observations indicate that flowering signals from vernalization and photoperiod pathways are integrated through the regulation of *FT* and *SOC1*.

## 11.3.1.4 Gibberellin Pathway

*Arabidopsis* eventually flowers under non-inductive short days, despite an absence of *FT* signaling. Genetic studies indicate that gibberellins (GA) control flowering under short days, therefore compensating for the absence of *FT* signaling. For example, the *GA1* gene is involved in GA biosynthesis, and a mutation in this gene (*ga1-3*) results in plants that are severely dwarfed, unable to flower under short days, and strongly enhances the *co2* mutation under long days because the *co2 ga1-3* double mutant never flowers (Wilson et al. 1992; Reeves and Coupland 2001). *ga1-3* mutants carry a deletion of the gene encoding the enzyme *ent*-copalyl diphosphate synthase (formerly *ent*-kaurene synthetase A) that catalyzes the first step in GA biosynthesis (Sun and Kamiya 1994). Overexpression of *LFY* and *SOC1* restores flowering of the *ga1-3* mutants under short days (Blazquez et al. 1998; Moon et al. 2003). This suggests that GAs promote flowering in *Arabidopsis* through a pathway that controls *LFY* and *SOC1* transcription (Blazquez et al. 1998). The *LFY* promoter contains *cis*-elements (e.g., the 8-base-pair CAACTGTC motif) involved in GA$_3$ response (Balazquez and Weigel 2000). *GAMYB*-like genes (e.g., *AtMYB33*) bind to the *LFY* promoter.

   *GIBBERELLIC ACID INSENSITIVE (GAI), REPRESSOR OF GA1-3 (RGA), RGA-LIKE 1 (RGL1), RGL2,* and *SPINDLY (SPY)* negatively regulate the GA signaling pathway and play a role in control of flowering (Jacobsen and Olszewski 1993; Dill and Sun 2001; Cheng et al. 2004; Tyler et al. 2004). RGL1 is predicted to function in repressing GA responses in the inflorescence, given that in the absence of the DELLA domain of RGL1, sepals, petals, and stamens are underdeveloped and the flowers are male sterile (Wen and Chang 2002). The DELLA domain is a conserved sequence near the N-termini of RGA, GAI, and RGL1, and plays a role in GA response (Wen and Chang 2002). If the DELLA domain is removed, GAI is insensitive to GA (Peng et al. 1997). This causes repression of shoot growth and flowering in the presence of GA. The *spy* mutant shows an early flowering phenotype (Jacobsen and Olszewski 1993), possibly because of the increased activity in the GA signaling pathway. The *SPY* gene is highly similar to Ser/Thr O-linked N-acetylglucosamine transferases in rats and humans (Olszewski et al. 2002). This suggests that SPY may play a role in post-translational modification of unknown downstream proteins.

### 11.3.2 Poplar

The molecular basis of "first-time" and "seasonal" reproduction is poorly understood in poplar. Using the protein sequences of *Arabidopsis* flowering-time genes, a search in the poplar genome database was conducted (*http://genome.jgi-psf.org/Poptr1_1/Poptr1_1.home.html*). Each of these *Arabidopsis* genes was found to have at least one corresponding poplar homolog (Table 11-1) and in some cases many poplar homologs. For example, some transcription factor homologs such as *FWA*, *GAI/RGA1*, *CO*, and MADS-box proteins consist of large families of genes in poplar. To help resolve the phylogenetic relationships between well characterized flowering control

**Table 11-1** Genes involved in transition to flowering in *Arabidopsis thaliana* (At) and their closest homologs in *Populus trichocarpa* (Pt).

| Annotation | At (gene ID) | Pt (protein ID) | Pt (gene ID) |
|---|---|---|---|
| PHYTOCHROME A (PHYA) | At1g09570 | 729311 | estExt_Genewise1_v1.C_LG_XIII0395 |
| PHYTOCHROME B (PHYB) | At2g18790 | 832686 | estExt_fgenesh4_pm.C_LG_VIII0434 |
| | | 1091155 | estExt_Genewise1Plus.C_LG_X3762 |
| CRYPTOCHROME 1 (CYR1) | At4g08920 | 559103 | eugene3.00050718 |
| | | 830225 | estExt_fgenesh4_pm.C_LG_II0442 |
| CRYPTOCHROME 2 (CYR2) | At1g04400 | 1119034 | estExt_Genewise1Plus.C_2730024 |
| | | 803751 | fgenesh4_pm.C_LG_VIII000706 |
| ZEITLUPE (ZTL) | At5g57360 | 809263 | fgenesh4_pm.C_LG_XVIII000281 |
| | | 580505 | eugene3.01210044 |
| FLAVIN-BINDING KELCH DOMAIN F BOX PROTEIN (FKF1) | At1g68050 | 822050 | estExt_fgenesh4_pg.C_LG_X0958 |
| | | 564672 | eugene3.00081267 |
| ELF3 | At2g25920 | 1099051 | estExt_Genewise1Plus.C_LG_XIV1951 |
| | | 409304 | gw1.II.639.1 |
| PSEUDO-RESPONSE REGULATOR 1 (TOC1) | At5g61380 (TOC1; APRR1) | 784463 | fgenesh4_pg.C_scaffold_129000038 |
| | At5g60100 (APRR3) | 824063 | estExt_fgenesh4_pg.C_LG_XIV0468 |
| | At5g24470 (APRR5) | 832516 | estExt_fgenesh4_pm.C_LG_VIII0151 |

*Table 11-1 contd....*

*Table 11-1 contd....*

| Annotation | At (gene ID) | Pt (protein ID) | Pt (gene ID) |
|---|---|---|---|
| | At2g46790 (APRR9) | 755476 | fgenesh4-pg.C_LG_II001656 |
| | At5g02810 (APRR7) | 227771 | gw1.X.2468.1 |
| | At4g31920 (ARR10) | 422771 | gw1.XII.1231.1 |
| | At3g16857 (ARR1) | 574629 | eugene3.00150024 |
| | | 725513 | estExt_Genewise1_v1.C_LG_X3573 |
| | | 231691 | gw1.X.6388.1 |
| | | 419669 | gw1.VIII.1097.1 |
| | | 825773 | estExt_fgenesh4_pg.C_LG_XVIII0466 |
| | | 732723 | estExt_Genewise1_v1.C_LG_XV0053 |
| | | 575747 | eugene3.00151142 |
| | | 230318 | gw1.X.5015.1 |
| | | 419184 | gw1.VIII.612.1 |
| | | 763573 | fgenesh4-pg.C_LG_VI001883 |
| | | 415998 | gw1.VI.371.1 |
| | | 262782 | gw1.XVIII.3323.1 |
| CIRCADIAN CLOCK ASSOCIATED 1 (CCA1) | At2g46830 (CCA1) | 552368 | eugene3.00021683 |
| LATE ELONGATED HYPOCOTYL (LHY) | At1g01060 (LHY) | 731468 | estExt_Genewise1_v1.C_LG_XIV1950 |
| | | 552367 | eugene3.00021682 |
| | | 198884 | gw1.IV.3973.1 |
| | | 569412 | eugene3.0012065 |
| | | 784079 | fgenesh4-pg.C_scaffold_122000043 |
| | | 256693 | gw1.XVI.2632.1 |

| | | | |
|---|---|---|---|
| GIGANTEA (GI) | At1g22770 | 289426 | gw1.44.639.1 |
| | | 281820 | gw1.28.394.1 |
| | | 409823 | gw1.II.1158.1 |
| | | 572303 | eugene3.00140348 |
| | | 198916 | gw1.IV.4005.1 |
| | | 818606 | estExt_fgenesh4-pg.C_LG_V1131 |
| | | 551288 | eugene3.00020603 |
| FLOWERING LOCUS T (FT) | At1g65480 (FT) | 582519 | eugene3.14090001 |
| | At4g20370 (TSF) | 765657 | fgenesh4-pg.C_LG_VIII000671 |
| | At1g18100 (MFT) | 805395 | fgenesh4-pm.C_LG_X000701 |
| | | 790700 | fgenesh4-pg.C_scaffold_1444000001 |
| | | 775913 | fgenesh4-pg.C_LG_XV000341 |
| | | 573457 | eugene3.00141502 |
| TERMINAL FLOWER 1 (TFL1) | At5g03840 | 827246 | estExt_fgenesh4-pg.C_660171 |
| | At2g2755 (ATC) | 648937 | grail3.0001004901 |
| | At5g62040 (BFT) | 575797 | eugene3.00151192 |
| TERMINAL FLOWER 1 (TFL2) | At5g17690 | 738591 | estExt_Genewise1_v1.C_LG_XIX1329 |
| | | 571324 | eugene3.00130688 |
| FD | At4g35900 | 424191 | gw1.57.158.1 |
| | At2g17770 | 818828 | estExt_fgenesh4-pg.C_LG_V1569 |
| | | 642918 | grail3.0003013801 |
| SUPPRESSION OF OVEREXPRESSION OF CONSTANS 1 (SOC1) | At2g456609 (SOC1) | 730942 | estExt_Genewise1_v1.C_LG_XIV0937 |

*Table 11-1 contd....*

*Table 11-1 contd....*

| Annotation | At (gene ID) | Pt (protein ID) | Pt (gene ID) |
|---|---|---|---|
|  | At5g62165 (AGL42) | 644373 | grail3.0033033502 |
|  | At4g11880 (AGL14) | 640573 | grail3.0008056401 |
|  | At4g22950 (AGL19) | 554289 | eugene3.00030922 |
|  | At5g51870 (AGL71) |  |  |
|  | At5g51860 (AGL72) |  |  |
| LUMINIDEPENDENS (LD) |  |  |  |
| FVE | At4g02560 | 572730 | eugene3.00140775 |
|  | At2g19520 | 592758 | eugene3.00440093 |
|  | At4g29730 | 271079 | gw1.145.113.1 |
|  | At2g16780 | 589194 | eugene3.02850001 |
|  | At4g35050 | 200694 | gw1.IX.1159.1 |
|  | At5g58230 | 818199 | estExt_fgenesh4_pg.C_LG_IV1464 |
|  | At2g19540 | 653315 | grail3.0112007001 |
|  |  | 816770 | estExt_fgenesh4_pg.C_LG_II1945 |
|  |  | 824422 | estExt_fgenesh4_pg.C_LG_XIV1179 |
|  |  | 207387 | gw1.V.2788.1 |
|  |  | 217471 | gw1.VII.1776.1 |
| FCA | At4g16280 | 225956 | gw1.X.653.1 |
|  | At2g47310 | 755673 | fgenesh4_pg.C_LG_II001853 |
|  |  | 799273 | fgenesh4_pm.C_LG_II000962 |
|  |  | 246020 | gw1.XIV.2763.1 |
|  |  | 572009 | eugene3.00140054 |
|  |  | 829855 | estExt_fgenesh4_pm.C_LG_I1075 |
| FY | At5g13480 | 178488 | gw1.I.7088.1 |

| | | | |
|---|---|---|---|
| | | 415574 | gw1.III.2677.1 |
| | | 725686 | estExt_Genewise1_v1.C_LG_X3939 |
| | | 835833 | estExt_fgenesh4_pm.C_LG_XVIII0164 |
| | | 561983 | eugene3.00061942 |
| | | 803047 | fgenesh4_pm.C_LG_VIII000002 |
| FPA | At2g43410 | 292764 | gw1.64.416.1 |
| | At4g12640 | 802569 | fgenesh4_pm.C_LG_VII000055 |
| | At1g27750 | 247179 | gw1.XIV.3922.1 |
| | | 243881 | gw1.XIV.624.1 |
| SHORT VEGETATIVE PHASE (SVP) | At2g22540 (SVP) | 719775 | estExt_Genewise1_v1.C_LG_VII4001 |
| MADS-box protein (AGL24) | At4g24540 (AGL24) | 643871 | grail3.0003092401 |
| | | 651191 | grail3.0083000102 |
| | | 871682 | e_gw1.VII.1214.1 |
| | | 838041 | APOLLO_200 |
| | | 763951 | fgenesh4_pg.C_LG_VII000296 |
| | | 258177 | gw1.XVII.168.1 |
| | | 778182 | fgenesh4_pg.C_LG_XVII000061 |
| REDUCED VERNALIZATION RESPONSE 1 (VRN1) | At3g18990 | 267709 | gw1.130.31.1 |
| | At4g33280 | 267711 | gw1.130.33.1 |
| | | 791723 | fgenesh4_pg.C_scaffold_2789000001 |
| | | 649491 | grail3.0001051501 |
| | | 784500 | fgenesh4_pg.C_scaffold_130000008 |
| | | 744205 | estExt_Genewise1_v1.C_1300033 |

*Table 11-1 contd.....*

*Table 11-1 contd....*

| Annotation | At (gene ID) | Pt (protein ID) | Pt (gene ID) |
|---|---|---|---|
| | | 770235 | fgenesh4_pg.C_LG_X001725 |
| | | 783236 | fgenesh4_pg.C_scaffold_8600096 |
| | | 591695 | eugene3.00400040 |
| | | 581458 | eugene3.01300012 |
| | | 764626 | fgenesh4_pg.C_LG_VII000971 |
| | | 779701 | fgenesh4_pg.C_LG_XVIII001069 |
| | | 583676 | eugene3.01520080 |
| | | 286795 | gw1.40.555.1 |
| | | 755009 | fgenesh4_pg.C_LG_II001189 |
| | | 781471 | fgenesh4_pg.C_scaffold_40000221 |
| | | 551891 | eugene3.00021206 |
| | | 551892 | eugene3.00021207 |
| REDUCED VERNALIZATION RESPONSE 2 (VRN2) | At4g16845 (VRN2) | 830729 | estExt_fgenesh4_pm.C_LG_III0191 |
| | At5g51230 (EMF2) | 814985 | estExt_fgenesh4_pg.C_LG_I0694 |
| | At2g35670 (FIS2) | 412944 | gw1.III.47.1 |
| | At4g16810 | 554592 | eugene3.00031225 |
| | At4g16807 | | |
| FRIGIDA (FRI) | At4g00650 | 253112 | gw1.XV.2548.1 |
| | | 710332 | estExt_Genewise1_v1.C_LG_II1377 |
| | | 664960 | grail3.0012039001 |
| | | 552423 | eugene3.00021738 |
| FLOWERING LOCUS C (FLC) | At5g10140 (FLC) | 647039 | grail3.0047013603 |
| | At1g77080 (FLM) | 680171 | grail3.0690000103 |

| | | | |
|---|---|---|---|
| | At5g65050 (AGL31) | 647042 | grail3.0047013701 |
| | At5g65060 (AGL70) | 840315 | e_gw1.I.7774.1 |
| | At5g65070 (AGL69) | 288710 | gw1.4342.4.1 |
| | At5g65080 (AGL68) | 294451 | gw1.690.3.1 |
| FWA | At4g25530 (FWA) | 774292 | fgenesh4_pg.C_LG_XIV000202 |
| | At5g52170 (HDG7) | 552123 | eugene3.00021438 |
| | At3g61150 (HDG1) | 575800 | eugene3.00151195 |
| | At4g00730 (ANL2) | 773381 | fgenesh4_pg.C_LG_XII001156 |
| | | 591580 | eugene3.04030001 |
| EMBRYONIC FLOWER 1 (EMF1) | At5g11530 | 763364 | fgenesh4_pg.C_LG_VI001674 |
| | | 825684 | estExt_fgenesh4-pg.C_LG_XVIII0291 |
| EMBRYONIC FLOWER 2 (EMF2) | At5g51230 (EMF2) | 830729 | estExt_fgenesh4_pm.C_LG_III0191 |
| | At4g16845 (VRN2) | 814985 | estExt_fgenesh4-pg.C_LG_I0694 |
| | At2g35670 (FIS2) | 412944 | gw1.III.47.1 |
| | At4g16810 | 554592 | eugene3.00031225 |
| | At4g16807 | | |
| GA INSENSITIVE (GAI) | At1g14920 (GAI) | 803616 | fgenesh4_pm.C_LG_VIII00571 |
| REPRESSOR OF GA1 (RGA1) | At2g01570 (RGA1) | 726855 | estExt_Genewise1_v1.C_LG_X6481 |
| | At1g66350 (RGL1) | 411207 | gw1.II.2542.1 |
| | At3g03450 (RGL2) | 831148 | estExt_fgenesh4_pm.C_LG_IV0241 |
| | At5g17490 (RGL3) | | |
| APETALA 1 (AP1) | At1g69120 (AP1) | 719996 | estExt_Genewise1_v1.C_LG_VIII0399 |
| | At1g26310 (CAL) | 659188 | estExt_Genewise1_v1.C_LG_VIII0399 |

*Table 11-1 contd....*

*Table 11-1 contd....*

| Annotation | At (gene ID) | Pt (protein ID) | Pt (gene ID) |
|---|---|---|---|
| | At5g60910 (FUL) | 661810 | grail3.0042013901 |
| | At3g30260 (AGL79) | 1076378 | estExt_Genewise1Plus.C_LG_IV3240 |
| | | 745710 | estExt_Genewise1_v1.C_1550090 |
| LEAFY (LFY) | At5g61850 | 835248 | estExt_fgenesh4_pm.C_LG_XV0337 |
| CONSTANS (CO) | At5g15840 (CO) | 266027 | gw1.123.49.1 |
| | At5g15850 (COL1) | 831202 | estExt_fgenesh4_pm.C_LG_IV0339 |
| | At3g02380 (COL2) | | |

genes and their poplar homologs, we conducted phylogenetic analyses. Based on the Neighbor-Joining method (Saitou and Nei 1987), three genes closely cluster with *FWA* in *Arabidopsis*, but five genes are present in the same clade with *FWA* in poplar (Fig. 11-2). RGA1 and GAI belong to the GRAS family of transcription factors and cluster with three other RGL proteins in *Arabidopsis*, whereas there are four DELLA domain poplar proteins in this cluster (Fig. 11-3). Interestingly, a group of six poplar proteins form a sister group to the DELLA protein group, but lack a DELLA domain. A total of 16 CO-like (COL) proteins (including CO) are present in *Arabidopsis*, and COL1 and COL2 closely cluster with CO (Fig. 11-3). Two zinc finger-containing proteins in poplar show high similarity to the *Arabidopsis* CO protein in the same clade (Yuceer et al. 2002; Fig. 11-4). An analysis of the evolutionary relationship among the MADS-box proteins in poplar (Leseberg et al. 2006) showed that many poplar gene families have expanded due in part to gene duplications occurring after the divergence of *Arabidopsis* and poplar (Tuskan et al. 2006). The closest poplar homologs of *Arabidopsis* FLC, SVP, SOC1, and AP1 proteins are individually grouped in Fig. 11-5.

### 11.3.2.1 LFY May Play a Role in Poplar Flowering Time

Overexpression of the *Arabidopsis LFY* gene regulated by the CaMV 35S promoter (*Pro$_{35S}$:LFY*) caused early flowering in a male poplar (*P. tremula* x *P. tremuloides*) (Weigel and Nilsson 1995). However, *Pro$_{35S}$:LFY* did not consistently produce early flowering in other poplar genotypes (Rottmann et al. 2000), nor did it produce normal inflorescences and viable gametes. Four of seven lines flowered within six months, but flowering was observed primarily in males (*P. tremula* x *P. tremuloides*). Only two of 19 lines of a female poplar clone (*P. tremula* x *P. alba*) transformed with this construct flowered, doing so after two years of growth. Single flowers in these lines also formed anthers, suggesting that *LFY* may promote male flowering in poplar. A *LFY-like* gene, *PTLF*, is the only copy of a gene with substantive resemblance to *LFY* in the poplar genome (Rottmann et al. 2000). A construct with the native poplar *LFY* homolog under the 35S promoter (*Pro$_{35S}$:PTLF*) did not cause early flowering in the female clone, and only one of 16 transformed males produced unitary flowers without evidence of viable pollen production (Rottmann et al. 2000).

### 11.3.2.2 FT1 and FT2 Control Flowering Time

*FLOWERING LOCUS T1* and *T2* (*FT1* and *FT2*) are major players in "first-time" and "seasonal" reproduction in poplar, and their transcription is controlled by developmental and environmental factors

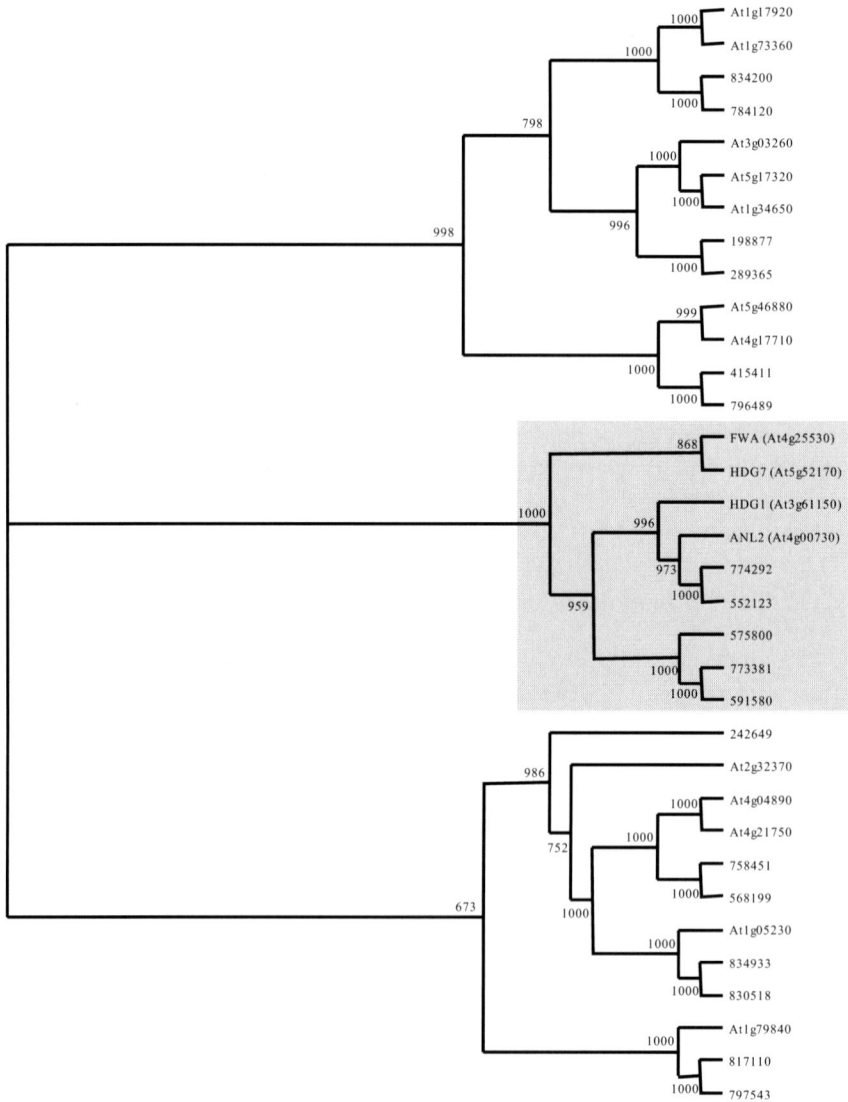

**Figure 11-2** Phylogenetic analysis of the FWA family proteins in *Arabidopsis thaliana* (At) and *Populus trichocarpa* using the Neighbor-Joining method. Gray shading indicates the clade with close homologs of FWA. Bootstrap analysis was conducted to estimate nodal support using 1,000 replicates.

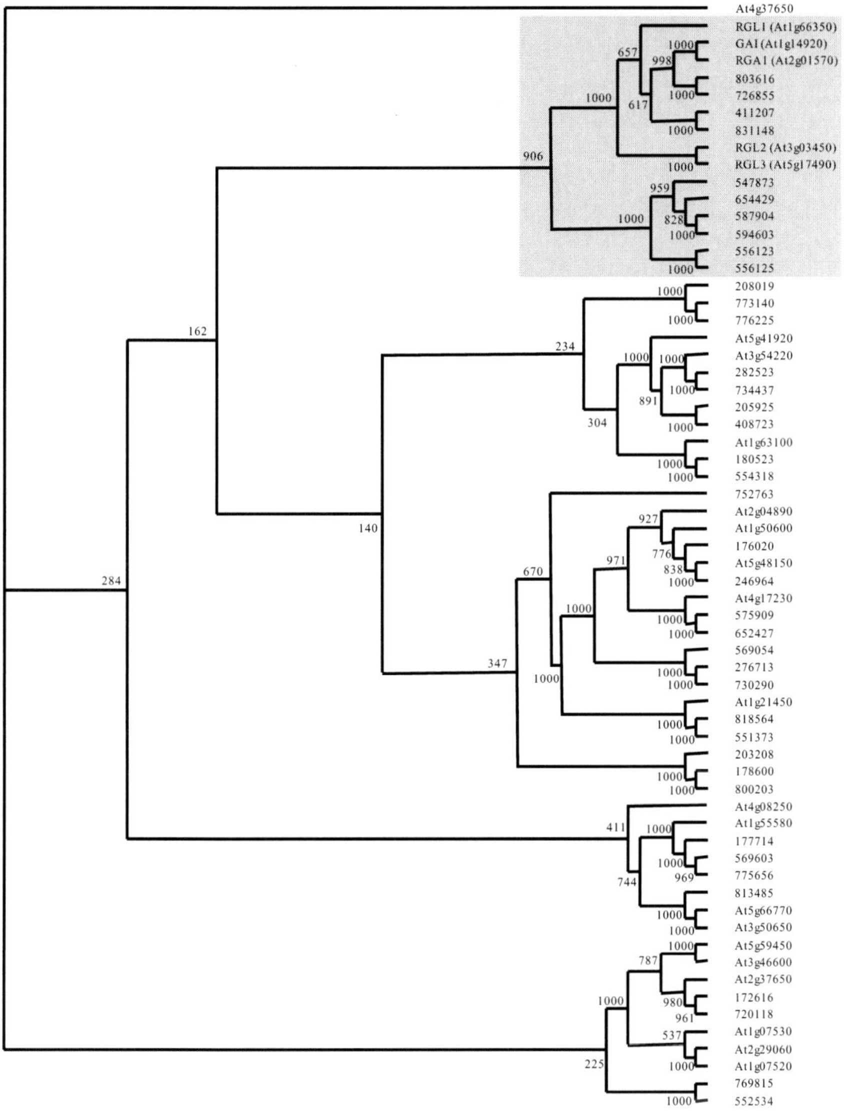

**Figure 11-3** Phylogenetic analysis of GRAS family proteins in *Arabidopsis thaliana* (At) and *Populus trichocarpa* using the Neighbor-Joining method. Gray shading indicates the clade with close homologs of GAI/RGA1. Bootstrap analysis was conducted to estimate nodal support using 1,000 replicates.

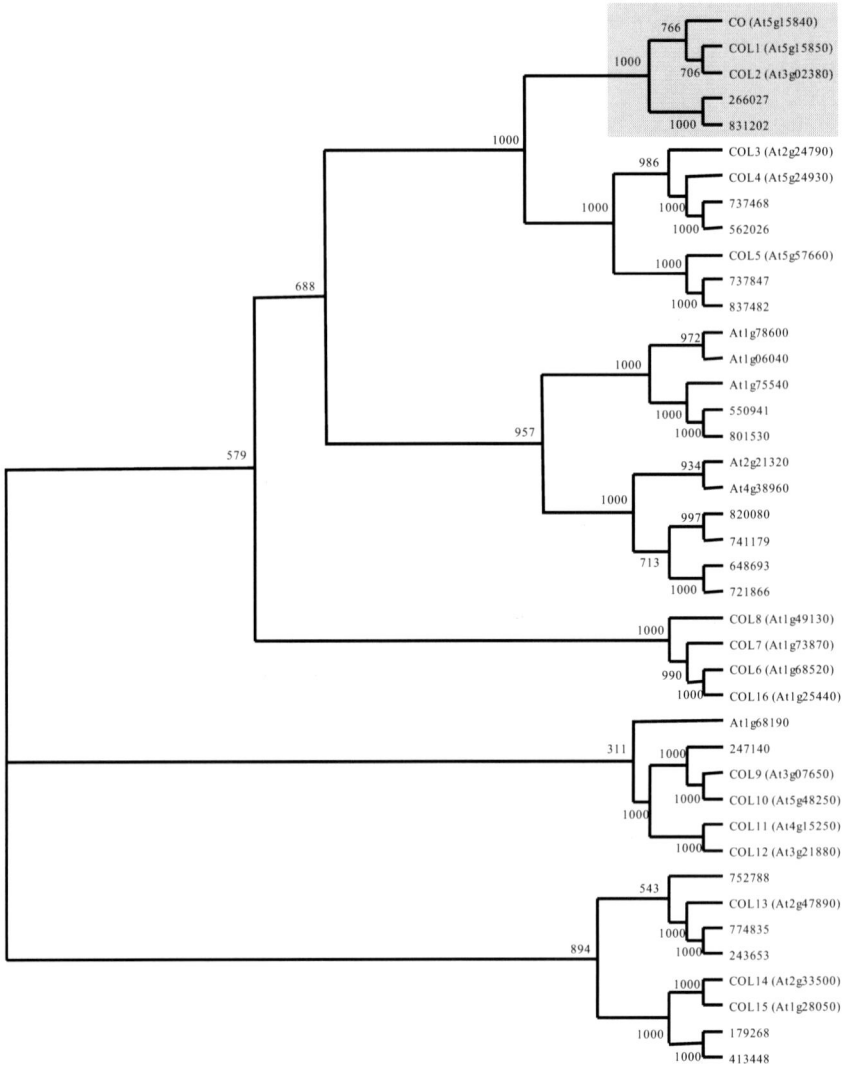

**Figure 11-4** Phylogenetic analysis of the CONSTANS (CO) family proteins in *Arabidopsis thaliana* (At) and *Populus trichocarpa* using the Neighbor-Joining method. Gray shading indicates the clade with close homologs of CO. Bootstrap analysis was conducted to estimate nodal support using 1,000 replicates.

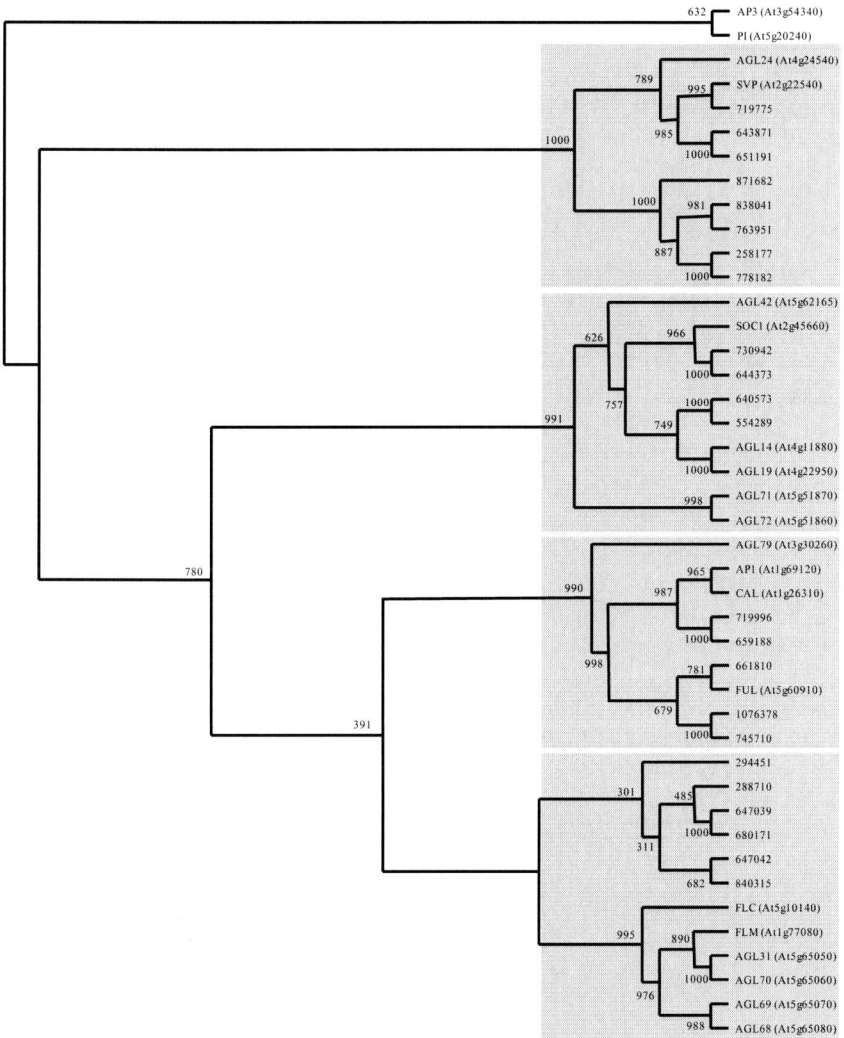

**Figure 11-5** Neighbor-Joining phylogenetic analysis of *Arabidopsis thaliana* (At) and *Populus trichocarpa* MADS domain family proteins involved in flowering. The clades with close homologs of FLC, SVP, SOC1, and AP1 subfamilies in both *Arabidopsis* and poplar are individually gray-boxed. Bootstrap analysis was conducted to estimate nodal support using 1,000 replicates.

(Bohlenius et al. 2006; Hsu et al. 2006). *FT1* and *FT2* are in the same gene family with 91% amino acid sequence similarity. *FT2* mRNA was detected at background levels in roots and the shoot apex (Hsu et al. 2006). However, its expression was most abundant in leaf 11 (from the base of the shoot) and in the bud in its axil that was destined to be reproductive, suggesting that *FT2* expression is upregulated in leaves and buds. The abundance of *FT2* transcripts in leaf 11 increased from the juvenile to reproductive developmental phases, suggesting that *FT2* might play a role in juvenile to mature transition. When $Pro_{35S}:FT1$ and $Pro_{35S}:FT2$ constructs were separately inserted into juvenile poplar, trees produced flowers within several months. The $Pro_{35S}:FT1$ trees were not, however, induced to enter dormancy under short days or cold temperatures such as in wild type trees (Bohlenius et al. 2006). This suggests that flowering and dormancy induction share common regulatory elements.

The abundance of *FT2* transcript in leaf 11 was low from February to April, but was high in mid-May (Hsu et al. 2006). During this time, leaves developed from a primordial preformed leaf to a fully expanded leaf. Beginning in mid-May, *FT2* transcript was abundant in bud 11 which formed an inflorescent shoot and floral meristems on its flanks. Potential factors involved in the increase of *FT2* transcript in leaves include temperature, development, and photoperiod. Poplar trees were treated under two temperature regimes (23°C and 38°C) to determine if this affected *FT2* transcript abundance. No change, however, was observed in the expression pattern of *FT2* under either temperature regime, suggesting that temperature is not a factor in the expression pattern of *FT2* (Hsu et al. 2006). When poplar trees were grown under long (14 hours) and short (8 hours) days for 14 days, *FT2* transcripts were abundant under long days throughout the experiment, whereas they were either at background levels (first 7 days) or undetectable after 14 days under short days (Hsu et al. 2006). These results suggest that long days promote the abundance of *FT2* transcript. The poplar genome contains at least two *FD* orthologs and all transgenic lines overexpressing *PtFD1* flowered when grown in a long day-length greenhouse, but flowering was not observed when transgenics were grown under short day-lengths (G. Coleman, pers. comm.). The *FT2* and *PtFD1* results suggest that long photoperiods promote floral bud formation in poplar.

Photoperiod controls many aspects of poplar growth and development including growth cessation and winter dormancy (Pauley and Perry 1954; Howe et al. 1995, 1996; Olsen et al. 1997). Reports indicate that photoperiod is a physiological stimulus that triggers flower bud initiation in woody perennial plants (Junttila 1980; Rivera and Borchert 2001). In the related species, *Salix pentandra*, flower bud formation was maximally promoted by photoperiods of 18 to 22 hours (Junttila 1980). However, detailed molecular studies have yet to be conducted to complete understanding of how photoperiod controls flowering in poplar. A major barrier has been the lack of naturally occurring early-flowering poplar genotypes that can be easily moved and studied in various controlled environments, but as the *FT* and *PtFD1* results show, use of early-flowering transgenics is likely to be useful in circumventing this problem.

Study of *PopCEN1*, a poplar homolog of snapdragon *CENTRORADIALIS (CEN)* and *TFL1* from *Arabidopsis* revealed a conserved role in repressing flowering (Mohamed et al. 2010). Downregulation of *PopCEN1* via RNAi did not induce the extreme early flowering seen in poplar transgenics overexpressing *FT1*, *FT2*, or *PtFD1*, but a multi-year field study revealed that suppression of *PopCEN1* did promote an earlier onset of flowering and a markedly increased number of lateral inflorescences.

These few examples show the power of transgenic manipulation in poplar and of comparative genomics, especially when whole genome sequences are available. Moreover, combining transgenesis, microarray expression analysis, protein-protein interaction studies, and other -omics approaches should reveal the transcription-based regulatory networks controlling flowering in poplar and thus, how these genes and pathways are modified to yield the dramatically different flowering habits of poplar and *Arabidopsis*.

## 11.4 Practical Applications

The rationale, projected benefits, and mechanisms for the manipulation of flowering via genetic engineering have been widely discussed, most recently in an extensive review by Brunner et al. (2007). They are: 1) Improved vegetative growth by removal or reduction of inflorescences, floral organs, and fruits as sinks for carbon and nutrients. The evidence that this could be substantial in some species and circumstances was discussed in depth by Strauss et al. (1995); 2) Containment of genes or exotic organisms by suppression of floral onset, floral organ or fruit function, or by transgene removal during gametogenesis. A very wide variety of options have been shown to work in *Arabidopsis*, tobacco, or other model annual plants. The only evidence that these kinds of genes can be effective in substantially reducing fertility in a field environment was presented by Brunner et al. (2007) with reference to poplars containing a gene for male sterility ($Pro_{TA29}$:*barnase*); Finally, 3) Acceleration of flowering to speed breeding or research has been a long sought goal in tree breeding, for which hormone treatments have been highly effective in conifers and some other woody species, but not in poplars (Meilan 1997). However, as discussed above, the transgenic approaches attempted to date have given unsatisfying results with respect to consistent production of viable gametes and seeds. The more normal appearance of catkins with *FT1* induction of flowering in poplar

(Bohlenius et al. 2006), and the graft-transmissibility of the *FT* signal protein, have inspired hope that transgenic rootstocks might be useful for inducing rapid flowering of grafted scions. The *FT*-associated inductive signal can be transmitted from leaves to shoot apical meristems, as demonstrated using intra- and inter-specific grafting experiments (Imaizumi and Kay 2006; Zeevaart 2006). Such a tactic could avoid the regulatory or environmental concerns of transgene deployment in production forests. However, graft induction of *FT*-associated flowering has yet to be demonstrated in woody species, and at least in Germany—where labeling of transgenic associated products is required by the EU whether transgenes persist or not—such a tactic would not be likely to obviate regulatory oversight of derived non-trangenic seeds and forests (M. Fladung, pers. comm.).

The process-based regulatory oversight of all transgenic products in the USA and most other countries, where transgenes are assumed to be dangerous until proven otherwise on a case by case basis, makes it extremely difficult to do the required field research evaluations to assess the level of fertility reduction, postponement, or precocious induction under conditions relevant to commercial forestry programs. This is because genetic dispersal of even minute amounts of as little as fertility-reducing genes is not permitted, yet it is very difficult to fully guarantee this during the course of multiple year research in large, flowering trees. Until there is substantial regulatory reform that takes into account the risks of specific classes of genes, as has been proposed earlier many times and in many ways (e.g., Hancock 2003; Strauss 2003a, b; Bradford 2005)—and is now under active consideration in the USA (USDA 2007)—research to develop practical applications for trees that have been genetically engineered for modified flowering characteristics will proceed very slowly and at great expense, if it can proceed at all.

## Acknowledgments

The authors thank the NSF Plant Genome Research Program, USDA-NRI Plant Biology Program, USDA Biotechnology Risk Assessment Grants Program, NSF Industry/University Centers, and the industrial members of the Tree Biosafety and Genomics Research Cooperative based at Oregon State University for supporting our research on flowering.

# References

Abe M, Kobayashi Y, Yamamoto S, Daimon Y, Yamaguchi A, Ikeda Y, Ichinoki H, Notaguchi M, Goto K, Araki T (2005) FD, a bZIP protein mediating signals from the floral pathway integrator FT at the shoot apex. Science 309: 1052–1056.

Ahn JH, Miller D, Winter DJ, Banfield MJ, Lee JH, Yoo SY, Henz SR, Brady RL, Weigel D (2006) A divergent external loop confers antagonistic activity on floral regulators FT and TFL1. EMBO J 25: 605–614.

Alvarez J, Guli CL, Yu X-H, Smyth DR (1992) *TERMINAL FLOWER*: A gene affecting inflorescence development in *Arabidopsis thaliana*. Plant J 2: 103–116.

An H, Roussot C, Suarez-Lopez P, Corbesier L, Vincent C, Pineiro M, Hepworth S, Mouradov A, Justin S, Turnbull C, Coupland G (2004) *CONSTANS* acts in the phloem to regulate a systemic signal that induces photoperiodic flowering of *Arabidopsis*. Development 131: 3615–3626.

Ausin I, Alonso-Blanco C, Jarillo JA, Ruiz-Garcia L, Martinez-Zapater JM (2004) Regulation of flowering time by FVE, a retinoblastoma-associated protein. Nat Genet 36: 162–166.

Ayre BG, Turgeon R (2004) Graft transmission of a floral stimulant derived from *CONSTANS*. Plant Physiol 135: 2271–2278.

Bastow R, Mylne JS, Lister C, Lippman Z, Martienssen RA, Dean C (2004) Vernalization requires epigenetic silencing of *FLC* by histone methylation. Nature 427: 164–167.

Bernier G, Perilleux C (2005) A physiological overview of the genetics of flowering time control. Plant Biotechnol J 3: 3–16.

Bernier G, Havelange A, Houssa C, Petitjean A, Lejeune P (1993) Physiological signals that induce flowering. Plant Cell 5: 1147–1155.

Blazquez MA, Weigel D (2000) Integration of floral inductive signals in *Arabidopsis*. Nature 404: 889–892.

Blazquez MA, Green R, Nilsson O, Sussman MR, Weigel D (1998) Gibberellins promote flowering of *Arabidopsis* by activating the *LEAFY* promoter. Plant Cell 10: 791–800.

Boes TK, Strauss SH (1994) Floral phenology and morphology of black cottonwood, *Populus trichocarpa* (*Salicaceae*). Am J Bot 81: 562–567.

Bohlenius H, Huang T, Charbonnel-Campaa L, Brunner AM, Jansson S, Strauss SH, Nilsson O (2006) *CO/FT* regulatory module controls timing of flowering and seasonal growth cessation in trees. Science 312: 1040–1043.

Borner R, Kampmann G, Chandler J, Gleissner R, Wisman E, Apel K, Melzer S (2000) A MADS domain gene involved in the transition to flowering in *Arabidopsis*. Plant J 24: 591–599.

Boss PK, Bastow RM, Mylne JS, Dean C (2004) Multiple pathways in the decision to flower: enabling, promoting, and resetting. Plant Cell 16 (suppl): S18–S31.

Braatne JH, Rood SB, Heilman PE (1996) Life history, ecology, and conservation of riparian cottonwoods in North America In: RF Stettler, HD Bradshaw, PE Heilman, TM Hincley (eds) Biology of *Populus*. NRC Research Press, Ottawa, Canada, pp 57–85.

Bradford K, Van Deynze A, Gutterson N, Parrott W, Strauss SH (2005) Regulating transgenic crops sensibly: lessons from plant breeding, biotechnology and genomics. Nat Biotechnol 23: 439–444.

Bradley D, Ratcliffe O, Vincent C, Carpenter R, Coen E (1997) Inflorescence commitment and architecture in *Arabidopsis*. Science 275: 80–83.

Brunner AM, Li J, DiFazio SP, Shevchenko O, Montgomery BE, Mohamed R, Wei H, Ma C, Elias AA, VanWormer K, Strauss SH (2007) Genetic containment of forest plantations. Tree Genet Genomes 3: 75–100.

Cashmore AR, Jarillo JA, Wu YJ, Dongmei L (1999) Cryptochromes, blue light photoreceptors for plants and animals. Science 284: 760–765.

Cheng H, Qin L, Lee S, Fu X, Richards DE, Cao D, Luo D, Harberd NP, Peng J (2004) Gibberellin regulates *Arabidopsis* floral development via suppression of DELLA protein function. Development 131: 1055–1064.

Conti L, Bradley D (2007) TERMINAL FLOWER1 is a mobile signal controlling *Arabidopsis* architecture. Plant Cell 19: 767–778.

Corbesier L, Coupland G (2005) Photoperiodic flowering of *Arabidopsis*: integrating genetic and physiological approaches to characterization of the floral stimulus. Plant Cell Environ 28: 54–66.

Corbesier L, Vincent C, Jang S, Fornara F, Fan Q, Searle I, Giakountis A, Farrona S, Gissot L, Turnbull C, Coupland G (2007) FT protein movement contributes to long-distance signaling in floral induction of *Arabidopsis*. Science 316: 1030–1033.

Critchfield WB (1960) Leaf dimorphism in *Populus trichocarpa*. Am J Bot 47: 699–711.

Cseke LJ, Cseke SB, Podila GP (2007) High efficiency poplar transformation. Plant Cell Rep 26: 1529–1538.

Devlin PF, Kay SA (2000) Cryptochromes are required for phytochrome signalling to the circadian clock but not for rhythmicity. Plant Cell 12: 2499–2509.

Dickson RE (1986) Carbon fixation and distribution in young *Populus* trees In: T Fujimori, D Whitehead (eds) Proceedings: Crown and Canopy Structure in Relation to Productivity. Forestry and Forest Products Research Institute, Ibaraki, Japan, pp 409–426.

Dill A, Sun T-P (2001) Synergistic derepression of gibberellin signaling by removing RGA and GAI function in *Arabidopsis thaliana*. Genetics 159: 777–785.

Fitter, AH, Fitter RS (2002) Rapid changes in flowering time in British plants. Science 296: 1689–1691.

Fowler S, Lee K, Onouchi H, Samach A, Richardson K, Morris B, Coupland G, Putterill J (1999) *GIGANTEA*: A circadian clock-controlled gene that regulates photoperiodic flowering in *Arabidopsis* and encodes a protein with several possible membrane-spanning domains. EMBO J 18: 4679–4688.

Gendall AR, Levy YY, Wilson A, Dean C (2001) The *VERNALIZATION 2* gene mediates the epigenetic regulation of vernalization in *Arabidopsis*. Cell 107: 525–535.

Goffinet MC, Larson PR (1981) Structural changes in *Populus deltoides* terminal buds and in the vascular transition zone of the stems during dormancy induction. Am J Bot 68: 118–129.

Goto N, Kumagai T, Koornneef M (1991) Flowering responses to light-break in photomorphogenic mutants of *Arabidopsis thaliana*, a long-day plant. Physiol Plant 83: 209–215.

Guo H, Yang H, Mockler TC, Lin C (1998) Regulation of flowering time by *Arabidopsis* photoreceptors. Science 279: 1360–1363.

Han K, Meilan R, Ma C, Strauss SH (2000) An *Agrobacterium tumefaciens* transformation protocol effective on a variety of cottonwood hybrids (genus *Populus*). Plant Cell Rep 19: 315–320.

Hancock JF (2003) A framework for assessing the risk of transgenic crops. BioScience 53: 512–519.

He Y, Michaels S, Amasino RM (2003) Regulation of flowering time by histone acetylation in *Arabidopsis*. Science 302: 1751–1754.

Henderson IR, Liu F, Drea S, Simpson GG, Dean C (2005) An allelic series reveals essential roles for FY in plant development in addition to flowering-time control. Development 132: 3597–3607.

Hicks KA, Millar AJ, Carre IA, Somers DE, Straume M, Meeks-Wagner DR, Kay SA (1996) Conditional circadian dysfunction of the *Arabidopsis early-flowering 3* mutant. Science 274: 790–792.

Howe GT, Hackett WP, Furnier GR, Klevorn RE (1995) Photoperiodic responses of a northern and southern ecotype of black cottonwood. Physiol Plant 93: 695–708.

Howe GT, Gardner G, Hackett WP, Furnier GR (1996) Phytochrome control of short-day-induced bud set in black cottonwood. Physiol Plant 97: 95–103.

Hsu C-Y, Liu Y, Luthe DS, Yuceer C (2006) Poplar *FT2* shortens the juvenile phase and promotes seasonal flowering. Plant Cell 18: 1846–1861.

Imaizumi T, Kay SA (2006) Photoperiodic control of flowering: not only by coincidence. Trends Plant Sci 11: 550–558.

Jacobsen SE, Olszewski NE (1993) Mutations at the *SPINDLY* locus of *Arabidopsis* alter gibberellin signal transduction. Plant Cell 5: 887–896.

Jaeger KE, Wigge PA (2007) FT protein acts as a long-range signal in *Arabidopsis*. Curr Biol 17: 1050–1054.

Johnson E, Bradley M, Harberd NP, Whitelam GC (1994) Photoresponse of light-grown phyA mutants of *Arabidopsis*. Plant Physiol 105: 141–149.

Junttila O (1980) Flower bud differentiation in *Salix pentandra* as affected by photoperiod, temperature, and growth regulators. Physiol Plant 49: 127–134.

Kardailsky I, Shukla VK, Ahn JH, Dagenais N, Christensen SK, Nguyen JT, Chory J, Harrison MJ, Weigel D (1999) Activation tagging of the floral inducer *FT*. Science 286: 1962–1965.

Knott JE (1934) Effect of a localized photoperiod on spinach. Proc Soc Hort Sci 31: 152–154.

Kobayashi Y, Kaya H, Goto K, Iwabuchi M, Araki T (1999) A pair of related genes with antagonistic roles in mediating flowering signals. Science 286: 1960–1962.

Larson PR (1975) Development and organization of the primary vascular system in *Populus deltoides* according to phyllotaxy. Am J Bot 62: 1084–1099.

Larson PR, Pizzolato TD (1977) Axillary bud development in *Populus deltoides*. I. Origin and early ontogeny. Am J Bot 64: 835–848.

Lee H, Suh S-S, Park E, Cho E, Ahn JH, Kim S-G, Lee JS, Kwon YM, Lee I (2000) The AGAMOUS-LIKE 20 MADS domain protein integrates floral inductive pathways in *Arabidopsis*. Genes Dev 14: 2366–2376.

Lee I, Aukerman MJ, Gore SL, Lohman KN, Michaels SD, Weaver LM, John MC, Feldmann KA, Amasino RM (1994) Isolation of *LUMINIDEPENDENS*: a gene involved in the control of flowering time in *Arabidopsis*. Plant Cell 6: 75–83.

Leseberg CH, Li A, Kang H, Duvall M, Mao L (2006) Genome-wide analysis of the MADS-box gene family in *Populus trichocarpa*. Gene 378: 84–94.

Levy YY, Dean C (1998) The transition to flowering. Plant Cell 10: 1973–1989.

Levy YY, Mesnage S, Mylne JS, Gendall AR, Dean C (2002) Multiple roles of *Arabidopsis VRN1* in vernalization and flowering time control. Science 297: 243–246.

Lim M-H, Kim J, Kim Y-S, Chung K-S, Seo Y-H, Lee I, Kim J, Hong CB, Kim H-J, Park C-M (2004) A new *Arabidopsis* gene, *FLK*, encodes an RNA binding protein with K homology motifs and regulates flowering via *FLOWERING LOCUS C*. Plant Cell 16: 731–740.

Lin C (2000) Photoreceptors and regulation of flowering time. Plant Physiol 123: 39–50.

Lin C, Yang H, Guo H, Mockler T, Chen J, Cashmore AR (1998) Enhancement of blue-light sensitivity of *Arabidopsis* seedlings by a blue light receptor cryptochrome 2. Proc Natl Acad Sci USA 95: 2686–2690.

Macknight R, Bancroft I, Page T, Lister C, Schmidt R, Love K, Westphal L, Murphy G, Sherson S, Cobbett C, Dean C (1997) *FCA*, a gene controlling flowering time in *Arabidopsis*, encodes a protein containing RNA-binding domains. Cell 89: 737–745.

Mandel MA, Yanofsky MF (1995) A gene triggering flower formation in *Arabidopsis*. Nature 377: 522–524.

Martinez-Garcia JF, Virgos-Soler A, Prat S (2002) Control of photoperiod regulated tuberization in potato by the *Arabidopsis* flowering-time gene *CONSTANS*. Proc Natl Acad Sci USA 99: 15211–15216.

Mas P, Devlin PF, Panda S, Kay SA (2000) Functional interaction of phytochrome B and crytochrome 2. Nature 408: 207–211.

Mathieu J, Warthmann N, Kuttner F, Schmid M (2007) Export of FT protein from phloem companion cells is sufficient for floral induction in *Arabidopsis*. Curr Biol 17: 1055–1060.

Meilan R (1997) Floral induction in woody angiosperms. New For 14: 179–202.

Metzger E, Wissmann M, Yin N, Muller JM, Schneider R, Peters AH, Gunther T, Buettner R, Schule R (2005) LSD1 demethylates repressive histone marks to promote androgenreceptor-dependent transcription. Nature 437: 436–439.

Michaels SD, Amasino RM (1999) *FLOWERING LOCUS C* encodes a novel MADS domain protein that acts as a repressor of flowering. Plant Cell 11: 949–956.

Michaels SD, Amasino RM (2001) Loss of *FLOWERING LOCUS C* activity eliminates the late-flowering phenotype of *FRIGIDA* and autonomous-pathway mutations, but not responsiveness to vernalization. Plant Cell 13: 935–941.

Mizoguchi T, Wright L, Fujiwara S, Cremer F, Lee K, Onouchi H, Mouradov A, Fowler S, Kamada H, Putterill J, Coupland G (2005) Distinct roles of *GIGANTEA* in promoting flowering and regulating circadian rhythms in *Arabidopsis*. Plant Cell 17: 2255–2270.

Mockler TC, Guo H, Yang H, Duong H, Lin C (1999) Antagonistic actions of *Arabidopsis* cryptochromes and phytochrome B in the regulation of flowering. Development 126: 2073–2082.

Mockler TC, Yu X, Shalitin D, Parikh D, Michael TP, Liou J, Huang J, Smith Z, Alonso JM, Ecker JR, Chory J, Lin C (2004) Regulation of flowering time in *Arabidopsis* by K homology domain proteins. Proc Natl Acad Sci USA 101: 12759–12764.

Mohamed, R, Wang C-T, Ma C, Shevchenko O, Dye SJ, Puzey JR, Etherington E, Sheng X, Meilan RSH, Strauss, Brunner AM (2010) *Populus CEN/TFL1* regulates first onset of flowering, axillary meristem identity and dormancy release in *Populus*. Plant J 62: 674–688.

Moon J, Suh SS, Lee H, Choi KR, Hong CB, Paek NC, Kim SG, Lee I (2003) The *SOC1* MADS-box gene integrates vernalization and gibberellin signals for flowering in *Arabidopsis*. Plant J 35: 613–623.

Mouradov A, Cremer F, Coupland G (2002) Control of flowering time: interacting pathways as a basis for diversity. Plant Cell 14 (suppl): S111–S130.

Ohshima S, Murata M, Sakamoto W, Ogura Y, Motoyoshi F (1997) Cloning and molecular analysis of the *Arabidopsis* gene *Terminal Flower 1*. Mol Gen Genet 254: 186–194.

Onouchi H, Igeno MI, Perilleux C, Graves K, Coupland G (2000) Mutagenesis of plants overexpressing *CONSTANS* demonstrates novel interactions among *Arabidopsis* flowering-time genes. Plant Cell 12: 885–900.

Olsen JE, Junttila O, Nilsen J, Eriksson ME, Martinussen L, Olsson O, Sandberg G, Moritz T (1997) Ectopic expression of oat phytochrome A in hybrid aspen changes critical daylength for growth and prevents cold acclimatization. Plant J 12: 1339–1350.

Olszewski N, Sun T-P, Gubler F (2002) Gibberellin signaling: Biosynthesis, catabolism, and response pathways. Plant Cell 14 (suppl): S61–S80.

Park DH, Somers DE, Kim YS, Choy YH, Lim HK, Soh MS, Kim HJ, Kay SA, Nam HG (1999) Control of circadian rhythms and photoperiodic regulation of flowering by the *Arabidopsis GIGANTEA* gene. Science 285: 1579–1582.

Pauley SS, Perry TO (1954) Ecotypic variation of the photoperiodic response in *Populus*. J Arnold Arbor 35: 167–188.

Peng J, Carol P, Richards DE, King KE, Cowling RJ, Murphy GP, Harberd NP (1997) The *Arabidopsis GAI* gene defines a signaling pathway that negatively regulates gibberellin responses. Genes Dev 11: 3194–3205.

Pizzolato TD, Larson PR (1977) Axillary bud development in *Populus deltoides* II Late ontogeny and vascularization. Am J Bot 64: 849–860.

Putterill J, Lee K, Simon R, Coupland G (1995) The *CONSTANS* genes of *Arabidopsis* promotes flowering and encodes a protein showing similarities to zinc finger transcription factors. Cell 80: 847–857.

Ratcliffe OJ, Amaya I, Vincent CA, Rothstein S, Carpenter R, Coen ES, Bradley DJ (1998) A common mechanism controls the life cycle and architecture of plants. Development 125: 1609–1615.

Ratcliffe OJ, Bradley DJ, Coen ES (1999) Separation of shoot and floral identity in *Arabidopsis*. Development 126: 1109–1120.

Reed JW, Nagpal P, Poole DS, Furuya M, Chory J (1993) Mutations in the gene for the red/ far-red light receptor phytochrome B alter cell elongation and physiological responses throughout *Arabidopsis* development. Plant Cell 5: 147–157.

Reeves PH, Coupland G (2001) Analysis of flowering time control in *Arabidopsis* by comparison of double and triple mutants. Plant Physiol 126: 1085–1091.

Rivera G, Borchert R (2001) Induction of flowering in tropical trees by a 30-min reduction in photoperiod: evidence from field observations and herbarium specimens. Tree Physiol 21: 201–212.

Roden LC, Song HR, Jackson S, Morris K, Carre IA (2002) Floral responses to photoperiod are correlated with the timing of rhythmic expression relative to dawn and dusk in *Arabidopsis*. Proc Natl Acad Sci USA 99: 13313–13318.

Rottmann WH, Meilan R, Sheppard LA, Brunner AM, Skinner JS, Ma C, Cheng S, Jouanin L, Pilate G, Strauss SH (2000) Diverse effects of overexpression of *LEAFY* and *PTLF*, a poplar (*Populus*) homolog of *LEAFY/FLORICAULA*, in transgenic poplar and *Arabidopsis*. Plant J 22: 235–245.

Saitou N, Nei M (1987) The neighbor-joining method: A new method for reconstructing phylogenetic trees. Mol Biol Evol 4: 406–425.

Samach A, Onouchi H, Gold SE, Ditta GS, Schwarz-Sommer Z, Yanofsky MF, Coupland G (2000) Distinct roles of *CONSTANS* target genes in reproductive development of *Arabidopsis*. Science 288: 1613–1616.

Schaffer R, Ramsay N, Samach A, Corden S, Putterill J, Carre IA, Coupland G (1998) The *late elongated hypocotyl* mutation of *Arabidopsis* disrupts circadian rhythms and the photoperiodic control of flowering. Cell 93: 1219–1229.

Schomburg FM, Patton DA, Meinke DW, Amasino RM (2001) *FPA*, a gene involved in floral induction in *Arabidopsis*, encodes a protein containing RNA-recognition motifs. Plant Cell 13: 1427–1436.

Searle I, He Y, Turck F, Vincent C, Fornara F, Krober S, Amasino RM, Coupland G (2006) The transcription factor *FLC* confers a flowering response to vernalization by repressing meristem competence and systemic signaling in *Arabidopsis*. Genes Dev 20: 898–912.

Shannon S, Meeks-Wagner DR (1991) A mutation in the *Arabidopsis TFL1* gene affects inflorescence meristem development. Plant Cell 3: 877–892.

Sheldon CC, Burn JE, Perez PP, Metzger J, Edwards JA, Peacock WJ, Dennis ES (1999) The *FLF* MADS box gene. A repressor of flowering in *Arabidopsis* regulated by vernalization and methylation. Plant Cell 11: 445–458.

Sheldon CC, Finnegan EJ, Rouse DT, Tadege M, Bagnal DJ, Helliwell CA, Peacock WJ, Dennis ES (2000) The control of flowering by vernalization. Curr Opin Plant Biol 3: 418–422.

Sheldon CC, Conn AB, Dennis ES, Peacock WJ. (2002) Different regulatory regions are required for the vernalization-induced repression of *FLOWERING LOCUS C* and for the epigenetic maintenance of repression. Plant Cell 14: 2527–2537.

Shi Y, Lan F, Matson C, Mulligan P, Whetstine JR, Cole PA, Casero RA, Shi Y (2004) Histone demethylation mediated by the nuclear amine oxidase homolog LSD1. Cell 119: 941–953.

Simon R, Igeno MI, Coupland G (1996) Activation of floral meristem identity genes in *Arabidopsis*. Nature 384: 59–62.

Simpson GG, Dijkwel PP, Quesada V, Henderson I, Dean C (2003) FY is an RNA end-processing factor that interacts with *FCA* to control the *Arabidopsis* floral transition. Cell 113: 777–787.

Saitou N, Nei M (1987) The neighbor-joining method: A new method for reconstructing phylogenetic trees. Mol Biol Evol 4: 406–425.

Soltis PS, Soltis DE, Chase MW (1999) Angipsperm phylogeny inferred from multiple genes as a tool for comparative biology. Nature 402: 402–404.

Somers DE, Webb AAR, Pearson M, Kay SA (1998) The short-period mutant, *toc1-1*, alters circadian clock regulation of multiple outputs throughout development in *Arabidopsis thaliana*. Development 125: 485–494.

Somers DE, Schultz TF, Milnamow M, Kay SA (2000) *ZEITLUPE* encodes a novel clock-associated PAS protein from *Arabidopsis*. Cell 101: 319–329.

Somerville C, Koornneef M (2002) A fortunate choice: The history of *Arabidopsis* as a model plant. Nature 3: 883–889.

Song J, Lu S, Chen Z-Z, Lourenco R, Chiang V (2006) Genetic transformation of *Populus trichocarpa* genotype Nisqually-1: A functional genomic tool for woody plants. Plant Cell Physiol 47: 1582–1589.

Strauss SH (2003a) Genomics, genetic engineering, and domestication of crops. Science 300: 61–62.

Strauss SH (2003b) Regulation of biotechnology as though gene function mattered. BioScience 53: 453–454.

Strauss SH, Rottmann WH, Brunner AM, Sheppard LA (1995) Genetic engineering of reproductive sterility in forest trees. Mol Breed 1: 5–26.

Strayer C, Oyama T, Schultz TF, Raman R, Somers DE, Mas P, Panda S, Kreps JA, Kay SA (2000) Cloning of the *Arabidopsis* clock gene *TOC1*, an autoregulatory response regulator homolog. Science 289: 768–771.

Suarez-Lopez P, Wheatley K, Robson F, Onouchi H, Valverde F, Coupland G (2001) *CONSTANS* mediates between the circadian clock and the control of flowering in *Arabidopsis*. Nature 410: 1116–1120.

Sun T-P, Kamiya Y (1994) The *Arabidopsis ga1* locus encodes the cyclase ent-kaurene synthetase-A of gibberellin biosynthesis. Plant Cell 6: 1509–1518.

Sung S, Amasino RM (2004) Vernalization in *Arabidopsis thaliana* is mediated by the PHD finger protein *VIN3*. Nature 427: 159–164.

Takada S, Goto K (2003) TERMINAL FLOWER 2, an *Arabidopsis* homolog of HETEROCHROMATIN PROTEIN 1, counteracts the activation of FLOWERING LOCUS T by CONSTANS in the vascular tissues of leaves to regulate flowering time. Plant Cell 15: 2856–2865.

Tuskan GA, DiFazio SP, Hellsten U, Jansson S, Rombauts S, Putnam N, Sterck L, Bohlmann J, Schein J, Ralph S, Aerts A, Bhalerao RR, Bhalerao RP, Blaudez D, Boerjan W, Brun A, Brunner AM, Busov V, Campbell M, Carlson J, Chalot M, Chapman J, Chen G-L, Cooper D, Coutinho PM, Couturier J, Covert SF, Cunningham R, Davis J, Degroeve S, Dejardin A, dePamphilis C, Detter J, Dirks B, Dubchak I, Duplessis S, Ehlting J, Ellis BE, Gendler K, Goodstein D, Gribskov M, Grigoriev I, Grimwood J, Groover A, Gunter L, Hamberger B, Heinze B, Helariutta Y, Henrissat B, Holligan D, Holt R, Islam-Faridi N, Jones S, Jones-Rhoades M, Jorgensen R, Joshi C, Kangasjarvi J, Karlsson J, Kelleher C, Kirkpatrick K, Kirst M, Kohler A, Kalluri U, Larimer FW, Leebens-Mack J, Leple JC, Locascio PF, Lucas S, Martin F, Montanini B, Napoli C, Nelson DR, Nelson CD, Nieminen KM, Nilsson O, Peter G, Philippe R, Pilate G, Poliakov A, Richardson P, Rinaldi C, Ritland K, Rouze P, Ryaboy D, Salamov A, Schmutz J, Schrader J, Segerman B, Shin H, Siddiqui A, Sterky F, Terry A, Tsai C, Unneberg P, Wall K, Wessler S, Yang G, Yin T, Douglas CJ, Marra M, Sandberg G, Van de Peer Y, Rokhsar D (2006) The genome of black cottonwood, *Populus trichocarpa* (Torr. & Gray ex Brayshaw). Science 313: 1596–1604.

Tyler L, Thomas SG, Hu J, Dill A, Alonso JM, Ecker JR, Sun TP (2004) DELLA proteins and gibberellin-regulated seed germination and floral development in *Arabidopsis*. Plant Physiol 135: 1008–1019.

USDA (2007) Introduction of Genetically Engineered Organisms, USDA APHIS Draft Programmatic Environmental Impact Statement: *http://www.aphis.usda.gov/newsroom/content/2007/07/content/prin`/complete_eis.pdf* (last viewed May 12, 2008).

Wang Z-Y, Tobin EM (1998) Constitutive expression of the *CIRCADIAN CLOCK ASSOCIATED* (*CCA1*) gene disrupts circadian rhythms and suppresses its own expression. Cell 93: 1207–1217.

Weigel D, Nilsson O (1995) A developmental switch sufficient for flower initiation in diverse plants. Nature 377: 495–500.

Weigel D, Alvarez J, Smyth DR, Yanofsky MF, Meyerowitz EM (1992) *LEAFY* controls floral meristem identity in *Arabidopsis*. Cell 69: 843–859.

Wen C-K, Chang C (2002) *Arabidopsis RGL1* encodes a negative regulator of gibberellins responses. Plant Cell 14: 87–100.

Wigge PA, Kim MC, Jaeger KE, Busch W, Schmid M, Lohmann JU, Weigel D (2005) Integration of spatial and temporal information during floral induction in *Arabidopsis*. Science 309: 1056–1059.

Wikstrom N, Savolainen V, Chase MW (2001) Evolution of the angiosperms: calibrating the family tree. Proc Roy Soc Lond 268: 2211–2220.

Wilson RN, Heckman JW, Somerville CR (1992) Gibberellin is required for flowering in *Arabidopsis thaliana* under short days. Plant Physiol 100: 403–408.

Valverde F, Mouradov A, Soppe W, Ravenscroft D, Samach A, Coupland G (2004) Photoreceptor regulation of CONSTANS protein in photoperiodic flowering. Science 303: 1003–1006.

Yamaguchi A, Kobayashi Y, Goto K, Abe M, Araki T (2005) *TWIN SISTER OF FT (TSF)* acts as a floral pathway integrator redundantly with *FT*. Plant Cell Physiol 46: 1175–1189.

Yanovsky MJ, Kay SA (2002) Molecular basis of seasonal time measurement in *Arabidopsis*. Nature 419: 308–312.

Yoo SK, Chung KS, Kim J, Lee JH, Hong SM, Yoo SJ, Yoo SY, Lee JS, Ahn JH (2005) CONSTANS activates *SUPPRESSOR OF OVEREXPRESSION OF CONSTANS 1* through *FLOWERING LOCUS T* to promote flowering in *Arabidopsis*. Plant Physiol 139: 770–778.

Yuceer C, Harkess RL, Land Jr SB, Luthe DS (2002) Structure and developmental regulation of *CONSTANS-LIKE* genes isolated from *Populus deltoides*. Plant Sci 163: 615–625.

Yuceer C, Land SB, Kubiske ME, Harkess RL (2003) Shoot morphogenesis associated with flowering in *Populus deltoides* (Salicaceae). Am J Bot 90: 194–204.

Zagotta MT, Hicks KA, Jacobs CI, Young JC, Hangarter RP, Meeks-Wagner DR (1996) The *Arabidopsis ELF3* gene regulates vegetative morphogenesis and the photoperiodic induction of flowering. Plant J 10: 691–702.

Zeevaart JAD (1976) Physiology of flower formation. Ann Rev Plant Physiol 27: 321–348.

Zeevaart JAD (2006) Florigen coming of age after 70 years. Plant Cell 18: 1783–1789.

# 12

# Phenylpropanoid and Phenolic Metabolism in *Populus*: Gene Family Structure and Comparative and Functional Genomics

*Carl J. Douglas,[1],\* Jürgen Ehlting[2] and Scott A. Harding[3]*

## ABSTRACT

The completion and annotation of the *Populus trichocarpa* genome has allowed, for the first time in a tree, analyses of the complete sets of genes involved in shikimate, phenylpropanoid and phenolic metabolism. This is particularly important in *Populus*. Not only is wood a huge sink tissue for carbon derived from the shikimate pathway via phenylpropanoid metabolism into lignin biosynthesis, but the large biosynthetic commitment of leaves and other tissues to soluble phenolics and condensed tannins make this another very large sink for photosynthetically derived carbon. This chapter describes the currently understood repertoires of poplar phenylpropanoid and lignin biosynthetic genes, as well as the less well characterized sets of shikimate pathway genes required for phenylalanine biosynthesis that feeds directly into phenylpropanoid metabolism. The chapter concludes with an overview of the *Populus* genes that are involved in soluble phenolic and condensed tannin biosynthesis, and the ecological and biomass implications of *Populus'* commitment to these foliar compounds. Throughout the chapter, we focus on genes that appear to have clear functions in phenylpropanoid and phenolic metabolism, those for which more information is needed, and the evolutionary and genomic context of these gene sets.

[1]Department of Botany, University of British Columbia, Vancouver BC V6T 1Z4, Canada; e-mail: *cdouglas@interchange.ubc.ca*
[2]Centre for Forest Biology and Department of Biology, University of Victoria, Victoria BC V8W 3N5, Canada; e-mail: *je@uvic.ca*
[3]Warnell School of Forestry, University of Georgia, Athens, GA 30602-2152, USA; e-mail: *sharding@uga.edu*
\*Corresponding author

**Keywords:** phenylpropanoid metabolism, lignin, shikimate pathway, phenolic metabolism, gene duplication, enzyme function

## 12.1 Introduction

The completion and initial analysis of the *Populus trichocarpa* (black cottonwood) genome (Tuskan et al. 2006) was a milestone in plant and tree biology. The foundation for this accomplishment was laid by years of research on the genetics, physiology, and molecular biology of members of the *Populus* genus (poplars, cottonwoods, aspens; referred to as "poplar"), and by the development of complementary genomic resources such as expressed sequence tag (EST) collections, microarrays, and a physical map (Jansson and Douglas 2007). In addition to its status as a model tree, poplars are commercially important for pulp and paper, solid wood, potentially for biofuel production, and are ecologically important in their native ranges (Brunner et al. 2004). Phenylpropanoid and phenolic metabolism strongly affects many poplar traits that influence their suitability for such uses as well as the biology of poplars, and is the focus of this chapter. Secondary wall formation is a hugely important process in forest ecosystems, because the bulk of the biomass of trees is cellulose and encrusting lignin in secondary walls. Lignin, whose monomeric constituents are derived from the phenylpropanoid pathway, is a major component of wood, and has a profound impact on the properties of wood with respect to its use in pulp and paper production and conversion of ligno-cellulose into ethanol. Phenolic and phenylpropanoid derived compounds appear to play important roles in poplar defenses against pathogens and insect pests, and provide a potentially rich source of renewable organic compounds. These pathways of phenylpropanoid and phenolic natural product biosynthesis require the coordinated activity of the shikimate pathway to supply pathway intermediates and, especially, phenylalanine as a starter molecule for phenylpropanoid metabolism. Completion of the poplar genome and continued genome-level, biochemical, genetic, and comparative genomics analyses are providing new insights into phenylpropanoids and phenolics and their biosynthesis in these trees. This information will be of critical importance in improvement of poplar genotypes for deployment for practical purposes, and in understanding the biology of poplar as an ecological keystone species.

## 12.2 Phenylpropanoid Metabolism

*Populus* species elaborate a rich array of plant natural products, and one of the largest classes of such products is phenylpropanoids and phenolics derived from phenylalanine via phenolic and phenylpropanoid metabolism. The elaboration of the diversity and quantity of such products found in poplar and other plants relies on the unique ability of plants, as well as some fungi and microbes, to divert carbon from the shikimate pathway, which leads to the biosynthesis of the aromatic amino acids phenylalanine, tyrosine, and tryptophan, into aromatic natural product biosynthesis. The evolution of such biosynthetic ability is considered to have been a key innovation during land plant evolution 450–600 million years ago (Bowman et al. 2007), since such products afford protection from UV-light, structural support, and protection from pathogen attack.

The enzymes of general phenylpropanoid metabolism that channel carbon away from the shikimate pathway of primary metabolism into phenylpropanoid natural product biosynthesis have been studied in numerous plants and are depicted in Fig. 12-1. These enzymes include phenylalanine ammonia-lyase (PAL), which catalyzes the deamination of phenylalanine to produce cinnamic acid, cinnamate-4-hydroxylase (C4H), which generates 4-hydroxycinnamate (4-coumarate), 4-coumarate-CoA ligase (4CL), which generates activated CoA esters of 4-coumarate and its hydroxylated derivatives (Hahlbrock and Scheel 1989). Poplar tissues and organs such as buds, leaves, bark, and xylem are all rich sources of various classes of soluble natural products derived from phenylalanine, including phenylpropanoids, flavonoids, condensed tannins, and phenolic glycosides (Fig. 12-1). Some of these compounds accumulate to high levels, for example the phenolic glycosides may constitute a very high percentage of leaf dry weight in some species, as discussed in more depth below. Relative to short lived annuals such as *Arabidopsis*, chemical defenses, including those involving phenolics and phenylpropanoid compounds, may be more important in trees such as poplar (Ralph et al. 2006; Miranda et al. 2007).

## 12.3 Lignin

Lignin is a major phenylpropanoid natural product and a major structural component of secondary cell walls in the wood of poplar and other trees. Lignin is a polymer of hydroxycinnamyl alcohols (monolignols) and other phenylpropanoids (Boerjan et al. 2003; Raes et al. 2003 and references therein), and is an essential constituent of secondarily thickened plant cell walls such as tracheary elements (water conducting cells) and fibers of xylem. Lignin constitutes up to 30% of the weight of secondary xylem (wood) in trees. Thus, lignin is the second most abundant biological polymer in the biosphere after cellulose, and its biosynthesis requires a considerable metabolic commitment on the part of woody perennial plants such as *Populus*. Biosynthesis of the monolignol building blocks of the lignin polymer requires enzymes of the phenylpropanoid pathway. These

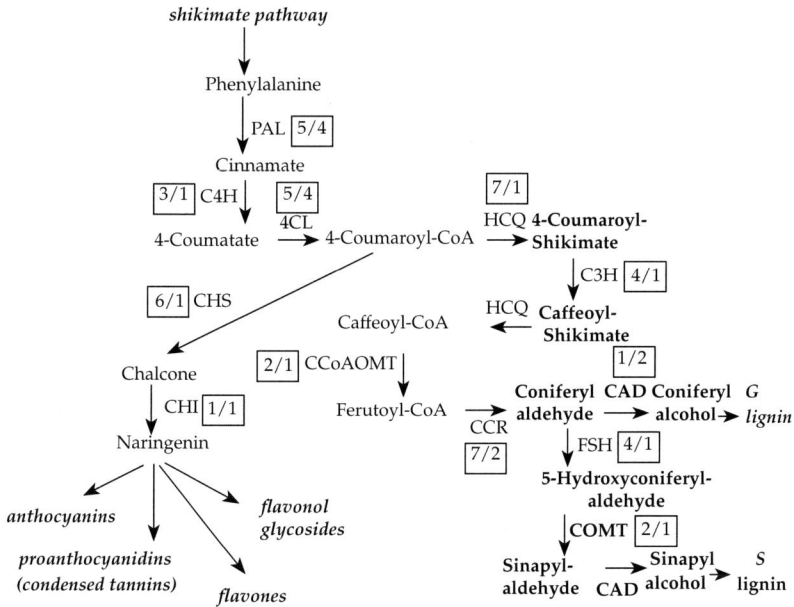

**Figure 12-1** Simplified representation of the phenylpropanoid, monolignol and flavonoid biosynthetic pathways. The numbers of genes encoding enzymes of phenylpropanoid and monolignol biosynthesis in *Populus* and *Arabidopsis* are shown. Enzyme names are annotated with gene numbers. Numbers in boxes refer to the number of *Populus* genes relative to the number of *Arabidopsis* genes for each enzymatic step. Chemical intermediates and product are also indicated. Names in italics refer to additional precursor or end-product pathways. Abbreviations: (*4CL*) 4-couumarate:CoA ligase, (*CAD*) cinnamyl alcohol dehydrogenase, (*C4H*) cinnamate-4-hydroxylase gene family, (*C3H*) 4-coumaroyl-shikimate/quinate-3-hydroxlase (*CCR*) cinnamyl CoA reductase, (*CHI*) chalcone isomerase, (*CHS*) chalcone synthase, (*COMT*) caffeic acid O-methyltransferase, (*CCoAOMT*) caffeoyl-CoA O-methyltransferase, (*HCQ*), hydroxycinnamyl-quinate transferase, (*F5H*) ferulate-5-hydroxylase, (*PAL*) phenylalanine ammonia-lyase.

*Color image of this figure appears in the color plate section at the end of the book.*

reactions lead from 4-coumarate via a series of intermediates (generated by hydroxylation, methylation and reduction) to coniferyl or sinapyl alcohols (monolignols for G and S lignin subunits, respectively; Fig. 1; Humphreys and Chapple 2002); These lignin precursors are exported across the plasma membrane into secondary cell walls, where they undergo oxidative polymerization (Boerjan et al. 2003). The enzymes involved in phenylpropanoid biosynthesis are usually encoded by gene families, which often vary in size according to plant species (Hahlbrock and Scheel 1989; Dixon et al. 2002; Hamberger et al. 2006).

The genes encoding core enzymes in phenylpropanoid, flavonoid, and lignin metabolism have been cloned from many species, and the completed *Arabidopsis* genome allowed the first description of the full set of phenylpropanoid and lignin biosynthetic genes to be identified in a single species, using a combination of phylogenetic, functional, and transcriptional profiling approaches (Costa et al. 2003; Goujon et al. 2003; Raes et al. 2003). In addition to bona fide phenylpropanoid enzymes, these and other studies have revealed large numbers of "phenylpropanoid-like" enzymes related to members of phenylpropanoid gene families but of largely unknown biochemical function, in the *Arabidopsis* genome (AGI 2000), and these have been described in recent reports (Cukovic et al. 2001; Costa et al. 2003; Raes et al. 2003; Shockey et al. 2003; de Azevedo Souza et al. 2008).

The completed poplar genome sequence allows genome-wide comparisons of phenylpropanoid and phenylpropanoid-like gene families in angiosperms, allowing insights into the evolution and diversification of these families in lineages represented by poplar, *Arabidopsis* and rice. These studies have shown that genes encoding phenylpropanoid-like enzymes are conserved in poplar as well as rice and other plants (Tuskan et al. 2006; Hamberger et al. 2007; de Azevedo Souza et al. 2008). While most or all phenylpropanoid-like enzymes are unlikely to have functions in phenylpropanoid metabolism, their conservation in angiosperms, and in some cases the moss *Physcomitrella*, suggests that they play important and conserved roles in natural product biosynthesis in plants. Certain phenylpropanoid-like gene family clades have undergone expansion and diversification in poplar relative to *Arabidopsis* and rice, most prominently two clades encoding cinnamyl alcohol dehydrogenase-like (CADL) proteins (Tuskan et al. 2006; Hamberger et al. 2007) and a clade of genes encoding cinnamyl CoA reductase-like (CCRL) proteins (Hamberger et al. 2007). Functional analysis of this expanded repertoire of poplar phenylpropanoid-like enzymes should shed light on natural product pathways important to poplars.

Annotation of the poplar genome, phylogenetic analyses, and functional genomics approaches such as microarray enabled transcript profiling has allowed the "*Populus* lignin biosynthetic toolbox" (the set of poplar genes encoding enzymes required for developmental lignin monomer biosynthesis) to be described (Tuskan et al. 2006; Hamberger et al. 2007). BLAST searches of the poplar genome using *Arabidopsis* genes as queries, alignment of amino acid sequences, and phylogenetic reconstructions revealed sets of poplar phenylpropanoid genes encoding enzymes closely related to previously characterized enzymes from *Arabidopsis*, poplar, and other species known or inferred to be involved in lignin biosynthesis (Tuskan et al. 2006; Hamberger et al. 2007). Analysis of poplar microarray gene expression data and EST abundance data provided further evidence for the preferential expression of these gene sets during wood and secondary xylem development.

As shown in Fig. 12-1, the poplar "toolbox" of lignin biosynthetic genes is about twice as big as the corrollary set in *Arabidopsis*. The poplar genome has been profoundly shaped by the "salicoid duplication", a whole-genome duplication event that is inferred to have occurred 60–65 million years ago (Tuskan et al. 2006) and was followed by diploidization and chromomsome rearrangement, and a large proportion of duplicated genes have been retained in *Populus*, suggesting selection for retention of duplicated genes due to subfunctionalization and neofunctionalization (Tuskan et al. 2006).

The high number of lignin biosynthetic genes is likely to be at least partially due to retention of such segmentally duplicated genes (Hamberger et al. 2007). Especially interesting is the high number of poplar genes relative to those in *Arabidopsis* encoding cytochrome P450-mediated enzymes required for key hydroxylation steps of the phenylpropanoid phenolic ring at the 4', 3', and 5' positions, required for H, G, and S lignin monomer biosynthesis (Fig. 12-1). P450 enzymes often catalyze rate-limiting steps in plant natural product biosynthesis, suggesting a greater need for maintaining flux into monolignol biosynthesis during the massive commitment to secondary wall biosynthesis, or reflecting evolutionary pressure for subfunctionalization. A good example of subfunctionalization is found in the poplar cinnamate-4-hydroxylase (*C4H*) gene family which have distinct and largely non-overlapping expression patterns (Lu et al. 2006; Hamberger et al. 2007).

Many studies in the pre-genomic era have shown that manipulation of lignin biosynthsesis can be achieved in poplar by antisense and/or sense expression of biosynthetic genes, leading to gene silencing and resulting changes in lignin depostion (for example, Pilate et al. 2002). With the lignin biosynthetic tool box now well defined by analysis of the poplar genome, it will become possible to define more precisely the roles of individual genes in the biosnynthesis of lignin and other phenylpropanoids. For example, a recent report shows that RNAi-mediated suppression of one of the poplar *C3H* genes leads to profound changes in lignin amount and composition in trees with supressed expression. These trees display not only greatly reduced lignin deposition, but also strong enrichment of *p*-hydroxyphenyl (H) subunits at the expense of guaiacyl (G) subunits (Coleman et al. 2008), consistent with C3H enzymatic function in vitro. Pool levels of soluble phenolic glycosides were also affected, consistent with a block in the entry into the monolignol biosynthetic pathway, supporting a role for C3H in catalyzing a rate limiting step in lignin monomer biosynthesis.

## 12.4 Shikimate Pathway

The shikimate pathway connects primary sugar metabolism with the biosynthesis of the aromatic amino acids tryptophan (Trp), tyrosine (Tyr), and phenylalanine (Phe), the latter being the precursor of phenylpropanoids. Thus, regulation of carbon flux through this pathway is important for the regulation of phenylpropanoid and phenolic metabolism. The shikimate pathway precursors, phosphoenolpyruvate (PEP) and erythrose-4-phosphate (E4P), are ultimately derived from photosynthetic carbon fixation in the Calvin cycle and are fed into the pathway via glycolysis (PEP) and the pentose phosphate shunt (E4P). In plants, especially in woody species, the phenylpropanoid pathway and in particular the branch leading to lignin is the primary carbon sink and at least 20% of the carbon fixed is channeled through the shikimate pathway (Coruzzi and Last 2000). Almost every step of the shikimate pathway may also be considered a branching point leading to a distinct set of natural products. Besides lignin, two other classes of shikimate pathway derivatives accumulate to high levels in *Populus*, namely quinate esters of hydroxycinnamic acids and salicin-containing phenolic glycosides, whose biosynthetic routes will be covered later in more detail.

Most plant shikimate pathway genes have been originally identified based on their ability to complement *Escherichia coli* or yeast mutants deficient in the respective step (Klee et al. 1987; Keith et al. 1991; Schaller et al. 1991; Schmid et al. 1992; Eberhard et al. 1993; Bonner and Jensen 1994; Bischoff et al. 1996; Mobley et al. 1999; Hsieh and Goodman 2002). Subsequently, these enzymes and putative orthologs from other plants have been characterized biochemically by heterologous expression, mainly in *E. coli*, and by analysis of the corresponding recombinant proteins. The pathway has been most intensely studied in members of the Solanaceae including *Nicotiana tabacum* (tobacco), *Solanum tuberosum* (potato) and, in particular, *Solanum lycopersicon* (formerly *Lycopersicon esculentum*, tomato). Despite the importance of this pathway particularly in woody species, to date no shikimate pathway enzyme from *Populus* has been characterized at the molecular and biochemical level. However, employing comparative genomics studies, likely *Populus* orthologs have been annotated and, based on EST abundance data, initial expression characterizations have been performed (Hamberger et al. 2006; Tsai et al. 2006).

The gene families encoding individual enzymatic steps in the shikimate pathway in poplar and other plants are generally less complex than those encoding phenylpropanoid pathway enzymes (Hamberger et al. 2006). With some exceptions, all isoforms in a given family share high degrees of similarity and only in some cases related divergent genes are present. This is in contrast to the phenylpropanoid enzyme gene families present in poplar and other genomes. In addition to encoding bona fide phenylpropanoid enzymes, such families also encode enzymes that are related to phenylpropanoid enzymes but likely have divergent functions (see above). It thus seems that generally purifying selection is acting on shikimate pathway genes and that duplicated genes have rarely been recruited to act in divergent or newly evolving biochemical pathways.

The main branch of the shikimate pathway leads to chorismate biosynthesis and is initiated by the condensation of PEP and 4EP to form 3-deoxy-D-*arabino*-heptulosonate-7-phosphate (DAHP), catalyzed by DAHP synthase (DHS). The poplar genome encodes four DHS isoforms (Fig. 12-2) and a likely pseudogene. In comparison, the *Arabidopsis*, *Ricinus* (castor bean), and *Sorghum* genomes each encode three isoforms, while the *Oryza* (rice) and *Vitis* (grape vine) genomes each contain five. Surprisingly, the moss *Physcomitrella* contains seven genes encoding DHS while the lycophyte *Selaginella moellendorfii* and green algae contain only a single gene (not shown). Plant DHS isoforms are highly conserved, both within species and among lineages. When excluding the variable 5'-chloroplast targeting signals, the *Populus* proteins are 82–95% identical to each other and share 80–96% identity to other angiosperms; even the moss and green algal proteins are 72–79% identical to the poplar proteins. A total of 31 DHS full-length sequences from vascular plants were retrieved from sequence depositories and all of these are predicted to be targeted to the chloroplast, based on subcellular targeting predictions. Despite high levels of sequence similarity within *DHS* genes, differences in transcript abundance among gene family members have been observed, with members from one phylogenetic subclass being expressed in correlation with enhanced phenylpropanoid biosynthesis in poplar and other plants (Gorlach et al. 1995; Ehlting et al. 2005; Hamberger et al. 2006). It thus appears plausible that different DHS isoforms have been recruited for specialized physiological roles while maintaining their plastid localization and biochemical functions. However, DHS isoforms with distinct biochemical properties and subcellular localizations (plastids and cytoplasm) have been described in *Nicotiana sylvestris*, and both isoforms have been detected in a wide range of plants (Ganson et al. 1986). Only systematic functional characterization of the whole *DHS* family in poplar and other plants can clarify this apparent contradiction.

In contrast to DHS, poplar and most other plants investigated appear to contain only a single gene encoding 3-dehydroquinate synthase (DHQS), which catalyzes the dephosphorylation of DAHP to form 3-dehydroquinate (Fig. 12-2). A total of 27 full-length reading frames were identified in diverse databases: 26 from vascular land plants, mosses, and green algae including 10 species with completely elucidated genomes. We found that only *Zea mays* (maize) contains two isoforms, and these are highly similar (95% identity on the amino acid level), which can be explained with the likely tetraploid origin of maize (Swigonova et al. 2004). One of these cDNAs contains a frame shift mutation, suggesting it is likely the product of a pseudogene. It thus appears that strong selection pressure exists to maintain the single-isoform status of DHQS in poplar and other plants, all of which are clearly predicted

Phosphoenolpyruvate
+
Erythrose-4-P

| 4/3 |
- DHS1 (Chl)
- DHS2 (Chl)
- DHS3 (Chl)
- DHS4 (Chl)
- DHS5P (Chl)

3-Deoxy-D-arabino-
heptulosonate-7-P

| 1/1 | DHQS (Chl)

3-dehydroquinate

| 5/1 |
- DHQD1 (Chl)
- DHQD2 (Chl/Mit?)
- DHQD3 (Sec)
- DHQD4 (Cyt)
- DHQD5 (Cyt)

quinate

shikimate

| 4/4 |
- SK1 (Chl)
- SK2 (Chl)
- SK3 (Chl)
- SK4P (Cyt)
- SKL (Chl)

quinate /
shikimate
esters

shikimate-3-P

| 2/2 |
- EPSPS1 (Chl)
- EPSPS2 (Chl)

5-enolpyruvylshikimate-3-P

| 2/1 |
- CS1 (Chl)
- CS2 (Chl)

tryptophane ◀--- chorismate

| 3/3 |
- CM1 (Chl)
- CM2 (Chl)
- CM3 (Cyt)
- CM4P (Sec)

tyrosine ◀─── prephenate

| PNT [?]

arogenate

| 5/6 |
- ADT1 (Chl/Mit)
- ADT2 (Chl/Mit)
- ADT3 (Chl/Mit)
- ADT4 (Chl)
- ADT5P (Chl)

phenylalanine

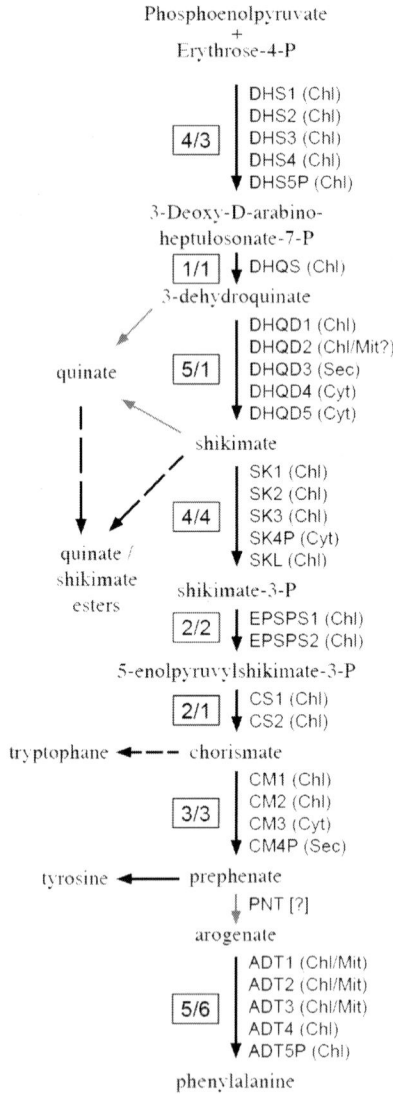

**Figure 12-2** Simplified representation of the shikimate pathway and post-chorismate pathway leading to phenylalanine. The numbers of genes encoding enzymes of phenylalanine biosynthesis in *Populus* and *Arabidopsis* are shown in boxes. Annotated *Populus* enzyme names and predicted subcellular localization (in brackets) are shown. Chemical intermediates and product are given; solid arrows indicate single enzymatic steps, dashed arrows represent multiple steps. Abbreviations: (*DHS*) 3-deoxy-arabino-heptulosonate-7-phosphate synthase, (*DHQS*) 3-dehydroquinate synthase, (*DHQD*) 3-dehydro-quinate dehydratase/shikimate dehydrogenase, (*SK*) shikimate kinase (*EPSPS*) 5-enolpyruvylshikimate-3-phosphate synthase, (*CS*) chorismate synthase, (*CM*) chorismate mutase, (*PNT*) prephenate aminotransferase (not characterized at the gene level), (*ADT*) arogenate dehydratase.

to reside in the plastid. In agreement with a broader function in diverse shikimate sub-pathways, *DHQS* appears to be more broadly expressed compared to other genes in the pathway both in poplar and *Arabidopsis* (Hamberger et al. 2006). Taken together, this suggests that DHQS could be the initial control point regulating flux into the shikimate pathway.

The subsequent step in the pathway, dehydrogenation and reduction of 3-dehydroquinate to shikimate, is catalyzed by 3-dehydroquinate dehydratase/shikimate dehydrogenase (DHQD/SDH), which appears to be encoded by five genes in poplar, relative to the single *Arabidopsis* gene (Hamberger et al. 2006). This suggests subfunctionalization of DHQD/SDH function in poplar, and evidence for differential expression of DHQD/SDH gene family members supports this hypothesis (Hamberger et al. 2006).

While it appears that the first two steps of the shikimate pathway are strictly confined to the chloroplast (based on subcellular localization predictions of full-length sequences), Ding et al. (2007) recently showed, by means of transient expression of green fluorescence protein tagged DHQD/SDH versions, that one of two tobacco DHQD/SDH isoforms (NtDHD/SHD1) is localized in the cytoplasm, while the other (NtDHD/SHD2) is localized in the cytoplasm. Both enzymes do have NADP$^+$ dependent activity with shikimate when transiently over-expressed in tobacco leaves (Ding et al. 2007), but kinetic properties of recombinant enzymes could only be obtained for NtDHD/SHD1. Among the five poplar DHQD/SDH isoforms, only DHQD/SDH1 is clearly predicted to be localized in plastids.

We generated phylogenetic reconstructions of aligned DHQD/SDH sequences from numerous higher plants and found two distinct monophyletic DHQD/SDH clusters, which appear to have arisen close to the divergence of angiosperm/gymnosperm lineages (not shown). The plastid localized isoforms from poplar (PoptrDHQD/SDH1) and tobacco (NictaDHD/SHD2) reside in one cluster together with a second isoform from Poplar (PoptrDHQD/SDH5), while the cytosolic tobacco isoform groups together with the remaining four poplar enzymes in the second cluster. The second cluster is anchored by two conifer sequences and contains *Ricinus* and *Vitis* isoforms, but does not contain monocot proteins, or any isoforms from *Arabidopsis* or *Medicago*, all of which group in the first cluster (not shown). It thus appears that DHQD/SDH isoforms from this class were lost in these two lineages. The likely plastidial poplar isoform shares 65–85% identity with other dicot isoforms in the same cluster and identity remains high at 62% in comparison with conifer sequences at the base of the same clade, thus resembling the identities found for DHS and DHQD enzymes. In contrast, protein identity between PoptrDHQD/SDH1 to conifer, dicot, and poplar sequences of the second cluster ranges from 49 to 51%, similar to the identity found in comparison to moss and green algal sequences.

Taken together, these data suggest that DHQD/SDH genes duplicated early in land plants to acquire novel functions in cellular compartments other than the chloroplast, thus becoming subfunctionalized. Since over-expression of both tobacco isoforms results in enhanced DHQD/SDH activity (Ding et al. 2007), both isoform types may retain the same biochemical function. Their different subcellular locations, however, may serve to channel the shikimate intermediate (the product of the DHQD/SDH catalyzed reaction) into distinct branches of secondary metabolism in the lineages including *Populus* that contain both isoform types. Both poplar and tobacco accumulate high levels of quinate and shikimate esters of hydroxycinnamates, such as chlorogenic acid (see below). Since quinate may be derived from either shikimate or 3-dehydroquinate (the substrate and product of DHQD/SDH, respectively (Hermann and Weaver, 1999)), it appears plausible that the cytosolic isoforms may have been recruited to channel carbon into these pathways. However, transcriptional down-regulation of the plastidial DHQD/SDH1 isoform in tobacco also results in decreased chlorogenic acid accumulation indicating an involvement of this isoform in quinate ester biosynthesis (Ding et al. 2007).

A similar example of gene duplication and functional divergence can be found in the families of genes encoding shikimate kinase (SK). Poplar contains four *SK* gene family members including a likely pseudogene, and these isoforms were placed in distinct phylogenetic clades (Hamberger et al. 2006). PoptrSK1, PoptrSK3 and the likely pseudogene were placed in the same clade as the biochemically characterized SK from tomato (Schmid et al. 1992). Further analysis showed that this clade also contains sequences from monocots and is anchored by a sequence from the moss *Physcomitrella* (not shown). The poplar sequences are 51 to 84% identical to other proteins in this clade, but share only 33% identity on average to members of a sister clade that contains the PoptrSK2 isoform alongside with single isoforms from *Arabidopsis*, *Vitis*, *Ricinus*, and *Physcomitrella*. A third clade also contains a single SK-related isoform from poplar (PoptrSKL), with single representatives from each of the other fully sequenced dicot, monocot, and *Physcomitrella* genomes. Members from this divergent clade are even more divergent and share only 13% identity on average with PoptrSK1 or PoptrSK3, and it is not known if the enzymes have SK activity. It thus appears that the poplar *SK* gene family contains three functionally divergent sets of SK or SK-like isoforms that are shared in common with other land plant lineages, suggesting common evolutionary origins. However, except for the likely pseudogene, all poplar SK family members are predicted to be localized in the chloroplast and thus appear to have retained plastidial functions.

The organization of genes in the poplar and other genomes that encode the enzyme catalyzing the next step in the shikimate pathway,

5-enolpyruvylshikimate-3-phosphate synthase (EPSPS), well known as the target for the herbicide glyphosate, is again much simpler than that of phenylpropanoid genes, and resembles that of the first enzymes in the pathway. Poplar contains two genes likely encoding this function with one being a likely pseudogene (Hamberger et al. 2006). Only in *Arabidopsis* and *Physcomitrella* are two copies present while all other plants contain a single gene encoding EPSPS. All isoforms from higher plants with fully sequenced genomes are predicted to be located in the chloroplast. However, biochemical evidence exists for cytoplasmic localization of EPSPS in diverse plants (Mousdale and Coggins 1986). In each case where duplicate gene copies exist they appear to have been independently duplicated in each lineage since they form paralogous sister pairs in phylogenetic reconstructions (not shown). Isoforms from divergent plants are also highly conserved, with the poplar enzyme being 81–91% identical to isoforms from monocots or dicots. Even the *Physcomitrella* proteins share 71% identity with the poplar enzyme. It thus seems that *EPSPS* genes are under strict purifying selection with a tendency to maintain a single copy status. Similarly, analysis of the family encoding the enzyme catalyzing the last step of the pre-chorismate pathway, chorismate synthase (CS), in fully sequenced genomes shows that the enzyme is encoded by a either one or two genes in *Physcomitrella*, monocots, and dicots, all of which encode plastidial proteins. Only poplar and tomato contain two genes, which appear to have arisen by independent duplication in each lineage.

The first step of the post-chorismate pathway branch that leads to the biosynthesis of Phe and Tyr, catalyzed by chorismate mutase (CM), is encoded by a small family in poplar comprising three members and a pseudogene (Hamberger et al. 2006). Similar gene family structure (three *CM* genes) is found in other higher plants with completed genome sequences. Based on initial phylogenetic reconstructions, it appears that two duplication events, one prior to the monocot-dicot divergence, and a second one early within the dicot radiation led to the creation of these isoforms (data not shown). Subcellular localization predictions suggest that the more divergent isoforms (including PoptrCM3) are not localized to the plastid, while most isoforms from the other two clades are predicted to be plastidial. Sequence identities are in the range of 70% within a clade, but drop down to only 51% between the divergent PoptrCM1 and PoptrCM3 proteins.

The last two steps of Phe biosynthesis branch from prephenate, the product of the CM catalyzed reaction. This Phe-specific branch is of interest with respect to the regulation of carbon channeling into phenylpropanoids and other Phe-derived phenolics, and is of special interest in poplar, given the large commitment to lignin, phenolic glycoside, condensed tannin, and other phenolic compounds in this plant. In most bacteria and fungi, Phe is synthesized via phenylpyruvate, while biochemical evidence indicates that in plants, Phe is synthesized via arogenate. In both cases an aminotransferase reaction (prephenate to arogenate in plants and phenylpyruvate to Phe in bacteria) and a dehydratase reaction (arogenate to Phe in plants and prephenate to phenylpyruvate in bacteria) are involved albeit in reversed succession. The plant prephenate aminotransferase (PNT) has not been characterized on the gene level, but *Arabidopsis* candidates from the large aminotransferase gene family have been identified based on co-expression analysis with shikimate and phenylpropanoid pathway genes (Ehlting et al. 2005) and poplar orthologs with congruent expression preferences are present in the poplar genome (Hamberger et al. 2006). A smaller number of poplar and *Arabidopsis* ADT candidates have been identified based on sequence similarity with fungal and bacterial prephenate dehydratases (Cho et al. 2007; Hamberger et al. 2007), and the enzymes encoded by the six putative *Arabidopsis ADT* genes, *AtADT1-AtADT6*, have been shown to have ADT activity (Cho et al. 2007), providing biochemical support for their provisional designation as ADTs.

The poplar genome contains five putative *ADT* gene family members (Hamberger et al. 2007), which is within the range found in other plants with completely sequenced genomes, ranging from three in *Selaginella* to six in *Arabidopsis*. Recently, a rice mutant has been characterized as being impaired in Phe and Tyr metabolism and the affected gene, *OrysaADT3*, was shown to encode an enzyme with primarily ADT and minor PNT activity (Yamada et al. 2008). Interestingly, this gene groups in a divergent phylogenetic clade compared to a set of previously described *Arabidopsis*, poplar, and rice genes that had been previously identified based on expression patterns related to tissues or organs with high demand for phenylpropanoids in *Arabidopsis* and poplar (Ehlting et al. 2005; Hamberger et al. 2007). Proteins within this set are highly similar to the PoptrADT1 protein core sequence (89% identical), but the only poplar ADT sequence that groups with the divergent *OrysaADT3*, PoptrADT5p, is a likely pseudogene. It thus seems possible that despite the large evolutionary distance and fairly low sequence similarity (61% between PoptrADT1 and OrysaADT3) that separates them, PoptrADT1 and OrysaADT3 biochemical functions have remained conserved.

Analysis of the poplar genome relative to other fully sequenced plant genomes has allowed detailed analysis of the complete gene families encoding enzymes of the shikimate pathway, leading to Phe biosynthesis. Considering these complete sets of genes, it is intriguing to note that each gene family is characterized by its own distinct evolutionary history. Some gene families are apparently under strict purifying selection, in some cases limiting its members to a single gene per species, in others allowing lineage specific multiplications. Apparently (based on conservation of gene family structure from moss through angiosperms), some shikimate pathway gene families were duplicated already in the common Viridiplantae (green plant) ancestor that first colonized land, while others diverged only within the vascular plant clade or in species specific branches. Together, these analyses highlight both evolutionary pressure for conservation of enzyme function within the pathway, and the flexibility to (sub-) divert duplicated genes to novel, sometimes lineage specific, functions. It is important to note though that the vast majority of genes identified in large-scale genome and transcriptome sequencing projects, including all genes from poplar described in this section, still await functional characterization, and that only this will show the levels of biochemical and physiological diversification suggested.

It has long been known that the shikimate pathway is plastid localized. The genome-aided identification of shikimate pathway enzyme isoforms predicted to be cytoplasmically localized suggests that parallel reactions also occur outside the plastid, increasing the potential biochemical complexity of shikimate and phenolic metabolism. However, based on the sum of subcellular localization predictions, it appears unlikely that the complete shikimate pathway exists in the cytosol in any plant species, but rather that certain isoforms from individual families, which vary among plant species, have cytosolic variants. Either these isoforms have evolved to fulfill divergent functions in the cytosol, or they catalyze the same reaction and thereby channel carbon flow into cytosolic side branches to produce lineage specific natural compounds derived from the shikimate pathway. The examples highlighted here show the importance of comparative genomic analyses to distinguish between common pathways present in all plants and lineage specific variations.

## 12.5 Condensed Tannins and Phenolic Glycosides

The major non-structural secondary-metabolite end products in *Populus* are the condensed tannins (CT), the salicin-containing phenolic glycosides (PG), and the esters of quinic and caffeic acid that comprise chlorogenic acid (reviewed in Tsai et al. 2006). CT's arise from proanthocyanidin monomer end-products of the flavonoid pathway, exhibit wide structural and functional diversity within *Populus* and form up to 35% of *Populus* leaf dry matter (Lindroth and Hwang 1996; Ayres et al. 1997; Harding et al. 2005). The genes and metabolites of the flavonoid pathway are well characterized (Winkel-Shirley 2001), and specific examples in support of their function in *Populus* defense and anti-herbivory have been described (Peters and Constabel 2002). Foliar CT levels are also strongly affected by nutrient and light stress and may have the beneficial effect of reducing nitrogen cycling in leach-prone or nutrient poor *Populus* habitats (Northup et al. 1995; Schweitzer et al. 2004).

A recent overview of the expanded network of *Populus* flavonoid and CT biosynthetic pathway genes, their expression in various organs and their response to wounding is available (Tsai et al. 2006). This analysis provides a framework for investigation of the roles of individual genes in flavonoid, condensed tannin, and phenolic glycoside biosynthesis in poplar. Chalcone synthase (CHS) catalyzes the entry point into flavonoid biosynthesis, and, interestingly, poplar has an expanded repertoire of genes encoding CHS enzymes (Tsai et al. 2006; Fig. 12-1) relative to *Arabidopsis*, similar to other plants such as legumes. The increased number of poplar *CHS* genes appears to be the result of both the salicoid whole-genome duplication, and of tandem duplication of the genes on linkage groups I and III (Tsai et al. 2006). Consistent with their retention in the poplar genome, duplicated *CHS* genes have distinct developmental expression patterns (Tsai et al. 2006). The annotation of these genes will be useful in ascribing functions to individual gene family members in developmental and stress-related flavonoid biosynthesis in poplar. While structural pathway genes for flavonoid and CT biosynthesis are expressed in all shoot and root organs of *Populus* (Tsai et al. 2006), long-distance movement of flavonoids in *Arabidopsis* has recently been reported (Buer et al. 2007) suggesting participation of additional gene suites in flavonoid and CT control.

Based on sub-cellular distribution, CTs and their flavonoid precursors probably have fundamentally different roles in plant protection. While proanthocyanidin CT monomers accumulate in the vacuoles and cell wall, flavonoid pathway intermediates important for UV-B protection are found in plant nuclei (Saslowsky et al. 2005). Details of the regulation of CTs in vegetative tissues of *Populus* continue to emerge. The accumulation of anthocyanins during autumnal leaf senescence in relation to photosynthetic decline and nutrient resorption has been reported in *Populus* and other woody taxa (Hoch et al. 2003; Keskitalo et al. 2005). CT induction during short day-induced bud formation of *Populus* and its possible regulation by abscisic acid has also been reported (Ruttink et al. 2007). The participation of flavonoids in processes as fundamental as reproductive development and mycorrhizal association have been described in many angiosperm species (Scervino et al. 2005) but not *Populus*.

PGs are as abundant in *Populus* foliage as CTs, but PG abundance appears to be regulated in an agonistic or competitive manner with respect to CT abundance (Lindroth and Hwang 1996; Harding et al. 2005). The apparently agonistic pattern continues throughout the ontogeny of quaking aspen, during which there is a shift toward increased CT and decreased PG abundance in foliage (Donaldson et al. 2006b). The inheritance of CT and PG abundance traits does not appear, however, to be strictly additive (Hardig et al. 2000). The chemistry and diversity of the major salicin-containing *Populus* PGs has been thoroughly documented (Pearl and Darling 1971), but the PG biosynthetic pathway remains unsolved (Zenk 1967; Morse et al. 2007). Two of the most abundant *Populus* PGs are salicortin and tremulacin (reviewed in (Lindroth and Hwang 1996)). To date there is no clear evidence that the core salicin moiety of those PGs is derived from salicylic acid, as radio-labeled salicylic acid is not incorporated into salicin, nor do salicortin or tremulacin levels exhibit any change in abundance in *nahG* transgenic poplar (Zenk 1967; Morse et al. 2007). Therefore, although the assumption supported by the experimental use of PAL enzyme inhibitors is that PG biosynthesis depends on the phenylpropanoid pathway (Ruuhola and Julkunen-Tiitto 2003), the biosynthetic relationship of PG to CT and lignin remains unknown. We have recently found through stable isotope feeding, that benzoates, but not salicylates, are likely precursors of higher order PGs in *Populus* (B.A. Babst, S. Harding and C.J. Tsai, unpublished results). PGs appear to participate in a complex suite of *Populus* fitness traits including generalist herbivore deterrence, attraction of specific insects, and UV-B screening (Lindroth and Hwang 1996; Warren et al. 2003). The biosynthetic pathway of coniferyl benzoate, an important phenylpropanoid-derived anti-nutritive in flower buds of quaking aspen (Jakubas et al. 1993), has also not been described.

With their large contribution to leaf dry mass in *Populus*, PGs and CTs shape carbon flow and community diversity in habitats dominated by these species (Schweitzer et al. 2004; Fischer et al. 2006; Fischer et al. 2007). The potential growth cost of carbon allocation to CT and PG has prompted the application of a number of trade-off based theoretical mechanisms to describe growth and fitness in *Populus* and other pioneering tree taxa (reviewed in Stamp 2003). CT and PG function as sinks, since they accumulate throughout leaf growth and exhibit slower turnover than labile carbohydrates (Kleiner et al. 1999; Ruuhola and Julkunen-Tiitto 2003). Calculated on the basis of the biosynthetic steps of CT, its metabolic cost is comparable to that of proteins (Kandil et al. 2004). The biosynthetic cost of PG cannot be calculated with certainty, but glucose comprises 30% to more than 40% of PG mass. It is not very surprising therefore, that genetic modulation of aspen growth under sub-optimal nutrient conditions is at least partially due to costs of PG accrual (Osier and Lindroth 2006). Feeding of phenylpropanoid pathway inhibitors to investigate effects of PG biosynthesis on growth and on other phenylpropanoid branch pathways have yielded mixed results in willow species (Ruuhola and Julkunen-Tiitto 2003; Keski-Saari et al. 2007). In part, the varied responses are due to additional regulation of phenylpropanoid homeostasis by carbon that enters the biosynthetic network downstream of PAL (e.g., Mattson et al. 2005; Keski-Saari et al. 2007).

There is enormous plasticity in the control of CT and PG accrual in *Populus*. There is also evidence for strong genetic control of CT among cottonwood genotypes inhabiting elevational gradients (Schweitzer et al. 2004; Fischer et al. 2006). The metabolic basis for CT and PG accrual, and thus the potential cost of their biosynthesis to growth is likely to depend on the mode of induction, whether by nutrient, light, pathogen or herbivore attack. Nitrogen deficiency reliably decreases growth and eventually photosynthesis in *Populus* and stimulates the accrual of CT (Donaldson et al. 2006a). Here, the control of phenylpropanoid flux depends on nitrate sensing and involves a decrease in starch utilization and carbohydrate metabolism in response to N deficiency (e.g., Fritz et al. 2006). Consistent with this, the trade-off models used to conceptualize phenolic metabolism and growth in pioneering tree species predict passive utilization of carbon for secondary metabolism when growth is already nutrient limited (Bryant et al. 1983; Herms and Mattson 1992; Glynn et al. 2007). Signifying a more active and metabolically costly induction, insect feeding and wounding cause changes in transcript levels of phenlypropanoid as well as genes of primary metabolism (Major and Constabel 2006). Sink strength, sucrose transport to, and invertase activity at sites of CT accrual are enhanced during the systemic response of *Populus* to wounding signals (Arnold et al. 2004). Therefore, active accrual of phenylpropanoid metabolites during a wound response might be more costly to plant growth than passive accrual during nutrient deficiency.

   Much remains unknown about the control of carbon partitioning between phenylpropanoid pathway branches. Although distinct phenylpropanoid pathway gene paralogs are expressed in CT and lignin accumulating cells of *Populus* (Harding et al. 2002; Kao et al. 2002), there are no definitive reports on transcription factor regulation of phenylpropanoid composition in *Populus* tissues. However, various members of the Myb family of transcription factors may coordinately regulate partitioning between phenylpropanoid branches in *Populus* as they do in other species (Tamagnone et al. 1998; Karpinska et al. 2004). In addition to the complex genetic additivity of CT and PG accrual, evidence for feedback control of PG accumulation has been observed in transgenic *Populus* with altered lignin biosynthesis or polymerization (Ranocha et al. 2002; Coleman et al. 2008). It is also important to note that post transcriptional regulation plays an important role in the flavonoid pathway network through metabolic channeling and multi-functional enzymes (Turnbull et al. 2004; Winkel 2004).

   The combination of genomics resources, metabolic profiling techniques and pedigreed populations promises to expand the known inventory of functions for CTs, PGs and other non-structural phenylpropanoids in *Populus* growth and development. Flavonoids alone have evolved a complex assortment of roles in plant function that will be challenging to dissect in tree species. The hallmark function of certain flavonoids as UV-irradiation filters in terrestrial plants was probably preceded by roles as internal regulatory agents in aquatic ancestors (Stafford 1991). Demonstrations of flavonoid regulatory function in signaling and auxin transport exemplify the continued evolution of this pathway in land plants (Winkel-Shirley 2001). A dense linkage map created using *P. fremontii* and *P. angustifolia* should lead to the mapping of quantitative trait loci (QTLs) associated with CT and PG function at the ecosystem level (Woolbright et al. 2008). The feasibility of genetical metabolomics approaches to identify QTLs associated with specific flavonoid pathway intermediates has also been demonstrated in *P. deltoides* and *P. nigra* (Morreel et al. 2006). The likelihood that continued changes in atmospheric $CO_2$, $O_3$, and nitrogen will affect CT and PG levels in complex ways as already reported for *Populus*, *Betula* and *Salix* species (e.g., Gupta et al. 2005; Karonen et al. 2006) should provide fertile ground for future applications of QTL and genetical metabolics approaches (Morreel et al. 2006), which will be greatly facilitated by the poplar genome and genomic tools.

## Acknowledgements

Work by the authors relevant to this chapter was supported as follows: CJD, by a Discovery Grant from the Natural Sciences and Engineering Research Council of Canada (NSERC) and by Genome Canada, and the Province of BC; JE, by an NSERC Discovery Grant; SH, by the Office of Science (BER), US Department of Energy, Grant No. DE-FG02-05ER64112 and by NSF Plant Genome Project DBI-0421756.

## References

AGI (2000) Analysis of the genome sequence of the flowering plant *Arabidopsis thaliana*. Nature 408: 796–815.

Arnold T, Appel H, Patel V, Stocum E, Kavalier A, Schultz J (2004) Carbohydrate translocation determines the phenolic content of *Populus* foliage: a test of the sink-source model of plant defense. New Phytol 164: 157–164.

Ayres MP, Clausen TP, MacLean SF, Redman AM, Reichardt PB (1997) Diversity of structure and antiherbivore activity in condensed tannins. Ecology 78: 1696–1712.

Bischoff M, Rosler J, Raesecke HR, Gorlach J, Amrhein N, Schmid J (1996) Cloning of a cDNA encoding a 3-dehydroquinate synthase from a higher plant, and analysis of the organ-specific and elicitor-induced expression of the corresponding gene. Plant Mol Biol 31: 69–76.

Boerjan W, Ralph J, Baucher M (2003) Lignin biosynthesis. Annu Rev Plant Biol 54: 519–546.

Bonner CA, Jensen RA (1994) Cloning of cDNA encoding the bifunctional dehydroquinase. shikimate dehydrogenase of aromatic-amino-acid biosynthesis in *Nicotiana tabacum*. Biochem J 302: 11–14.

Bowman JL, Floyd SK, Sakakibara K (2007) Green genes-comparative genomics of the green branch of life. Cell 129: 229–234.

Brunner AM, Busov VB, Strauss SH (2004) Poplar genome sequence: functional genomics in an ecologically dominant plant species. Trends Plant Sci 9: 49–56.

Bryant JP, Chapin FS, Klein DR (1983) Carbon nutrient balance of boreal plants in relation to vertebrate herbivory. Oikos 40: 357–368.

Buer CS, Muday GK, Djordjevic MA (2007) Flavonoids are differentially taken up and transported long distances in *Arabidopsis*. Plant Physiol 145: 478–490.

Cho MH, Corea OR, Yang H, Bedgar DL, Laskar DD, Anterola AM, Moog-Anterola FA, Hood RL, Kohalmi SE, Bernards MA, Kang C, Davin LB, Lewis NG (2007) Phenylalanine biosynthesis in *Arabidopsis thaliana*. Identification and characterization of arogenate dehydratases. J Biol Chem 282: 30827–30835.

Coleman HD, Park JY, Nair R, Chapple C, Mansfield SD (2008) RNAi-mediated suppression of p-coumaroyl-CoA 3 '-hydroxylase in hybrid poplar impacts lignin deposition and soluble secondary metabolism. Proc Natl Acad Sci USA 105: 4501–4506.

Coruzzi G, Last R (2000) Amino acids In: B Buchanan ,W Gruissem, R Jones (eds) Biochemistry and Molecular Biology of Plants. American Society of Plant Physiologists, Rockville, MD, USA, pp 379–395.

Costa MA, Collins RE, Anterola AM, Cochrane FC, Davin LB, Lewis NG (2003) An in silico assessment of gene function and organization of the phenylpropanoid pathway metabolic networks in *Arabidopsis thaliana* and limitations thereof. Phytochem. 64: 1097–1112.

Cukovic D, Ehlting J, VanZiffle JA, Douglas CJ (2001) Structure and evolution of 4-coumarate: coenzyme A ligase (4CL) gene families. Biol Chem 382: 645–654.

de Azevedo Souza C, Barbazuk B, Ralph SG, Bohlmann J, Hamberger B, Douglas CJ (2008) Genome-wide analysis of a land plant-specific acyl:coenzyme A synthetase (*ACS*) gene family in *Arabidopsis*, poplar, rice and *Physcomitrella*. New Phytol 179: 987–1003.

Ding L, Hofius D, Hajirezaei MR, Fernie AR, Bornke F, Sonnewald U (2007) Functional analysis of the essential bifunctional tobacco enzyme 3-dehydroquinate dehydratase/shikimate dehydrogenase in transgenic tobacco plants. J Exp Bot 58: 2053–2067.

Dixon R, Achnine L, Kota P, Liu C-J, Reddy M, Wang L(2002) The phenylpropanoid pathway and plant defence—a genomics perspective. Mol Plant Pathol 3: 371–390.

Donaldson JR, Kruger EL, Lindroth RL (2006a) Competition- and resource-mediated tradeoffs between growth and defensive chemistry in trembling aspen (*Populus tremuloides*). New Phytol 169: 561–570.

Donaldson JR, Stevens MT, Barnhill HR, Lindroth RL (2006b) Age-related shifts in leaf chemistry of clonal aspen (*Populus tremuloides*). J Chem Ecol 32: 1415–1429.

Eberhard J, Raesecke HR, Schmid J, Amrhein N (1993) Cloning and expression in yeast of a higher plant chorismate mutase Molecular cloning, sequencing of the cDNA and characterization of the *Arabidopsis thaliana* enzyme expressed in yeast. FEBS Lett 334: 233–236.

Ehlting J, Mattheus N, Aeschliman DS, Li E, Hamberger B, Cullis IF, Zhuang J, Kaneda M, Mansfield SD, Samuels L, Ritland K, Ellis BE, Bohlmann J, Douglas CJ (2005) Global transcript profiling of primary stems from *Arabidopsis thaliana* identifies candidate genes for missing links in lignin biosynthesis and transcriptional regulators of fiber differentiation. Plant J 42: 618–640.

Fischer DG, Hart SC, Rehill BJ, Lindroth RL, Keim P, Whitham TG (2006) Do high-tannin leaves require more roots? Oecologia 149: 668–675.

Fischer DG, Hart SC, LeRoy CJ, Whitham TG (2007) Variation in below-ground carbon fluxes along a *Populus* hybridization gradient. New Phytol 176: 415–425.

Fritz C, Palacios-Rojas N, Feil R, Stitt M (2006) Regulation of secondary metabolism by the carbon-nitrogen status in tobacco: nitrate inhibits large sectors of phenylpropanoid metabolism. Plant J 46: 533–548.

Ganson RJ, D'Amato TA, Jensen RA (1986) The two-Isozyme system of 3-deoxy-d-arabino-heptulosonate 7-phosphate synthase in *Nicotiana silvestris* and other higher plants. Plant Physiol 82: 203–210.

Glynn C, Herms DA, Orians CM, Hansen RC, Larsson S (2007) Testing the growth-differentiation balance hypothesis: dynamic responses of willows to nutrient availability. New Phytol 176: 623–634.

Gorlach J, Raesecke HR, Rentsch D, Regenass M, Roy P, Zala M, Keel C, Boller T, Amrhein N, Schmid J (1995) Temporally distinct accumulation of transcripts encoding enzymes of the prechorismate pathway in elicitor-treated, cultured tomato cells. Proc Natl Acad Sci USA 92: 3166–3170.

Goujon T, Sibout R, Eudes A, MacKay J, Jouanin L (2003) Genes involved in the biosynthesis of lignin precursors in *Arabidopsis thaliana*. Plant Physiol Biochem 41: 677–687.

Gupta P, Duplessis S, White H, Karnosky DF, Martin F, Podila GK (2005) Gene expression patterns of trembling aspen trees following long-term exposure to interacting elevated CO2 and tropospheric O-3. New Phytol 167: 129–142.

Hahlbrock K, Scheel D (1989) Physiology and molecular biology of phenylpropanoid metabolism. Annu. Rev Plant Physiol. Plant Mol Biol 40: 347–369.

Hamberger B, Ehlting J, Barbazuk B, Douglas C (2006) Comparative genomics of the shikimate pathway in Arabidopsis, *Populus trichocarpa* and *Oryza sativa*: shikimate pathway gene family structure and identification of candidates for missing links in phenylalanine biosynthesis In: J Romeo (ed) Recent Advances in Phytochemistry, vol 40. Integrative Plant Biochemistry. Elsevier Ltd., Amsterdam, The Netherlands, pp 85–113.

Hamberger B, Ellis M, Friedmann M, Souza CDA, Barbazuk B, Douglas CJ (2007) Genome-wide analyses of phenylpropanoid-related genes in *Populus* trichocarpa, *Arabidopsis thaliana*, and *Oryza sativa*: the *Populus* lignin toolbox and conservation and diversification of angiosperm gene families. Can J Bot 85: 1182–1201.

Hardig TM, Brunsfeld SJ, Fritz RS, Morgan M, Orians CM (2000) Morphological and molecular evidence for hybridization and introgression in a willow (*Salix*) hybrid zone. Mol Ecol 9: 9–24.

Harding SA, Leshkevich J, Chiang VL, Tsai CJ (2002) Differential substrate inhibition couples kinetically distinct 4-coumarate: coenzyme A ligases with spatially distinct metabolic roles in quaking aspen. Plant Physiol 128: 428–438.

Harding SA, Jiang HY, Jeong ML, Casado FL, Lin HW, Tsai CJ (2005) Functional genomics analysis of foliar condensed tannin and phenolic glycoside regulation in natural cottonwood hybrids. Tree Physiol 25: 1475–1486.

Herrmann KM, Weaver LM (1999) The shikimate pathway. Annu Rev Plant Physiol Plant Mol Biol 50: 473–503.

Herms DA, Mattson WJ (1992) The dilemma of plants—to grow or defend. Quart Rev Biol 67: 283–335.

Hoch WA, Singsaas EL, McCown BH (2003) Resorption protection. Anthocyanins facilitate nutrient recovery in autumn by shielding leaves from potentially damaging light levels. Plant Physiol 133: 1296–1305.

Hsieh MH, Goodman HM (2002) Molecular characterization of a novel gene family encoding ACT domain repeat proteins in *Arabidopsis*. Plant Physiol 130: 1797–1806.

Humphreys JM, Chapple C (2002) Rewriting the lignin roadmap. Curr Opin Plant Biol 5: 224–229.

Jakubas WJ, Wentworth BC, Karasov WH (1993) Physiological and behavioral effects of coniferyl benzoate on avian reproduction. J Chem Ecol 19: 2353–2377.

Jansson S, Douglas CJ (2007) *Populus*: a model system for plant biology. Annu Rev Plant Biol 58: 435–458.

Kandil FE, Grace MH, Seigler DS, Cheeseman JM (2004) Polyphenolics in *Rhizophora mangle* L. leaves and their changes during leaf development and senescence. Trees-Struct Funct 18: 518–528.

Kao YY, Harding SA, Tsai CJ (2002) Differential expression of two distinct phenylalanine ammonia-lyase genes in condensed tannin-accumulating and lignifying cells of quaking aspen. Plant Physiol 130: 796–807.

Karonen M, Ossipov V, Ossipova S, Kapari L, Loponen J, Matsumura H, Kohno Y, Mikami C, Sakai Y, Izuta T, Pihlaja K (2006) Effects of elevated carbon dioxide and ozone on foliar proanthocyanidins in *Betula platyphylla*, *Betula ermanii*, and *Fagus crenata* seedlings. J Chem Ecol 32: 1445–1458.

Karpinska B, Karlsson M, Srivastava M, Stenberg A, Schrader J, Sterky F, Bhalerao R, Wingsle G (2004) MYB transcription factors are differentially expressed and regulated during secondary vascular tissue development in hybrid aspen. Plant Mol Biol 56: 255–270.

Keith B, Dong XN, Ausubel FM, Fink GR (1991) Differential induction of 3-deoxy-D-arabino-heptulosonate 7-phosphate synthase genes in *Arabidopsis thaliana* by wounding and pathogenic attack. Proc Natl Acad Sci USA 88: 8821–8825.

Keski-Saari S, Falck M, Heinonen J, Zon J, Julkunen-Tiitto R (2007) Phenolics during early development of *Betula pubescens* seedlings: inhibition of phenylalanine ammonia lyase. Trees-Struct Funct 21: 263–272.

Keskitalo J, Bergquist G, Gardestrom P, Jansson S (2005) A cellular timetable of autumn senescence. Plant Physiol 139: 1635–1648.

Klee HJ, Muskopf YM, Gasser CS (1987) Cloning of an *Arabidopsis thaliana* gene encoding 5-enolpyruvylshikimate-3-phosphate synthase: sequence analysis and manipulation to obtain glyphosate-tolerant plants. Mol Gen Genet 210: 437–442.

Kleiner KW, Raffa KF, Dickson RE (1999) Partitioning of C-14-labeled photosynthate to allelochemicals and primary metabolites in source and sink leaves of aspen: evidence for secondary metabolite turnover. Oecologia 119: 408–418.

Lindroth RS, Hwang SY (1996) Diversity, redundancy and multiplicity in chemical defense systems of aspen In: J Romero, J Saunders, PP Barbarosa (eds) Phytochemical Diversity and Redundancy in Ecological Interactions. Plenum Press, New York, pp 25–56.

Lu S, Zhou Y, Li L, Chiang VL (2006) Distinct roles of cinnamate 4-hydroxylase genes in *Populus*. Plant Cell Physiol 47: 905–914.

Major IT, Constabel CP (2006) Molecular analysis of poplar defense against herbivory: comparison of wound- and insect elicitor-induced gene expression. New Phytol 172: 617–635.

Mattson W, Julkunen-Tiitto R, Herms D (2005) CO2 enrichment and carbon partitioning to phenolics: Do plant responses accord better with the protein competition or the growth differentiation balance models? Oikos 111: 337–347.

Miranda M, Ralph SG, Mellway R, White R, Heath MC, Bohlmann J, Constabel CP (2007) The transcriptional response of hybrid poplar (*Populus trichocarpa* x *P. deltoides*) to infection by *Melampsora medusae* leaf rust involves induction of flavonoid pathway genes leading to the accumulation of proanthocyanidins. Mol Plant-Microb Interact 20: 816–831.

Mobley EM, Kunkel BN, Keith B (1999) Identification, characterization and comparative analysis of a novel chorismate mutase gene in *Arabidopsis thaliana*. Gene 240: 115–123.

Morreel K, Goeminne G, Storme V, Sterck L, Ralph J, Coppieters W, Breyne P, Steenackers M, Georges M, Messens E, Boerjan W (2006) Genetical metabolomics of flavonoid biosynthesis in *Populus*: a case study. Plant J 47: 224–237.

Morse AM, Tschaplinski TJ, Dervinis C, Pijut PM, Schmelz EA, Day W, Davis JM (2007) Salicylate and catechol levels are maintained in nahG transgenic poplar. Phytochemistry 68: 2043–2052.

Mousdale DM, Coggins JR (1986) Rapid chromatographic purification of glyphosate-sensitive 5-enolpyruvylshikimate 3-phosphate synthase from higher plant chloroplasts. J Chromatogr 367: 217–222.

Northup RR, Yu ZS, Dahlgren RA, Vogt KA (1995) Polyphenol control of nitrogen release from pine litter. Nature 377: 227–229.

Osier TL, Lindroth RL (2006) Genotype and environment determine allocation to and costs of resistance in quaking aspen. Oecologia 148: 293–303.

Pearl IA, Darling SF (1971) The structures of salicortin and tremulacin. Phytochemistry 10: 3161–3166.

Peters DJ, Constabel CP (2002) Molecular analysis of herbivore-induced condensed tannin synthesis: cloning and expression of dihydroflavonol reductase from trembling aspen (*Populus tremuloides*). Plant J 32: 701–712.

Pilate G, Guiney E, Holt K, Petit-Conil M, Lapierre C, Leple JC, Pollet B, Mila I, Webster EA, Marstorp HG, Hopkins DW, Jouanin L, Boerjan W, Schuch W, Cornu D, Halpin C

(2002) Field and pulping performances of transgenic trees with altered lignification. Nat Biotechnol 20: 607–612.

Raes J, Rohde A, Christensen JH, Van de Peer Y, Boerjan W (2003) Genome-wide characterization of the lignification toolbox in *Arabidopsis*. Plant Physiol 133: 1051–1071.

Ralph S, Oddy C, Cooper D, Yueh H, Jancsik S, Kolosova N, Philippe RN, Aeschliman D, White R, Huber D, Ritland CE, Benoit F, Rigby T, Nantel A, Butterfield YS, Kirkpatrick R, Chun E, Liu J, Palmquist D, Wynhoven B, Stott J, Yang G, Barber S, Holt RA, Siddiqui A, Jones SJ, Marra MA, Ellis BE, Douglas CJ, Ritland K, Bohlmann J (2006) Genomics of hybrid poplar (*Populus trichocarpa* x *deltoides*) interacting with forest tent caterpillars (*Malacosoma disstria*): normalized and full-length cDNA libraries, expressed sequence tags, and a cDNA microarray for the study of insect-induced defences in poplar. Mol Ecol 15: 1275–1297.

Ranocha P, Chabannes M, Chamayou S, Danoun S, Jauneau A, Boudet AM, Goffner D (2002) Laccase down-regulation causes alterations in phenolic metabolism and cell wall structure in poplar. Plant Physiol 129: 145–155.

Ruttink T, Arend M, Morreel K, Storme V, Rombauts S, Fromm J, Bhalerao RP, Boerjan W, Rohde A (2007) A molecular timetable for apical bud formation and dormancy induction in poplar. Plant Cell 19: 2370–2390.

Ruuhola TR, and Julkunen-Tiitto MRK (2003) Trade-off between synthesis of salicylates and growth of micropropagated *Salix pentandra*. J Chem Ecol 29: 1565–1588.

Saslowsky DE, Warek U, Winkel BSJ (2005) Nuclear localization of flavonoid enzymes in *Arabidopsis*. J Biol Chem 280: 23735–23740.

Scervino JM, Ponce MA, Erra-Bassells R, Vierheilig H, Ocampo JA, Godeas A (2005) Arbuscular mycorrhizal colonization of tomato by *Gigaspora* and *Glomus* species in the presence of root flavonoids. J Plant Physiol 162: 625–633.

Schaller A, Schmid J, Leibinger U, Amrhein N (1991) Molecular cloning and analysis of a cDNA coding for chorismate synthase from the higher plant Corydalis sempervirens Pers. J Biol Chem 266: 21434–21438.

Schmid J, Schaller A, Leibinger U, Boll W, Amrhein N (1992) The *in vitro* synthesized tomato shikimate kinase precursor is enzymatically active and is imported and processed to the mature enzyme by chloroplasts. Plant J 2: 375–383.

Schweitzer JA, Bailey JK, Rehill BJ, Martinsen GD, Hart SC, Lindroth RL, Keim P, Whitham TG (2004) Genetically based trait in a dominant tree affects ecosystem processes. Ecol Lett 7: 127–134.

Shockey JM, Fulda MS, Browse J (2003) Arabidopsis contains a large superfamily of acyl-activating enzymes. Phylogenetic and biochemical analysis reveals a new class of acyl-coenzyme A synthetases. Plant Physiol 132: 1065–1076.

Stafford HA (1991) Flavonoid evolution—an enzymatic approach. Plant Physiol. 96: 680–685.

Stamp N (2003) Out of the quagmire of plant defense hypotheses. Quart Rev Biol 78: 23–55.

Swigonova Z, Lai J, Ma J, Ramakrishna W, Llaca V, Bennetzen JL, Messing J (2004) On the tetraploid origin of the maize genome. Comp Funct Genom 5: 281–284.

Tamagnone L, Merida A, Parr A, Mackay S, Culianez-Macia FA, Roberts K, Martin C (1998) The AmMYB308 and AmMYB330 transcription factors from antirrhinum regulate phenylpropanoid and lignin biosynthesis in transgenic tobacco. Plant Cell 10: 135–154.

Tsai CJ, Harding SA, Tschaplinski TJ, Lindroth RL, Yuan YN (2006) Genome-wide analysis of the structural genes regulating defense phenylpropanoid metabolism in *Populus*. New Phytol 172: 47–62.

Turnbull JJ, Nakajima J, Welford RW, Yamazaki M, Saito K, Schofield CJ (2004) Mechanistic studies on three 2-oxoglutarate-dependent oxygenases of flavonoid biosynthesis: anthocyanidin synthase, flavonol synthase, and flavanone 3beta-hydroxylase. J Biol Chem 279: 1206–1216.

Tuskan GA, Difazio S, Jansson S, Bohlmann J, Grigoriev I, Hellsten U, Putnam N, Ralph S, Rombauts S, Salamov A, Schein J, Sterck L, Aerts A, Bhalerao RR, Bhalerao RP, Blaudez

D, Boerjan W, Brun A, Brunner A, Busov V, Campbell M, Carlson J, Chalot M, Chapman J, Chen GL, Cooper D, Coutinho PM, Couturier J, Covert S, Cronk Q, Cunningham R, Davis J, Degroeve S, Dejardin A, Depamphilis C, Detter J Dirks B, Dubchak I, Duplessis S, Ehlting J, Ellis B, Gendler K, Goodstein D, Gribskov M, Grimwood J, Groover A, Gunter L, Hamberger B, Heinze B, Helariutta Y, Henrissat B, Holligan D, Holt R, Huang W, Islam-Faridi N, Jones S, Jones-Rhoades M, Jorgensen R, Joshi C, Kangasjarvi J, Karlsson J, Kelleher C, Kirkpatrick R, Kirst M, Kohler A, Kalluri U, Larimer F, Leebens-Mack J, Leple JC, Locascio P, Lou Y, Lucas S, Martin F, Montanini B, Napoli C, Nelson DR, Nelson C, Nieminen K, Nilsson O, Pereda V, Peter G, Philippe R, Pilate G, Poliakov A, Razumovskaya J, Richardson P, Rinaldi C, Ritland K, Rouze P, Ryaboy D, Schmutz J, Schrader J, Segerman B, Shin H, Siddiqui A, Sterky F, Terry A, Tsai CJ, Uberbacher E, Unneberg P, Vahala J, Wall K, Wessler S, Yang G, Yin T, Douglas C, Marra M, Sandberg G, Van de Peer Y, Rokhsar D (2006) The genome of black cottonwood, *Populus trichocarpa* (Torr. & Gray). Science 313: 1596–1604.

Warren JM, Bassman JH, Fellman JK, Mattinson DS, Eigenbrode S (2003) Ultraviolet-B radiation alters phenolic salicylate and flavonoid composition of *Populus trichocarpa* leaves. Tree Physiol 23: 527–535.

Winkel BSJ (2004) Metabolic channeling in plants. Annu Rev Plant Biol 55: 85–107.

Winkel-Shirley B (2001) Flavonoid biosynthesis. A colorful model for genetics, biochemistry, cell biology, and biotechnology. Plant Physiol 126: 485–493.

Woolbright SA, DiFazio SP, Yin T, Martinsen GD, Zhang X, Allan GJ, Whitham TG, Keim P (2008) A dense linkage map of hybrid cottonwood (*Populus fremontii* x *P. angustifolia*) contributes to long-term ecological research and comparison mapping in a model forest tree. Heredity 100: 59–70.

Yamada T, Matsuda F, Kasai K, Fukuoka S, Kitamura K, Tozawa Y, Miyagawa H, Wakasa K (2008) Mutation of a rice gene encoding a phenylalanine biosynthetic enzyme results in accumulation of phenylalanine and tryptophan. Plant Cell 20: 1316–1329.

Zenk M (1967) Pathways of salicyl alcohol and salicin formation in *Salix purpurea* L. Phytochemistry 6: 245–252.

# 13

# Ecogenomics of Mycorrhizal Interactions Mediated by the Aspen Genome under Elevated Ozone and Carbon Dioxide[#]

*Gopi K. Podila,[1,a,†] R. Michael Miller,[2,*] Leland J. Cseke,[1]*
*Holly L. White[3] and Victoria Allison[4]*

## ABSTRACT

The molecular mechanisms controlling the response of keystone organism responses to environmental perturbations can provide insights to system responses at higher scales of organization. This is especially true for responses to elevated atmospheric levels of the greenhouse gas $CO_2$ ($eCO_2$) and the atmospheric pollutant ozone ($eO_3$). Mycorrhizae are a biotrophic association between roots and fungi that form an essential component of terrestrial ecosystems. The kinds of molecular interactions between mycorrhizal fungi and their host roots cause strong feedbacks with cascading implications for host function. In return, host functional responses to this symbiotic association have profound implications to mycorrhizal fungal and rhizosphere community composition. Understanding these molecular determinants

---

[#]The submitted manuscript has been created by the University of Chicago as operator of Argonne National Laboratory under Contract No. W-31-109-ENG-38 with the U.S. Department of Energy. The U.S. government retains for itself, and others acting on its behalf, a paid-up, nonexclusive, irrevocable worldwide license in said article to reproduce, prepare derivative works, distribute copies to the public, and perform publicly and display publicly, by or on behalf of the government.

[1]Dept. of Biological Sciences, University of Alabama, Huntsville, AL 35899, USA.
[a]e-mail :*podilag@uah.edu*
[2]Biosciences Division, 9700 South Cass Ave, Argonne National Laboratory, Argonne, IL 60439, USA; e-mail: *rmmiller@anl.gov*
[3]DIATHERIX Laboratories, Inc. Huntsville, AL 35806.
[4]Ministry of Agriculture and Forestry Auckland, New Zealand; e-mail: *victoria.allison@maf.govt.nz*
[†]Dr. Podila passed away on February 12, 2010.
[*]Corresponding author

will lead to an improved mechanistic understanding of this keystone association where the identification of dysfunctional expression of critical metabolic pathways has the potential for being an early warning transponder to systems under stress. By using phenotypic, physiological and gene expression differences among aspen (*Populus tremuloides*) clones having a competitive advantage when grown in a FACE (Free Air Carbon Dioxide Environment) environment supplemented with ozone, we have a powerful tool with which to elucidate the genetic basis of the mycorrhizal association controlling or influencing host growth and survival. The experiment offers a unique opportunity to study the responses of specific tree genotype and associated mycorrhizal fungi to $eCO_2$ and $eO_3$. The experimental design enabled us to determine how the above-ground responses of three aspen genotypes to $eCO_2$, $eO_3$ and the interaction of $eCO_2+eO_3$ influenced below-ground mycorrhizal fungal growth and allocation, as well as rhizosphere dynamics. Here we describe some of the physiological and molecular mechanisms associated with aspen genotype responses to $eCO_2$ and $eO_3$ by using microarray expression profiles and metabolic profiling of the AM fungus, *Glomus intraradices* that is associated with aspen roots as a mycorrhizal partner.

**Key words:** mycorrhiza, *Glomus intraradices*, fatty acids, *Populus tremuloides*, elevated $CO_2$, elevated ozone, expression profiles

## 13.1 Introduction

Human activities have greatly accelerated the rates of global environmental change (Vitousek et al. 1997). Understanding the consequences of these changes for forest ecosystems is one of the more pressing challenges confronting the scientific community today, given the importance of forests in global net primary production (NPP), carbon sequestration, human economies, and as repositories of biodiversity (Hassan et al. 2005). The effects of environmental change will occur first and foremost in the activation or suppression of genes within the species of these systems. Altered gene expression drives plant function, and gene expression at all levels is the first transponder that receives the signal that the atmosphere has changed. Any alteration of gene expression sequence that controls the allocation of photosynthate to below-ground structures also has the potential of influencing root function via effects on processes associated with membrane permeability and transporter function, membrane repair, and the production of signaling molecules important in regulating mycorrhizal and pathogen interactions (Garrett et al. 2006). It is important then, in order to better anticipate the consequences of environmental change, to identify those genes that are activated by an alteration of environmental conditions. More specifically, it is important to be able to identify those genes or gene sequences that are good surrogates of the metabolic state of an individual but can also be used to scale from the individual to higher organization levels as a robust indicator of a change in the community or ecosystem properties.

The use of genetic information on the gene groups that are associated with key metabolic and structural processes is critical to this endeavor. It has already been demonstrated for various *Populus* species that there is considerable genetic variation in how processes respond to elevated levels of atmospheric $CO_2$ and $O_3$ (Gupta et al. 2005; Di Baccio et al. 2008; Cseke et al. 2009). The recent advances in transcriptome analysis, genome sequencing and associated genomics and bioinformatics tools provide an unprecedented opportunity to identify the key components of intraspecific and organism-environment interactions that modulate ecosystem responses to global change. By examining patterns of gene expression and metabolic changes associated with the target organisms, identification of the genetic control points regulating plant responses to changing atmospheric chemistry is possible, including how these responses influence below-ground interactions with mycorrhizal fungi.

The mycorrhizal symbiosis is a key regulator of the feedbacks controlling the partitioning of photosynthate and mineral nutrients (Miller and Kling 2000; Graham and Miller 2005; Smith and Read 2008). Because mycorrhizal fungi require host-supplied carbohydrates for growth, any forcing factor or pollutant that alters photoassimilate production and allocation could affect the functional balance of the mycorrhizal symbiosis and, in turn, alter the nutritional status of the plant and its ability to respond to environmental changes (Allen et al. 2003). Hence, accurate predictions about whole-plant responses to a changing environment, especially changes driven by multiple forcing factors, will require a better understanding of how the mycorrhizal fungus responds to changes in host allocation of assimilated carbohydrates and soil nutrients.

Members of the genus *Populus* appear to be an ideal model tree system for elucidating such feedbacks between host and mycorrhizal fungal partners as they respond either directly or indirectly to multiple environmental and anthropogenic forcing factors. However, a major obstacle to predicting plant responses to multiple environmental forcing factors is the current lack of knowledge about the trade-offs between plant carbon allocation and nutrient acquisition. Thus, understanding the interactive effects of elevated atmospheric concentrations of carbon dioxide ($eCO_2$) and ozone ($eO_3$) on the growth and allocation of mycorrhizal fungi requires investigation of the responses of mycorrhizal fungi to multiple-factor stressors.

Much of the research on enhanced atmospheric $CO_2$ levels suggests a potential fertilizer effect on plant growth, whereas the rise in tropospheric $O_3$ levels could easily nullify these predicted growth enhancements through adverse effects on leaf tissue and metabolism. Although the potential importance of interactions among environmental factors is well recognized, the majority of research to date on climate change and anthropogenic factors has mainly addressed the effects of single factors. Therefore, we took on the challenge of identifying mycorrhizal responses in three different genotypes of trembling aspen (*Populus tremuloides* Michx) to FACE delivered $eCO_2$, $eO_3$ and $eCO_2 + eO_3$ treatments. Additionally, we determined the responses not only for aspen roots, but also for their associated arbuscular mycorrhizal (AM) and ectomycorrhizal (ECM) fungi.

Mycorrhizal associations are comprised of several distinctive types that can be phylogenetically grouped for plant and fungus that for the most part are consistent with the structures that form on root systems. *Populus* species are unusual in that their root systems may possess two of the more common types of mycorrhizae, arbuscular mycorrhiza and ectomycorrhiza. Previous research has demonstrated that the levels of root colonization by these fungi are a good indicator for C demand by the mycorrhizal fungus (Graham 2000; Wright et al. 2000; Miller et al. 2002). The costs of the symbiosis are usually expressed in terms of the amount of photosynthate allocated to the fungus and associated root tissue and suggest up to 30% of the plant's total C budget can be assimilated by the fungus (Graham 2000). Respiration of mycorrhizal roots and associated extramatrical mycelia can account for more than 50% of total soil respiration (Bhupinderpal-Singh et al. 2003). Benefits of the symbiosis are usually recognized as improved access for a host to limiting soil resources where the fungus, through exploitation of the soil environment, can deliver up to 80% of a plant's P requirements and up to 25% of its N requirements (Marschner and Dell 1994). Other benefits to the host include improved water relations, increased resistance to pathogens, and modification of the soil environment via improvement of soil structure (Daniell et al. 1999; Miller and Jastrow 2000; Augé 2001). In addition, along with the quantity and quality of root inputs, the extraradical mycelia of mycorrhizal fungi are important contributors to the mechanisms associated with carbon sequestration in soils (Langley and Hungate 2003; Zhu and Miller 2003).

Although many plants grown under $eCO_2$ have demonstrated biomass allocation that favors root growth at the expense of shoots (Fitter et al. 2000; Norby et al. 2004; Alberton et al. 2005), data on the responses of mycorrhizal fungi to $eCO_2$ under field or free-air carbon dioxide enrichment (FACE) growing conditions have been limited by their short duration (Lukac et al. 2003; Gamper et al. 2004; Staddon et al. 2004; Pritchard et al. 2008). Reported studies indicate a range of observed responses from increased amounts of fungal growth to no clear effects with ectomycorrhizal fungi being more responsive to enhanced levels of atmospheric $CO_2$ than arbuscular mycorrhizal fungi (Treseder 2004; Alberton et al. 2005; Andrew and Lilleskov 2009).

In contrast to the generally observed stimulation of plant growth caused by enhanced atmospheric $CO_2$, increased atmospheric concentrations of ozone is toxic to plants (Andersen 2003; Karnosky et al. 2007; Lindroth 2010). Ozone exposure has been shown to be detrimental to the growth of many North American trees, including trembling aspen. Natural selection for $O_3$ tolerance has also been observed in populations of aspen trees experiencing long term exposure to higher background levels of $O_3$ (Berrang et al. 1996). A primary consequence of enhanced background exposure to $O_3$ is a cascade of physiological responses that are a direct response of the highly reactive $O_3$ damage to cellular lipids and proteins (Lindroth 2010). Plant response to $O_3$ damage is a greater allocation of growth to above-ground structures at the expense of below-ground growth. The effects of $eO_3$ on mycorrhizal fungi are just as variable as the results reported for $eCO_2$, ranging from decreases in mycorrhizal colonization to no obvious effects, with even a few studies showing an increase in colonization (Andersen 2003). More recent studies have demonstrated that different ectomycorrhizal fungal communities appear to develop with ozone fumigation (Grebenc and Kraigher 2007; Andrew and Lilleskov 2009) and that these changes can affect nutrient cycling (Haberer et al. 2007; Zeleznik et al. 2007).

We report on a study of three *Populus tremuloides* genotypes conducted at the Rhinelander Aspen FACE facility that is also capable of $O_3$ fumigation. The facility is located on a simulated regenerating forest ecosystem composed of rapidly growing, shade-intolerant and slower growing, shade-tolerant species of trees. Taking advantage of an ongoing experiment that includes genetically well-characterized aspen genotypes (see Table 13-1), has allowed us to evaluate the effects of $eCO_2$, $eO_3$ and the interaction effects of $eCO_2$ and $eO_3$ on the mycorrhizae of aspen genotypes demonstrating a range of responsiveness to the fumigations.

**Table 13-1** *Populus tremuloides* clones used in this study[1].

| Clone | Origin (County) | $O_3$ Tolerance | $CO_2$ Responsiveness |
|-------|-----------------|-----------------|------------------------|
| 271 | Michigan (Emmet) | Very Tolerant | High |
| 216 | Wisconsin (Bayfield) | Tolerant | Low |
| 259 | Indiana (Porter) | Sensitive | High |

[1]Isebrands et al. 2001; Karnosky et al. 2003.

## 13.2 Methods

### 13.2.1 Study Site

The Aspen FACE facility (*http://aspenface.mtu.edu*) is a 32-ha site located in Rhinelander, WI (45°40.5′N, 89°37.5′E) (See Dickson et al. (2000) for a full facility description). The facility consists of 12 circular, 30-m-diameter rings spaced 100 m apart. Each ring is split into three community types in sections planted with equal tree densities. Half of each ring is planted with five aspen clones of various sensitivities to $O_3$, a quarter is planted with paper birch and aspen in a 2:1 ratio, and the remaining quarter is planted with sugar maple and aspen in a 2:1 ratio. A total of 670 trees were planted in each ring at 1 m × 1 m spacing in June 1997. The experimental design is a split-plot, randomized, complete block design with three replicates of factorial atmospheric $CO_2$ and $O_3$ treatments; the three tree community types split the $CO_2$–$O_3$ main plots. The $CO_2$ and $O_3$ treatments have been applied during each growing season since 1998. The $eCO_2$ rings are maintained at 560 μL $L^{-1}$, which is 200 μL $L^{-1}$ above ambient atmospheric $CO_2$. The $e$O3 rings are fumigated with ≈ 55 nL $L^{-1}$, which is 20 nL $L^{-1}$ above average ambient atmospheric $O_3$ levels at this site (Karnosky et al. 2003). For this study sampling was limited to the three aspen clones growing in the aspen side of the ring (Table 13-1). Site soils are mixed, frigid, coarse loamy Alfic Haplorthods. The sandy loam A-horizon (≈ 15 cm thick) grades into a loamy B-horizon (≈ 30 cm thick). In general, soil properties vary little across the site (see Table 1 in Dickson et al. 2000).

### 13.2.2 Sampling

In July 2002, a whole-tree harvest was conducted to derive an allometry for determining tree biomass allocation (King et al. 2005). Trees were selected to represent a range in above-ground biomass, based on tree height and bole diameter at 10-cm above soil surface. The harvest allowed us to investigate the relationship between biomass allocation and the quantity of mycorrhizal fungi. Roots were harvested at the center of the severed tree stem using a 25.4-cm diameter corer to a depth of 25 cm. Only coarse and fine roots that were attached to the tap root were collected for analysis. Roots were washed quickly with half of the sample frozen at −20°C, and then freeze-dried. The other portion of roots was placed in liquid $N_2$ until further analyzed for RNA and gene expression analyses (the time period between field extraction of roots and their placement in liquid $N_2$ was usually 30–45 minutes). Identity of AMF partner is determined by PCR using specific primers for *G. intraradices* as described by Maldonado-Mendoza et al. (2002).

## 13.2.3 Determination of Signature Fatty Acids

Phospholipid fatty acids (PLFA)s and neutral lipid fatty acids (NLFA)s were quantified using the freeze-dried root subsample. Lipids were extracted from a 30 mg root sample that had been ground in a Spex mill (Spex-Certiprep Inc., Metuchen, NJ) in a single-phase mixture of chloroform, methanol, and phosphate buffer (pH 7.4) in a ratio of 1:2:0.8, by an adaptation of the method described by Bligh and Dyer (1959). After 3 hours, water and chloroform were added to separate the mixture into polar and nonpolar fractions, and total lipids were extracted from the nonpolar chloroform phase. The PLFAs were separated from other lipid classes by using silicic acid column chromatography (Vestal and White 1989; Allison and Miller 2005). The PLFAs were then methylated by using a mild-alkaline solution, and the samples frozen until analysis.

Prior to analysis, both PLFA and NLFA samples were thawed and dissolved in a 20 ng μL$^{-1}$ solution of FAME 19:0 (Matreya Inc., PA) in hexane, as an internal standard. PLFA separation was by high-resolution fused-silica capillary gas chromatography (GC), using an HP 6890 GC, with an HP7683 autosampler (Agilent Technologies, Palo Alto, CA). A 30 m HP-5MS column was used, with hydrogen as the carrier gas at a constant flow rate of 4.0 mL min$^{-1}$. A 1 μL splitless injection was made for each sample, with the inlet temperature set at 230°C, and the inlet purged at 47.0 ml min$^{-1}$, 0.75 minutes after injection. The oven temperature was held at 80°C for 1 minute, increased at a rate of 20°C min$^{-1}$ to 155°C, and then increased at 5°C min$^{-1}$ to a final temperature of 270°C and held for 5 minutes. Detection of fatty acids was by flame ionization at 350°C. PLFA and NLFA markers including 16:1ω5c and ergosterol were identified by retention time in comparison to known standards, and quantified using the 19:0 fatty acid as internal standard.

## 13.2.4 Mycorrhiza Analysis

Intraradical AMF biomass was converted to a root and area basis by using a conversion factor of 3.92 nmols PLFA 16:1ω5c mg$^{-1}$ AMF biomass (Allison and Miller 2004) and scaling by fine root biomass. Root associated ECM biomass was converted to a root and area basis by using a conversion factor of 4.1 μg ergosterol mg$^{-1}$ ECM fungus and also scaled by multiplying by fine root biomass. The responses of the fungus and tree clones to single and multiple factors and their interactions in association with the various hosts were evaluated using analysis of variance (ANOVA), Pearson product-moment correlations, and a statistical comparison of response curves (SAS, SAS Institute, Version 8.2).

### 13.2.5 Preparation of Oligo Arrays from Glomus Intraradices

Oligo probes (70-mer) were designed from publicly available NCBI EST sequences (~ 4,500) from *G. intraradices* deposited by our laboratory and others. From approximately 4,000 EST sequences available, oligo probes were designed using UOLIGO tool from the Oligospawn suite (Zheng et al. 2004), which produced 1,000 unique oligomers. The source sequences from Genbank were annotated according to their Gene Ontology (GO) function (The Gene Ontology Consortium 2000) by searching for homologies with WU-BLAST version 2.0 (*http://blast.wustl.edu*) using the non-redundant database from the Protein Information Resource (PIR-NREF) release 1.46 (Wu et al. 2003). Accessions that did not return a result were annotated as "Unknown". The oligomer selections were verified by using WU-BLAST to search for redundancies within the oligomers and to search for oligomers that failed to match any EST. Each oligomer was found to be unique and to match at least one EST. Arrays were constructed by the Genome Technologies Support Facility at Michigan State University (*http://www.genomics.msu.edu/*) spotted in quadruplicate format per each slide. These oligos represented genes from signaling, membrane biogenesis, metabolism, defense, protein synthesis, structural, hypothetical as well as unknown categories.

### 13.2.6 RNA Extraction, cDNA Preparation, and Labeling of cDNA

Total RNA from 50–100 mg of mycorrhizal roots was extracted using the RNAeasy kit (Qiagen Inc., CA) and treated with the TURBO DNA-free™ kit (Ambion Inc., Austin, TX) to remove genomic DNA contamination. Preparation of Cy3 and Cy5 labeled targets was done from 1 µg of each RNA sample using the Amino Allyl messageAmp II aRNA kit (Ambion, TX). The yield and dye incorporation were evaluated using a NanoDrop ND-1000 Spectrophotometer (NanoDrop Technologies, Wilmington, DE).

### 13.2.7 Hybridization of Microarrays

The widely accepted MIAME guidelines for microarray analysis and verification (Zimmermann et al. 2006) were followed. For each of the two-channel arrays, RNA collected from mycorrhizal roots grown in the ambient (control) ring was compared directly to RNA derived from the same genotype grown under $eCO_2$ or $eO_3$ or $eO_3 + eCO_2$ following methods found in Cseke et al. 2009. Two independent biological replicates derived from trees grown in three independent replicate FACE rings were used. In addition, each clone and time point included dye swap reciprocal two-color experiments for each biological replicate (Churchill 2002; Allison and Coffey 2003; Allison et al. 2006). Thus, eight data points per cDNA (two biological replicates with four technical replicates each) were used.

All microarray slides were checked using a dissecting microscope for uniformity of spots. Slides were prehybridized for 30 minutes on a slow rotary shaker in 5 × SSC, 0.1% SDS, and 1% BSA, followed by rinsing 20 times in ddH$_2$O. DNA was denatured by incubating slides in 95°C ddH$_2$O for 1 minute, followed by washing in 100% ice-cold ethanol for 15 seconds. Slides were dried by centrifugation (4°C, 2,500 rpm, 2 minutes) and stored at 4°C. Hybridization was performed within 1 hour of prehybridization.

Targets were resuspended in 55 μl hybridization solution (50% formamide, 5 × SSC, 0.1% SDS, and 0.1% BSA) and centrifuged at 14,000 × g for 1 minute before denaturation at 45°C for 5–10 minutes. Hybridization was performed in Corning hybridization chambers (Corning Inc., Acton, MA) with Lifter slips (Erie Scientific Co., Portsmouth, NH) and incubated in a 43°C water bath in a hybridization oven for 36 hours (Sartor et al. 2004). Slides were rinsed in wash solution I (1 × SSC, 0.2% SDS), prewarmed to 45°C, to remove the lifter slip. Slides were washed once in wash solution I for 15 minutes at 45°C with gentle agitation, once in wash solution II (0.1 × SSC, 0.2% SDS) for 10 minutes, and twice in wash solution III (0.1 × SSC) for 2 minutes each. Finally, slides were immersed in ddH$_2$O for 10 seconds, 100% ethanol for 10 seconds, then dried by centrifugation (4°C, 2,500 rpm, 2 minutes) and stored in darkness at room temperature.

### 13.2.8 Slide Scanning and Data Collection

Slides were scanned using a VersArray ChipReader™ scanner (BioRad, Hercules, CA) at 5 μm resolution (Cseke et al. 2009). Cy3 and Cy5 images were aligned and spots flagged using VersArray Analyzer 5.0 (BioRad, CA). After local background subtraction, signal intensity was log transformed and normalized by the LOWESS algorithm with a smoothing parameter of 0.2, using GeneGazer software (Bio-Rad, CA). Normalized intensity values were filtered by a coefficient of variance (CV) cutoff of 0.25, and spots having intensities below 100 in both channels were also excluded from further analysis. Filtered gene lists were subjected to *t*-test with a false discovery rate at 0.1 ($p < 0.1$). Principal component analysis (PCA) was conducted using standard correlation and Hierarchical cluster analysis (HCA) was performed using the Euclidean distance metric and average cluster linkage.

### 13.2.9 Quantitative Real-Time RT-PCR

Expression of selected genes from oligo arrays was validated using QRT-PCR (Table 13-2). DNA-free total RNA was used for cDNA synthesis with oligo-dT primers or gene- or group-specific primers flanking 137 to 290 bp amplicons near the 3′-UTRs were designed corresponding GenBank *G. intraradices* EST sequences. QRT-PCR and data analysis was performed following the well-established protocols for *G. intraradices* (Govindarajulu et al. 2005).

**Table 13-2** Gene expression analysis of *G. intraradices* from roots of aspen clones 216, 259 and 271 exposed to *e*CO$_2$ or *e*O$_3$ or *e*CO$_2$ + *e*O$_3$. Expression levels are calculated against expression levels in the ambient air rings as described in the methods section. Upregulated expression is indicated by '↑' and down regulated expression is indicated by "↓". Fold level of expression are indicated by number of arrows.

| Gene ID | IEA GO terms | 216 eCO$_2$ | 259 eCO$_2$ | 271 eCO$_2$ | 216 eO$_3$ | 259 O$_3$ | 271 eO$_3$ | 216 eCO$_2$+O$_3$ | 259 eCO$_2$+O$_3$ | 271 eCO$_2$+O$_3$ | QRT-PCR |
|---|---|---|---|---|---|---|---|---|---|---|---|
| BI451891 | Amino acid biosynthesis | ↑ | ↓ | ↑ | ↑ | ↓ | | ↓↓ | ↓ | ↑ | |
| BE604005 | Acyltransferase activity | | ↑ | ↑ | | | | ↑ | | ↑↑↑ | 3 |
| BI451906 | DNA binding | | ↓ | | | | | | | ↓↓ | |
| BI451872 | Serine carboxypeptidase | | | | | | | ↑↑ | | ↑↑ | 3 |
| AF260996 | chitin synthase activity | ↑↑ | | | | | | | | ↑↑ | |
| BI451987 | Tryptophan catabolism | | | | | | | | | ↑↑↑ | 3 |
| BM959086 | DNA binding | ↓ | | | ↓ | | | ↓ | | ↓ | |
| BM959485 | DNA-directed RNA polymerase | | | | | | | | ↓ | | |
| CG432119 | DNA binding | ↑↑ | ↓ | | | | | | | | |
| AJ574744 | Gamma-glutamyltransferase | | ↓ | | | | | | | ↓↓ | |
| AU098274 | Glutamate-ammonia ligase | | | | ↓ | ↓ | ↓ | ↓ | | ↓ | |
| BM026972 | glutamine biosynthesis | | | | ↓ | ↓ | ↓ | ↓ | ↑↑↑ | ↑↑ | 3 |
| BI246179 | GTPase activity | | | | | | | | ↑ | ↑↑↑ | |
| BI246191 | Microtubule-based movement | ↓ | | | | | | | | ↓↓↓ | |
| CG431656 | Guanylate cyclase; signaling | | | | | | | ↑↓ | | ↑↓ | |
| CG431645 | Guanylate kinase | | | | | | | | ↓ | ↓ | 3 |
| BM959400 | Integral membrane Protein | | | | | | | | | ↑↑↑ | |
| CG431714 | Clathrin coated vesicle | | | | | | | ↑↓ | ↓ | ↑↓ | |
| CF200419 | Intracellular transport | ↑↑ | | | | | | ↑↓ | | | |
| BM958976 | Cation transport | ↓ | | | | | | ↓ | ↓ | ↓↓ | 3 |
| BI246182 | one-carbon metabolism | | | | | | | ↓ | ↑↓ | ↑↑ | |

| Gene ID | Function | | | | | | | | | | |
|---|---|---|---|---|---|---|---|---|---|---|---|
| BM959166 | Membrane dipeptidase activity | ↓ | ↓ | ↓ | | ↓ | | | ↑↑ | ↑↑↑ | 3 |
| BM959418 | SAM methyltransferase activity | ↓ | ↓ | | | | | | | ↓↓ | |
| BM027046 | Mitotic chromosome condensation | ↓ | ↑ | ↓ | | ↓↑ | | | | | |
| BM959317 | Monooxygenase activity | | | | | | ↓↑ | | ↑↑ | ↑↑ | 3 |
| BE604002 | Unknown | | | | | | | | | ↑↑ | |
| BI452184 | Hypothetical protein | | | | ↑↑ | | | | | ↓↑ | |
| BM027189 | Hypothetical protein | | | | | | ↓↑ | ↓↑ | | ↓↑ | |
| BM439297 | Unknown | | | | | | ↓↑ | | | ↑↑ | |
| BM958918 | Unknown | | | | ↑↑ | | ↑↓ | ↑↓ | ↑↓ | ↑↑ | |
| CF200353 | Hypothetical protein | ↓ | ↓ | | ↓ | | | | | | |
| CG431746 | RNA dependent RNA polymerase | | | | ↓ | | | | | ↓ | 3 |
| CG431891 | Hypothetical protein | ↑↑ | ↓↑ | | | | | | | | |
| CG432123 | Hypothetical protein | ↓↑ | | | | | | | | ↓ | |
| CG431797 | Hypothetical protein | ↓↑ | | | | | | ↓ | ↓ | | |
| BM959245 | Unknown | | ↓↑ | | ↑↑ | | | | ↑↑↑ | ↑↑↑ | 3 |
| CF200246 | Unknown | | | | | | ↑↑ | ↑↑ | ↑↑ | ↓↓ | 3 |
| BM959461 | None found | | | | | | | | | | |
| BM959457 | Hypothetical protein | | | | ↑↑ | ↑↑↑ | ↑ | ↑↑↑ | ↑↑↑ | ↑↑ | 3 |
| BM959574 | Unknown | | | | ↑↑ | ↑ | | ↑↑ | ↑↑ | ↑↑ | |
| BM959464 | Unknown | | | | ↓↑ | | ↓↑ | | ↑↑↑ | ↑↑↑ | 3 |
| BM959332 | Unknown | | | | ↑ | ↑↑ | | | ↑↑ | ↑ | 3 |
| BM959283 | Unknown | | ↓ | | ↑↑↑ | | | | | ↑↑ | |
| BM959287 | Unknown | | ↓↑ | | ↑↑ | | | | | ↑↑ | |
| BM959255 | Hypothetical protein | | ↓↑ | | ↑↑ | | | | | ↑↑ | |
| BI452015 | Unknown | | | | | | | | | ↑↑ | |

*Table 13-2 contd....*

*Table 13-2 contd....*

| Gene ID | IEA GO terms | 216 eCO$_2$ | 259 eCO$_2$ | 271 eCO$_2$ | 216 eO$_3$ | 259 O$_3$ | 271 eO$_3$ | 216 eCO$_2$+O$_3$ | 259 eCO$_2$+O$_3$ | 271 eCO$_2$+O$_3$ | QRT-PCR |
|---|---|---|---|---|---|---|---|---|---|---|---|
| BM958862 | Unknown | | | | | ↑↑ | | | | ↑↑ | |
| BM027139 | Unknown | | | | | | | | | ↑↓ | |
| BM027800 | Unknown | ↑ | ↑↓ | | | | | | | | |
| BM027138 | Unknown | | | | | | | | ↑↑ | | |
| CF200266 | Hypothetical protein | | ↑↓ | | | | | | | | 3 |
| AU098252 | Hypothetical protein | | | | | | ↑ | | ↑↑ | → | |
| BI452331 | Hypothetical protein | | | | | ↑↑ | | | | ↑↓ | |
| BE603813 | Putative Zn transporter | | | | | | | | | ↑↓ | |
| CF619491 | Hypothetical protein | | | | | ↑↑↑↑ | ↑↑ | | ↑↑↑↑ | ↑↑↑ | 3 |
| CG432148 | Hypothetical protein | → | ↑ | | | | | | | ↑↓ | |
| CG432190 | Hypothetical protein | → | | | | | | | | ↑↓ | |
| BM959247 | Helicase activity | | ↑ | | | | | ↑↓ | | → | |
| BI452026 | RNA binding | ↑↑↓ | ↑↓ | | | ↑ | | | | | |
| BM439216 | GTP biosynthesis | ↑↑↓ | | | | ↑ | | | | | |
| BE604019 | Chromosome biogenesis | | | | | | | | | ↑↑ | |
| CG431872 | nucleotide binding | | | | | | | ↑↑ | → | ↑↑ | |
| CG431989 | 3'-5' exonuclease activity | | | | | | | ↑↑ | | | |
| BM959423 | Nucleotide-excision repair | | | | ↑↑ | | | ↑↑ | | | |
| BM027338 | Cyclophilin | | | | | ↑↑↑ | | | ↑↑↑↑ | ↑↑↑ | 3 |
| BM027140 | Protein folding | | | | | | | | | ↑↑ | |
| BI452079 | Protein modification | | | | | | | ↑↓ | ↑ | | |
| BI451974 | Regulation of nitrogen utilization | | | | | | | ↑↑ | | → | 3 |
| BM026944 | RNP reductase | | | | | ↑↓ | | | ↑↓ | | |
| AU082823 | mRNA metabolism | ↑ | | | | | | | | ↑↑ | |
| BI451933 | L5b ribosomal protein | | | | | ↑ | | | → | ↑↑↑ | |
| BM959282 | peptidase activity | | | | | → | | ↑↓ | | ↑↓ | |
| BM027222 | stearoyl-CoA 9-desaturase | | ↑↓ | | | | | ↑↓ | | | 3 |

| Accession | Protein |
|---|---|
| BM027273 | Hypothetical protein |
| BI452277 | Ribosomal protein S8 |
| CF200502 | Ribosome 60S L6 |
| AU082792 | S19 ribosomal protein |
| BM027541 | Ribosome 60S L7 |
| BI451956 | Ras2 like protein |
| CF200424 | 40S ribosomal protein S3a |
| BM027149 | S-phase specific ribosomal protein |
| BM027500 | S2 ribosomal protein |
| BM958906 | Two-component sensor kinase |

## 13.3 Results and Discussion

In July of 2002, roots were collected in association with the whole-tree harvest at the Aspen FACE site. The primary purpose of the harvest was to derive an allometry for determining biomass allocation for the aspen clones (King et al. 2005). Trees of various diameters and heights were selected to represent a range in above-ground biomass. Importantly, the harvest allowed investigation of the relationship between host biomass allocation and the quantity of mycorrhizal fungi, as well as the response of mycorrhizal fungi gene expression to exposure of the host plant aspen to $eCO_2$, or $eO_3$ combination of $eCO_2 + eO_3$.

### *13.3.1 Biochemical and Biomass Analyses*

A primary interest was to determine whether a difference exists between the aspen clones for AM fungal and ECM fungal colonization. Through the use of the AM fungal marker fatty acid 16:1ω5c, a difference for PLFA 16:1ω5c concentration was found to exist for the tested aspen genotype fibrous roots. Because the genotypes were grown in the same environmental setting, these differences between genotypes suggest diversity for this trait in these genotypes. Fig. 13-1A shows that aspen clone 259 possessed a greater concentration of the AMF marker PLFA, suggesting that a host factor contributes to colonization. A significant clone effect was also found for ECM using ergosterol content where clone 216 had significantly less ergosterol concentration than the other clones (Fig. 13-1B). When evaluated for total root PLFA concentration, clone 216 had significantly lower biovolume than the other clones (Fig. 13-1C).

We initially hypothesized that the amount of photosynthate allocated to below-ground structures determines the amount of root-associated mycorrhizal fungus biomass. In support of this hypothesis is the strong positive relationship between mycorrhizal fungal biomass and fine root biomass (Fig. 13-2). In addition, significant positive relationships exist between aspen whole-tree biomass and the amount of root-associated mycorrhizal fungal biomass (AM fungal biomass + ECM fungal biomass, $r^2 = 0.80$, $P \leq 0.0001$), and for mycorrhizal fungal biomass and total leaf dry mass ($r^2 = 0.69$, $P < 0.0001$).

Further, the relationship between mycorrhizal biomass and tree biomass is influenced by treatment. We hypothesized that the mycorrhizal fungi associated with $O_3$-fumigated hosts would have depleted energy reserves that could, in turn, lead to a loss of active mycorrhizal fungus biomass. The greatest amount of root associated AMF biomass was associated with clones grown with $eCO_2$ fumigation ($P \leq 0.0469$) (Fig. 13-2), whereas $eO_3$ had little influence on AMF biomass concentration ($P \leq 0.4055$) (Fig. 13-2). No clone differences were observed for the production of the associated 16:1ω5c

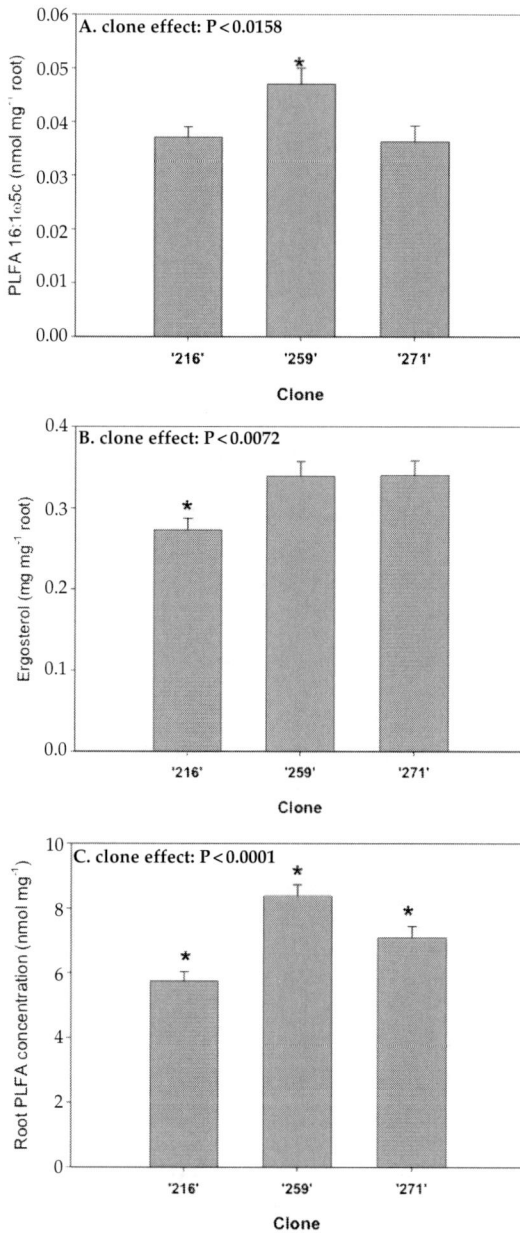

**Figure 13-1** An ANOVA demonstrated an aspen clone effect was evident for concentration of AM fungal marker PLFA 16:1ω5c (A); the ECM fungal marker ergosterol (B); and total root PLFA concentration, a measure of root biovolume (C). Clones with (*) are significantly different at the $P < 0.05$ level using Tukey.

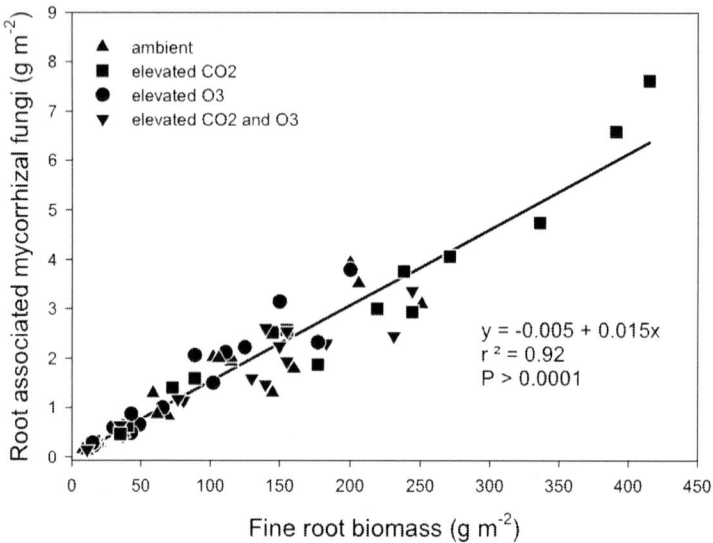

Figure 13-2 A significant positive linear relationship was found for the association between fine-root biomass and root associated mycorrhizal fungal biomass (AM fungal biomass + ECM fungal biomass expressed as g $m^{-2}$ to a depth of 20 cm).

NLFA ($P \le 0.2290$), indicating that the growth environments for the fungi were most likely similar (Fig. 13-3). Likewise, the NLFA 16:1ω5c increased under $eO_3$ ($P \le 0.0001$), suggesting that AMF allocate a greater portion of the carbon they receive from their host to storage reserves when aspen is grown under $eO_3$. This effect is clearly demonstrated by the increase in the ratio of NLFA:PLFA 16:1ω5c under $eO_3$ and is supported by a concomitant increase in vesicle structures in the roots (Fig. 13-3).

There is a clear correlation in the association of ECM and AM fungal species under $eCO_2$ versus $eO_3$. Under $eCO_2$ both ECM and AM fungal species thrived, and ECM fungi were dominant (Fig. 13-4). However, under $eO_3$ or $eCO_2 + eO_3$ conditions, the amount of ECM fungal biomass was reduced to a greater extent than the biomass of associated AM fungi. Because of the surprising abundance of AM fungi encountered in the aspen roots, this study focused on further analysis of how the treatment effects in combination with aspen genotypes influence the AMF below ground.

In addition to the effects of $eO_3$, the treatment effects of $eCO_2 + eO_3$ also were more pronounced in the ratio of NLFA:PLFA 16:1ω5c (Fig. 13-5). This suggests that the AM fungus is channeling lipids for storage rather than building new cell membranes necessary for growth. This change in resource use would in turn have implications for the plants ability to respond to environmental stressors, as preferential allocation to resources over active growth would in turn limit nutrient acquisition for the plant. The treatment

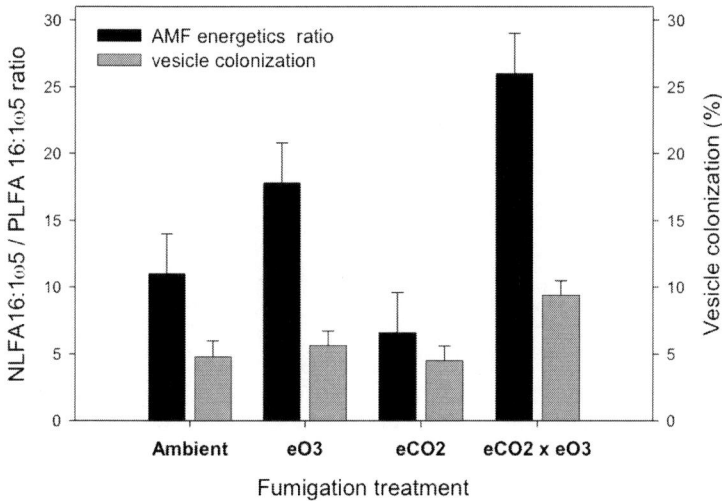

**Figure 13-3** The effects of fumigation treatments on the relationship between AM fungal energetics-to-biomass ratio (NLFA/PLFA for 16:1ω5c) and the occurrence of AM fungal vesicle structures in aspen fine roots. An ANOVA found $eO_3$ and $eCO_2 \times eO_3$ treatments to significantly enhance the energetics ratio (P < 0.05). A significant $eCO_2 \times eO_3$ treatment effect for the increase in vesicle structures (P < 0.05) indicates that $eCO_2 \times eO_2$ fumigation favors the accumulation of fungal storage lipids over active growth and that the mechanism appears to be independent of AM fungal growth.

effect impact on the elevated NLFA:PLFA ratios indicates there are changes in the overall metabolism of AMF in addition to reallocation of carbon by these fungi depending on the treatment (Fig. 13-6).

## 13.3.2 *Gene Expression Analysis*

We were interested in how the above-ground effects of $eCO_2$, $eO_3$ and $eCO_2 + eO_3$ on different aspen genotypes influence the below-ground gene expression in mycorrhizae. DNA analysis of fine roots showed aspen is colonized by both ECM and AM fungi. Due to the abundance of AMF colonization of fine roots, the effects on gene expression by AMF were tested. PCR analyses of aspen fine-roots (< 2.0 mm diameter) demonstrated that the majority of the roots were colonized by the AM fungus *G. intraradices,* as well as other *Glomus* species (data not shown). Gene expression analysis of aspen roots was then performed using a 1000-element custom oligo array prepared using available EST sequences of *G. intraradices.*

It is clear from the gene expression data obtained that the above-ground responses of aspen to the $eCO_2$, $eO_3$ and $eCO_2 + eO_3$ fumigation treatments trickle down to below-ground structures with dramatic consequences on the growth and metabolism of the mycorrhizal fungi inhabiting aspen

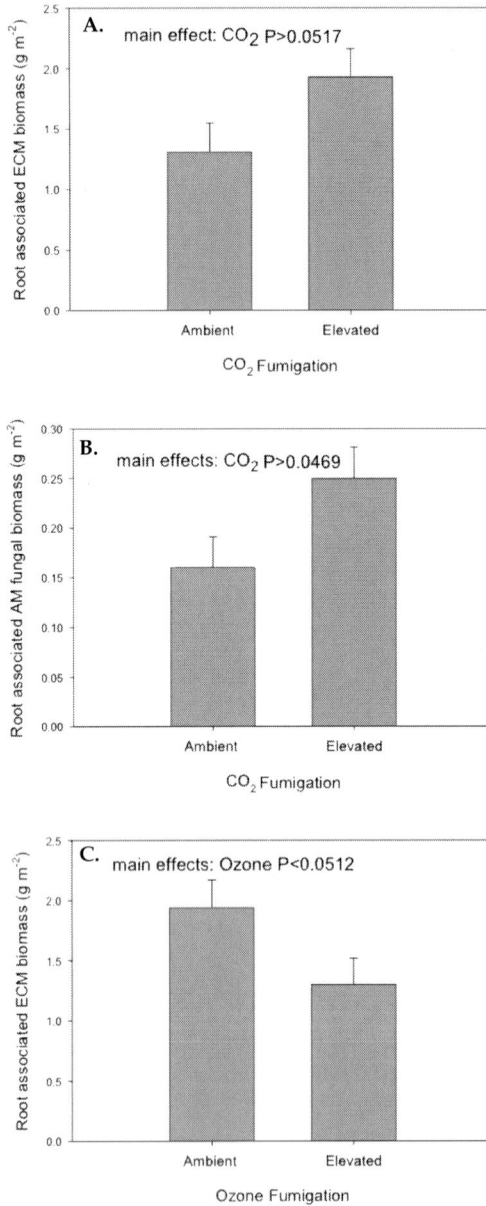

**Figure 13-4** The fumigation treatment main effects on root associated mycorrhizal fungal biomass (g m$^{-2}$ to a depth of 20 cm) indicates that the $eCO_2$ fumigation treatment favors an increase in both AM fungal and ECM fungal biomass (**A** and **B**), whereas, only ECM fungal biomass was reduced when the host experienced ozone fumigation (**C**).

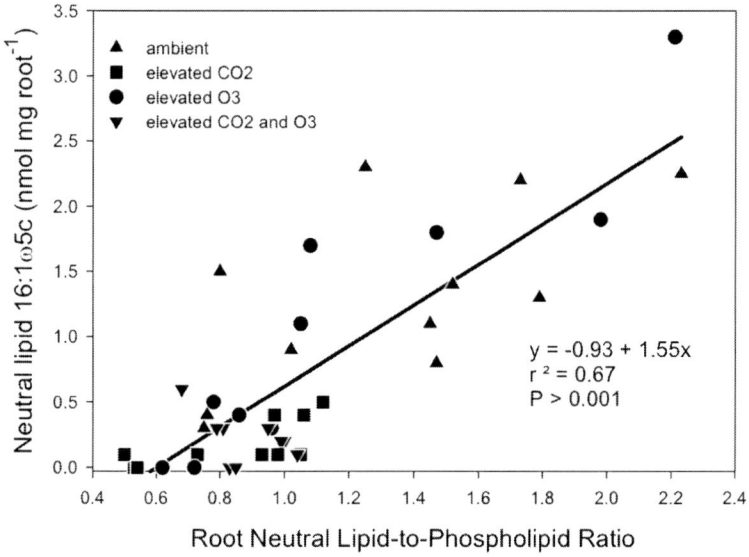

**Figure 13-5** A significant linear relationship was observed between root total neutral lipid-to-phospholipid ratio and the AM fungal storage lipid 16:1ω5c. This association indicates the observed increase in root total neutral lipid-to-phospholipid ratio is most likely in response to an increase in AM fungal storage lipids.

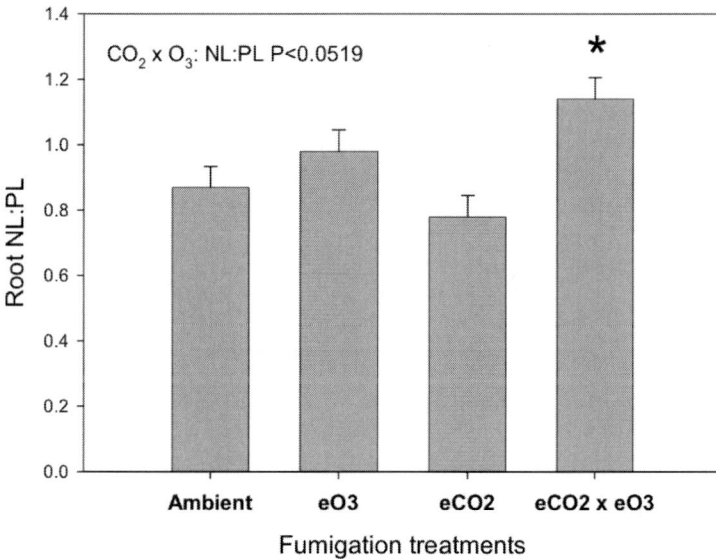

**Figure 13-6** An ANOVA demonstrated a significant $eCO_2 \times eO_3$ treatment effect was evident for root neutral lipid-to-phospholipid (NL: PL) ratio.

roots. Gene expression analysis showed that there is a distinct clone-specific response followed by a treatment-specific response. The HCA analysis derived from this gene expression data demonstrates this response, showing significant differences in the *Glomus* gene expression patterns according to both aspen genotype and according to treatment (see Fig. 13-7). This follows similar observations made by using fatty acid analysis.

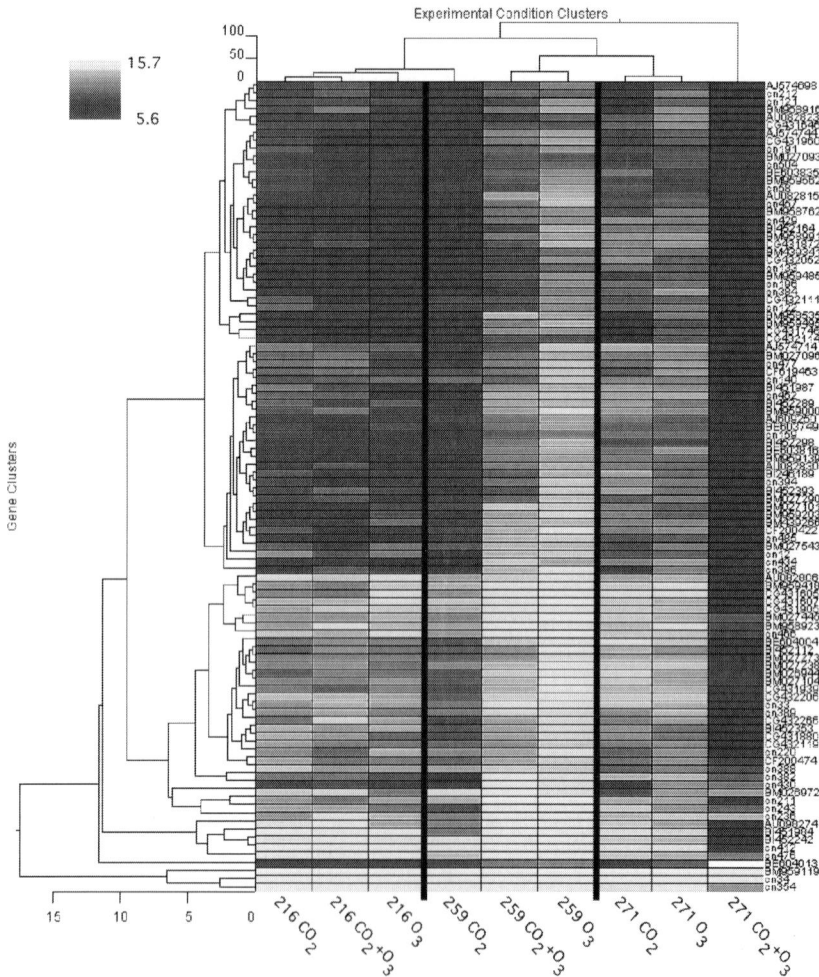

**Figure 13-7** Hierarchical Cluster Analysis (HCA) of relative normalized gene expression data from *G. intraradices*. Gene expression data obtained from the oligoarray of *G. intraradices* associated with three different aspen clones (216, 259 and 271) exposed to $eCO_2$, $eO_3$ and $eCO_2 + eO_3$ was analyzed as described in methods section. Blue indicates low relative gene expression. Yellow indicates high relative gene expression.

*Color image of this figure appears in the color plate section at the end of the book.*

By using Principal Component Analysis (PCA) of normalized transcript ratios of *G. intraradices* show that the PCA that correlate to the majority of the observed array variation are related to both clone and treatment. Principal component axis 1 represents 39.0% of the variation and is associated with the effect of $eCO_2 + eO_3$ on all three clones. Principal components 2 and 4, representing 17.4% and 9.7% of the variation respectively, is associated with both ECM vs. AMF effects on gene expression. Principal component 3 accounts for 14.2% of the variation, and is associated with the genotype response.

Out of the 1000 genes from *G. intraradices* tested on the oligoarrays, 169 genes showed statistically significant differences in expression, with more than 85 showing fold changes between 1.74 and 11.7 in one or more of the clones and in one or more treatments (see Table 13-2). Of the three clones, *Glomus* gene expression levels were most significantly impacted by both $eO_3$ and $eCO_2 + eO_3$ on clone 259, where overall expression levels are seen to increase from $eCO_2$ to $eO_3$ with $eCO_2 + eO_3$ having intermediate expression levels. This most likely is due to the significant $O_3$ sensitivity seen in clone 259 that is somewhat overcome under $eCO_2 + eO_3$ conditions (Fig. 13-7). It is also evident from using the oligoarrays that the interactions of fumigation with both $CO_2$ and $O_3$ are able to alter gene expression in *G. intraradices* at higher levels than either $eCO_2$ or $eO_3$ treatments on these aspen genotypes. This is especially true for clone 271 where *Glomus* expression levels are seen to drop in response to $eCO_2 + eO_3$ compared to the other treatments alone. The expression levels of the same *Glomus* genes on clone 259 have a significant increasing trend as more $eO_3$ is included, yet the expression levels on clone 216 have much more minor alteration in response to each treatment (Fig. 13-7). It has previously been demonstrated that the interacting effects of $eCO_2 + eO_3$ are different from the separate effects of $eCO_2$ or $eO_3$ in various aspen genotypes and that the damage resulting from $eO_3$ is not ameliorated or compensated by $eCO_2$ (Wustman et al. 2001). Similarly, global gene expression profiling in aspen genotypes 216, 259 and 271 has shown that the effects of these imposed fumigation treatments manifest differently (Gupta et al. 2005). Thus, it is interesting to see that the above-ground fumigations by $CO_2$ and $O_3$ influence root metabolism and in turn lead to different responses by the AM fungi based on treatment and aspen genotype.

As seen from the summary of gene expression differences presented in Table 13-2, the greatest number of differential gene expressions of AMF was observed by fungi growing in roots of clone 271 exposed to the $eCO_2 + eO_3$ fumigation treatment. This enhanced expression profile is not unexpected in that clone 271 has been shown to be responsive to $eCO_2$ and is highly tolerant to $eO_3$ compared to the other two aspen genotypes (see Table 13-1). On the other hand, clone 216 is not responsive to $eCO_2$ and is only moderately tolerant to $eO_3$, whereas clone 259 is responsive to $eCO_2$ and also is sensitive to $eO_3$. Gene expression changes in AMF were found for all major groups of genes including those involved in carbohydrate metabolism, lipid biosynthesis, protein synthesis and regulation and oxidative stress. The most commonly up-regulated groups of genes included the acyltransferases, which are involved in shifts in neutral fatty acid biosynthesis. Similarly, up-regulation was observed in chitin synthase genes indicating changes in cell wall biogenesis as a response to treatment induced changes in carbohydrate transfer from the host. Up-regulation of expression was also found in both a clone-specific and treatment-specific manner for genes such as Cylcophilin, which is involved in a variety of functions including response to stresses and as protein chaperones (Joseph et al. 1999). Monooxygenase genes were also up-regulated in a treatment-specific manner indicating the response of AMF to the oxidative stress perceived from the plant roots. Many unknown or hypothetical proteins also showed significant up-regulation of expression in either a clone-specific or treatment-specific manner (Table 13-2).

For the most part, genes associated with nitrogen metabolism, protein synthesis and biosynthesis of unsaturated fatty acids (stearoyl-CoA 9-destaurase), as well as several genes for unknown proteins, were down-regulated in a clone-specific manner (Table 13-2). Stearoyl-CoA 9-destaurases are involved in the biosynthesis of unsaturated fatty acids (Stukey et al. 1990). The reduction in gene expression for unsaturated fatty acid biosynthesis suggests reorganization of membrane structure as well as channeling of lipids towards storage as seen from the AMF fatty acid analysis (Fig. 13-3). This is likely in response to the effect of ozone on the plant host rather than a direct effect on the fungus. The down-regulation of nitrogen metabolism and protein synthesis genes indicates the change in reallocation of carbon metabolism under different treatments, especially due to the interacting effects imposed by the $eCO_2 + eO_3$ treatment. These results mirror some of the effects of $eCO_2 + eO_3$ on the plant host. A total of 18 selected genes from the list of over 85 genes that showed significant differences in expression from *G. intraradices* arrays were checked using QRT-PCR (Table 13-2). The QRT-PCR results were consistent with the array data.

## 13.4 Conclusions

This study demonstrates that the above-ground fumigation of aspen to $eCO_2$ and $eO_3$ leads to significant below-ground effects to the growth and metabolism of the root associated mycorrhizal fungi, specifically in regulating their gene expression. The gene expression profile differences observed need to be placed within the context of the experiment. Above-ground fumigation with enhanced levels of $CO_2$ and $O_3$ does not result in an increase in the concentration of these gases belowground, rather the reported up- and down-regulation for gene expression of *G. intraradices* is an indirect effect primarily driven by the aspen genotype responses to the imposed fumigation treatments. More importantly, the fumigation $O_3$ exposure level presented to the aspen genotypes is at a concentration below levels identified as safe to human health by EPA standards. This study reveals that the gene expression in AMF is regulated both in a clone-specific (aspen genotype) and fumigation treatment-specific manner. The facts that aspen clone 259 is sensitive to $eO_3$ and clone 271 is responsive to $eCO_2$ and tolerant of $eO_3$ are also reflected in the gene expression responses of its mycorrhizal associate *G. intraradices*. It is also interesting to note that many of the changes in mycorrhizal fungal gene expression and metabolism reflect the changes in carbon transfer from the host to the mycorrhizal fungi. In the specific example of *G. intraradices*, there is a clear reallocation and reorganization of carbon metabolism from one that is involved in growth to one involved in carbon conservation and storage. Also, there are clear differences in the response of ECM fungi versus AM fungi when the aspen trees are exposed to $eCO_2$ or $eO_3$. While $eCO_2$ tends to increase the colonization of roots and extraradical growth by both ECM and AM fungi, $eO_3$ seems to be associated with a reduction in colonization and ECM sheath formation. This will have significant impact on the mycorrhizal biodiversity and soil fertility over the long term for trees exposed to ozone. Future studies with a complete set of gene arrays for the AMF and ECM fungi commonly found with aspen trees will provide a more comprehensive outlook on the below-ground effects of elevated $eCO_2$ and $eO_3$ and their interaction. These types of studies will also allow discovery of specific gene markers to determine the right combination of tree genotypes and mycorrhizal symbionts to improve forest productivity, as well as for anticipating future forest growth in response to increased atmospheric $CO_2$ and associated insult from atmospheric pollutants like ozone.

## Acknowledgements

Part of this research was supported through a grant 4F-02201 to GKP from Argonne National Laboratory. RMM's participation was funded by the U.S. Department of Energy, Office of Science, Office of Biological and Environmental Research, Climate and Environmental Research Division under contract DE-AC02-06CH11357 at Argonne. The authors would like to thank Pooja Gupta for helping with the field preparation of root samples. We also thank the Aspen Face tree harvesting crew of 2002 for their help in extracting roots in a timely manner.

# References

Alberton O, Kuyper TW, Gorissen A (2005) Taking mycocentrism seriously: mycorrhizal fungal and plant responses to elevated $CO_2$. New Phytol 167: 859–868.

Allen MF, Swenson W, Querejeta IJ, Egerton-Warburton LM, Treseder KK (2003) Ecology of mycorrhizae: A Conceptual Framework for Complex Interactions Among Plants and Fungi. Ann Rev Phytopath 41: 271–303.

Allison DB, Coffey CS (2003) Two stage testing in microarray analysis: What is gained? J Gerontol Biol Sci 57: 189–192.

Allison DB, Page G, Beasley TM, Edwards JW (2006) DNA Microarrays and Related Genomics Techniques. Chapman and Hall/CRC Taylor and Francis Group, Boca Raton, Florida, USA.

Allison VJ, Miller RM (2004) Using fatty acids to quantify arbuscular mycorrhizal fungi In: G Podila, A Varma (eds) Basic Research and Applications of Mycorrhizae. IK International, New Delhi, India, pp 141–161.

Allison VJ, Miller RM, Jastrow JD, Matamala R, Zak DR (2005) Changes in soil microbial community structure in a tallgrass prairie chronosequence. Soil Science Soc Am J 69: 1412–1421.

Anderson CP (2003) Source-sink balance and carbon allocation below ground in plants exposed to ozone. New Phytol 157: 213–228.

Andrew C, Lilleskov EA (2009) Productivity and community structure of ectomycorrhizal fungal sporocarps under increased atmospheric $CO_2$ and $O_3$. Ecol Letters 12: 813–822.

Augé RM (2001) Water relations, drought and VA mycorrhizal symbiosis. Mycorrhiza 11: 3-42.

Berrang P, Karnosky DF, Mickler RA, Bennet JP (1996) Natural selection for ozone tolerance in *Populus tremuloides*. Can J For Res 16: 1214–1216.

Bligh EC, Dyer WJ (1959) A rapid method of total lipid extraction and purification. Can J Biochem Physiol 37: 911–917.

Bhupinderpal-Singh NA, Lofvenius MO, Högberg MN (2003) Tree root and soil heterotrophic respiration as revealed by girdling of boreal Scots pine forests: extending observations beyond the first year. Plant Cell Environ 26: 1287–1296.

Cseke LJ, Tsai C-J, Rogers A, Nelsen MP, White HL, Karnosky DF, Podila GK (2009) Transcriptomic comparison in the leaves of two aspen genotypes having similar carbon assimilation rates but different partitioning patterns under elevated [CO2]. New Phytol 182: 891–911.

Di Baccio D, Castagna A, Paoletti E, Sebastiani L, Ranieri A (2008) Could the differences in $O_3$ sensitivity between two poplar clones be related to a difference in antioxidant defense and secondary metabolic response to $O_3$ influx? Tree Physiol 28: 1761–1772.

Dickson RE, Lewin KF, Isebrands JG, Coleman MD, Heilman WE, Riemenschneider DE, Sober J, Host GE, Zak DR, Hendrey GR, Pregitzer KS, Karnosky DF (2000) Forest Atmosphere Carbon Transfer and Storage (FACTS-II) The Aspen Free-Air $CO_2$ and $O_3$ Enrichment (FACE) Project: An Overview. Forest Service- USDA, St Paul, MN, USA.

Fitter AH, Heinemeyer A, Staddon PL (2000) The impact of elevated $CO_2$ and global climate change on arbuscular mycorrhizas: a mycocentric approach. New Phytol 147: 179–187.

Gamper H, Peter M, Jansa J, Lüscher A, Hartwig UA, Leuchtmann A (2004) Arbuscular mycorrhizal fungi benefit from 7 years of free air CO2 enrichment in well-fertilized grass and legume monocultures. Global Change Biol 10: 189–199.

Garrett KA, Dendy SP, Frank EE, Rouse MN, Travers SE (2006) Climate Change Effects on Plant Disease: Genomes to Ecosystems Ann Rev Phytopath 44: 489–509.

Govindarajulu M, Pfeffer PE, Jin H, Abubaker J, Douds DD, Allen JW, Bücking H, Lammers PJ, Shachar-Hill Y (2005) Nitrogen transfer in the arbuscular mycorrhizal symbiosis. Nature 435: 819–823.

Graham JH (2000) Assessing costs of arbuscular mycorrhizal symbiosis in agroecosystems. In: GK Podila, DD Douds (eds) Current Advances in Mycorrhizal Research. The American Phytopathological Society, St. Paul, MN, USA, pp 127–140.

Graham JH, Miller RM (2005) Mycorrhizas: Gene to Function. Plant Soil 274: 79–100.

Grebenc T, Kraigher H (2007) Types of ectomycorrhiza of mature beech and spruce at ozone-fumigated and control forest plots. Environ Monit Assess 128: 47–59.

Gupta P, Duplessis S, White H, Karnosky DF, Martin F, Podila GK (2005) Gene expression patterns of trembling aspen trees following long-term exposure to interacting elevated $CO_2$ and tropospheric $O_3$. New Phytol 167: 129–41.

Haberer K, Grebenc T, Alexou M, Gessler A, Kraigher H, Rennenberg H (2007) Effects of long-term free-air ozone fumigation on $\delta^{15}N$ and total N in *Fagus sylvatica* and associated mycorrhizal fungi. Plant Biol 9: 242–252.

Hassan R, Scholes R, Ash N (2005) Ecosystems and Human Wellbeing: Current State and Trends, vol 1. Island Press, Washington DC, USA.

Isebrands JG, McDonald EP, Kruger E, Hendrey G, Pregitzer K, Percy K, Sober J, Karnosky DF (2001) Growth responses of *Populus tremuloides* clones to interacting carbon dioxide and tropospheric ozone. Environ Pollut 115: 359–371.

James TY, Kauff F, Schoch CL, Matheny PB, Hofstetter V, Cox CJ, Celio G, Gueidan C, Fraker E, Miadlikowska J, Lumbsch HT, Rauhut A, Reeb V, Arnold AE, Amtoft A, Stajich JE, Hosaka K, Sung GH, Johnson D, O'Rourke B, Crockett M, Binder M, Curtis JM, Slot JC, Wang Z, Wilson AW, Schussler A, Longcore JE, O'Donnell K, Mozley-Standridge S, Porter D, Letcher PM, Powell MJ, Taylor JW, White MM, Griffith GW, Davies DR, Humber RA, Morton JB, Sugiyama J, Rossman AY, Rogers JD, Pfister DH, Hewitt D, Hansen K, Hambleton S, Shoemaker RA, Kohlmeyer J, Volkmann-Kohlmeyer B, Spotts RA, Serdani M, Crous PW, Hughes KW, Matsuura K, Langer E, Langer G, Untereiner WA, Lucking R, Budel B, Geiser DM, Aptroot A, Diederich P, Schmitt I, Schultz M, Yahr R, Hibbett DS, Lutzoni F, McLaughlin DJ, Spatafora JW, Vilgalys R (2006) Reconstructing the early evolution of fungi using a six-gene phylogeny. Nature 443: 818–822.

Joseph JD, Heitman J, Means AR (1999) Molecular cloning and characterization of *Aspergillus nidulans* Cyclophilin B. Fungal Genetics Biol 27: 55–66.

Karnosky DF, Zak DR, Pregitzer KS, Awmack CS, Bockheim JG, Dickson RE, Hendrey GR, Host GE, King JS, Kopper BJ, Kruger EL, Kubiske ME, Lindroth RL, Mattson WJ, McDonald EP, Noormets A, Oksanen E, Parsons WFJ, Percy KE, Podila GK, Riemenschneider DE, Sharma P, Thakur RC, Sober A, Sober J, Jones WS, Anttonen S, Vapaavuori E, Mankovska B, Heilman WE, Isebrands JG (2003) Tropospheric $O_3$ moderates responses of temperate hardwood forests to elevated $CO_2$: A synthesis of molecular to ecosystem results from the Aspen FACE project. Funct Ecol 17: 289–304.

Karnosky DF, Skelly JM, Percy KE, Chappelka AH (2007) Perspectives regarding 50 years of research effects on tropospheric ozone air pollution on US forests. Environ Pollut 147: 489–506.

King JS, Kubiske ME, Pregitzer KS, Hendrey GR, McDonald EP, Giardina CP, Quinn VS, Karnosky DF (2005) Tropospheric $O_3$ compromises net primary production in young stands of trembling aspen, paper birch and sugar maple in response to elevated atmospheric $CO_2$. New Phytol 168: 623–636.

Lammers P, Tuskan GA, DiFazio SP, Podila GK, Martin F (2004) Mycorrhizal symbionts of *Populus* to be sequenced by the United States Department of Energy's Joint Genome Institute. Mycorrhiza 14: 63–64.

Langley JA, Hungate BA (2003) Mycorrhizal controls on below-ground litter quality. Ecology 84: 2302–2312.

Lindroth RL (2010) Impacts of elevated atmospheric CO2 and O3 on forests: Phytochemistry, trophic interactions, and forest dynamics. J Chem Ecol 36: 2–21.

Lopez R, Silventoinen V, Robinson S, Kibria A, Gish W (2003) WU-Blast2 server at the European Bioinformatics Institute. Nucl Acids Res 31: 3795–3798.

Lukac M, Calfapietra C, Godbold DL (2003) Production, turnover and mycorrhizal colonization of root systems of three *Populus* species grown under elevated $CO_2$ (POPFACE). Glob Chang Biol 9: 838–848.

Maldonado-Mendoza IE, Dewbre GR, van Buuren ML, Versaw WK, Harrison MJ (2002) Methods to estimate the proportion of plant and fungal RNA in an arbuscular mycorrhiza. Mycorrhiza 12: 67–74.

Marschner H, Dell B (1994) Nutrient uptake in mycorrhizal symbiosis. Plant Soil 159: 89–102.

Meesters PA, Springer J, Eggink G (1997) Cloning and expression of the delta 9 fatty acid desaturase gene from *Cryptococcus curvatus* ATCC 20509 containing histidine boxes and a cytochrome b5 domain. Appl Microbiol Biotechnol 47: 663–667.

Miller RM, Jastrow JD (2000) Mycorrhizal fungi influence soil structure In: Y Kapulnik, D Douds (eds) Arbuscular Mycorrhizas: Physiology and Function. Kluwer Academic Publ, Dordrecht, The Netherlands, pp 4–18.

Miller RM, Kling M (2000) The importance of integration and scale in the arbuscular mycorrhizal symbiosis. Plant Soil 226: 295–309.

Miller RM, Miller SP, Jastrow JD, Rivetta CB (2002) Mycorrhizal mediated feedbacks influence net carbon gain and nutrient uptake in *Andropogon gerardii*. New Phytol 155: 149–162.

Smith SE, Read DJ (2008) *Mycorrhizal symbiosis*. Academic Press, New York USA.

Staddon PL, Jakobsen I, Blum H (2004) Nitrogen input mediates the effect of free-air $CO_2$ enrichment on mycorrhizal fungal abundance. Glob Chang Biol 10: 1678–1688.

Stukey JE, McDonough VM, Martin CE (1990) The OLE1 gene of *Saccharomyces cerevisiae* encodes the delta 9 fatty acid desaturase and can be functionally replaced by the rat stearoyl-CoA desaturase gene. J Biol Chem 265: 20144–20149.

Treseder KK (2004) A meta-analysis of mycorrhizal responses to nitrogen, phosphorus, and atmospheric $CO_2$ in field studies. New Phytol 164: 347–355.

Vestal JR, White DC (1989) Lipid analysis in microbial ecology: quantitative approaches to the study of microbial communities. BioScience 39: 535–541.

Wright DP, Scholes JD, Read DJ, Rolfe SA (2000) Changes in carbon allocation and expression of carbon transporter genes in *Betula pendula* Roth colonized by the ectomycorrhizal fungus *Paxillus involutus* (Batsch) Fr. Plant Cell Environ 23: 39–49.

Wu CH, Huang H, Yeh L-S, Barker WC (2003) Protein family classification and functional annotation. Comp Biol Chem 27: 37–47.

Wustman BA, Oksanen E, Karnosky DF, Sober J, Isebrands JG, Hendrey GR, Pregitzer KS, Podila GK (2001) Effects of elevated $CO_2$ and $O_3$ on aspen clones varying in $O_3$ sensitivity: Can $CO_2$ ameliorate the harmful effects of $O_3$? Environ Pollut 115: 473–481.

Zeleznik P, Hrenko M, Then C, Koch N, Grebenc T, Lavanic T, Kraigher H (2007) CASIROZ: Root parameters and types of ectomycorrhiza of young beech plants exposed to different ozone and light regimes. Plant Biol 9: 298–308.

Zelles L (1997) Phospholipid fatty acid profiles in selected members of soil microbial communities. Chemosphere 35: 275–294.

Zheng J, Close TJ, Jiang T, Lonardi S (2004) Efficient selection of unique and popular oligos for large EST databases. Bioinformatics 20: 2101–2112.

Zhu Y-G, Miller RM (2003) Carbon cycling by arbuscular mycorrhizal fungi in soil—plant systems. Trends Plant Sci 8: 407–409.

Zimmermann P, Schildknecht B, Craigon D, Garcia-Hernandez M, Gruissem W, May S, Mukherjee G, Parkinson H, Rhee S, Wagner U, Hennig L (2006) MIAME/Plant—adding value to plant microarray experiments. Plant Meth 2:1 doi: 10.1186/1746-4811-2-1.

# Index

# Color Plate Section

## Dedication

**Dr. Gopi Krishna Podila**

**Chapter 1**

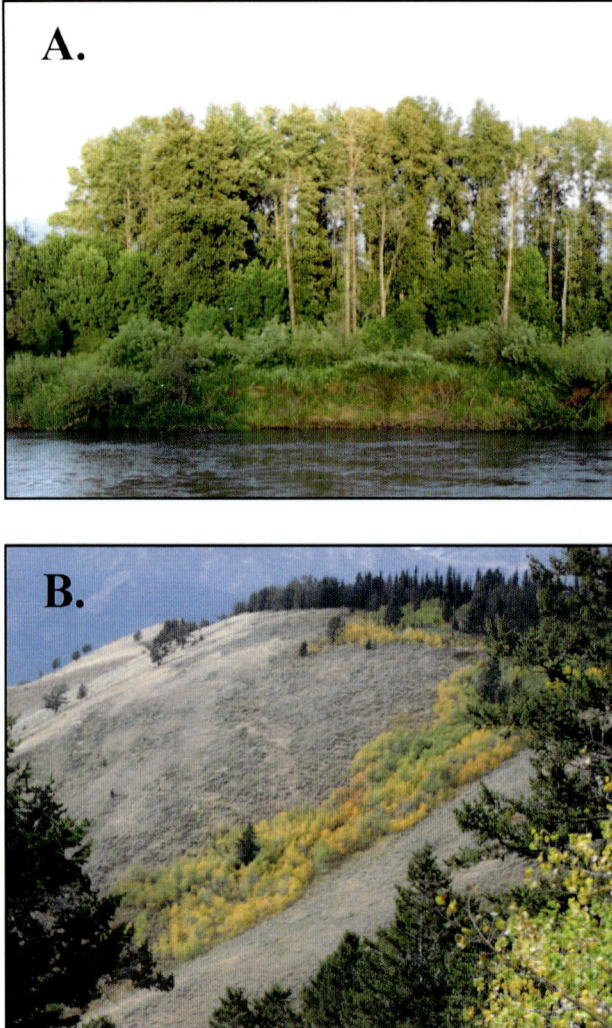

**Figure 1-3** Photos of *Populus* habitats and reproductive structures. **A.** A typical *P. trichocarpa* stand on the Willamette River near Salem, Oregon. **B.** Stands of *P. tremuloides* in Grand Teton National Park, Wyoming. Different genets can be distinguished based on variation in fall leaf colors. All photos by S. DiFazio.

# Chapter 3

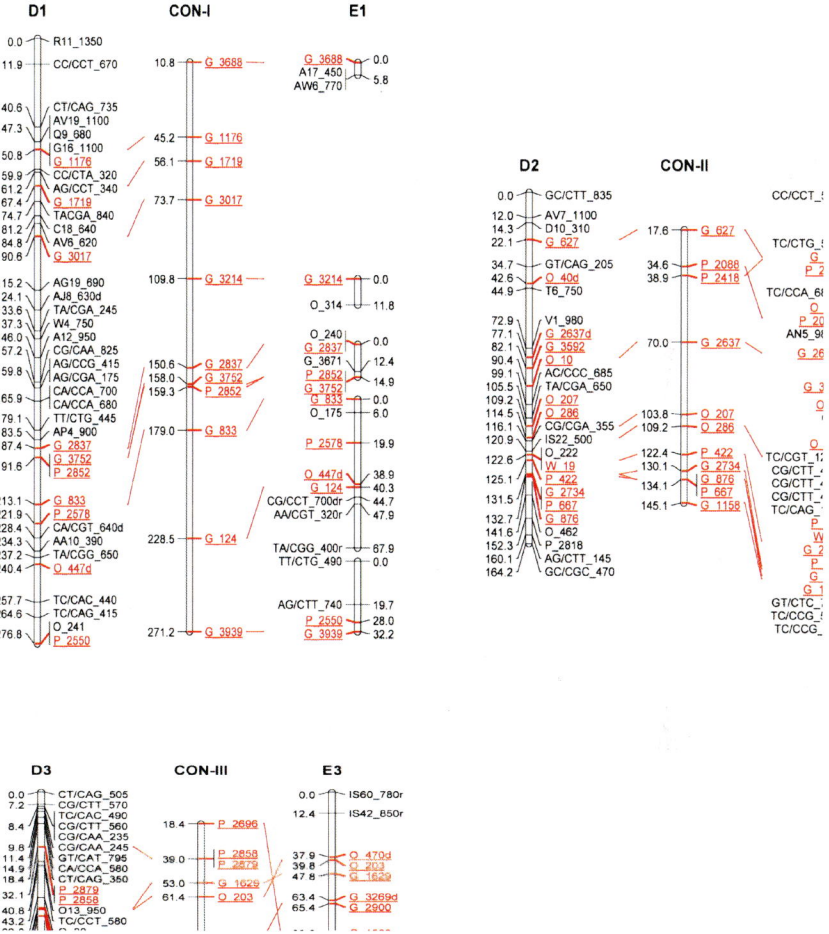

**Figure 3-3** Part of two linkage maps for interspecific hybrids between P. deltoides cv. "I-69" (D) and P. euramericana cv. "I-45" (E). The D and E maps are aligned to the reference linkage map of Populus (CON). Markers in red are homologous loci among the three maps and red lines link the homologous locations of the three genomes.

# Chapter 6

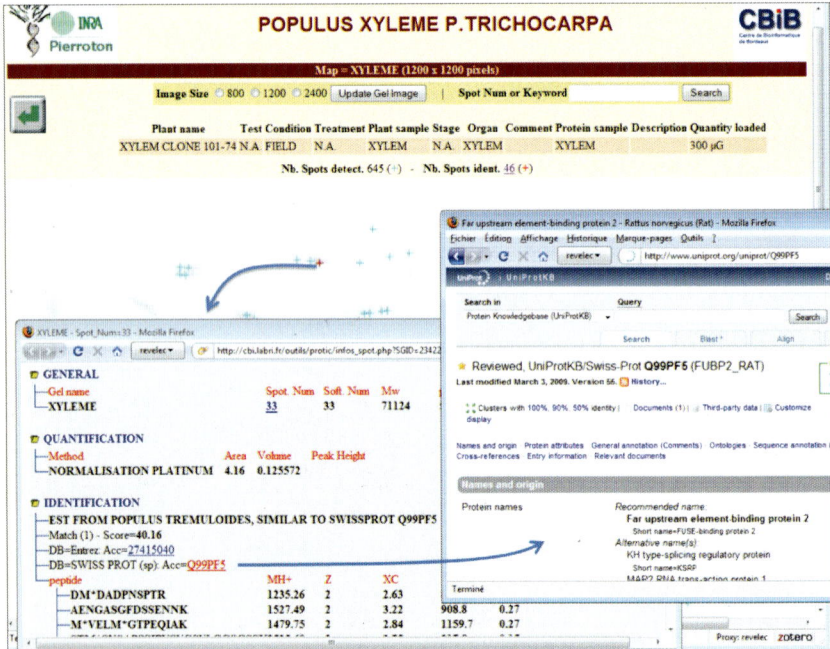

**Figure 6-3** The PROTICdb gel browser displays 2D-PAGE annotated images. Complete dataset for a spot including quantitative data and mass spectrometry identification are available as well as links to external sequence databases.

# Chapter 9

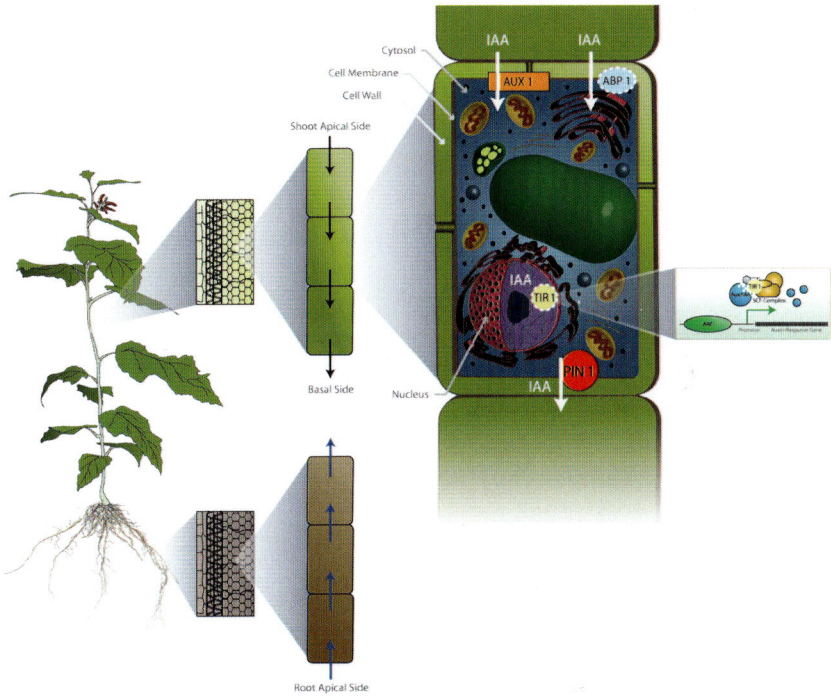

**Figure 9-2** Depiction of auxin transport mechanism at whole plant, tissue and cellular levels. The polar transport of auxin facilitated by AUX1-like influx and PIN1-like efflux proteins can occur in acropetal [from shoot apex to roots] and basipetal manner [mainly roots, upwards from root apex]. The cell structure conceptually illustrates membrane localized auxin pathway proteins, AUX1, ABP1 and PIN1, and the zoomed view of the nucleus conceptually illustrates the action of auxin-receptor, TIR-mediated control of auxin response transcription. Note: The various components in the cartoon are not drawn to scale. Auxin transport occurs in Basipetal (from lateral root cap through root epidermis, depicted here) as well as acropetal (through central tissues of the root toward the root tip) directions in roots. ABP1 is also known to be localized in endoplasmic reticulum).

# Chapter 10

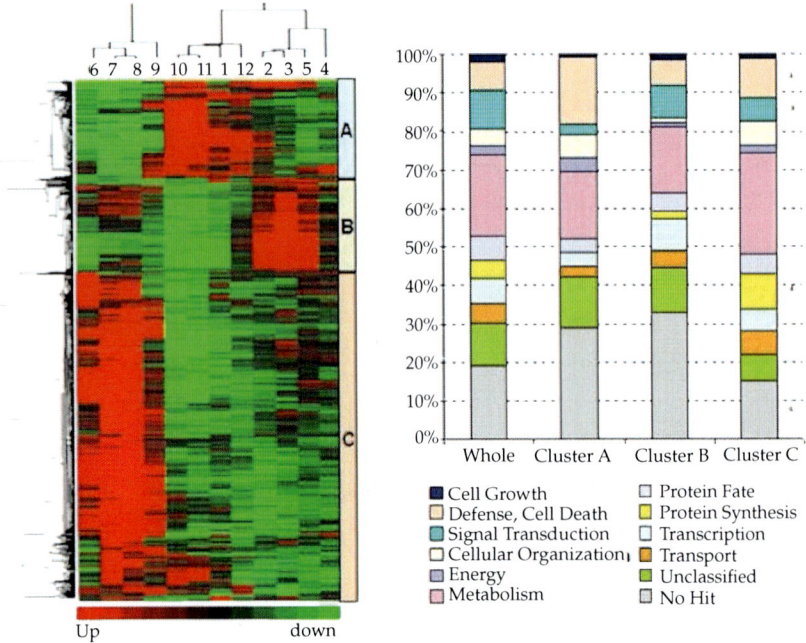

**Figure 10-1** Hierarchical clustering (left) and functional categorization (right) of 714 differentially expressed genes in bark tissue. In the hierarchical clustering, months are indicated numerically on each column. In functional characterization, the differentially expressed genes are presented as a percentage of the 1,953 unigenes on the array. The asterisk indicates statistically significant functional category (contingency test, p< 0.001) (Park et al. 2008).

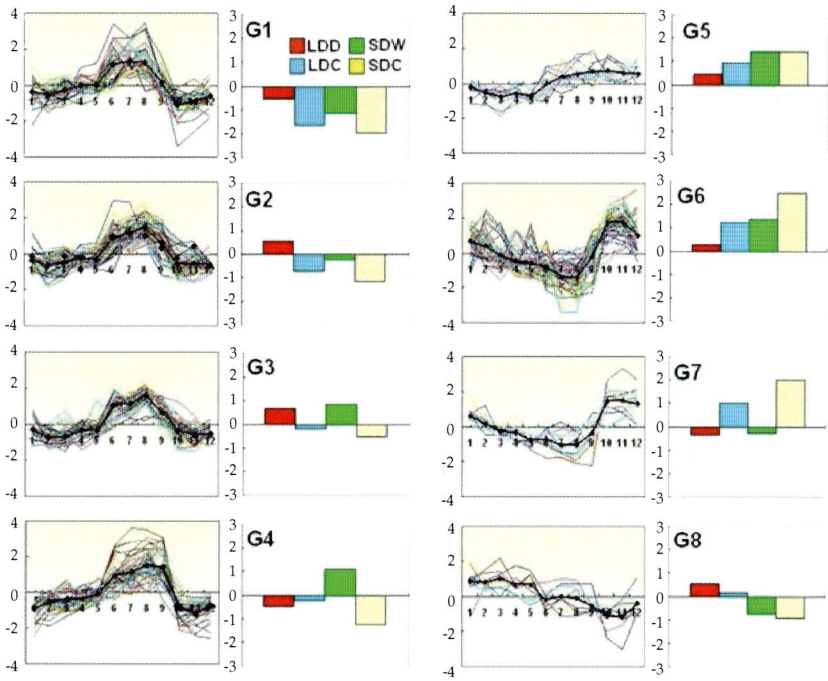

**Figure 10-2** K-means clustering of 274 differentially expressed genes. The line graphs on the left show seasonal expression changes, where the x-axis represents 12 months (1=Jan, 2=Feb, etc.) and the y-axis represents the fold change on a log2 scale (compared to the annual average). The bar graphs on the right show their environmental responsiveness, where mean fold changes (log2 scale) for each of four treatments (LDD, LDC, SDW, and SDC) over LDW are shown on the y-axis (Park et al. 2008).

# Chapter 11

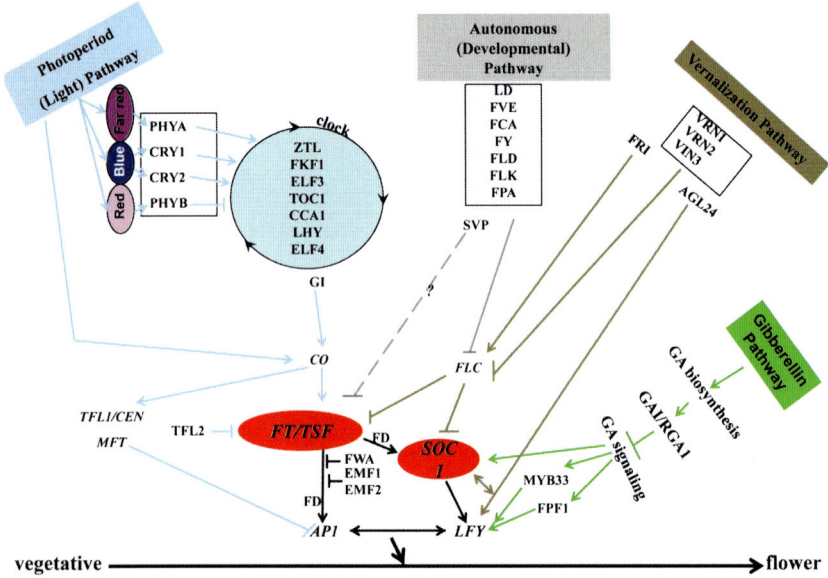

**Figure 11-1** A simplified genetic network that controls flowering time in the annual plant *Arabidopsis*. *FT/TSF* and *SOC1* are floral integrators. Arrows indicate promotion and bars indicate repression.

# Chapter 12

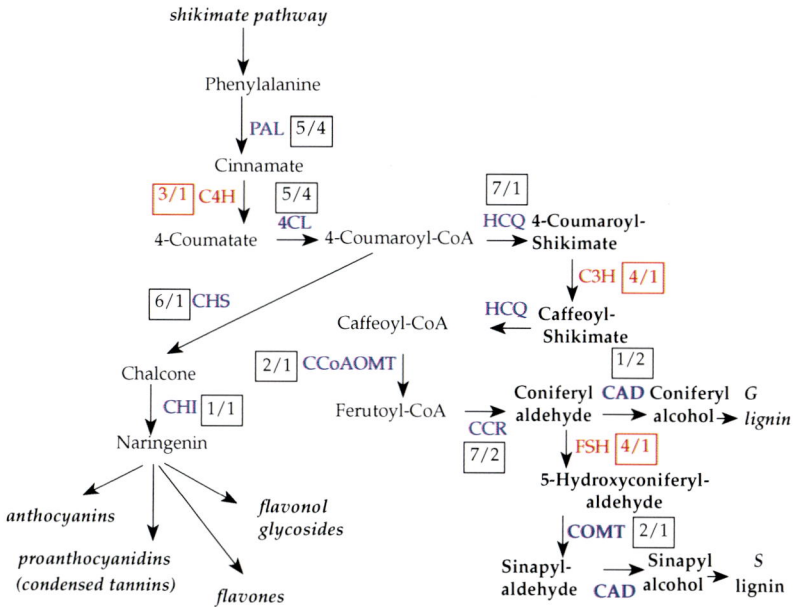

**Figure 12-1** Simplified representation of the phenylpropanoid, monolignol and flavonoid biosynthetic pathways. The numbers of genes encoding enzymes of phenylpropanoid and monolignol biosynthesis in *Populus* and *Arabidopsis* are shown. Enzyme names are annotated with gene numbers. Numbers in boxes refer to the number of *Populus* genes relative to the number of *Arabidopsis* genes for each enzymatic step. Chemical intermediates and product are also indicated. Names in italics refer to additional precursor or end-product pathways. Abbreviations: (4CL) 4-couumarate:CoA ligase, (CAD) cinnamyl alcohol dehydrogenase, (C4H) cinnamate-4-hydroxylase gene family, (C3H) 4-coumaroyl-shikimate/quinate-3-hydroxlase (CCR) cinnamyl CoA reductase, (CHI) chalcone isomerase, (CHS) chalcone synthase, (COMT) caffeic acid O-methyltransferase, (CCoAOMT) caffeoyl-CoA O-methyltransferase, (HCQ), hydroxycinnamyl-quinate transferase, (F5H) ferulate-5-hydroxylase, (PAL) phenylalanine ammonia-lyase.

# Chapter 13

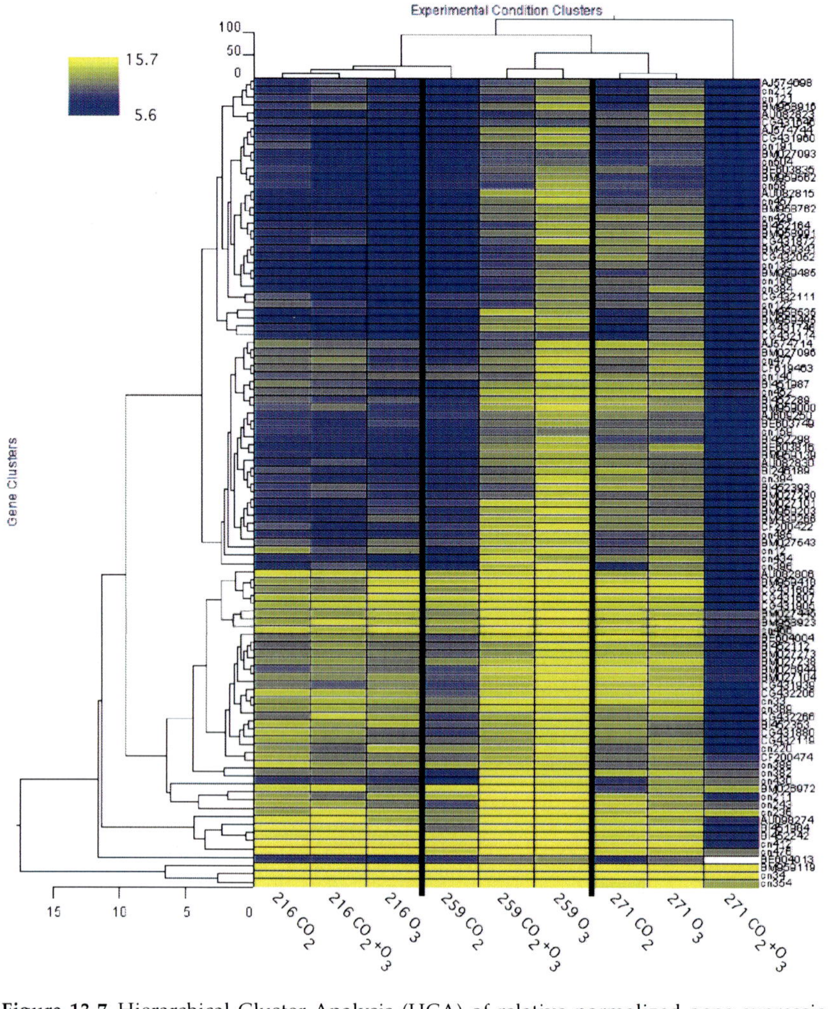

**Figure 13-7** Hierarchical Cluster Analysis (HCA) of relative normalized gene expression data from *G. intraradices*. Gene expression data obtained from the oligoarray of *G. intraradices* associated with three different aspen clones (216, 259 and 271) exposed to $eCO_2$, $eO_3$ and $eCO_2+eO_3$ was analyzed as described in methods section. Blue indicates low relative gene expression. Yellow indicates high relative gene expression.